Saas-Fee Advanced Course 26
Lecture Notes 1996

Springer
Berlin
Heidelberg
New York
Barcelona
Budapest
Hong Kong
London
Milan
Paris
Santa Clara
Singapore
Tokyo

R. C. Kennicutt, Jr. F. Schweizer
J. E. Barnes

Galaxies: Interactions and Induced Star Formation

Saas-Fee Advanced Course 26
Lecture Notes 1996
Swiss Society for Astrophysics and Astronomy

Edited by D. Friedli, L. Martinet, and D. Pfenniger

With 214 Figures

Springer

Professor R. C. Kennicutt, Jr.
Steward Observatory, University of Arizona,
Tucson, AZ 87721, USA

Professor F. Schweizer
Carnegie Institution of Washington, Department of Terrestrial Magnetism,
5241 Broad Branch Road, N.W., Washington, DC 20015-1305, USA

Professor J. E. Barnes
Institute of Astronomy, University of Hawaii,
2680 Woodlawn Drive, Honolulu, HI 96822, USA

Volume Editors:

Dr. D. Friedli
Professor L. Martinet
Dr. D. Pfenniger
Observatoire de Genève, ch. des Maillettes 51, CH-1290 Sauverny, Switzerland

This series is edited on behalf of the Swiss Society for Astrophysics and Astronomy:

Société Suisse d'Astrophysique et d'Astronomie
Observatoire de Genève, ch. des Maillettes 51, CH-1290 Sauverny, Switzerland

Cover photograph: A detailed view of the central part of the Antennae (NGC 4038/4039) taken by the Hubble Space Telescope (B. Whitmore, NASA/ESA)

ISBN 978-3-642-08342-6 e-ISBN 978-3-540-31630-5

Library of Congress Cataloging-in-Publication Data
Galaxies: interactions and induced star formation : lecture notes 1996 / Saas Fee Advanced Course 26.
R. C. Kennicutt ; F. Schweizer ; J. E. Barnes. Swiss Society for Astrophysics and Astronomy. Ed. by D. Friedli ...-
Berlin ; Heidelberg ; New York ; Barcelona ; Budapest ; Hong Kong ; London ; Milan ; Paris ; Santa Clara ;
Singapore ; Tokyo : Springer, 1998
(Saas Fee Advanced Courses ; 26)

Cover design: *design & production* GmbH, Heidelberg

Preface

The 26th Advanced Course of the Swiss Society of Astronomy and Astrophysics (SSAA) took place in Les Diablerets, a small resort in the Swiss Alps, from March 25 to 30 1996. It brought together 95 participants from 17 countries worldwide. The format of the course was traditional, i.e. 28 one-hour lectures given by three outstanding scientists in the field being studied.

The 6th Advanced Course of the SSAA was entitled *Galaxies*; twenty years later, the much longer title *Galaxies: Interactions and Induced Star Formation* clearly indicates in what direction the most remarkable progress has been made since then. Galaxies are no longer seen as immutable objects: they evolve, interact, merge, blaze, reshape, etc.

This course was quite innovative since we succeeded in inviting a very special guest: comet Hyakutake! On the first evening of the course all the participants enjoyed a marvelous and impressive view of the many degrees long tail of the comet in a clear mountain night sky. As live illustration for the course of a two-body interaction, this show was the largest one that the SSAA could ever afford...

Our deep gratitude goes to the three lecturers for their commitment, before, during, and above all(!), after the course for delivering high-quality manuscripts.

Thanks are also due to our secretary, Irène Scheffre, for her very efficient help in making everything runs smoothly. The kind hospitality as well as the excellent and beautiful closing banquet offered by the hotel Eurotel-Victoria have been highly appreciated. The financial support from the Swiss Academy of Natural Sciences was essential to the organization of these lectures.

To conclude, we strongly hope that this book will provide invaluable astronomical knowledge to many astronomers for the next twenty years, i.e. until the 46th Advanced Course of the SSAA!

Geneva
October 1997

Daniel Friedli
Louis Martinet
Daniel Pfenniger

Table of Contents

List of Previous Saas-Fee Advanced Courses

!! 1995 Stellar Remnants
S.D. Kawaler, I. Novikov, G. Srinivasan

!! 1994 Plasma Astrophysics
J.G. Kirk, D.B. Melrose, E.R. Priest

!! 1993 The Deep Universe
A.R. Sandage, R.G. Kron, M.S. Longair

* 1992 Interacting Binaries
S.N. Shore, M. Livio, E.J.P. van den Heuvel

!! 1991 The Galactic Interstellar Medium
W.B. Burton, B.G. Elmegreen, R. Genzel

* 1990 Active Galactic Nuclei
R. Blandford, H. Netzer, L. Woltjer

! 1989 The Milky Way as a Galaxy
G. Gilmore, I. King, P. van der Kruit

! 1988 Radiation in Moving Gaseous Media
H. Frisch, R.P. Kudritzki, H.W. Yorke

! 1987 Large Scale Structures in the Universe
A.C. Fabian, M. Geller, A. Szalay

! 1986 Nucleosynthesis and Chemical Evolution
J. Audouze, C. Chiosi, S.E. Woosley

! 1985 High Resolution in Astronomy
R.S. Booth, J.W. Brault, A. Labeyrie

! 1984 Planets, Their Origin, Interior and Atmosphere
D. Gautier, W.B. Hubbard, H. Reeves

! 1983 Astrophysical Processes in Upper Main Sequence Stars
A.N. Cox, S. Vauclair, J.P. Zahn

* 1982 Morphology and Dynamics of Galaxies
J. Binney, J. Kormendy, S.D.M. White

! 1981 Activity and Outer Atmospheres of the Sun and Stars
F. Praderie, D.S. Spicer, G.L. Withbroe

* 1980 Star Formation
J. Appenzeller, J. Lequeux, J. Silk

* 1979 Extragalactic High Energy Astrophysics
F. Pacini, C. Ryter, P.A. Strittmatter

* 1978 Observational Cosmology
J.E. Gunn, M.S. Longair, M.J. Rees

* 1977 Advanced Stages in Stellar Evolution
I. Iben jr., A. Renzini, D.N. Schramm

* 1976 Galaxies
K. Freeman, R.C. Larson, B. Tinsley

* 1975 Atomic and Molecular Processes in Astrophysics
A. Dalgarno, F. Masnou-Seeuws, R.V.P. McWhirter

* 1974 Magnetohydrodynamics
L. Mestel, N.O. Weiss

* 1973 Dynamical Structure and Evolution of Stellar Systems
G. Contopoulos, M. Hénon, D. Lynden-Bell

* 1972 Interstellar Matter
N.C. Wickramasinghe, F.D. Kahn, P.G. Mezger

* 1971 Theory of the Stellar Atmospheres
D. Mihalas, B. Pagel, P. Souffrin

---

* Out of print

! May be ordered from **Geneva Observatory**
Saas-Fee Courses
Geneva Observatory
CH-1290 Sauverny
Switzerland

!! May be ordered from **Springer-Verlag**

Induced Star Formation

Robert C. Kennicutt, Jr.

Steward Observatory, University of Arizona, Tucson AZ 87721, USA

1 Overview: Induced Star Formation and Interactions

1.1 Introduction

Star forming galaxies in the local universe provide vital clues to the evolutionary properties of galaxies and the physical processes which drive that evolution. Of these, the strongly interacting and merging galaxies tell an especially important story, providing unique insights into the galaxy formation process, and the keys to understanding such diverse phenomena as starbursts, active nuclei, quasars, and perhaps even the origins of present-day stellar populations and the Hubble sequence. What was regarded by most as a fringe subject 20 years ago has become one of the most active and exciting subfields of extragalactic astronomy.

As the title indicates, the main theme of this lecture series is star formation in interacting galaxies. This is a rich subject and I will interpret it broadly, to encompass starbursts in general, and the interconnections between interactions, starbursts, and nuclear activities, as one cannot fully describe or understand any one of these three subjects in isolation from the other two. Although I will emphasize observational results, I will also address current ideas about the physical triggering and regulation of star formation in these systems, a subject of crucial importance to galaxy formation and evolution in general. I will deliberately take a broad view of interactions themselves, describing not only the mergers that are the main focus of the other lectures, but also the dominant population of less spectacular interactions, so that the reader can better place the mergers in the broader context of galaxy evolution.

These lectures are aimed at readers who seek a broad introduction to star formation in interacting galaxies and the relevant astrophysical issues, with emphasis on basic concepts and an introduction to the relevant literature. The lectures presume only an undergraduate-level background in extragalactic astronomy and star formation. With that in mind, I begin this chapter with a historical overview of star formation and nuclear activity in interacting galaxies, and introduce the basic phenomena and terminology that will be discussed in the remaining lectures. I then highlight some of the key astrophysical questions and issues that will be addressed, and briefly outline the organization of the lecture series.

1.2 Historical Background: First Hints

Hints that interactions might play an important role in triggering star formation and other activities go back at least 40 years. One of the first quantitative indications was discovered by Holmberg (1958), as a byproduct of his famous photometric survey of nearby galaxies. When comparing the photographic colors of paired galaxies, he found a significant correlation, with blue galaxies tending to have blue companions, etc. This phenomenon has since been referred to as the "Holmberg effect". As discussed in Lecture 3, the physical explanation of the Holmberg effect is complex, reflecting in part a tendency for similar types of galaxies to form together (morphological segregation), but also reflects mutual tidal triggering of star formation (Kennicutt et al. 1987).

Holmberg's insight was no accident; it required a large database of photographic photometry of normal galaxies and a comparable data set on galaxy pairs, two of Holmberg's enduring contributions to extragalactic astronomy. Likewise, any subsequent progress in understanding the properties of interacting galaxies (as opposed to paired galaxies in general) would require the compilation of catalogs of strongly interacting and disturbed systems. These were provided in the famous photographic catalogs of Vorontsov-Velyaminov (1959; 1977) and Arp (1966). The catalog photographs themselves provided numerous anecdotal evidences of induced star formation. The catalogs stimulated a campaign of broadband photoelectric photometry that was subsequently compiled in the Second Reference Catalog of Bright Galaxies (de Vaucouleurs, de Vaucouleurs, & Corwin 1976; RC2). During this same period, the numerical experiments of Toomre & Toomre (1972) and Toomre (1977) began to attract the attention of theorists to the subject.

Further hints of the importance of interactions for triggering activity came with the release of the Markarian survey of UV-bright galaxies (Markarian 1967). Statistical analyses by several investigators showed that the fraction of galaxies with companions and multiple nuclei was several times higher among the Markarian galaxies than in the general field (e.g., Heidmann & Kalloghlian 1973; Karachentsev & Karachentseva 1974).

1.3 Systematic Studies: Starbursts

The first comprehensive study of the evolutionary properties of interacting galaxies was published in a seminal paper by Larson & Tinsley (1978; LT). Many of our current ideas about star formation bursts in interacting galaxies, including many that were "discovered" years later, are introduced in this paper, and the paper stimulated my own interest in this field. LT first compared the UBV colors of peculiar galaxies from the Arp atlas with those of normal galaxies drawn from the Hubble Atlas (Sandage 1961), using the wealth of new photometry that had been compiled in the RC2. Figure 1, taken from LT, shows the resulting color distributions. The Arp atlas galaxies exhibit both a larger dispersion in color and a much bluer envelope in the UBV plane. This alone provided circumstantial evidence for some kind of abnormal stellar population in the Arp atlas sample, which was dominated by interacting systems.

Larson and Tinsley tentatively interpreted the peculiar colors as being due to bursts of star formation (the term "starbursts" had not yet been introduced)

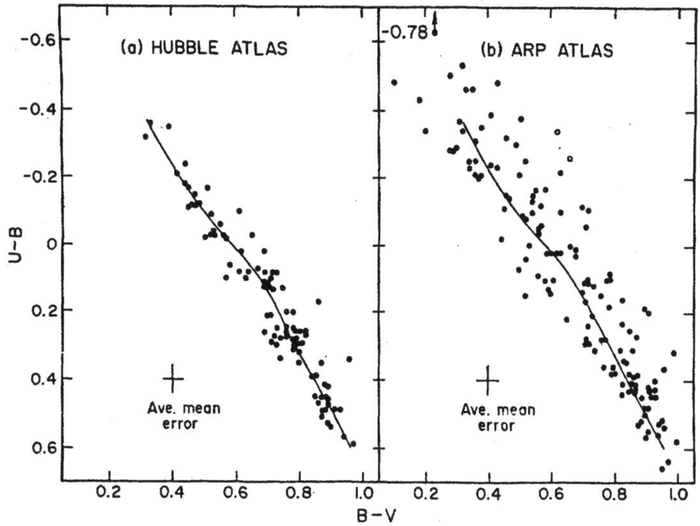

Fig. 1. UBV colors of normal (left) and peculiar galaxies (right), from Larson & Tinsley (1978).

triggered by the tidal interactions. To quantify their case, LT computed evolutionary synthesis models for galaxies with a range of star formation histories and burst strengths, and applied them to the observed color distributions (Lecture 6). The tight color locus for normal galaxies could be explained with a simple model of coeval stellar populations differing only in the e-folding time of the past star formation rate, confirming earlier work by Tinsley (1972) and Searle, Sargent, & Bagnuolo (1973). However the colors of interacting and peculiar galaxies could only be explained by adding to these models bursts of star formation with ages and durations of order $10^7 - 10^8$ yr. By fitting their models to the observed range of colors, they were able to make the first reliable estimates of the strengths, duration, and masses of the star formation bursts. Hence this single paper provided both the observational and the theoretical foundations for our current understanding of starbursts in interacting and peculiar galaxies.

The LT paper stimulated a series of follow-up studies of the evolutionary properties of interacting galaxies. An important development in the 1980's came with calibration of more direct methods of measuring absolute star formation rates (SFRs) in galaxies, using for example Balmer emission-line fluxes together with the integrated colors (Kennicutt 1983). This technique was applied to several samples of paired, interacting, and merging galaxies (Keel et al. 1985; Bushouse 1987; Kennicutt et al. 1987), confirming the conclusions of LT and providing better constraints on the SFRs and burst properties. Radio continuum surveys provided independent evidence for enhanced star formation and nuclear activities (Sulentic 1976; Stocke 1978; Hummel 1981; Condon et al. 1982). These investigations revealed a vast range in the SFRs, from virtually unperturbed systems to galaxies with dominant young populations formed in the interactions. The term "starburst" was coined by Weedman et al. (1981) to characterize NGC 7714, a prototype system dominated by a luminous nuclear star formation burst.

1.4 Interactions and Nuclear Activity

At about the same time, several lines of evidence established a clear link between tidal interactions and nuclear activity. Hummel (1981) showed that most of the excess of radio continuum emission observed in interacting pairs originated in the central regions, and Kennicutt & Keel (1984) and Keel et al. (1985) found that interacting galaxies exhibited an excess of nuclear line emission as well as a higher frequency of Seyfert activity. Likewise Dahari (1984) observed that Seyfert galaxies possess an apparent excess of companions relative to normal spirals. Strong infrared excesses in the centers of strongly interacting systems were also identified (e.g., Joseph et al. 1984; Cutri & McAlary 1985), observations which presaged an explosion of discoveries from IRAS (next section). Further evidence came from imaging studies of low-redshift quasars, which revealed a preponderance of interacting and disturbed hosts (e.g., Stockton 1978; Hutchings et al. 1982; Heckman et al. 1984).

1.5 IRAS and Ultraluminous Starburst Galaxies

In the mid-1980's the Infrared Astronomical Satellite (IRAS) provided the first deep all-sky survey in the $10 - 100\,\mu$m region, and led to a series of spectacular discoveries that have stimulated much of the current interest in this subject. The survey revealed a class of luminous galaxies which emit most of their bolometric luminosity in the infrared, with absolute luminosities ranging from those of normal galaxies to the most luminous optical quasars (Houck et al. 1984; Rowan-Robinson et al. 1984; Soifer et al. 1984). Nearby examples of such objects (e.g., M 82, Mrk 231) had been identified previously from the ground (e.g., Rieke & Low 1972; Joseph & Wright 1985), but the IRAS results firmly established the ubiquity and importance of this class of activity.

Subsequent studies at other wavelengths soon established a strong link between the most luminous infrared galaxies and strong interactions and mergers (e.g., Soifer et al. 1984; Joseph & Wright 1985; Sanders et al. 1988; Melnick & Mirabel 1990). As reviewed in Lectures 5 and 7, the ultraluminous galaxies appear to be a distinct class of objects in several respects, distinguished not only in terms of their high luminosities and heavy extinction, but moreover in the interstellar gas distributions and in terms of the physical processes leading to the starbursts. While many IRAS galaxies appear to be powered by scaled-up starbursts, perhaps with an anomalous stellar population, others appear to be powered by dust-enshrouded AGNs, or by a combination of a starburst and AGN (Lectures 5, 6).

1.6 The 1990's: HST, Supercomputers, and the Distant Universe

The most exciting recent observations in this field are the extraordinary images of the distant universe provided by the Hubble Space Telescope (HST). Two decades of ground-based observations provided circumstantial evidence that interactions and mergers were much more common in the past (e.g., Zepf & Koo 1989), but such data were severely hampered by the limited resolution and sensitivity of the data. Images of galaxy clusters in the redshift range $z = 0.4 - 2.4$ and of deep fields such as the Medium Deep Survey or the Hubble Deep Field provide manifest evidence that interactions were a common phenomenon in the past (e.g., Burkey et al. 1994; Dressler et al. 1994; Abraham et al. 1996). Such data can be used in principle to measure the cosmic history of the galaxy merger rate.

This decade has also witnessed dramatic improvements in computational capabilities, which have made it possible to model the formation of galaxies and large scale structure numerically, using the techniques that are described in Josh Barnes' lectures. In the framework of the most popular cold dark matter (CDM) cosmology, galaxy merging is a pervasive and ongoing process in the universe. If this hierarchical picture is correct, mergers may well be the dominant processes by which galaxies form and evolve, and the dominant mechanism for reproducing the Hubble sequence and forming stars in the universe. Although the global role of merger-induced star formation remains uncertain (see Lecture 9), we now recognize the importance of this process for galactic evolution on a much broader level.

1.7 Key Questions and Issues

In the course of this historical preview, several fundamental astrophysical questions and issues have emerged. Those motivating questions form the basis of the lectures that follow.

What is the relationship between galaxy interactions and global star formation? There are spectacular examples of massive starbursts in interacting systems, but how ubiquitous are these bursts? What is the range of responses of galaxies to tidal interactions? Are there instances of galaxies that do *not* form stars in mergers? How are the induced SFRs related to the global properties of the parent galaxies and the dynamical properties of the interactions?

What is the relationship between galaxy interactions and nuclear star formation? One might naively expect that the nuclear star formation would simply mimic the global response of a galaxy, but in fact the observations clearly show that the centers of galaxies often respond in a profoundly different manner. Hence we need to answer all of the questions raised above for the nuclear regions independently.

How does the character of the star formation in interacting galaxies compare to the steady-state star formation in normal galaxies? It is clear that strong interactions and mergers can increase the rate at which a galaxy converts its interstellar gas into stars, but can they also change the nature of the star formation and the resulting stellar populations? For example, is the initial mass function (IMF) different in starbursts? Is the spatial distribution of the star formation distinct, in terms of its nuclear fraction, spiral structure, or bar structure? Is the clustering of the star formation systematically different? The fact that our own Galaxy contains at least four distinct stellar components (disk, spheroid, nucleus, globular clusters) testifies to the diversity of star forming environments in galaxies, and we would like to know what role interactions may have played in forming these components.

What are the physical properties of starbursts in interacting galaxies? The most basic information we would like are the SFRs, durations, ages, and star formation efficiencies of the bursts, and how those properties are coupled to the dynamical properties of the interactions. In this context, interacting galaxies offer a powerful testing ground for diagnostic techniques that can then be applied to interpreting the star formation properties of more distant objects.

What are the properties and physical nature of the infrared ultraluminous galaxies? We would like to measure the same burst properties as listed above, but because of the highly obscured nature of the IRAS galaxies, the methods are somewhat different. More fundamentally, we need to identify the basic nature of the IR power source (star formation or AGN or both?) and understand how the ultraluminous galaxies relate to other interacting starburst galaxies.

How is the star formation in interacting galaxies triggered and regulated? This is one of the most important questions, and one that we would dearly like to answer for *any* galaxy! What determines the timescale for the starburst, and why is it frequently much shorter than the dynamical timescale of the interaction? Why is the efficiency of star formation $1-2$ orders of magnitude higher than it is in normal disks? Why is the star formation so highly confined in the most luminous starbursts? What mechanism transports the gas to the galactic center? How are the nuclear starburst and AGN activities connected? We are only beginning to confront these kinds of questions.

What is the cosmological significance of interaction-induced star formation? How are local interacting and merging galaxies related to primeval galaxies? What fraction of stars in the universe form in interactions today? How does this fraction change with cosmological lookback time? What constraints do the dynamical properties of galaxies provide on the recent merger history of the universe?

1.8 Organization of Lectures

Before we confront the issues raised above, I will begin with a brief tutorial on the star formation properties of normal galaxies, and the observational techniques that are used to measure these properties (Lecture 2). Before we address the properties of interacting galaxies we need to know something about star formation in isolated galaxies!

In Lecture 3, I review the global star formation properties of interacting galaxies, with emphasis on statistical studies of large samples, to provide an objective assessment of the full range of properties of this diverse class of objects.

I will also present several examples of prototype systems, to provide points of reference for interpreting the statistical results. Then in Lecture 4, I review the evidence for triggering of nuclear activity in interactions, both in terms of nuclear star formation bursts and nonthermal (AGN) activities. In Lecture 5, I summarize the properties of ultraluminous infrared galaxies. Although the properties of these galaxies overlap with those of other interacting systems, the enormous literature on these objects merits a separate discussion.

Lectures 6, 7, and 8 deal with the astrophysical interpretation of these observations. In Lecture 6, I describe the spectral synthesis techniques that are used to diagnose the star formation properties of galaxies at various wavelengths, and apply them to a few examples of nearby systems. These results provide us with a rudimentary picture of the range of strengths, timescales, and efficiencies of the bursts in different types of objects. I then apply these results to confront the problem of how the bursts are triggered and regulated in Lectures 7 and 8. Our understanding of the underlying physics of large-scale star formation— in interacting or noninteracting galaxies— is at a very primitive stage. I will discuss distinct physical regimes that are present in these galaxies (Lecture 7), and review some of the most promising ideas for explaining the star formation properties (Lecture 8).

Finally, Lecture 9 addresses the question of the cosmological significance of mergers and interactions for galaxy evolution. The case for a dominant role of mergers is presented in depth in the lectures by François Schweizer and Josh Barnes. For the sake of balance, I will present a more conservative assessment, and offer some devil's advocate arguments for why the role of mergers may in fact be less important, at least for the spiral and irregular galaxies that make up the bulk of the present-day field population.

2 Star Formation Properties of Normal Galaxies

Before we address the star formation properties of interacting galaxies, it is useful to review the spectral diagnostics of star formation and the properties of normal galaxies. Most galaxies manage to form copious numbers of stars without the benefit of a tidal interaction, and it is a common mistake to attribute everything in a disturbed galaxy to the interaction! I begin with a summary of the methods used to measure current star formation rates (SFRs) in galaxies, then briefly summarize the star formation properties of the Hubble sequence, and the underlying evolutionary interpretation of this sequence. The measurements of past SFRs in fossil starbursts is described separately in Lecture 6. For more complete reviews, I refer the reader to Kennicutt (1990a; 1992c).

2.1 Observational Techniques

Ideally, one would measure the SFR in a galaxy by directly counting the number of young stars in a certain mass range. Unfortunately this is impossible for all but the nearest galaxies, and even there the brightest stars are predominantly evolved supergiants, making it difficult to measure an accurate SFR. As a result, almost all measurements of SFRs in galaxies are based on measurements of the integrated spectrum. Several observational techniques have been developed and applied to interacting galaxies. Although I will describe each method separately, the most effective applications combine the results of two or more techniques, to improve the accuracy of the SFR measurement and minimize systematic errors.

Integrated Colors and SEDs. It has been recognized since the pioneering work of Morgan & Mayall (1957) that galaxies exhibit a large range of spectral types. A modern illustration of this diversity is displayed in Fig. 2, which shows the integrated spectra of 8 nearby galaxies (Kennicutt 1992a). Although the spectra are sorted by morphological type, a parallel spectral sequence is apparent. As one progresses from elliptical through spiral to irregular systems, the composite spectrum or spectral energy distribution (SED) evolves from one dominated by G and K-type giants, characteristic of stars several billion years old, to one increasingly dominated by a blue continuum characteristic of A-type main sequence stars.

Although the integrated spectra contain contributions from the full range of stellar spectral types and luminosities, it is easy to show that at visible wavelengths the dominant contributors are intermediate-type main sequence stars (mainly A to early F) and bright red (G–K) giants (Kennicutt 1986). The practical consequence is that normal galaxies fall on a relatively tight spectral sequence, with the spectrum of any given object dominated by the ratio of early to late-type stars, or alternatively by the ratio of young ($< 1\,\mathrm{Gyr}$) to old ($3-15\,\mathrm{Gyr}$) stars. This allows one to extract the stellar age ratio (and an estimate of the SFR) by modeling the spectrum with a relatively small number of colors or spectral indices.

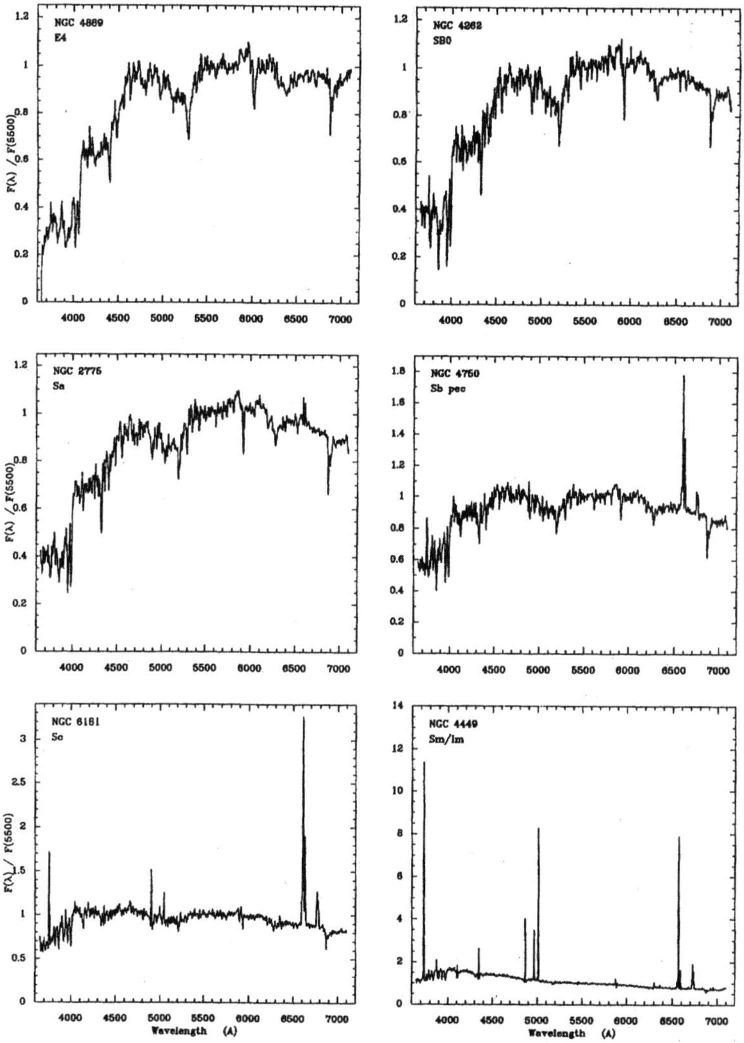

Fig. 2. Integrated spectra of elliptical, spiral, and irregular galaxies. The fluxes have been normalized to unity at 5500 Å (Kennicutt 1992a).

The most general form of this technique involves the synthesis of the full stellar spectrum of a galaxy (Lecture 6). A simpler application is to model the broadband colors of galaxies (Tinsley 1972; Searle, Sargent, & Bagnuolo 1973; Larson & Tinsley 1978). Briefly (see Lecture 6 for details), stellar evolution models are used to derive the effective temperatures and bolometric luminosities for various stellar masses as a function of time, and these are converted into broadband luminosities vs time using stellar atmosphere models. Those luminosities in turn are weighted by an IMF and summed to synthesize the integrated luminosities and colors of model star clusters as a function of time. These "single-burst models" by themselves may adequately describe the properties of a galaxies with a nearly coeval population, for example an old elliptical galaxy or a young starburst system. However most galaxies contain a more complex star formation history, and those can be modeled by summing the spectra of single-burst models of different ages, with the weighting function reflecting the presumed star formation history. For normal galaxies it is customary to adopt an exponential form:

$$R(t) = R_0 \, \mathrm{e}^{-t/\tau} \,, \tag{1}$$

where R is the SFR, R_0 is the initial SFR (at $t = 0$), and τ is the characteristic timescale for the star formation. The exponential form is little more than a convenient approximation, and often a grid of models with a range of timescales τ is calculated. Another useful parameterization is in terms of the ratio of the present-day SFR to the average past SFR (the birthrate parameter):

$$b \equiv \frac{R_{\mathrm{p}}}{\langle R \rangle_{\mathrm{past}}} \,. \tag{2}$$

Hence a galaxy which has ceased star formation would have $b = 0$, one with constant star formation $b = 1$, and a galaxy with a star formation burst $b \gg 1$. For the purposes of these lectures I will define a starburst as being an episode of star formation sufficiently strong that it cannot be sustained for longer than a small fraction of the Hubble time ($b \gg 1$), independent of the absolute luminosity or mass of the burst.

Although these models contain several parameters, including the star formation history, age, metallicity, and IMF, the colors of normal galaxies are remarkably well represented by a single parameter sequence of models with fixed age, composition and IMF, varying only in the time dependence of the SFR (Searle et al. 1973; Larson & Tinsley 1978). This is illustrated in Fig. 3, which shows the colors of normal galaxies from the RC2, along with a set of models from Kennicutt (1983). Each line corresponds to a set of coeval populations with fixed IMF, and exponential star formation histories with b ranging from 0 to 2. The best fitting model (middle line) is for roughly Salpeter IMF and age of 15 Gyr (Kennicutt 1983).

The models allow one to infer b directly for a given galaxy, simply by matching its colors to the closest point on a model sequence (e.g., Kennicutt, Tamblyn, & Congdon 1994). The models also predict the SFR per unit luminosity in a given bandpass as a function of b (Searle et al. 1978; Larson & Tinsley 1978; Tinsley & Danly 1980), and this can used to crudely estimate the absolute SFR.

The astute reader will immediately recognize several potential pitfalls in this method. The galaxy colors are logarithmic functions of b, so color errors will propagate exponentially into the inferred SFRs. An error of only a few hundredths of a magnitude in a red galaxy can easily correspond to an order-of-magnitude error in the SFR. For the same reasons, the SFR scale is sensitive to systematic error if the presumed metallicity, age, IMF, or reddening are incorrect (Larson & Tinsley 1978). Unfortunately most of these effects are nearly degenerate with the galaxy color sequence itself, so one cannot easily correct for

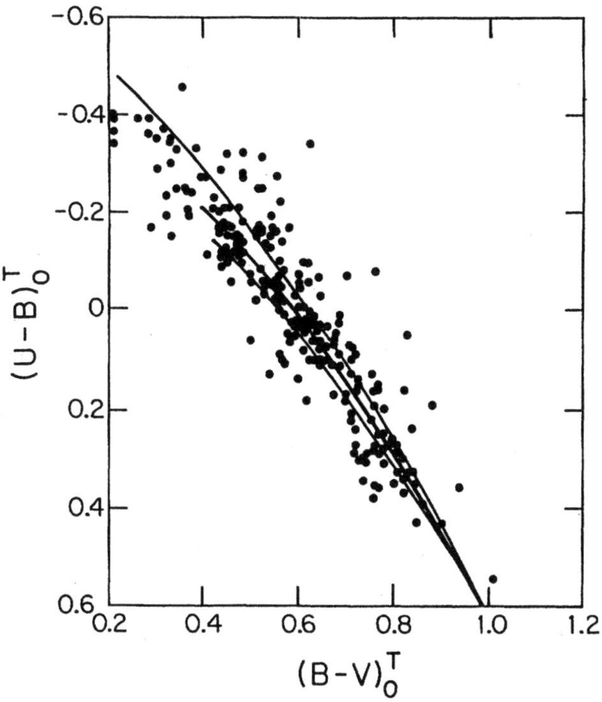

Fig. 3. *UBV* colors of spiral galaxies and simple synthesis models with constant age and IMF, and varying only in the star formation history (Kennicutt 1983).

these uncertainties. As a result, color synthesis is primarily useful in characterizing the general star formation properties of samples of galaxies, but not for providing an accurate absolute SFR for a given galaxy. Nevertheless modeling of integrated colors and SEDs remains a powerful technique for studying the systematics of star formation in distant galaxies, where only limited spectral information is available (e.g., Schade et al. 1995).

UV Continuum Fluxes. Many of the limitations encountered in modeling of the stellar continuum can be avoided if one observes at wavelengths where the continuum is dominated by young stars; in that case the SFR will scale linearly with the monochromatic flux. For active star-forming galaxies, the best continuum wavelengths are in the range 1300 – 2500 Å, longward of the Lyman absorption lines but short enough to minimize spectral contamination from older stellar populations. This wavelength range has the additional advantage that it is accessible from the ground in highly redshifted systems ($z \simeq 2-5$). A beautiful example is shown in Fig. 4, which displays near-UV spectra of an OB association in the Magellanic irregular NGC 4214 observed with HST (Leitherer et al. 1996) next to the redshifted spectra of a $z = 3.2$ starburst galaxy observed with the Keck telescope (Steidel et al. 1996).

The inaccessibility of this part of the spectrum for low-redshift galaxies has limited the application of this technique, but the situation is improving

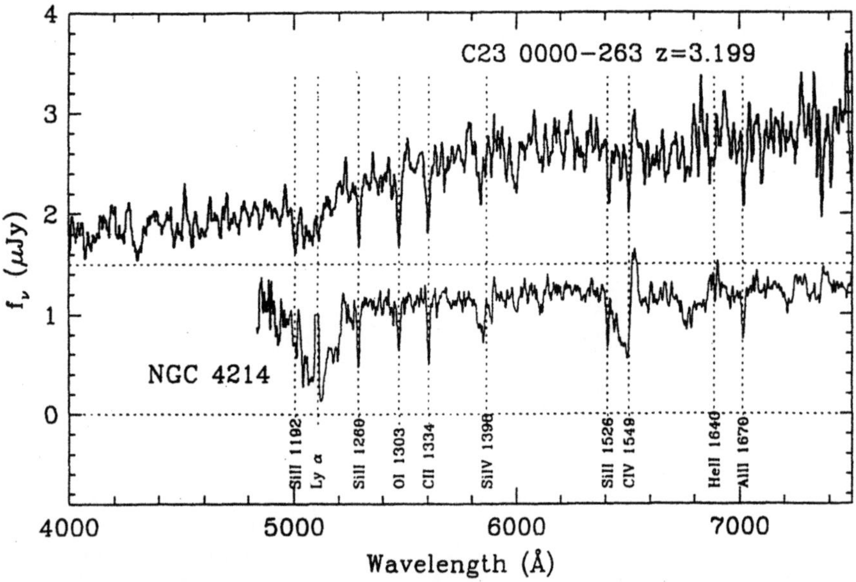

Fig. 4. Ultraviolet spectra of star forming galaxies near and far. The upper spectrum is a $z = 3$ object identified with the Keck telescope, while the lower spectrum is of an HII/OB association in the nearby galaxy NGC 4214, observed with HST (Leitherer et al. 1996). Figure from Steidel et al. (1996).

rapidly. Integrated fluxes of nearby galaxies are available from balloon experiments and observations with sounding rockets and small satellite payloads, most notably from groups at Marseille (Buat 1992 and references therein) and Goddard (Bohlin et al. 1990 and references therein). Recent imaging observations with ASTRO-B and HST have added significantly to the database in this area (e.g., Meurer et al. 1995; Maoz et al. 1996). A complete database of IUE spectra has also been compiled (McQuade, Calzetti, & Kinney 1995; Storchi-Bergmann, Kinney, & Challis 1995).

Calculating the conversion between the flux at an arbitrary UV wavelength and the SFR follows the same procedures as outlined earlier. SFR calibrations in terms of monochromatic UV fluxes are available from Buat, Donas, & Deharveng (1987) and Leitherer & Heckman (1995). Since the UV luminosity is directly coupled to the youngest population (< 20 Myr) the derived SFRs are insensitive to the star formation history (unlike broadband colors), but on the other hand they only trace the massive end of the IMF, several solar masses and higher. Hence one must either be content with measuring the SFR for massive stars only, or other methods or assumptions must be applied to extrapolate the IMF to lower stellar masses. In systems with very low SFRs, contamination from older post-AGB populations can become very important, and the UV continuum can only yield upper limits on the SFR.

The most serious limitation of this method is the very strong sensitivity of the UV flux to interstellar extinction. Because of the patchy structure of the extinction, a simple single-screen reddening model is inadequate, and a more general radiative transfer model is required. Such models have been applied to broadband UV data (e.g., Buat 1992; Buat & Xu 1996), and these indicate that corrections of order 1 mag are typical in normal spiral galaxies; in the dusty cores of many starburst galaxies much larger corrections are required, and the UV spectrum is of little practical value. For a detailed discussion of the UV extinction properties of galaxies, I refer the reader to papers by Calzetti et al. (1994). Despite these limitations, the UV spectrum remains a very powerful tool for characterizing the SFRs of galaxies, especially at high redshift where it is readily observed with ground-based telescopes.

Emission-Line Fluxes. If we look again at Fig. 2, we see that the most dramatic change in the integrated spectrum with galaxy type is the rapid increase in the strengths of the nebular emission lines. From basic Strömgren theory we know that the nebular lines effectively re-emit the ultraviolet luminosity of the galaxies shortward of the Lyman limit; for a radiation-bounded system the fluxes of the hydrogen recombination lines scale directly with the ionizing flux. Thus the emission lines provide a direct measure of the SFR for the massive, hot stars that dominate the ionizing continuum. This makes the recombination lines, Hα in particular, exceptionally sensitive tracers of the SFR for massive stars, and the most widely applied method for normal galaxies. Databases for normal galaxies are available from Kennicutt & Kent (1983) and Romanishin (1990). Studies of interacting galaxies in Hα are described in detail in Lecture 3.

The method is straightforward. The integrated Hα flux is usually measured with aperture photometers or nowadays with CCD imagers, using interference filters tuned to the emission line and the neighboring stellar continuum. The conversion factor between the Balmer flux and the SFR is then computed by constructing model star clusters as described in earlier, except that the stellar atmospheres are integrated shortward of the Lyman limit to derive the ionizing photon flux. These are then converted to Balmer emission-line fluxes using standard recombination theory. Widely used conversions are published by Kennicutt (1983), Gallagher, Hunter, & Tutukov (1984), Leitherer (1990), Kennicutt, Tamblyn & Congdon (1994), and Leitherer & Heckman (1995). The conversions are sensitive to the assumed IMF slope and stellar mass limits, and different authors adopt different conventions, so I refer the reader directly to the primary papers for details.

Although Hα is used most often (it is the strongest recombination line in unreddened objects), any hydrogen recombination line will suffice, and for dusty regions the Brγ or Brα lines are often applied in lieu of the Balmer lines. The thermal radio continuum provides another direct measure of the ionization rate, if contamination from synchrotron emission is removed (Israel & van der Hulst 1983). In the blue, higher-order Balmer lines can be used, but underlying stellar absorption often is difficult or impossible to remove. The [OII] doublet can serve as a substitute for the Balmer lines, and is especially useful for high-redshift galaxies, where Hα becomes unobservable. Kennicutt (1992a) provides a rough empirical calibration of the [OII] SFR scale.

The primary advantages of this technique are its sensitivity (nearly half of the ionizing photons produce an Hα photon), and the direct coupling between the nebular emission and the massive SFR. With CCD detectors and narrowband filters, the Hα emission can be mapped at high resolution with even a small telescope, so this technique has provided our most detailed information about the distribution and structure of the star forming regions in galaxies.

The chief disadvantages of the method are its sensitivity to extinction, the IMF, and the escape of ionizing photons from the galaxy. As with the UV continuum fluxes, the emission-line fluxes only trace the most massive stars ($M \geq 10\,\mathrm{M}_{\odot}$), and the SFR for lower-mass stars must be derived in some other way. Kennicutt (1983) used measurements of the visible continuum together with Hα to constrain the IMF down to $\sim 1\,\mathrm{M}_{\odot}$, but it is not clear whether this IMF is valid in extreme starburst environments (Lecture 6). Extinction is the other serious problem, especially in luminous starburst galaxies. The extinction in normal galaxies can be constrained by measuring the reddening of HII regions, or by comparing the integrated Hα and thermal radio fluxes of galaxies (e.g., Kennicutt 1983; van der Hulst et al. 1988). These results show that the typical extinction correction is about 1 mag. However it can be much higher in the nuclear regions of galaxies, and in the strongly concentrated starbursts found in many mergers. For those objects, the SFR is best determined instead from infrared or submillimeter recombination lines or from modeling of the far-infrared emission from dust (below).

Far-Infrared Continuum. A significant fraction of the bolometric luminosity of a galaxy is absorbed by interstellar dust, and this radiation is re-emitted in the thermal infrared, at wavelengths of roughly $10 - 300\,\mu$m. Since the absorption cross section of the dust is strongly peaked in the ultraviolet, this far-infrared (FIR) emission can be a sensitive tracer of the SFR. The IRAS survey has produced a database of over 30,000 galaxies (Moshir et al. 1992), offering a rich reward to those who can calibrate an accurate SFR scale from the $10 - 100\,\mu$m FIR emission.

Calibration of this SFR scale for normal galaxies has proven to be difficult, however. The dust grains absorb radiation with a broad range of wavelengths, and hence the heating function contains contributions from young and old stars alike. Modeling this heating, and separating the component associated with star formation is complicated by the different spatial distributions of young stars, old stars, and dust. Detailed modeling of the Galactic FIR spectrum reveals several distinct components (Cox, Krügel, & Mezger 1986), including a warm component peaking near $60\,\mu$m, associated mainly with young star forming regions, and a cooler "FIR cirrus" component, generally attributed to heating from the general interstellar radiation field. The latter probably includes contributions from young stars as well (Devereux & Young 1990), but in normal galaxies much of the heating may be due to older stars (e.g., Walterbos & Schwering 1987; Lonsdale & Helou 1987; Walterbos & Greenawalt 1996; Buat & Xu 1996). Uncertainty

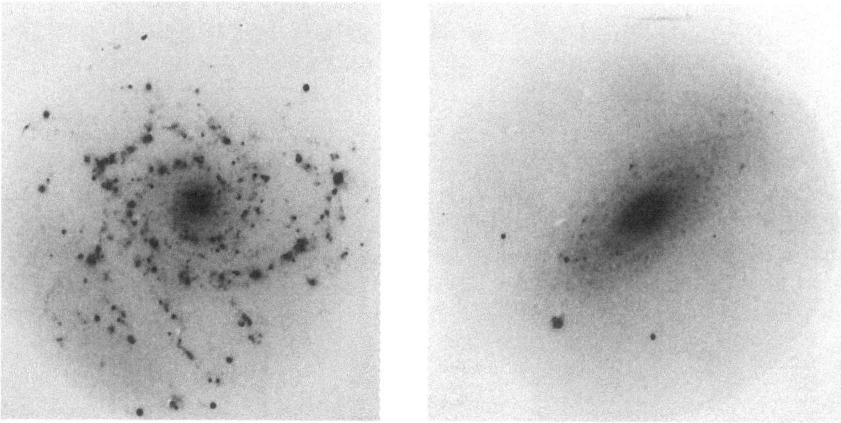

Fig. 5. Comparison of Hα images of NGC 628, an Sc galaxy, and NGC 2841, an early-type Sb galaxy. The bright objects in the latter image are foreground stars.

over the nature of this cool component has stymied efforts to construct a linear SFR calibration based on some combination of the IRAS spectral indices. Many galaxies also show shorter wavelength components that appear to be associated with nuclear sources and/or emission from small grains.

The physical situation is often much simpler in dusty starburst regions, however. There the stellar radiation field is dominated by a young stellar population, and the optical depth of the dust often is very high. In such objects the interpretation of the FIR emission is simple; it effectively measures the bolometric luminosity of the starburst. This in turn makes the FIR flux a very powerful probe of the total SFR, and often the only reliable probe available. We exploit this in Lectures 5 and 6.

2.2 Results: Star Formation in Normal Galaxies

Applications of these methods reveal an enormous diversity in the star formation properties of normal galaxies, and strong correlations with the Hubble types and other properties of the parent galaxies. Here I present a very brief synopsis, emphasizing those results which are the most useful for interpreting the properties of interacting galaxies. For simplicity I will focus mainly on results from Hα surveys, but most other studies lead to similar conclusions (with exceptions discussed below).

Absolute SFRs measured for nearby galaxies range over $\sim 0.001-10\,M_\odot\,yr^{-1}$, with the large range reflecting differences in both galaxy mass and morphological type (Kennicutt 1983; Gallagher et al. 1984; Caldwell et al. 1991, 1994; Kennicutt et al. 1994). The strong type dependence is already apparent in Fig. 2, and it is illustrated directly in Fig. 5, which compares Hα images of an early Sb galaxy (NGC 2841) to a comparable luminosity Sc galaxy (NGC 628). A more quantitative illustration is shown in Fig. 6, which shows the distribution of Hα

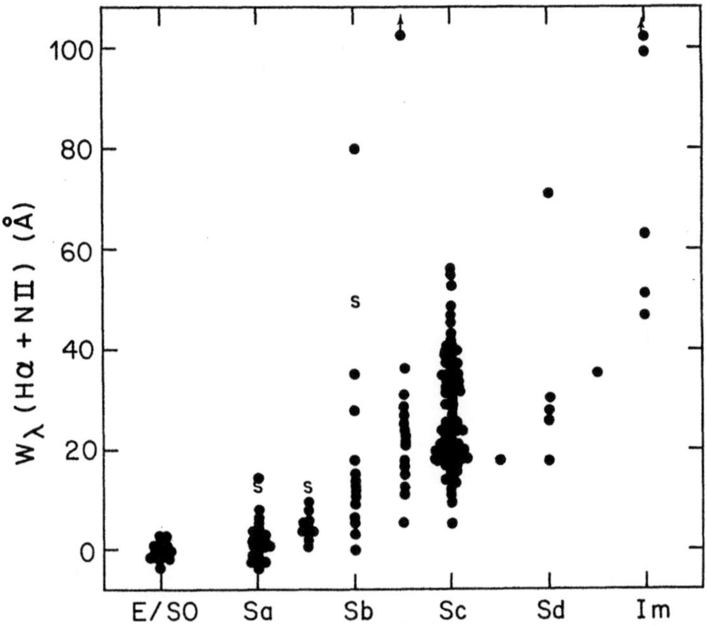

Fig. 6. Distribution of Hα + [NII] emission-line equivalent widths for a large sample of nearby galaxies (Kennicutt & Kent 1983).

+ [NII] equivalent widths in a sample of 150 nearby galaxies, subdivided by Hubble type. This index is defined as the emission-line luminosity normalized to the adjacent continuum flux, and hence is proportional to the SFR per unit (red) luminosity. In all but the strongest emission-line galaxies, the red continuum is dominated by old stars, so this provides an approximate index of the SFR per unit galaxy mass.

The SFR per unit luminosity is a strongly increasing function of Hubble type on average, increasing from zero in E/S0 galaxies (within the observational errors) to several solar masses per year for an Sc–Irr L^* galaxy (roughly comparable to the Milky Way), increasing severalfold with each successive spiral subtype. Also apparent in Fig. 6 is a very large dispersion in relative SFR among galaxies of the same type. The scatter, roughly an order of magnitude for most types, is much larger than would be expected from photometric errors or extinction effects, so most of it must reflect real variations in the SFR. Several mechanisms appear to be responsible, including variations in gas content (Kennicutt 1989), nuclear emission, interactions (Lecture 3), and most likely some level of temporal variation in the SFR within individual objects (Kennicutt et al. 1994).

A very different interpretation of these trends along the Hubble sequence has been proposed by Devereux & Young (1991), Devereux & Hameed (1997), and Tomita, Tomita, & Saito (1996), based primarily on the statistics of FIR emission. These authors point out that the average FIR luminosities and FIR/blue luminosity ratios of spirals show almost no change over the type range S0/a to

Im. If the FIR fluxes scale with the SFR, then this would imply that SFRs and star formation histories are virtually independent of morphological type. The flat inconsistencies between the interpretation of the FIR data and the optical colors, UV fluxes, and Hα results have never been completely explained. Devereux & Hameed (1997) and others have suggested that the optically-selected samples are biased in such a way as to produce a spurious Hubble type dependence, and that such samples systematically underestimate nuclear star formation in early-type galaxies (due to extinction).

The other possibility is that the FIR statistics are strongly affected by differences in dust heating across the Hubble sequence, with much of the FIR emission in early-type galaxies being produced by non-star forming populations (e.g., Lonsdale & Helou 1987; Kennicutt et al. 1994; Walterbos & Greenawalt 1995). Normalizing the infrared luminosities to the blue luminosity also may bias the FIR statistics, because the blue mass/light ratio itself is a strong function of Hubble type. Although I do not believe that selection effects in the optical data can erase the dramatic trends with Hubble type that are observed in Figs. 2–6, I do suspect that extinction effects in the centers of early-type galaxies, barred systems in particular, may have caused some of them to be overlooked in optically selected surveys.

High-resolution imaging of individual galaxies in Hα reveals that the order-of-magnitude changes in the total SFR along the Hubble sequence are produced in roughly equal parts by an increase in the total number of star forming regions per unit mass or area, and an increase in the characteristic masses of individual regions (Kennicutt, Edgar, & Hodge 1989). Both trends are nicely illustrated in Fig. 5. The changes in the mass spectra of the dominant star forming regions may have profound implications for the structure and evolution of their respective interstellar media. For example, in an Sa or early Sb spiral the average OB star forms in an HII region of roughly the size of the Orion nebula, in a cluster containing only a handful of massive stars (or singly). By contrast, the average massive star in a late-type Sc or Irr galaxy forms in a giant HII/OB association containing hundreds or thousands of OB stars, and often in dense clusters analogous to the "blue globular" clusters in the Large Magellanic Cloud. Clearly the physical or dynamical conditions in these gas-rich, late-type galaxies have altered not only the amounts but also the character of the star formation, and we should not be surprised to find even more dramatic differences in the star formation environments in the most active interacting systems.

2.3 Interpretation: Star Formation Histories

The strong trends observed in the relative SFRs of galaxies along the Hubble sequence mirror underlying trends in their past star formation histories (Kennicutt 1983; Gallagher et al. 1984; Sandage et al. 1986; Kennicutt et al. 1994). This is shown directly in Fig. 7, which plots the birthrate parameter b for a large set of spiral galaxies studied by Kennicutt et al. (1994). The b values were derived from a combination of Hα measurements and broadband color modeling,

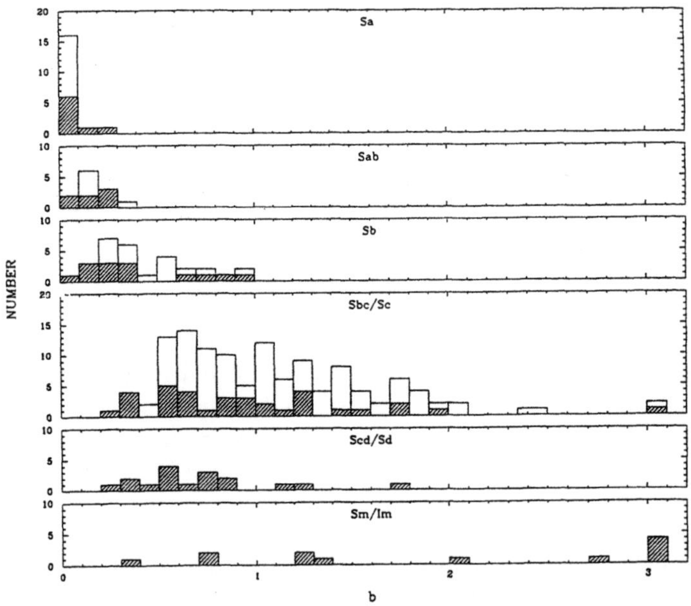

Fig. 7. Distribution of the ratio of the current SFR to the average past SFR for disks, subdivided by Hubble type (Kennicutt et al. 1994). Galaxies indicated by shading have individual bulge/disk measurements, while other data have been corrected statistically for bulge contamination.

and have been corrected for the effects of bulge contamination, so the numbers characterize the star formation histories of the disks alone.

Figure 7 shows that the typical late-type spiral has formed stars at a roughly constant rate ($b \sim 1$), consistent with direct measurements of the stellar age distribution in the Galactic disk (e.g., Scalo 1986). By contrast, early-type disks are characterized by rapidly declining SFRs, with $b \sim 0.01 - 0.1$. These changes in disk properties are much larger than those caused by changes in bulge/disk ratio, though differences in bulge structure may be important for driving the disk evolution. Unfortunately it is not possible with these kinds of measurements to characterize the detailed time dependence of the star formation histories across the Hubble sequence, but a schematic illustration of one such interpretation is shown in Fig. 8, taken from Sandage (1986). Note that other time dependences could fit the b constraints in Fig. 7, for example a high initial rate followed by a rapid decline, with the duration of the peak SFR increasing with later Hubble type. As more direct observations of distant galaxies become available, it should be possible to constrain the possibilities.

The main purpose of this discussion has been to provide a basis for comparing interacting and noninteracting galaxies, and to emphasize the range of properties in normal galaxies along the Hubble sequence. This complicates the interpretation of the evolutionary properties of interacting systems, because part of what we observe in the remnants may reflect the pre-existing populations (and ongoing

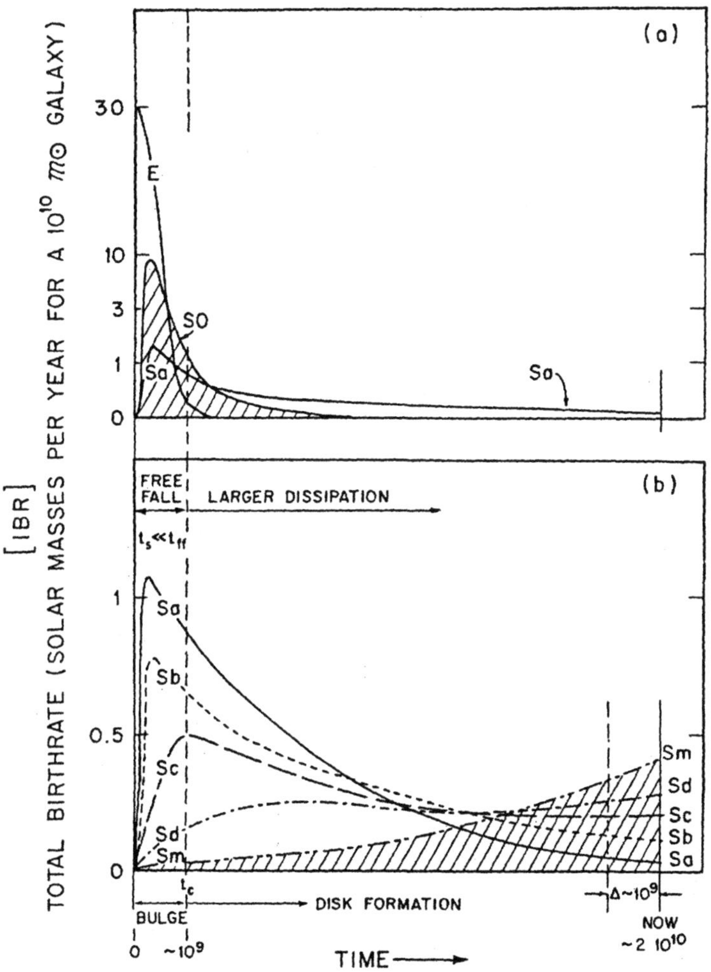

Fig. 8. A schematic illustration of how the stellar birthrate may have changed with galaxy type, based on data of the type shown in Fig. 7. Figure taken from Sandage (1986).

star formation) in the host galaxies, independent of the interactions themselves. Fortunately there are a number of unmistakable signatures of star formation among interacting galaxies, as we will see in the next lecture.

3 Global Star Formation in Interacting Galaxies

A preponderance of evidence points to a strong causal relationship between tidal interactions and large-scale bursts of star formation in gas-rich galaxies. As we will see, the effects of the interactions extend in scale from galactic nuclei to truly global triggering of star formation, often with very different responses on different spatial scales. This lecture concentrates on the large-scale star formation in interacting galaxies, with nuclear star formation covered in Lecture 4.

When confronted with the large body of statistical data on the star formation properties of interacting galaxies, it is easy to lose sight of the basic processes that take place in individual objects. For that reason, I begin by describing several case studies of interacting and merging galaxies, which illustrate the diversity of star formation properties. I then discuss systematic comparisons of the star formation properties of interacting vs isolated systems.

3.1 A Gallery of Interactions and Mergers

In order to illustrate the broad range of star formation properties in these systems, I have selected several examples, largely taken from the surveys done at Arizona (Kennicutt 1992b; Liu & Kennicutt 1995a, b; Turner 1997), including some of the famous objects that are discussed elsewhere in this volume. Most of the spectra shown are integrated spectra, made by drift scanning the galaxies across a spectrograph slit (Kennicutt 1992b).

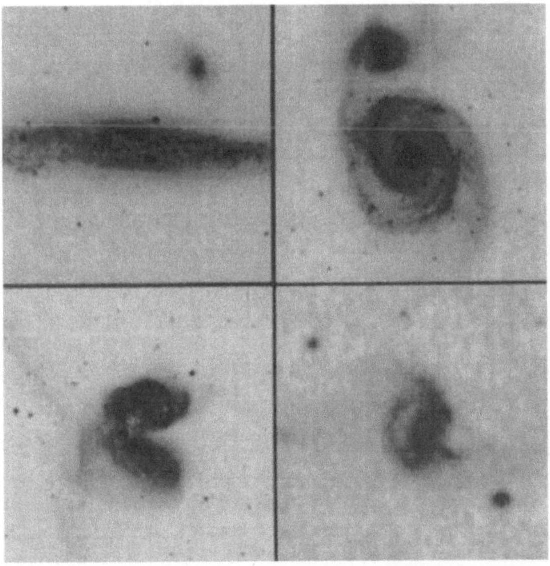

Fig. 9. Blue images of four examples of interacting galaxies, with a wide range in interaction strengths and star formation rates: **NGC 4631** (top left); **M51+NGC5195** (top right); **NGC 4038/9** (lower left); **NGC 7714/5** (lower right). For corresponding spectra see Figs. 10 and 11. Images are all from the Digitized Sky Surveys.[1]

[1] The Digitized Sky Surveys were produced at the Space Telescope Science Institute under U.S. government grant NAG W-2166. The images of these surveys are based

NGC 4631 (Arp 281) typifies the kind of weak interaction that is very common among paired and multiple spiral galaxies (Fig. 9a). Similar nearby examples would be M 31, M 101, or even the Galaxy. The only evidence of tidal interaction in NGC 4631 is a distorted HI distribution (Rand 1994), and as such it is representative of the weak end of the interaction scale. Thus it is not especially surprising that the integrated spectrum of NGC 4631 (Fig. 10a) is hardly abnormal either; it shows evidence of vigorous star formation, but at a level that is within the norm for late-type gas-rich spirals (compare with the spectra in Fig. 2). Not all interactions trigger a starburst; most weak interactions produce relatively modest effects on the star formation properties of their host galaxies.

M 51 + NGC 5195 (Arp 85) is one of the most famous and well studied interacting systems (Fig. 9b). The encounter clearly has distorted the disk of M 51 and produced a dramatic grand-design spiral pattern, but once again the integrated spectrum and SFR of the galaxy are quite typical for galaxies of its type, galaxies, as shown in Fig. 10b. This spectrum is of lower resolution than most shown in this section, which partly explains the weakness of the emission and absorption lines in the blue, but the SED and Hα emission are near the average for Sbc–Sc galaxies (Kennicutt 1983; 1992a). But how are we to physically interpret this result? If M 51 was a late-type spiral prior to its interaction, then the tidal encounter has hardly altered its global SFR at all.

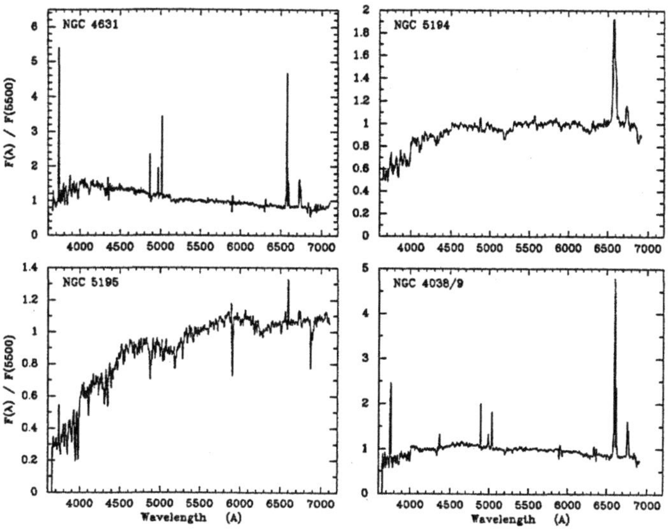

Fig. 10. Integrated spectra of four galaxies illustrated in Fig. 9, NGC 4631, M 51 (= NGC 5194), NGC 5195, and NGC 4038/9. The flux scales are normalized to unity at 5500 Å.

---

on photographic data obtained using the Oschin Schmidt Telescope on Palomar Mountain and the UK Schmidt Telescope. The plates were processed into the present compressed digital form with the permission of these institutions.

However several properties of M 51, including its relatively low gas fraction, its high metal abundance, its tight central spiral arms, and its rapidly rising rotation curve suggest that the galaxy might have been an earlier-type spiral prior to the interaction, perhaps an Sb galaxy (M 51 lacks a massive bulge, which precludes its being a very early-type system). In that case, the effect of the interaction would be quite profound, not only increasing the global SFR and exciting the spectacular spiral pattern but also transforming the Hubble type of the galaxy! This example illustrates the ambiguity we frequently face when trying to quantify the effects of interactions on individual galaxies; the interpretation of the observations often rests heavily on what we assume for the initial conditions.

The companion to M 51, **NGC 5195**, shows a completely different morphology (Fig. 9b), that of an amorphous or Irr II (I0) galaxy. Such galaxies are rare, and they almost always are associated with interactions or mergers (Cottrell 1978). The integrated spectrum of NGC 5195 is also peculiar (Fig. 10c). The stellar continuum shows a composite spectrum with very strong Balmer absorption lines, indicative of a recent starburst superimposed on an older population. The spectrum is heavily reddened and the strongest absorption features are interstellar lines (e.g., Ca II H+K, Na D). The galaxy possesses a very bright stellar nucleus that strongly influences the integrated spectrum.

NGC 4038/9 (Arp 244), The Antennae, is a well-studied example of an early-stage merger (Fig. 9c). The Hα images show a wealth of star formation, and HST observations reveal a rich population of massive young star clusters, perhaps even newly formed globular clusters (Whitmore & Schweizer 1995). The integrated spectrum (Fig. 10d) shows the characteristic features of an active star forming galaxy, with strong emission lines and a blue continuum. In this case the mean equivalent width of the Hα line is $2 - 3$ times higher than normal for late-type spirals and irregulars, so NGC 4038/9 can legitimately be categorized as a starburst galaxy, but again that the strength of the burst is at most a few times that in a steadily star forming disk. In most interactions, the increases in the SFR are relatively modest.

NGC 7714 (Arp 284A) typifies the upper end of the scale for optically selected starbursts; indeed the term "starburst" was first applied to this object (Weedman et al. 1981). It differs from the previous examples in two important respects. Its current star formation is very strongly concentrated to the nuclear regions (Figure 9d), and the starburst is extremely powerful, dominating the integrated spectrum (Fig. 11), which resembles that of a giant HII region! NGC 7714 is representative of an extreme class of interactions or mergers in which a substantial fraction of the stars in the galaxy are being formed over a dynamical timescale ($\sim 10^7 - 10^8$ yr). The properties of this class of nuclear starburst galaxies are described in more detail in Lecture 4.

NGC 5278/9 (Arp 239) represents an example of an interaction that has excited strong nonthermal activities in the nuclei of its galaxies. Both disks show high levels of star formation (Figure 12a), but the nuclei both show nonthermal spectral signatures, as shown in Fig. 13 (from Keel et al. 1985). The strong emis-

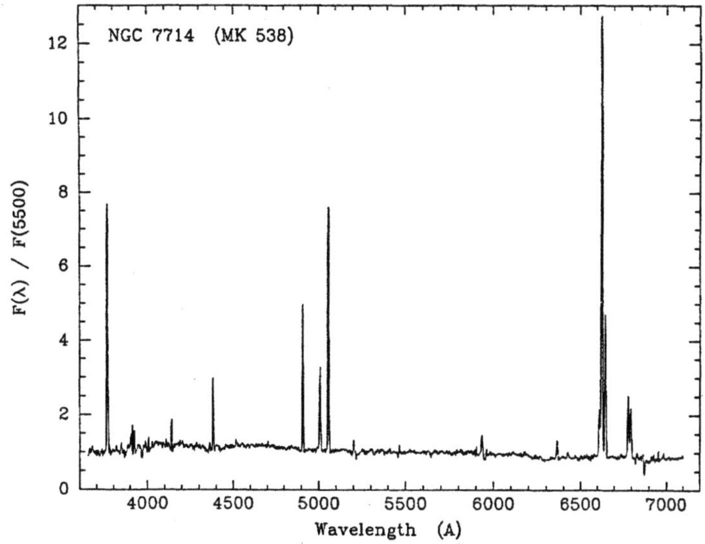

Fig. 11. Integrated spectrum of the prototype starburst galaxy NGC 7714 (Fig. 9).

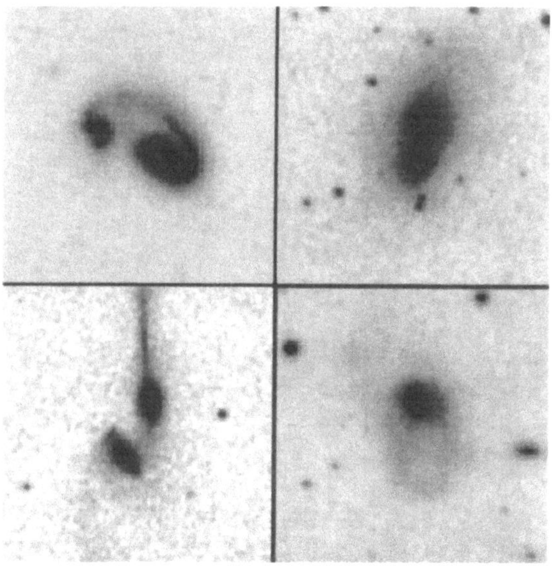

Fig. 12. Blue images of four strongly interacting or merging system:
NGC 5278/9 (top left), with AGNs in both galaxies;
NGC 750/1 (top right);
NGC 4676 (lower left);
NGC 3921 (lower right).
Corresponding spectra are shown in Figs. 13 and 14. Images taken from the Digitized Sky Surveys.

sion at [NII]λ6583, [OIII]λ4959,5007, and [OI]λ6300 is characteristic of Seyfert 2 or LINER spectra (Lecture 4), and Keel et al. (1985) classed both objects as borderline Seyfert/LINER nuclei. Another even more spectacular example of a double Seyfert interacting system is Markarian 463 (Mazzarella et al. 1991). Seyfert nuclei are rare, occurring in only $\sim 5\%$ of normal spirals, so the chances of a coincidental pairing of Seyfert galaxies are very small. The very existence of these systems offers evidence that interactions can trigger an AGN activity (Lecture 4).

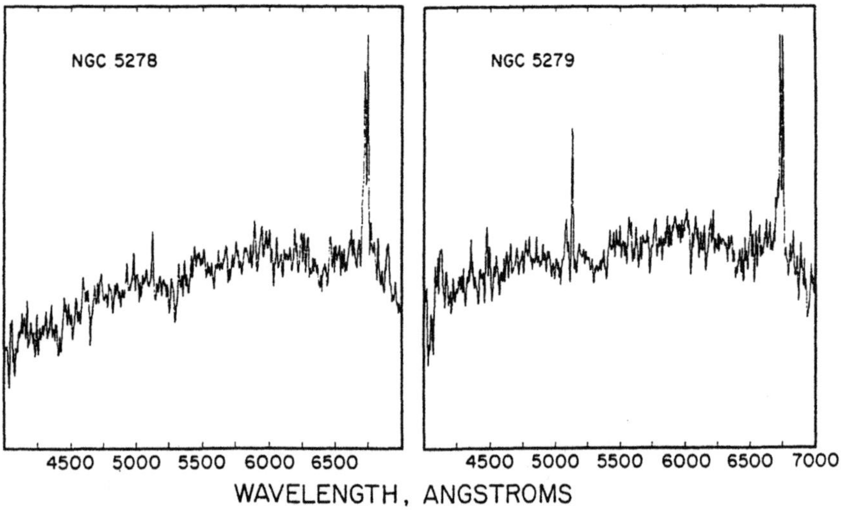

Fig. 13. Nuclear spectra of NGC 5278 and NGC 5279, showing non-thermal Seyfert 2 or LINER spectra. Figure taken from Keel et al. (1985).

NGC 750/1 (Arp 116) serves as a reminder that not all interactions or mergers need produce spectacular starbursts or AGN activities. As shown in Fig. 12b, it is an early merger of two elliptical or S0 galaxies. In this case the integrated spectrum (Fig. 14a) shows no evidence of star formation whatsoever; its spectrum is indistinguishable from that of normal E/S0 galaxies. The morphologies of NGC 750 and NGC 751 suggest that they are elliptical galaxies, and hence probably gas-poor objects, so the absence of star formation is not particularly surprising. Systems of this type may comprise $\sim 15-20\%$ of a morphologically selected sample of local mergers (Liu & Kennicutt 1995a), outnumbering those systems which produce ultraluminous starbursts.

Many other examples of evolved mergers *do* show evidence of strong starbursts, however. Three examples of such systems are **NGC 520** (Arp 157), **NGC 3921** (Arp 224), and **NGC 4676** (Arp 242), illustrated in Figs. 12 and 15. The spectra of these objects are quite distinct, as shown in Figs. 14. NGC 520 and NGC 3921 lack strong emission lines, indicating relatively low levels of current star formation (at visible wavelengths), but the spectra are dominated by Balmer absorption lines, a classic post-starburst signature. This young population has overwhelmed the light of any previous generations of stars; these galaxies clearly experienced very large starbursts that formed an appreciable fraction of their stellar mass, and we are observing the aftermath 10^7 to 10^9 years later. Such postburst spectra (sometimes called "E+A" spectra in analogy to postburst spectra of distant galaxies) are quite common among advanced mergers, but are extremely rare in other types of galaxies. In my survey of integrated spectra of 90 nearby galaxies the *only* examples of post-burst spectra were found in merging systems (Kennicutt 1992b).

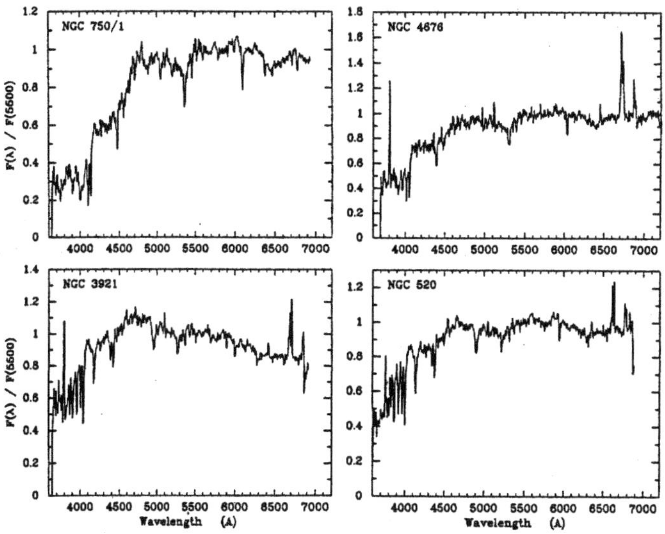

Fig. 14. Spectra of four examples of strongly interacting galaxies or evolved mergers. The spectra of NGC 520 and NGC 3921 are drift-scanned integrated spectra. The spectra of NGC 750 and NGC 4676 are integrated along a single long-slit position, but are representative of the spectra over most of the galaxies.

Not all advanced mergers fit this pattern. The spectrum of NGC 4676 (Fig. 14) shows abnormally strong Balmer lines from a postburst population, but the spectrum is composite, with a prominent older population as well. Apparently the starburst in this galaxy did not form as large a fraction of the stellar mass of the system. This example serves to illustrate the importance of observing a representative fraction of the galaxy before characterizing its entire stellar population.

The last class of interacting galaxies we consider here are the luminous infrared galaxies, illustrated in Fig. 15 by **M 82** (Arp 337), **Arp 220** (IC 4553), and **NGC 6240**. Here we must be very cautious when interpreting optical spectra, because we know that the bulk of the ultraviolet and optical luminosity of the galaxies is obscured by dust. Nevertheless the optical spectra provide ample evidence of a peculiar stellar population, and many parallels to the mergers just described.

M 82 is the prototype example of a luminous infrared starburst galaxy. It is not an interacting galaxy in the usual sense, in that there is no evidence of a merger of two stellar systems. However deep 21 cm HI maps show the presence of a tidal bridge of gas extending from M81 to M 82 (Yun, Ho, & Lo 1993), and it is likely that this interaction triggered the starburst in M 82. I present it as a example of an infrared-luminous starburst and as a probable interacting system.

The integrated spectrum of M 82 is shown in Figs. 16a-b. Despite the heavy extinction, the visible spectrum is dominated by Hα emission from a cluster of giant HII regions. At first glance the continuum appears anomalously red for such an active star forming galaxy, until one recognizes that the spectrum is severely reddened. An expanded plot of the blue spectrum is shown in Fig. 16b,

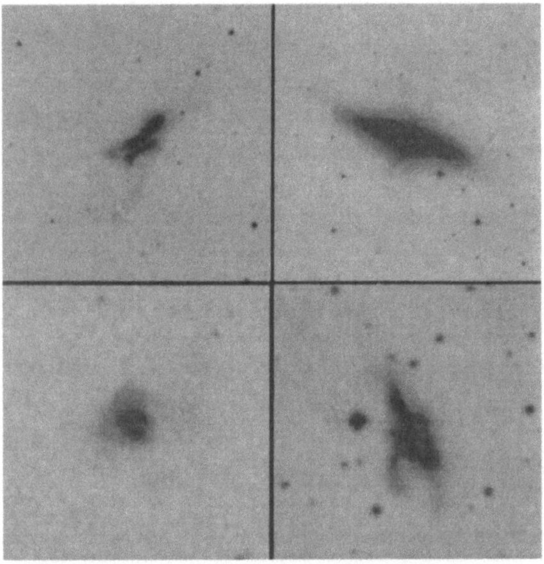

Fig. 15. Examples of four infrared-luminous mergers: **NGC 520** (top left); **M 82** (top right); **Arp 220** (lower left); **NGC 6240** (lower right). Images taken from the Digitized Sky Surveys.

and we now see the characteristic post-burst signature of strong Balmer lines (cf. Fig. 14). Note also the abnormally strong interstellar Ca and Na lines. The effects of the dust on the integrated SED are illustrated in Fig. 17, which compares the energy distribution of M 82 from the ultraviolet to the far-infrared with an optical emission-line starburst galaxy, NGC 1569. These objects share many physical similarities— enormous starbursts centered on massive central star clusters, strong bipolar galactic winds, and prominent young stellar populations–– but the presence of an obscuring gas and dust disk in M 82 has altered its observed spectrum entirely. It is interesting to speculate whether NGC 1569 represents the kind of galaxy that M 82 may evolve into over the next few tens of Myr.

Arp 220 is the type example of the ultraluminous infrared galaxies, with a bolometric luminosity in excess of $10^{12}\,L_{\odot}$. Although its primary power source is heavily obscured in the visible, its integrated spectrum (Fig. 16d) still shows the characteristics of a post-burst population superimposed on an older stellar component. The last spectrum shown is for NGC 6240 (Fig. 16c). It is one of the most peculiar integrated spectra I have ever observed for a galaxy, dominated by a nonthermal emission-line spectrum, not unlike that of an extraordinarily luminous supernova remnant! Otherwise the spectrum is very similar to that of M 82, with a heavily reddened continuum dominated by Balmer lines from young stars. Although dust severely limits the view of the central core sources in these three IR-luminous systems, the signatures of abnormal star formation and/or AGN activities are unmistakable.

I hope that these examples have conveyed the incredible diversity of star formation properties and stellar population in interacting galaxies, differences that reflect both the range of evolutionary stages and most of all a tremendous diversity of physical responses in individual systems. Although this leaves us

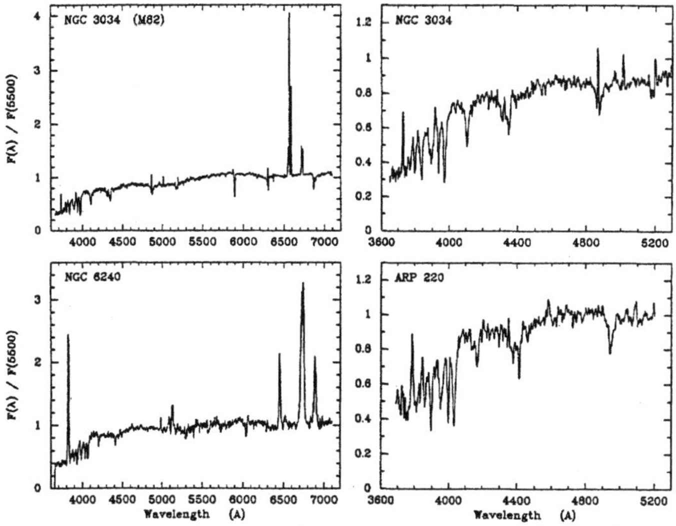

Fig. 16. Integrated spectra of the infrared-luminous galaxies M 82, NGC 6240, and Arp 220. Note the different wavelength scales for the spectra on the left and right sides of the figure.

with a daunting task of trying to integrate the gamut of properties within a single common astrophysical framework, the wide range of physical conditions provides important clues for constructing this common picture.

3.2 Star Formation Statistics: Guilt By Association Tests

We now turn to systematic comparisons of the star formation properties of interacting and noninteracting galaxies. One can approach this problem from two directions, either by measuring complete samples of interacting galaxies and comparing their SFRs with those of properly drawn control samples, or by comparing the frequencies of interactions in various samples of starburst galaxies. I first discuss examples of these latter "guilt by association" statistics. Although these results are more difficult to interpret in quantitative astrophysical terms, they provide an immediate indication of the importance of interactions in different types of starbursts.

As mentioned in Lecture 1, the first convincing association of interactions with starbursts came from the Byurakan surveys of UV-bright galaxies (Markarian et al. 1989 and references therein). Most of the galaxies identified in these objective prism selected catalogs are either Seyfert galaxies or starburst systems (Weedman & Khachikian 1969; Huchra 1977). From the early stages of the survey it became apparent that paired, interacting, and multiple-nucleus galaxies comprised an unusually high fraction of the systems (Heidmann & Kalloghlian 1973; Casini & Heidmann 1975; Karachentsev & Karachentseva 1974; Kazarian & Kazarian 1987). Similar studies are now being undertaken of the Kiso sample of ultraviolet excess galaxies (Comte et al. 1994).

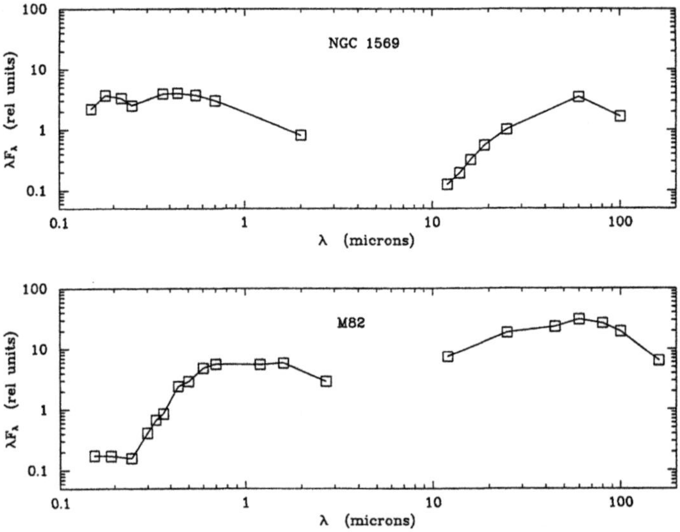

Fig. 17. Integrated spectral energy distributions (SEDs) for the infrared-luminous starburst galaxy M 82 and the optical starburst galaxy NGC 1569. The SEDs are plotted in terms of νF_ν, so that the area under any given portion of the plot is indicative of the contribution to the bolometric luminosity. Note the strong redistribution of energy from the UV-visible to the far-infrared in M 82.

The most complete study of the pairing properties of Markarian starburst galaxies was recently completed by Keel & van Soest (1992). They identified a sample of 516 non-Seyfert Markarian galaxies and found that 36% are members of interacting pairs, mergers, or double-nucleus systems. The Keel & van Soest study is especially valuable because a proper control sample was analyzed. When similar criteria are applied to a magnitude-limited sample of all disk galaxies, the fraction is only 7%. Thus interactions are clearly a triggering agent for starbursts, but most Markarian starbursts –over 60%– are associated with isolated systems. For a detailed study of the properties of multiple-nucleus Markarian galaxies see Mazzarella (1989).

Another major source of optically-selected starburst galaxies comes from objective prism surveys of emission-line galaxies. The resulting population of objects is similar to those revealed in the Markarian survey, but with a larger representation of dwarf galaxies. Studies of pairing fractions of various samples of emission-line galaxies have been carried out by Bothun et al. (1989), Taylor et al. (1995), and Telles & Terlevich (1995), yielding fractions of 25 – 75%, 57%, and 24%, respectively. The inconsistencies in these respective fractions reflect differences in the nature of the samples and in the criteria used to define companions, but in all cases they show an excess of companions over quiescent dwarf galaxies. However it is also clear that an interaction is not a precondition for a starburst; Telles & Terlevich emphasize that the most luminous starbursts in their HII galaxy sample are associated with isolated systems.

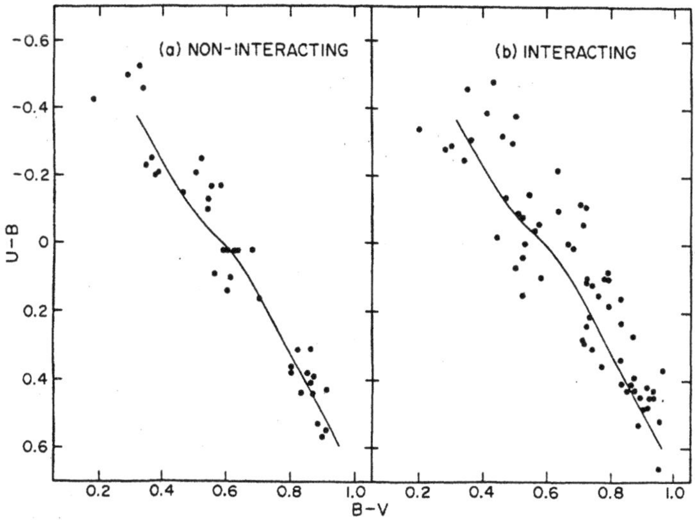

Fig. 18. Comparison of the *UBV* colors of interacting and noninteracting galaxies in the Arp atlas, from Larson & Tinsley (1978).

Perhaps the most intriguing "guilt by association" results apply to the statistics of luminous IRAS galaxies (LIGs). As discussed in Lecture 5, there is a very strong excess of interacting galaxies and mergers among LIGs, and the interaction rate is a strongly increasing function of infrared luminosity, rising from about 30% for systems brighter than $10^{10}\,L_\odot$ to $\sim 80-90\%$ for $L > 10^{12}\,L_\odot$ (Leech et al. 1994; Sanders & Mirabel 1996).

Taking these results together we can conclude that: (1) Interactions are an important triggering agent for starbursts over a wide range of galaxy types, luminosities, starburst strengths, and interaction types, accounting for at least 25% of all starburst events. (2) The importance of interactions increases sharply with the luminosity of the starburst. (3) All but the most powerful starbursts ($> 10^{12}\,L_\odot$) can be triggered *without* a visible interaction. Indeed a majority of low-luminosity starburst activities occur in noninteracting or very weakly interacting galaxies.

3.3 SFRs in Interacting vs Noninteracting Galaxies

Most of the diagnostic methods for measuring SFRs described in Lecture 2 have been applied to samples of interacting galaxies, and virtually all of the data point to a similar excess of star formation. Space does not allow for a complete discussion of the enormous literature on this subject; for more complete reviews see Kennicutt (1990b) and Keel (1991). Here I will describe several examples, with particular emphasis on the Hα studies.

Optical Colors. Analysis of the *UBV* colors of interacting and normal galaxies by Larson & Tinsley (1978) provided some of the first quantitative evidence

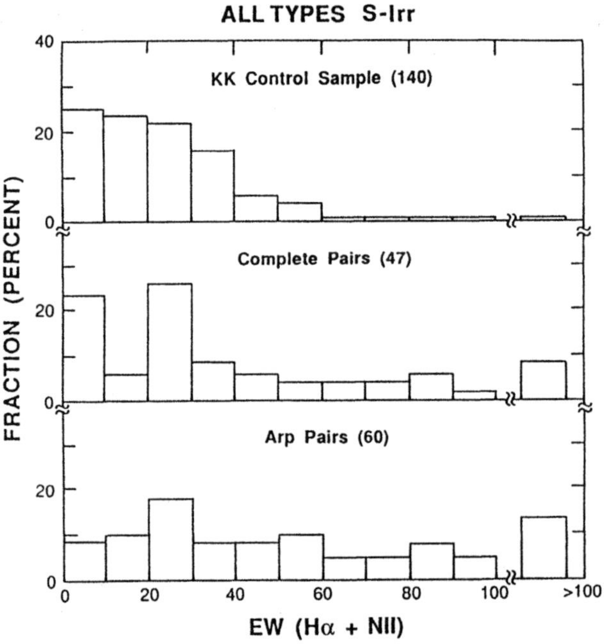

Fig. 19. Comparison of distributions of Hα+[NII] emission-line equivalent widths for two samples of interacting galaxies and a control sample of isolated galaxies, from Kennicutt et al. (1987).

for starbursts in interacting systems, as discussed already in Lecture 1. They showed that peculiar and interacting galaxies have a larger dispersion in color and an extended blue tail in the two-color diagram, and were able to model these results by adding starbursts to underlying normal stellar populations. Fig. 18, taken from Larson & Tinsley (1978), shows an explicit comparison for interacting galaxies in the Arp atlas. The interpretation of these data is described in detail in Lecture 6. The database of integrated photometry of interacting galaxies has been much improved in recent years, most notably with observations of southern galaxies by Bergvall & Johansson (1995 and references therein), and with observations of compact groups by Moles et al. (1994).

Multi-color CCD imaging has great potential for characterizing the distribution of star formation and the stellar populations in tidal features, but this potential is only beginning to be exploited. A notable study of this kind was published by Schombert, Wallin, & Struck-Marcell (1990). It provided comprehensive data on the colors of the tidal features in 25 interacting pairs, and analysis of the star formation properties of these features (see Lecture 2 by Schweizer). An extensive imaging study of Arp-Madore galaxies has been published by Smith & Hintzen (1991), and a detailed study of a small sample of mergers has been completed by Hibbard & van Gorkom (1996). A search of the literature will reveal numerous other imaging studies of individual objects.

Hα Emission. The sensitivity of Hα emission to the instantaneous SFR, and the relative ease with which data can be obtained have made this an ideal method for probing the systematics of star formation in interacting galaxies. The most

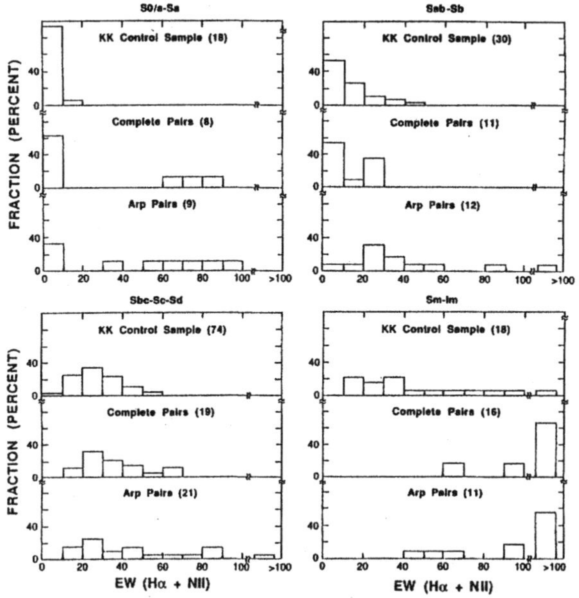

Fig. 20. Type-by-type comparison of Hα emission-line strengths of interacting and isolated galaxies, from Kennicutt et al. (1987).

extensive surveys to date were undertaken by Bushouse (1987) and Kennicutt et al. (1987; hereafter K87). Bushouse studied a sample of strongly interacting and merging systems, based on visual inspection of the POSS, while K87 studied two samples at opposite ends of the interaction spectrum, a complete sample of paired galaxies selected on the basis of a pairing algorithm (and independent of any visible evidence of interaction), and a second sample selected subjectively from the Arp (1966) atlas.

Figure 19 compares the distribution of SFRs in terms of the equivalent width of the Hα + [NII] emission lines (hereafter denoted "EW" for convenience), for both sets of paired galaxies as well as a control sample of noninteracting systems (Kennicutt & Kent 1983). The median EWs are 21 Å for the control sample, 27 Å for the complete pairs sample, and 46 Å for the Arp sample. Hence both samples of paired/interacting galaxies show excess star formation, but the average enhancement is quite modest. Bushouse's study (not shown) showed a similar enhancement of $\sim$ 2.5 times that in an isolated galaxy sample. The averages tell only part of the story, however, because the EW distributions of the interacting galaxy samples show enormous dispersions, ranging from galaxies with no enhancements to systems forming a substantial fraction of their total mass in new stars. The absolute SFRs in these latter systems reach levels as high as $\sim 100\,M_\odot\,yr^{-1}$ (K87; Bushouse 1987). However in the majority of interactions, the SFR enhancements are modest, a result that we will see repeatedly with the other star formation tracers.

It is important to be aware of two caveats when interpreting these results. One is that the Hα emission only measures the instantaneous SFR, produced by young stars with lifetimes of $< 10\,$Myr, so if an interaction has produced a major

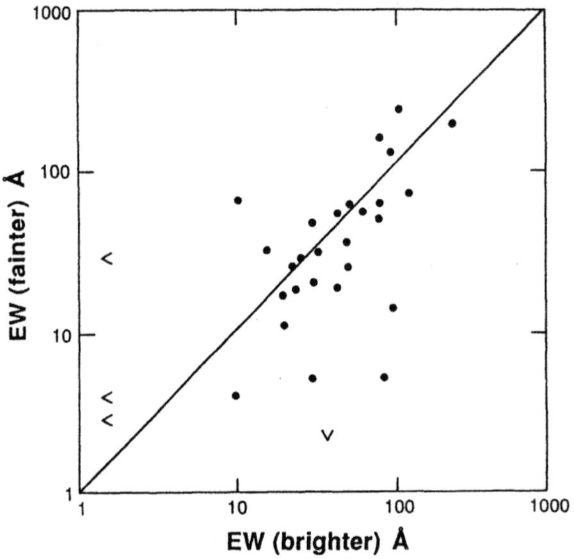

Fig. 21. Correlation between Hα emission-line strengths of individual members of galaxy pairs, from Kennicutt et al. (1987).

burst of star formation that subsided more than 10 Myr ago (e.g., NGC 520 or NGC 3921 in the previous section) this will not be reflected in the Hα EW. The Hα flux will also tend to underestimate the SFR in dusty starbursts. Both effects however act to diminish any actual star formation enhancements in the interacting galaxies.

Given the large dispersion observed in the SFRs of interacting galaxies, can we associate the SFRs with any physical property of the parent galaxies or the interactions themselves? In Lecture 2 we saw that the SFRs in normal galaxies are strongly correlated with properties such as Hubble type and gas content. However these correlations are virtually absent in interacting galaxies. Figure 20 shows the EW distributions from the previous figure broken down by the Hubble types of the component galaxies (K87), and demonstrates —quite remarkably— that the strong coupling between galaxy type and SFR that characterizes the Hubble sequence is completely absent among the interacting galaxies! Similarly a comparison of EW as a function of gas surface density shows little or no correlation (K87). Apparently the SFRs are being driven by processes that are virtually independent of the parent galaxy type, or alternatively the morphological types themselves have been completely altered by the interaction.

Are there any properties of the interactions that *do* correlate with the SFRs? Interestingly, one of the strongest correlations is between the SFRs of individual members of an interacting pair, the "Holmberg effect". Holmberg (1958) showed that the colors of paired galaxies are correlated, and a similar correlation is seen in the Hα EWs for the K87 sample, as shown in Fig. 21. This correlation holds whether the absolute EWs are compared, or if the EWs are scaled to the average for a given type (K87). Starburst galaxies tend to pair with other starburst galaxies, though exceptions are not uncommon. This strongly points to the mutual tidal interaction as being the primary triggering agent, overwhelming any effect of the initial Hubble types or gas contents of the galaxies (K87).

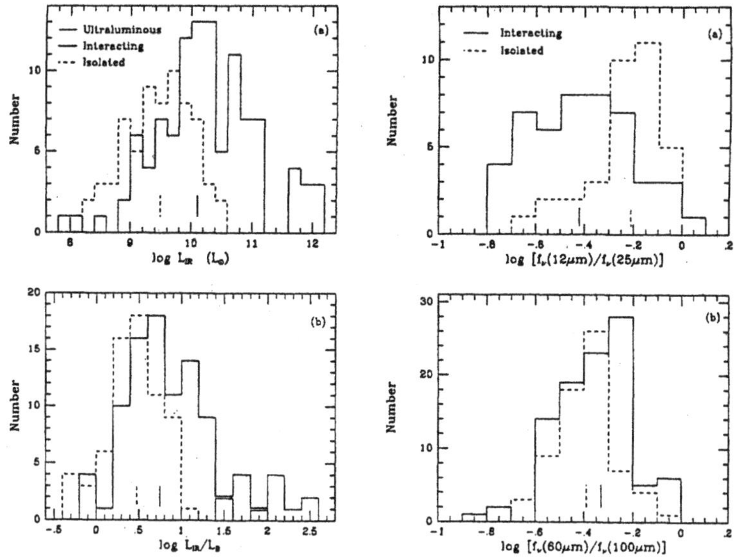

Fig. 22. Comparison of IRAS far-infrared luminosities and colors of interacting and isolated galaxies. Figure taken from Bushouse et al. (1988).

Given these results, one might expect to observe a strong correlation between the SFR and the pair separation or other dynamical property of the interaction. However several attempts to detect a correlation between projected separation and SFR have yielded negative results (e.g., Bushouse 1987; K87; Keel 1993). We return to this topic briefly in Sec. 3.4. The exception are the IR ultraluminous galaxies, which seem to be almost uniquely associated with evolved mergers, as discussed in Lecture 5.

Far-Infrared Emission. The IRAS survey has stimulated dozens of investigations of the FIR properties of interacting galaxies and mergers. I will discuss this subject in more depth in Lecture 5, and reviews can be found in Kennicutt (1990), Sanders (1992), and Sanders & Mirabel (1996). Here I restrict the discussion to global FIR properties of interacting galaxies, with nuclear starburst properties and IRAS ultraluminous galaxies covered in the next two lectures.

One of the first in-depth applications of the IRAS database, and still one of the most illuminating comparisons of interacting vs isolated galaxies was published by Bushouse, Lamb, & Werner (1988). Figure 22, taken from that paper, compares the FIR luminosities and colors of a set of interacting galaxies taken from Bushouse (1987), a small sample of ultraluminous objects from Sanders et al. (1988), and a control sample of isolated galaxies. The interacting galaxies show systematically stronger infrared emission, both in absolute terms and in terms of the FIR/blue ratios. Similar results were found by Kennicutt et al. (1987), Telesco, Wolstencroft, & Done (1988), Hintzen et al. (1989), Xu & Sulentic (1991), and Liu & Kennicutt (1995a). The FIR enhancements are about

3 times stronger than observed in Hα for the same sample (Bushouse 1987), which probably reflects a combination of the presence of evolved starbursts and extinction, the biases mentioned in the previous section. The basic conclusions are consistent with those seen in Hα, an enormous range of SFRs and a modest enhancement in SFR on average, ranging from 20 – 60% for objectively drawn samples of paired galaxies to factors of a few for the most disturbed systems.

The other major difference apparent in Fig. 22 is a tendency toward stronger 25 – 60 μm emission in the interacting galaxies (cf. Telesco et al. 1988). In normal galaxies, this warm dust component is associated with star forming regions, so the colors support the interpretation of the enhanced FIR emission being caused by an elevated SFR. However one should caution that enhanced nonthermal nuclear activity can cause similar excesses, especially in the 25 μm band. These points are addressed in more detail in Lecture 5.

Radio Continuum Emission and Supernova Rates. Some of the first evidence of anomalous activity in interacting galaxies came from radio continuum studies (Stocke 1978; Condon et al. 1982). Systematic studies of the global radio emission in paired and interacting galaxies have been carried out by Hummel (1981) and Smith & Kassim (1993), and for members of compact groups by Menon (1995). Interestingly, the integrated radio fluxes show quite modest enhancements, and are significantly *below normal* in the dense groups (Menon 1995). High resolution VLA mapping shows that the central radio sources are significantly enhanced in all of the interacting samples (Lecture 4).

The ultimate sources of the radio continuum emission are thought to be high-energy particles from supernovae interacting with the galactic magnetic field. In view of the previous results, we might expect to see an elevated supernova rate in interacting galaxies, at least in the central regions. Gathering adequate statistics for such a test is difficult, but an excess supernova rate has been reported by Smirnov & Tsvetkov (1981). The properties of these supernovae have been investigated further by Petrosian & Turatto (1995), who find a central radial concentration in interacting systems, consistent with radio continuum and Hα results (next section).

3.4 Kinematic Properties and Regulation of SFRs

Summarizing all of the results of these studies: (1) There is strong evidence for a systematic increase in star formation in interacting galaxies; (2) The average SFR enhancements are small, factors of a few or less; (3) There is a large range —at least two orders of magnitude— in the responses of individual systems.

It is unclear from these statistics alone how much of the dispersion in properties reflects the wide range of evolutionary stages present in the samples, and how much reflects differences in the actual responses of the galaxies themselves. As mentioned earlier, attempts to correlate the SFR with simple kinematic parameters such as pair separation or velocity difference have yielded negative results, but given projection effects and observational selection effects in the samples this

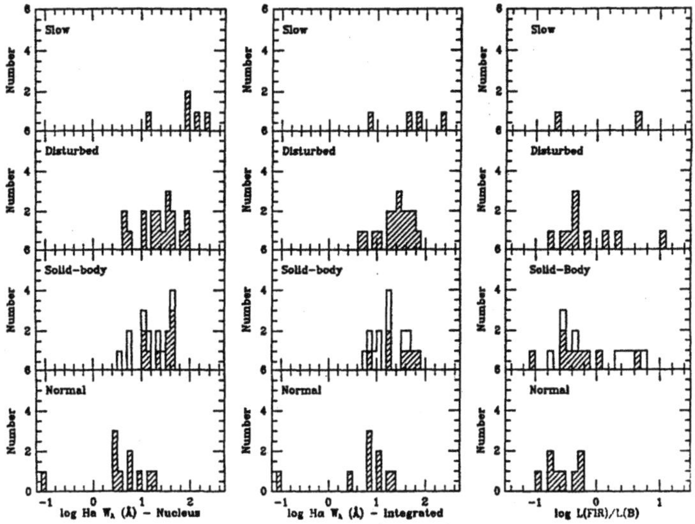

Fig. 23. Distributions of three star formation indicators for members of paired and interacting systems, sorted by rotation curve properties. Note the logarithmic flux scales. Figure taken from Keel (1993).

is not very surprising. Nevertheless we do expect the dynamics of the interactions to be critical in determining the disk responses (see lectures by Barnes), so the question remains whether we can observe any coupling of interaction kinematics with the resulting SFRs.

An ambitious attempt to isolate such effects was undertaken by Keel (1993). He selected 59 galaxy pairs from the Karachentsev catalogs with velocity separations of $150 - 400\,\mathrm{km\,s^{-1}}$ and various other criteria designed to constrain projection effects, and obtained Hα imaging and longslit spectra of the systems. The spectra provided rotation curves for the systems and information on the radial distribution of star formation and the level of nuclear activity.

Some of the results of this study are summarized in Fig. 23. Most of the tests for kinematic correlations were negative; the inferred SFRs, whether determined from Hα strength or FIR/optical flux ratios, were identical within errors for retrograde and direct encounters, and for close and distant encounters. However there are pronounced correlations between the nuclear and disk SFRs with the degree of kinematic disturbance in the disks, as measured by the rotation curves (Fig. 22). The strongest starbursts occur in galaxies with large regions of solid-body rotation or other kinds of kinematic disturbances. These results are in qualitative accord with the theoretical picture presented by Barnes elsewhere in this volume, and it will be interesting to see whether the causal link between starbursts and disk kinematics can be strengthened with more detailed observations of individual systems (e.g., Hibbard 1995; Turner 1997).

4 Induced Nuclear Activity and Star Formation

One of the most dramatic consequences of strong interactions in gas-rich systems is the transport of large masses of interstellar gas to the centers of the remnant galaxies. This process in turn can trigger nuclear starbursts or nonthermal nuclear activities (AGNs). I begin this lecture with a brief overview of nuclear spectra and activity classes, then review the evidence for interaction-triggered star formation and AGN activities. For reviews of this subject see Heckman (1990; 1991) and Sanders (1992). Several relevant reviews can be found in the proceedings of a recent conference devoted to this subject (Shlosman 1994).

The main emphasis of these lectures is on star formation in interacting galaxies, so why are we discussing AGN activity at all? The reason is that nuclear starbursts and AGN activities are closely intertwined; both processes occur simultaneously in many interacting (and normal) galaxies, and it is often difficult to separate them observationally (Lecture 5). Moreover, the two processes may be physically connected in some galaxies. Hence one cannot hope to fully understand nuclear star formation and starbursts in these objects without taking into account the properties of the nonthermal nuclei, and *vice versa*. However, the subject of nuclear activity could easily occupy an entire Saas-Fee school of its own (Courvoisier & Mayor 1990), so I will restrict my discussion to the rudimentary aspects of AGN phenomena that are most relevant to interactions and star formation, and avoid entirely the detailed physical processes that take place on the AU to parsec scale, as well as the problem of fuelling AGNs and quasars on those physical scales.

4.1 Background: Nuclear Spectra and Classification

The term "nuclear activity" encompasses a broad spectrum of phenomena ranging over at least 10 orders of magnitude in luminosity, at least 8 orders of magnitude in spatial scale, and physical processes ranging from circumnuclear star formation to accretion on to supermassive black holes. Interactions play important roles in all of these processes, and it is important to be able to distinguish them from one another. Fortunately, approximately 70% of galactic nuclei emit a nebular emission-line spectrum, and the spectra provide valuable diagnostics of the dominant physical processes. Fig. 24 shows examples of the four most common nuclear spectral classes. For a complete discussion see Osterbrock (1989).

Roughly 40% of spiral nuclei have spectra that are similar to those of star forming disk HII regions, so-called "HII region nuclei". An excellent example is in the interacting spiral NGC 2798 (Figure 24). The prototype starburst nucleus NGC 7714 (Fig. 11) is another example. There are subtle differences between the properties of the HII region nuclei and disk HII regions, mainly in terms of their continuum properties, nonthermal radio emission, density, and forbidden-line excitation (Kennicutt, Keel, & Blaha 1989), but the dominant ionization mechanism in both cases is photoionization from hot massive stars.

At the opposite end of the scale are Seyfert nuclei, which further subdivide into two main spectral classes. Seyfert 1 nuclei are characterized by broad

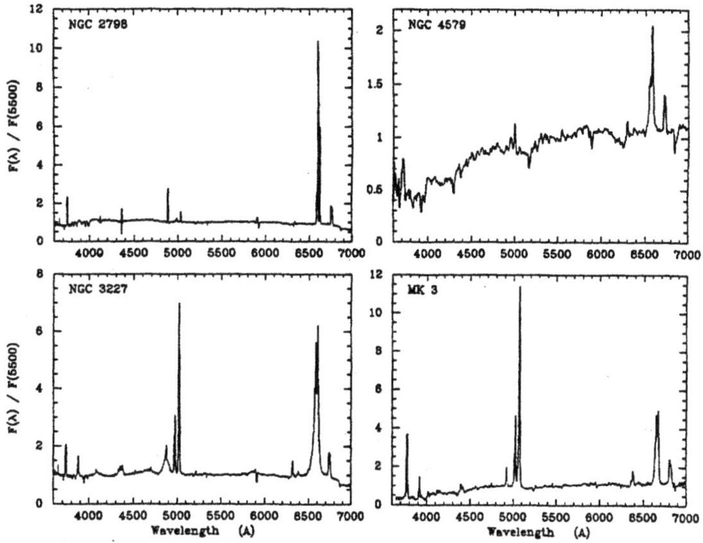

Fig. 24. Examples of four types of nuclear spectra: (a) an HII region nucleus, in NGC 2798; (b) a LINER nucleus, in NGC 4579; (c) a Seyfert 1 nucleus, in NGC 3227; (d) a Seyfert 2 nucleus, in Markarian 3. Spectrum of NGC 4579 kindly provided by Bill Keel. Other spectra are from the author.

Balmer and other permitted lines (linewidths hundreds to thousands of km s^{-1}) and strong emission in the forbidden lines of [OII], [OIII], [OI], [NII], and [SII]. Seyfert 2 nuclei show the anomalous forbidden-line excitation without the broad permitted lines (Fig. 24). These objects are also frequently associated with intense nonthermal radio cores and jets, hard X-ray emission, and other signatures of nonthermal emission from a compact core. The characteristics of Seyfert 1 nuclei extend to the more luminous broad-line radio galaxies and quasars. Seyfert nuclei are rare among normal galaxies, occurring in $\sim 5\%$ of nearby spirals, as discussed later.

Roughly half of all galactic nuclei fall into a third class, the low ionization nuclear emission regions, or LINERs (Heckman, Balick, & Crane 1980). These objects show excitation of [OIII]/Hβ that is typical of moderate-excitation HII regions, but also show anomalously strong forbidden lines of [OII], [OI], [NII], and other low-ionization species (Fig. 24). They are especially common among early-type galaxies. The ionization mechanism for LINERs is less well established than for the other nuclear types, with the most popular explanation being photoionization from a power-law continuum, similar to Seyfert nuclei (e.g., Ferland & Netzer 1983), possibly with some contributions from shock ionization. In this picture, LINERs represent a low-level form of AGN activity, probably powered by a central black hole. A minority view attributes many of the properties of LINERs and Seyfert galaxies to star formation in an unusual interstellar environment (e.g., Terlevich & Boyle 1993 and references therein).

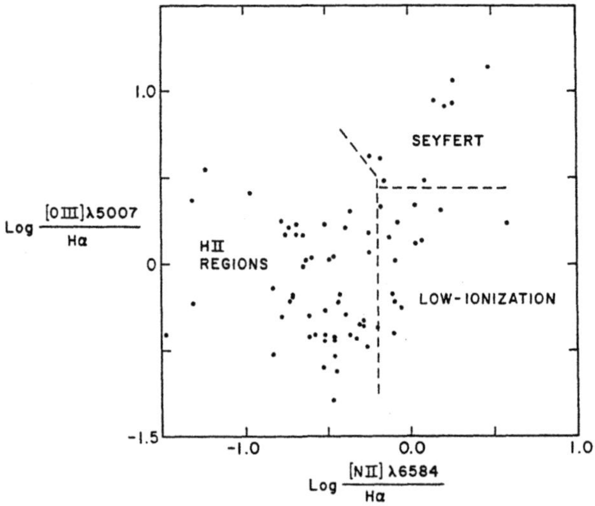

Fig. 25. Example of a diagnostic diagram of forbidden-line strengths to classify a sample of nuclei in interacting galaxies, from Keel et al. (1985).

The different nuclear types can be readily separated by correlating key diagnostic forbidden line ratios such as [OIII]/Hβ, [NII]/Hα, [SII]/Hα, and [OI]/Hα (Baldwin, Phillips, & Terlevich 1981; Veilleux & Osterbrock 1987). Figure 25 shows an example of the most popular diagram, plotting [OIII]/Hβ vs [NII]/Hα, for a sample of interacting galaxy nuclei from Keel et al. (1985). Although we have defined these three distinct nuclear classes, many nuclei are composite objects, with star forming HII regions superimposed on emission from a central AGN (e.g., Kennicutt, Keel, & Blaha 1989; Ho, Filippenko, & Sargent 1995).

4.2 Nuclear Star Formation and Starbursts

The realization that many bright galactic nuclei harbor luminous star forming HII regions led Weedman et al. (1981) to coin the term "star-burst" to describe them. The prototype starburst nucleus is in a strongly interacting galaxy, NGC 7714 (Figs. 9 and 11). The nuclear Hα luminosity in NGC 7714 is $\sim 10^{42}$ erg s^{-1}, comparable to the integrated Hα luminosities of giant Sc galaxies! Application of the SFR calibration of Kennicutt (1983) yields a corresponding nuclear SFR of $\sim 10\,M_\odot\,yr^{-1}$ for a normal IMF.

The properties of a large sample of starburst nuclei were measured by Balzano (1983). Her sample was selected from the Markarian catalog, and includes interacting and noninteracting systems. Figure 26, taken from Kennicutt et al. (1989), compares the Hα luminosity functions of the Balzano starburst sample with HII region nuclei of a local sample of spirals and with the first-ranked disk HII regions in 81 nearby spirals (Kennicutt 1988). The starburst nuclei are typically 30 times brighter than the most luminous disk HII regions, and at least an order of magnitude brighter than typical HII region nuclei. NGC 7714 is among the most luminous objects, even for the starburst sample. The Hα luminosities

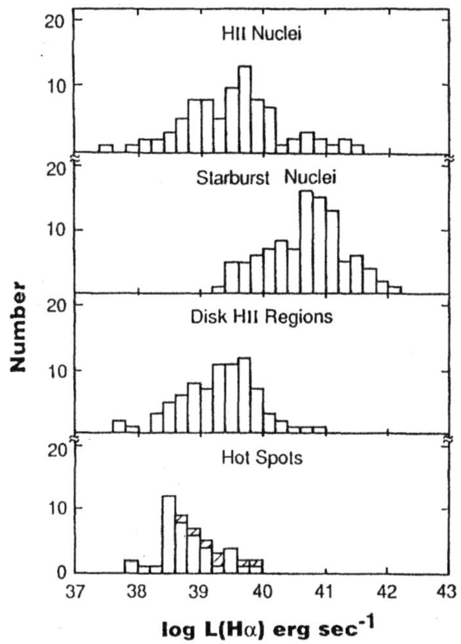

Fig. 26. A comparison of the Hα luminosities of Balzano's sample of starburst nuclei with those of a nearby sample of HII region nuclei, the brightest disk HII regions in nearby spiral galaxies, and a set of nearby "hotspot" circumstellar HII regions, from Kennicutt et al. (1989).

plotted in Fig. 26 are not corrected for extinction, and since the extinction tends to be higher in the nuclei, the actual contrast between the starburst nuclei and disk HII regions is even larger than indicated by the figure.

The ionized gas in most starburst nuclei is spatially extended, with typical FWHM diameters of 500 – 2000 pc (e.g., Balzano 1983; Kennicutt et al. 1989; Hummel et al. 1990). At high spatial resolution, many of the nuclear HII regions resolve into subclumps, and in barred galaxies they often resolve into a tight nuclear ring of "hotspots", probably associated with an inner Lindblad resonance in the bar. In many cases the hotspots surround a compact central Seyfert or LINER nucleus. Hence the "nuclear" starbursts should more correctly be regarded as a circumnuclear phenomenon in the inner kpc of a galaxy.

4.3 Nuclear Star Formation and Interactions

The links between interactions and nuclear star formation have been tested using the same kinds of statistical tests as described in Lecture 3 (see Heckman 1990, and references therein). Many of the general trends seen in the global SFRs of interacting galaxies are also seen in the nuclear SFRs. For example, Fig. 27 compares the distribution of Hα EWs for two samples of interacting galaxies with that of normal galaxies (Keel et al. 1985). In this case the distribution of EWs in the control sample was adjusted to reduce selection biases due to differences in galaxy type and luminosity distributions. The interacting galaxies have much stronger nuclear emission, with median EWs of 28 Å for a complete sample of nearby galaxy pairs and 24 Å for an Arp atlas pair sample, vs 8 Å

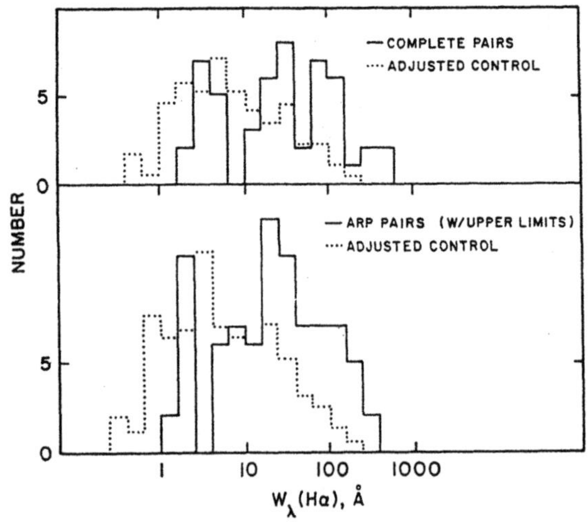

Fig. 27. Comparisons of distributions of Hα emission line equivalent widths for nuclei in two samples of interacting galaxies (complete and Arp samples), with an adjusted control sample of normal spiral nuclei, from Keel et al. (1985).

for a control sample of noninteracting spirals (see Section 3.3 for a description of the sample properties). This result is very reminiscent of what was observed for the global SFRs; an average enhancement of a factor of a few, with a large dispersion in the SFRs within individual objects.

Several properties of the nuclear star formation are quite distinct from those observed in the global SFRs, however. For example, several studies suggest that the strength of the nuclear emission is uncorrelated with the degree of tidal damage observed in the parent galaxy, and if anything it may be anticorrelated with the level of disturbance. In the Keel et al. (1985) study, objects selected from the Arp atlas show slightly *weaker* nuclear emission than a randomly selected sample of paired galaxies. Likewise Bushouse (1986) and Liu & Kennicutt (1995a) observed relatively low enhancements in nuclear emission in samples of highly disturbed systems and mergers. This result might reflect the tendency for the most disturbed interactions and mergers to be observed at late evolutionary stages, after a peak in nuclear star formation, or it may be that the dust obscures the star formation in the most violent interactions.

Similar effects of interactions have been observed in the radio continuum and infrared. Radio continuum studies of galaxies in interacting pairs and dense groups show enhancements in central radio luminosity of 5 – 10 times over that in control samples of isolated systems (Hummel et al. 1990; Menon 1995). Unfortunately, the low spatial resolution of IRAS does not allow one to make a similar test for the FIR emission, but groundbased mid-IR observations show systematic enhancements in interacting systems (e.g., Joseph et al. 1984; Lonsdale, Persson, & Mathews 1984; Curti & McAlary 1985). The typical IR enhancements are similar to those observed in Hα, but a subclass of advanced mergers show

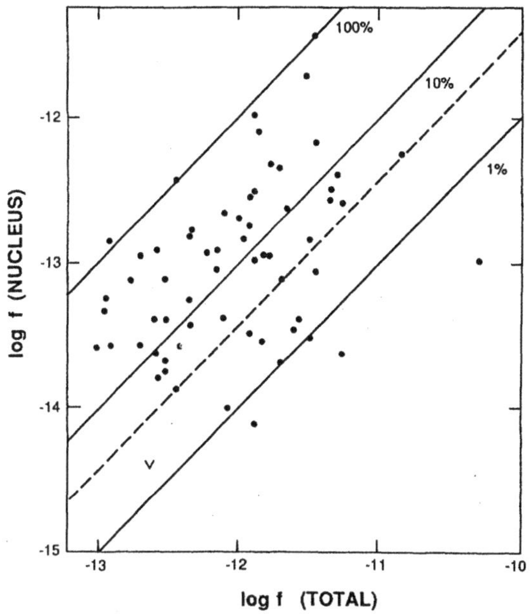

Fig. 28. Correlation of nuclear Hα + [NII] flux and total flux for interacting galaxies. Lines of constant nuclear flux fraction are superimposed. From Kennicutt et al. (1987).

much stronger emission, the "ultraluminous" infrared galaxies (Joseph & Wright 1985). I defer further discussion of those objects to Lecture 5.

These data reveal several other unique properties of the star formation in the interacting systems. Although enhancements are observed in both the global and nuclear SFRs, the star formation tends to be more centrally concentrated in the interacting galaxies (Bushouse 1987; Kennicutt et al. 1987). This is shown in Fig. 28, which plots the nuclear and total Hα + [NII] luminosities of the paired and interacting galaxies in the Keel et al. (1985) and Kennicutt et al. (1987) samples. The fraction of nuclear emission covers the full range from 0 – 100%, with a median fraction of 13%; the comparable fraction in normal spirals is only 4% (neither number is corrected for extinction so the actual nuclear fractions are probably higher). Bushouse (1987) studied the actual radial distributions of star formation in his sample and observed pronounced central concentrations in many interacting galaxies directly. Thus the strongest effects of the interactions are felt in the centers of the galaxies.

Given these results, one might expect the level of nuclear star formation to be highest in systems with the highest global SFRs, and this is indeed observed. Figure 29 shows the correlation between nuclear and total Hα + [NII] emission line EWs for interacting and noninteracting galaxies (Kennicutt et al. 1987). Among normal galaxies there is a very loose correlation, reflecting the fact that gas-rich (late-type) galaxies tend to have higher levels of disk star formation and nuclear emission. The interacting galaxies are displaced in both axes, and there is a much more pronounced correlation between the nuclear and disk EWs.

Finally, Fig. 30 compares the breakdown of nuclear spectral types for the interacting and normal galaxies in the samples of Keel et al. (1985). Among

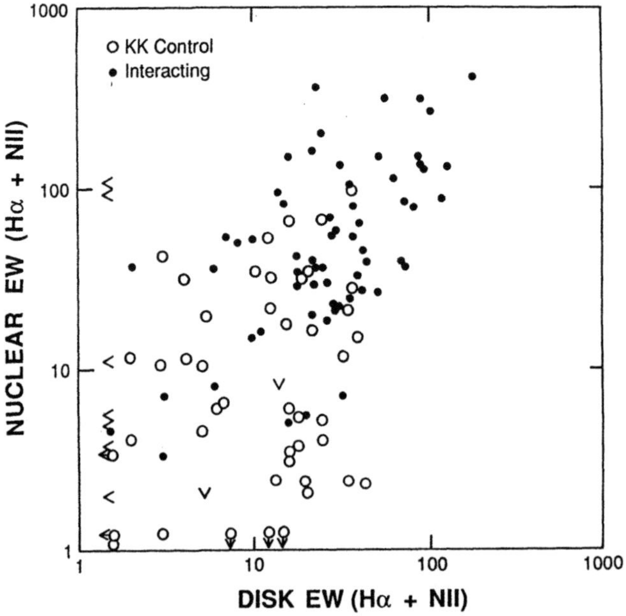

Fig. 29. Correlation of nuclear and disk Hα emission equivalent widths for interacting and noninteracting galaxies, from Kennicutt et al. (1987).

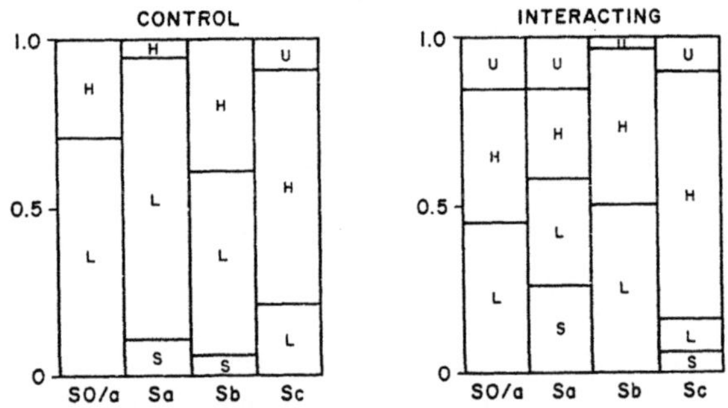

Fig. 30. Relative fractions of HII region, LINER, and Seyfert nuclei in interacting and noninteracting spirals, from the study of Keel et al. (1985).

normal galaxies, there is a strong dependence of nuclear emission type on parent galaxy morphological type, so the fractions are shown as a function of Hubble type. One immediately sees that HII region nuclei, those dominated by nuclear star formation, are much more prevalent in the interacting systems, with the strongest differences apparent in the early-type spirals. The same effect has been seen in a large sample of Arp-Madore systems studied recently by Sekiguchi & Wolstencroft (1993). This suggests that the predominant effect of gas inflow in interactions is to trigger star formation, and that in many cases this star formation may mask any effects of induced nuclear AGN activity.

4.4 Induced AGN Activity: Statistics of Seyfert Galaxies

Numerous studies have addressed the issue of whether interactions trigger AGN activity, Seyfert activity in particular. Although there are many prominent examples of Seyfert galaxies in interacting systems (e.g., NGC 3227, NGC 7469), the statistical case is very ambiguous. Excellent summaries of this work can be found in Heckman (1990) and Laurikainen & Salo (1995), which I use as the main sources for the following discussion.

As with tests for enhanced star formation, one can approach the problem of Seyfert triggering in two ways, either by comparing the frequency of Seyfert activity for well-defined samples of interacting and noninteracting systems, or by comparing the interaction rates for Seyfert and non-Seyfert nuclei, the "guilt by association" test. Statistics on Seyfert frequencies in interacting systems have been published by Kennicutt & Keel (1984), Keel et al. (1985), Dahari (1985), Bushouse (1986), and Sekiguchi & Wolstencroft (1992). The Keel et al. (1985) and Dahari (1985) samples show an excess of Seyfert galaxies, $8-13\%$ in complete samples of paired and strongly interacting systems, compared to a Seyfert frequency of $4-6\%$ in the control samples. However these results are contradicted by Bushouse (1986) and Sekiguchi & Wolstencroft (1992), who found no statistical excess in samples of highly disturbed systems; indeed there is marginal evidence for a *deficiency* of Seyfert nuclei in those interacting systems. Similar absences of Seyfert nuclei are seen among the most strongly disturbed galaxies in the Keel et al. and Dahari samples, so taken together, the data are consistent with a common picture in which Seyfert activity is mildly enhanced ($\sim 10\%$ vs 5%) in interacting galaxies as a class, but is notably absent in the most highly disturbed objects, perhaps due to a combination of evolutionary selection effects and a masking of nonthermal activity by dust and/or star formation. In any case, the excess of AGN activity in the interacting galaxies is very modest, a factor of $2-3$ at most.

Studies of the environments of Seyfert vs non-Seyfert galaxies also yield contradictory results. The first systematic studies suggested that Seyfert galaxies have roughly 3 times the number of companions as non-Seyfert hosts, with (Seyfert) companion frequencies ranging from 12% to 71% depending on the criteria used to define companions (Dahari 1984; MacKenty 1989). However these results were contradicted by Fuentes-Williams & Stocke (1988), who argued that selection biases due to differences in redshift distributions, luminosities, and morphological types between samples of Seyfert and normal galaxies could produce a false excess of companions in the Seyfert samples. When they took these biases into account, they found no significant excess of companions. Most recent studies show only marginal evidence for an excess of companions around Seyferts at best (Laurikainen & Salo 1995; McLeod & Rieke 1995; Maiolino et al. 1997), but Rafanelli, Violato, & Baruffolo (1995) do find evidence for an excess of companions, using many of the same databases as the other authors.

Most of the inconsistencies between these results can be traced to a combination of small number statistics and differences in the criteria used to define companions between the various studies. For instance, it is well known that

Seyfert galaxies participate in the same large scale clustering as other massive spirals, i.e., they tend to be located in groups or other high-density environments, and hence will show an excess of nearby galaxies but not necessarily an excess of physically interacting companions (e.g., Petrosian 1982; Monaco et al. 1994; Laurikainen & Salo 1995). Laurikainen & Salo (1995) find an excess in the number of companions around Seyfert galaxies but no increase in the frequency of companions. When statistics are compiled separately for Seyfert 1 and 2 galaxies, the latter show a higher frequency and number of companions, however.

What are we to conclude from this disparate set of results? Probably the main lesson we can draw is the importance of having large enough samples and proper treatment of control samples and background corrections! Given the rarity of Seyfert nuclei, the statistics in these studies usually are poor, and they are subject furthermore to a host of observational and physical selection effects: AGN frequency is a strong function of the type and luminosity of the parent galaxy; the importance of interactions probably increases sharply with the level of activity (Lecture 5); nuclear star formation –pervasive in interactions– probably masks any low-level AGN activity; other physical variables, such as the presence of bars and oval distortions, are important and will mask interaction effects; and viewing geometry may be very important as well, especially for studies that attempt to differentiate between Seyfert 1 and 2 systems.

My own assessment of this problem has not changed much from 10 years ago. I still believe that: (1) Some fraction of Seyfert activities are triggered by tidal interactions, but; (2) Most Seyfert nuclei are triggered by something else, for example gas inflow from a bar. Indeed the case for bar-triggered Seyfert activity is considerably stronger than for interaction-triggered activity (Simkin, Su, & Schwarz 1980; Moles, Márquez, & Pérez 1995; Mailoino et al. 1996). If these conclusions are correct, then the subject of interaction-triggered Seyfert activity is of some interest to understanding and constraining gas inflow models (i.e., it may be a challenge to explain why more interacting systems *do not* harbor Seyfert nuclei), but the problem is of secondary importance for understanding AGN triggering in general. Studies which take a broader view of the conditions triggering Seyfert activity, most notably Moles et al. (1995), point strongly toward nonaxisymmetric perturbations as being a primary triggering mechanism, with internal dynamical perturbations being more important than companions or interactions.

4.5 Environments of Quasars

Over the past 20 years there have been numerous investigations of the link between interactions and the triggering of luminous radio galaxies or quasars. For reviews of the extensive literature on this subject, I refer the reader to Heckman (1990; 1991) and Djorgovski (1994). This problem is much more difficult than for Seyfert galaxies, because the quasars are so distant, and because the light of the quasars themselves make it difficult to resolve the surrounding galaxies. However

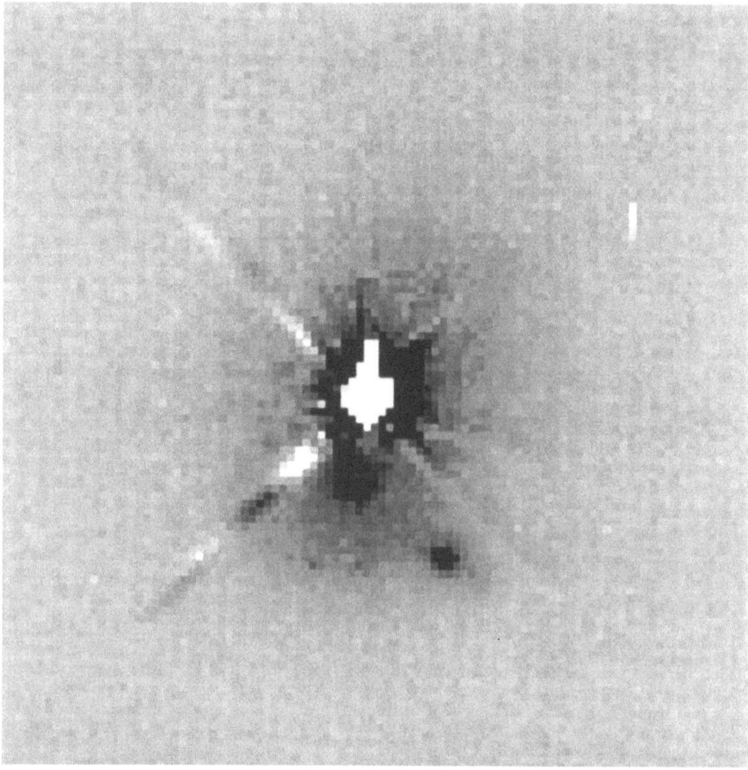

Fig. 31. Broadband HST image of the low-redshift quasar PKS 1302-102, showing evidence for interacting companions, from Bahcall, Kirhakos, & Schneider (1995).

HST and high-resolution groundbased imaging are opening new opportunities in this area.

Groundbased imaging already provides evidence that many low-redshift quasars possess physical companions or distorted morphologies (Heckman 1990 and references therein). The subject is being revolutionized by observations with HST. The most relevant study is a preliminary snapshot survey of low-redshift quasars by Bahcall, Kirhakos, & Schneider (1995). Much of the focus of this investigation is aimed at the structure and nature of quasar host galaxies, a controversial subject which we will not consider here, but the paper also contains a preliminary analysis of the companion statistics, which confirm a strong causal link between interactions and quasar activity.

Bahcall, Kirhakos, & Schneider (1995) present detailed observations for 8 quasars with $z < 0.30$ and $M_V < -22.9$ ($H_0 = 100$, $\Omega_0 = 1$), and preliminary companion statistics for 4 other quasars. Among this sample of 12 objects, they have identified a total of 14 companions around 9 quasars. For this application a companion is defined as any object with $M_V < -16.5$ within a projected radius of 25 kpc. Figure 31 shows one of the most dramatic examples, PKS 1302-102, with two companions at projected radii of 3.0 and 5.9 kpc. Since redshifts are not available for the companions, it is possible that not all of them are physically associated with the quasar host. However, a preliminary analysis of control fields surrounding the quasars suggests that the probability of chance superpositions is only $\sim 5\%$, so most of the companions are likely to be physically associated –if not interacting– with the quasars. There is also evidence of tidal disturbances in some cases, confirming the many groundbased studies cited earlier.

Spectroscopic follow-up of these observations would be very valuable, to confirm the redshifts and association of the apparent companions, and to reveal the nature of the stellar populations in the companions and the host galaxies. Groundbased spectroscopy of the "fuzz" around selected quasars and radio galaxies has sometimes revealed striking evidence of postburst spectra (e.g., Boroson et al. 1985), and longslit spectroscopy with HST would make it possible to test for similar populations in a large sample of quasars. The planned installation of the STIS spectrograph on HST in 1997 should open enormous new opportunities in this area.

These results suggest a much stronger link between interactions and quasars than is observed for the lower-luminosity Seyfert nuclei. It is important to confirm this conclusion with larger samples and proper treatment of background contamination; the many studies of Seyfert companions, if nothing else, have demonstrated the sensitivity of the inferred companion frequency on these parameters. However, this tendency for the role of interactions to be increasingly important in more luminous systems is consistent with the properties of ultraluminous infrared galaxies (Lecture 5). As will be discussed in Lecture 7, the result is also consistent with our rudimentary physical understanding of the gas fueling of starbursts and AGNs; it probably requires an event as catastrophic as a merger to fuel a major AGN outburst or starburst, whereas any number of other dynamical processes may suffice to trigger a lower-luminosity AGN.

4.6 Kinematic Clues to the Triggering of AGNs

In contrast to the glut of data on the statistics of interactions vs AGN activity, there are relatively few studies of the physical and dynamical properties of the host galaxies in interacting systems. Such data might shed light on why so few interactions result in AGN outbursts, and which conditions favor the formation of an AGN in an interacting system (or in any galaxy for that matter).

Keel (1996) recently published a study of the internal kinematics of Seyfert galaxies. His sample includes 29 Seyfert galaxies in interacting pairs, along with the necessary control samples of paired non-Seyfert galaxies, isolated Seyfert galaxies, and isolated non-Seyfert galaxies. The data are similar to those obtained for Keel's kinematic study of starbursts in interacting systems (Lecture 3). Subdividing the sample in this manner and considering the various selection effects makes it difficult to delineate robust statistical trends, but nevertheless, the data show one very strong trend, namely a preference for Seyfert nuclei to occur in host galaxies with small solid-body rotation regions (i.e., rapidly rising rotation curves). This condition may reflect more on the conditions needed to fuel an AGN in general (massive bulge, strong central mass concentration, inner Lindblad resonance) than a property that is uniquely connected to interactions. However Keel's work illustrates the approach that eventually will be needed to isolate the physical mechanisms that fuel the very luminous AGNs in strongly interacting systems.

5 Infrared Luminous Galaxies and Starbursts

One of the most exciting discoveries in this entire field has been the identification of a class of infrared luminous galaxies, which rank with quasars as the most luminous objects in the universe, and which are intimately linked to strong tidal interactions and mergers. Much of the research on galaxy interactions and star formation over the past decade has been motivated by the discovery of these objects and the mysteries surrounding their physical origins and power sources.

Hundreds of papers have been written on this subject over the past dozen years, and a thorough review is beyond the scope of these lectures. Instead I will highlight those aspects of infrared luminous galaxies that are most relevant to understanding starbursts and interactions. For more complete discussions, I highly recommend a recent review by Sanders & Mirabel (1996), along with earlier reviews by Soifer, Houck, & Neugebauer (1987a), Telesco (1988), Mirabel (1992), Sanders (1992), and Scoville et al. (1994).

I begin this lecture by providing an overview of the physical properties of IRAS galaxies, with special emphasis on the ultraluminous objects. These data provide important clues about the origin of the infrared luminosity and the processes that trigger the outbursts. I then review evidence for triggering by interactions, and briefly address the still unanswered question of what powers the far-infrared (FIR) emission. More details on the nature of the star formation in these objects are given in later lectures.

5.1 Background: IR Luminous Galaxies and IRAS

As mentioned in earlier lectures, the discovery of infrared luminous galaxies associated with strong interactions predates the IRAS survey (e.g., Rieke & Low 1972; Rieke et al. 1980; Gehrz et al. 1983; Joseph et al. 1984; Joseph & Wright 1985). Rieke & Low (1972) identified a class of "ultrahigh luminosity infrared galaxies" associated with interacting and merging galaxies, and Joseph & Wright (1985) applied the term "ultraluminous infrared galaxies" to a class of advanced mergers with very strong mid-IR emission. Henceforth I will use the conventional nomenclature of "luminous infrared galaxy" (LIG) for objects with $8 - 1000\,\mu$m luminosities $> 10^{11}\,L_\odot$ and "ultraluminous infrared galaxy" (ULIG) for $L_{IR} > 10^{12}\,L_\odot$. The ULIGs are quite extraordinary objects; if their energy were mostly emitted in the visible they would have absolute magnitudes of -25 or brighter!

The Infrared Astronomical Satellite (IRAS) provided a database of FIR fluxes for over 30,000 galaxies, and established the ubiquity of the LIG phenomenon. The satellite provided broadband flux densities in four passbands centered at 12, 25, 60, and $100\,\mu$m, with angular resolutions of $\sim 1 - 4.5'$. For descriptions of the survey parameters and references to the relevant IRAS databases see Soifer et al. (1987a) and Sanders & Mirabel (1996).

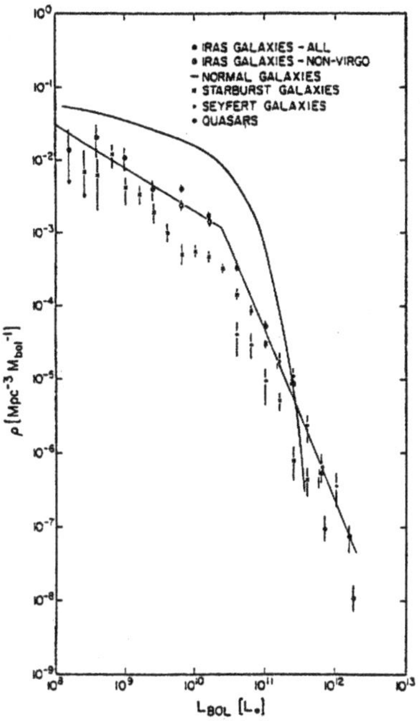

Fig. 32. Luminosity function of IRAS galaxies, compared to other samples of galaxies and AGNs, from Soifer et al. (1987b).

5.2 Infrared Luminosity Function and Spectra

The combination of the IRAS all-sky survey along with several deeper pointed observations have made it possible to construct an accurate infrared luminosity function (e.g., Soifer et al. 1987b; Saunders et al. 1990). Figure 32 compares the IRAS galaxy luminosity function with several other galaxy and AGN samples (Soifer et al. 1987a). At low luminosities the IR luminosity function roughly follows the optical function for normal galaxies, with a strong steepening of slope above $10^{10} - 10^{11}\,L_\odot$. At very high luminosities the IR luminosity function roughly parallels the optical luminosity function of quasars, but at a given power the space density of IRAS galaxies is higher than that of quasars, by factors of a few for $L_{\rm bol} > 10^{12}\,L_\odot$.

Figure 32 illustrates the enormous range of absolute luminosity among the IRAS galaxies. This diversity applies even among the famous examples of infrared-luminous galaxies, which range from systems such as M 82 ($L_{\rm IR} = 3 \times 10^{10}\,L_\odot$, $L_{\rm IR}/L_B = 10$) to Arp 220 ($L_{\rm IR} = 1.6 \times 10^{12}\,L_\odot$, $L_{\rm IR}/L_B = 80$). The range of optical luminosities diminishes at higher infrared powers; the ULIGs tend to be distinguished not only in terms of their high absolute luminosities but also in terms of their unusually high infrared/visible flux ratios.

Several investigators have studied the spectral energy distributions (SEDs) of LIGs (e.g., Sanders et al. 1988; Soifer & Neugebauer 1991; van den Broek 1993). More luminous sources exhibit higher dust temperatures, with the elevated $60\,\mu$m/$100\,\mu$m ratios that are characteristic of active star forming normal galaxies (Lecture 3). Many so-called 'warm' sources also show strong $25\,\mu$m emission, a feature that is often associated with AGNs.

5.3 Infrared Structure and Morphology

Most of the LIGs are considerably smaller than the $1 - 4'$ resolution of IRAS, so IRAS provides very little information on the spatial structure of the FIR emission. Limited information is available from mid-infrared and far-infrared imaging obtained from the ground or with the Kuiper Airborne Observatory (e.g., Matthews et al. 1987; Hill, Becklin, & Wynn-Williams 1988; Joy et al. 1989). These data show that most of the infrared emission is produced within very small regions in the centers of the galaxies, a few arcseconds in diameter or less. The corresponding linear diameters are typically $0.5 - 2\,\mathrm{kpc}$. A lower limit to the size of the emitting region can be computed for a blackbody from the observed luminosity and color temperature, and these sizes are of the same order, usually a few hundred parsecs. This is comparable to the sizes of the central molecular disks (below).

Telesco et al. (1993) mapped 21 nearby LIGs at $10.8\,\mu\mathrm{m}$ and found that on average 58% of the $12\,\mu\mathrm{m}$ IRAS emission was detected in the inner $20''$ ($1-2\,\mathrm{kpc}$ for these galaxies). The central emission is often resolved into multiple sources, circumnuclear rings, and other complex structures. Although these objects tend to be at the low end of the scale for LIGs they provide an indication of the diversity in infrared structures that is likely to characterize the LIGs and ULIGs as a class. The successful launch of the Infrared Space Observatory (ISO) promises to provide spectacular new mid-IR imaging observations of LIGs and normal star forming galaxies as well.

5.4 Interstellar Gas

Mapping of these galaxies in HI and CO has provided several important clues to the physical nature and origin of the LIGs (e.g., Scoville et al. 1994; Combes 1995). The integrated HI profiles of these galaxies are characterized by distorted 21 cm profiles and deep absorption (Mirabel & Sanders 1988). High-resolution HI mapping with the VLA reveals complex structures, with much of the atomic gas spread between long tidal tails and the main bodies of the galaxies (e.g., van der Hulst 1979; Yun, Ho, & Lo 1993; Hibbard & van Gorkom 1996). The HI sometimes shows central concentrations but often there is a central depression in HI where the gas is predominantly molecular (below). The HI in the tails provides an powerful diagnostic of the orbital kinematics of the interaction or merger (see lectures by Schweizer).

The most striking property of the interstellar medium (ISM) in the LIGs is the presence of massive central molecular disks in the host galaxies. Figure 33 shows the infrared luminosities of 89 galaxies in the IRAS Bright Galaxy Survey (BGS) plotted against (total) molecular gas mass, derived from from CO single-disk measurements by Sanders, Scoville, & Soifer (1991). The molecular masses were estimated from the intensity of the CO J=1–0 transition using the usual CO/H_2 conversion factor applicable to the solar neighborhood; this may systematically overestimate the mass of molecular hydrogen (e.g., Aalto et al. 1995). However this uncertainty does not alter the qualitative conclusions that

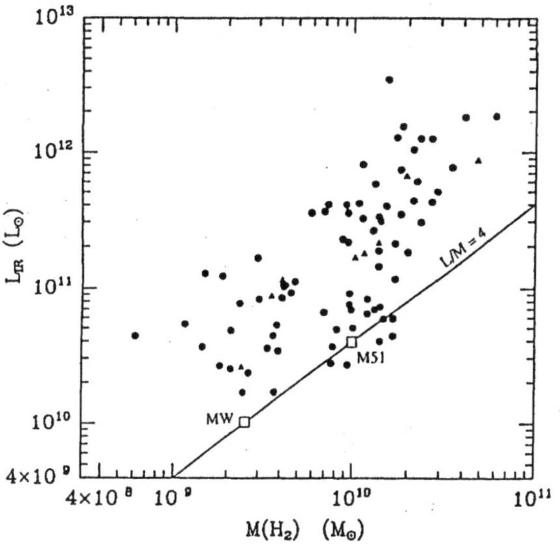

Fig. 33. Relationship between far-infrared luminosity and molecular gas mass for a sample of luminous infrared galaxies. The diagonal line indicates the mean relation for normal galaxies. Figure from Sanders et al. (1991).

the LIGs are characterized by very high molecular masses – typically of order $10^9 - 10^{11}\,M_\odot$, and a much higher ratio of FIR luminosity per unit gas mass relative to normal star forming galaxies (indicated by the line in Fig. 33).

Millimeter interferometers now make it possible to spatially resolve the molecular disks in these systems (Okamura et al. 1991; Scoville et al. 1994 and references therein). Over 30 LIGs and ULIGs have been mapped, and most show the emission concentrated in central disks with diameters of 0.5 – 3 kpc. This is a very important result because it provides *prima facie* evidence for the radial transport of massive amounts of interstellar gas into the centers of these systems. In some objects the structure and kinematics of the gas disks (or the underlying stellar components) are determined well enough so that the dynamical mass of the core can be measured. In such cases the inferred molecular gas masses comprise a substantial fraction of the total mass in the centers of the galaxies, typically 20 – 100% (Scoville et al. 1994). This tends to confirm the suspicion that the CO-inferred molecular masses are probably overestimated in some cases, but nevertheless the gas disks comprise a significant fraction of the total masses in the central regions. Adopting lower central H_2 masses would relieve the mass transport problem somewhat, but it would exacerbate the anomalously high FIR/H_2 ratios in the LIGs. In any case it is clear that most of the ISM of the merged galaxies has been transported to the central kiloparsec of the remnant.

An independent indication of extraordinary concentrations of molecular gas in these systems comes from the detection of OH megamasers in over 50 LIGs and ULIGs (Baan 1993 and references therein). The luminosities can reach over 10^6 times those observed for Galactic OH masers. The OH line profiles should provide valuable information about the kinematics and physical conditions of the dense molecular cores of the IRAS galaxies, and may help to discriminate whether starbursts or AGNs are the primary heating mechanisms for the dust.

Several observations of the interstellar gas in these objects show evidence for a very unusual, high-pressure ISM. This evidence comes from X-ray observations (e.g., Bregman, Schulman, & Tomisaka 1995), far-infrared and submillimeter spectroscopy of nebular lines (e.g., Carral et al. 1994), and multi-transition millimeter observations of the molecular gas (e.g., Aalto et al. 1991, 1995). Taken together these observations are consistent with a general picture in which the ISM is composed of very dense clouds or filaments ($> 10^4\,\mathrm{cm}^{-3}$) with a low filling factor, residing in an ionized intercloud medium with pressure $P/k = 3-30 \times 10^6$, roughly $100 - 1000$ times the pressure in the local ISM. As discussed in Lecture 8 this unusually high pressure may have important consequences for the star formation in these environments.

5.5 X-Ray Emission and Superwinds

IRAS galaxies have been detected in X-rays, both with Einstein (David et al. 1992) and in the ROSAT All Sky Survey (Boller et al. 1992). Most of the detected sources are associated with AGNs. A few objects have been mapped with ROSAT, including M 82 (e.g., Bregman et al. 1995) and Arp 220 (Heckman et al. 1996). Both objects show diffuse emission extending well into the halos of the galaxies, and are probably associated with outflows or "superwinds".

Emission-line imaging and spectroscopy of LIGs shows pervasive evidence of extended nebulosity associated with energetic outflows of interstellar material or "superwinds" (Heckman et al. 1990; Heckman 1996). The best studied example is M 82. Extended nebular gas along the minor axes of the galaxy are coincident with X-ray emission, and show both the velocity field and density signature of an outflow (McCarthy, Heckman, & van Breugel 1987). The energetics associated with the M 82 outflow are impressive, with a mechanical power of $\sim 2 \times 10^{42}\,\mathrm{erg\,s^{-1}}$, and total mass and energy depositions over the lifetime of the burst of order $10^8\,\mathrm{M_\odot}$ and 10^{57} erg, respectively. The superwinds are thought to be powered by the combined effects of stellar winds and supernovae from a starburst, though AGNs may play a role in some systems. The effect of an ULIG (e.g., Arp 220) on its surroundings can be substantial, with up to $10^9\,\mathrm{M_\odot}$ in metals and $10^{58} - 10^{59}$ erg of mechanical energy ejected into the intergalactic medium over the lifetime of a $10^7 - 10^8$ yr starburst (Heckman et al. 1990).

5.6 Optical, UV, and Near-Infrared Spectra

Because of the heavy extinction that is intrinsic to the LIGs, little if any of the central power source is likely to be detectable in the visible. Nevertheless, UV, optical, and near-infrared observations prove to be surprisingly useful for characterizing the circumnuclear environments of these objects, as well as the physical processes occuring in the cores.

Although some LIGs show very large reddening at visible wavelengths (e.g., M 82), most show surprisingly blue continua (Sanders et al. 1988; Leech et al. 1989; Veilleux et al. 1995; Liu & Kennicutt 1995a). This is illustrated in Fig. 34, which plots optical colors (4100 – 5000 Å) and [OII]λ3727 emission line equivalent

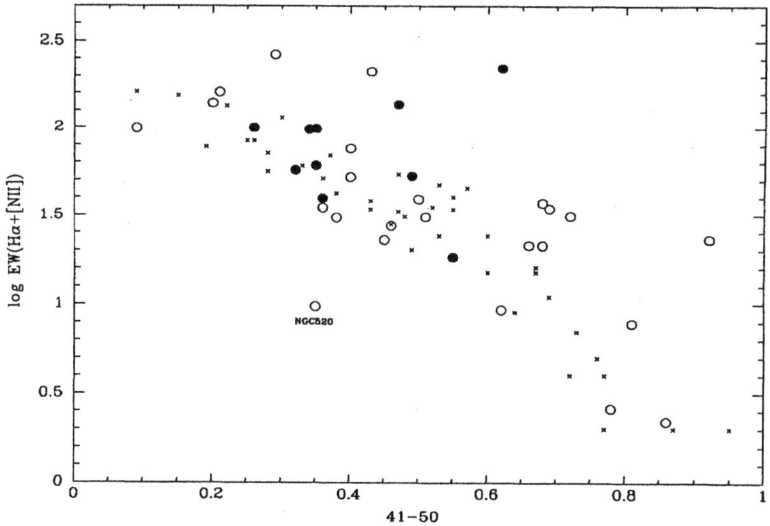

Fig. 34. Distributions of optical colors and [OII]λ3727 emission-line strengths for ULIGs (large points) along with optically selected merger remnants (open circles) and normal galaxies (small points), from Liu & Kennicutt (1995a).

widths for samples of mergers and normal galaxies from Liu & Kennicutt (1995a). The heavy symbols show the properties of 10 ULIGs from the BGS; although they are redder than the bluest morphologically selected mergers (open circles) they are still far bluer than average for normal galaxies (small points). This demonstrates that despite the heavy extinction there are relatively extinction-free patches which reveal an underlying blue stellar population. Indeed many LIGs show prominent UV continua with C IV and Si IV absorption lines (e.g., Augarde & Lequeux 1985; Sekiguchi & Anderson 1987).

The blue colors also suggest the presence of a dominant young stellar population at these wavelengths. Indeed the absorption spectra of the ULIGs tend to be dominated by young stars or by a Balmer-dominated post-burst population (Armus, Heckman, & Miley 1989; Veilleux et al. 1995). Figure 35 shows the example of UGC 5101 (Liu & Kennicutt 1995b), one of the original BGS ULIGs (see Fig. 16 for other examples). The Balmer-dominated spectra closely resemble those of classical Toomre (1977) merger remnants such as NGC 520, NGC 2623, NGC 3921, and NGC 7252 (Fig. 14), and are not unlike those of the "E+A" galaxies frequently observed in high-redshift galaxy clusters (e.g., Dressler & Gunn 1983). Regardless of what powers the central infrared source, these data show unambiguous evidence for massive bursts of star formation in the ULIG host galaxies. However, not all ULIGs show evidence of a stellar-dominated continuum. Figure 36 shows a counter-example in Markarian 231, with the characteristic emission and continuum spectrum of a broad-line AGN.

The emission-line spectra of the LIGs provide additional constraints on the underlying star formation and/or AGN activities (e.g., Sanders et al. 1988; Ar-

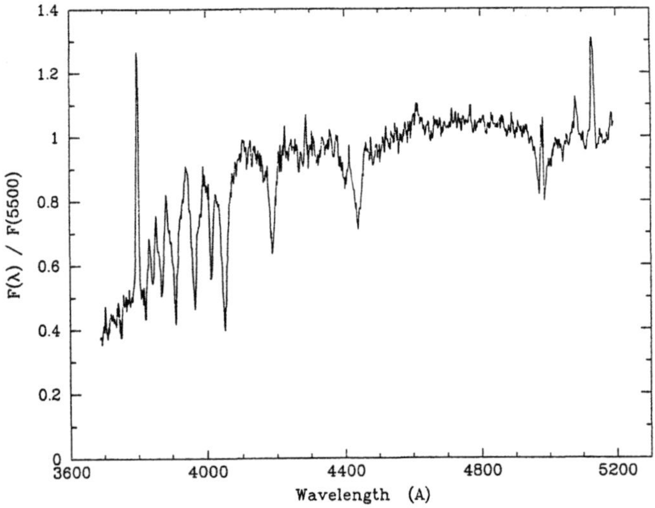

Fig. 35. Long-slit spectrum of the ultraluminous infrared galaxy UGC 5101. Note the surprisingly blue continuum and the strong Balmer absorption lines, typical of a post-starburst spectrum. From Liu & Kennicutt (1995b).

mus et al. 1989; Leech et al. 1989; Veilleux et al. 1995). Kim et al. (1995) and Veilleux et al. (1995) recently completed a comprehensive study of nuclear and long-slit spectra of 200 LIGs. All three major types of nuclear spectra –HII region, LINER, and Seyfert– are represented in their sample, along with many composite spectra. The breakdown of nuclear types is shown in Fig. 37, subdivided according to infrared luminosity (Veilleux et al. 1995). In low-luminosity galaxies ($< 10^{10}\,L_\odot$) the proportions of HII, LINER, and Seyfert nuclei are similar to those found for normal spirals, with HII region nuclei being the most common type. However the fraction of LINER and Seyfert nuclei increases sharply with IR luminosity, comprising 62% of the total for the ULIGs. The LINERs usually are associated with AGN activities, though one should bear in mind that other mechanisms can produce such a spectrum (e.g., Armus et al. 1990; Martin 1996). Broad-line nuclei are also detected with increasing frequency in the highest luminosity objects (e.g., Mrk 231, Fig. 36).

Veilleux et al. (1995) also obtained long-slit data for 23 LIGs in order to study the physical conditions outside the nuclei. Most of the objects, including those with AGN-like nuclear spectra, show extended emission with HII region-like line ratios. The extranuclear spectra also show the absorption features of young stellar populations. Together these results point to pervasive circumnuclear star formation in the LIGs, regardless of the predominant nuclear spectral type. Some objects show extended emission with a LINER-like spectrum (Armus et al. 1989, 1990; Heckman et al. 1990; Veilleux et al. 1995), suggesting perhaps the presence of a superwind.

Near-infrared spectroscopy offers a very powerful probe of the inner regions of these objects. Studies of this kind have been limited to date, but rapid im-

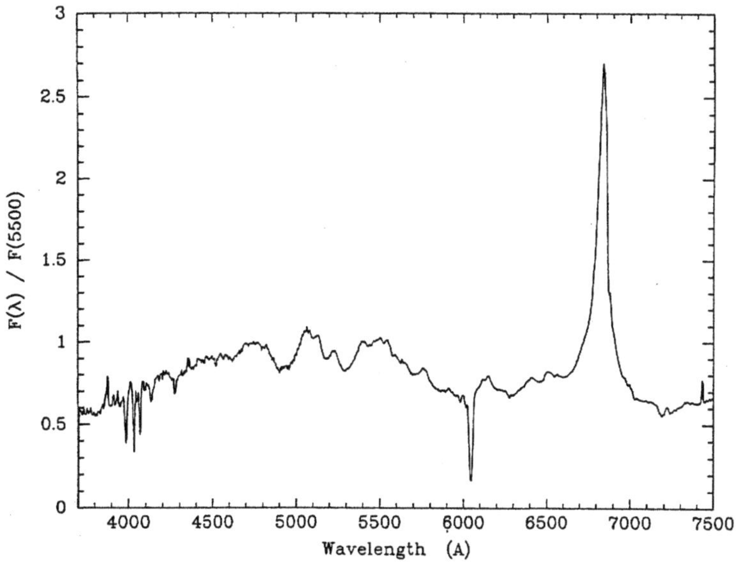

Fig. 36. Nuclear spectrum of the ultraluminous infrared galaxy Markarian 231, from Liu & Kennicutt (1995b). In this case the spectrum is dominated by a broad-line AGN.

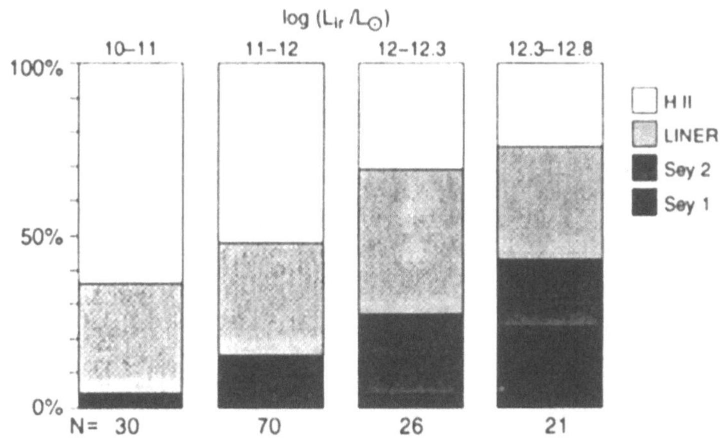

Fig. 37. Relative fractions of nuclear emission types for luminous infrared galaxies, shown as a function of total luminosity, from Veilleux et al. (1995).

provements in infrared instrumentation make this area with enormous potential. The continuum shapes and CO band spectra can help to distinguish between an AGN and starburst-dominated spectrum (e.g., Rieke et al. 1985; Lester, Harvey, & Carr 1988; Armus et al. 1995) and the velocity structure of the CO features can constrain the dynamics of the central regions (e.g., Lester & Gaffney 1994; Doyon et al. 1994; Armus et al. 1995; Shier 1995). Emission-line spectroscopy can provide useful constraints on the IMF in the central starbursts, as discussed in Lecture 6.

5.7 Radio Continuum Emission

The LIGs characteristically show very strong radio continuum emission that is tightly correlated with the infrared flux (Rieke & Low 1972). This strong radio–IR correlation appears to extend across a wide range of galaxy types and physical conditions, and its origin has never been fully explained, though several promising models have been proposed (e.g., Lisenfeld, Völk & Xu 1994 and references therein). If one assumes that the correlation also holds over a wide range of spatial scales, then in principle one can use interferometer maps in the radio to constrain the size of the infrared emitting region (e.g., Condon et al. 1991).

Large VLA surveys of the radio structure of LIGs have been carried out by Condon et al. (1990; 1991). The data show a gradual change in radio structure with increasing infrared luminosity; sources with $L_{\rm IR} < 10^{11}\,{\rm L}_\odot$ tend to show extended structure, while compact unresolved central sources become more dominant with increasing infrared luminosity. The radio sizes present a problem for a conventional starburst model, because the radio sources are frequently smaller that the minimum blackbody sizes of the infrared emitting regions, yet the objects follow the same IR-radio correlation observed in normal starburst galaxies. Condon et al. (1991) argue that this can be explained if the LIGs are ultra-compact starbursts, with densities so high that the regions are optically thick to all wavelengths shorter than $25\,\mu$m ($A_V > 1000$!). Most models of the FIR emission in these objects are based on the assumption that the sources are optically thin in the mid to far-infrared, and the Condon et al. study warns that such an assumption may not always be valid. The compact radio structure could also be produced by a central AGN, perhaps an AGN located within a more extended circumnuclear starburst, but in that case one needs to explain why the radio/FIR flux ratios are similar to those observed for starbursts.

High-resolution VLBI observations provide the most direct test for ultra-compact structure in the LIGs. Extensive LBI and VLBI observations have been reported by Norris et al. (1990) and Lonsdale et al. (1993). Cores with angular diameters $< 0''.25$ are detected in roughly half of the LIGs surveyed to date. The cores range in size from $5 - 150$ mas, and on average contain 12% of the total radio flux. It is interesting that the VLBI-scale emission *does not* appear to be correlated with the radio structure of the objects on larger angular scales (Lonsdale et al. 1993). Norris et al. (1990) show that compact cores are common among AGNs and those LIGs which show Seyfert-like optical spectra, but cores are rare among optically selected starburst galaxies or LIGs with HII region-like spectra. Taken together, these results provide perhaps the strongest evidence against starbursts being the sole heating source for the FIR emission in the LIGs, at least those which show strong VLBI components.

5.8 Evidence for Interactions and Mergers

From the first observations of LIGs, it was clear that the brightest objects were preferentially associated with strong tidal interactions and mergers; several prominent examples are illustrated throughout these lectures. The IRAS database has made it possible to analyze the connection between interactions and IRAS galaxies in a statistically rigorous manner. As with environmental studies of Seyfert galaxies (Lecture 4), the first estimates of the interaction frequencies of IRAS galaxies were quite controversial, with published estimates of the fraction of interacting LIGs and ULIGs ranging from $\sim$ 10% to 100%! Over the past few years a more consistent picture has begun to emerge. I refer the reader to the review by Sanders & Mirabel (1996) for a more complete discussion of this topic.

Most studies of interaction rates have been based on deep optical or near-infrared imaging of LIGs, to measure the frequency of tidal features, distorted morphologies, or interacting companions. Some of the inconsistencies in the early studies can be traced to different operational definitions of "interaction". Inferring the presence of a tidal interaction or merger on the basis of a peculiar optical morphology alone is very dangerous, because the visible appearance can easily be distorted by the heavy extinction or by asymmetric illumination and ejection of material from the starbursts, regardless of whether a tidal interaction took place. Stated another way, ULIGs are extraordinary, rare, peculiar objects by their very nature, and one would be surprised if even isolated objects *did not* appear peculiar! Consequently the most rigorous tests for interactions are the most conservative, insisting that one observe tidal signatures such as tails or bridges to a companion object. An excellent case study in these problems is the controversy over whether powerful radio galaxies are triggered by interactions and the so-called "alignment effect" (McCarthy 1993; Djorgovski 1994).

Investigations of interaction rates among LIGs include Sanders et al. (1988), Lawrence et al. (1989), Melnick & Mirabel (1990), Gallimore & Keel (1993), and Leech et al. (1994). One can summarize the conclusions of these studies as follows: (1) IRAS galaxies in general show an excess of companions and/or other tidal signatures over optically selected galaxies of comparable luminosity. (2) The frequency of interaction increases strongly with IR luminosity, from $\sim 20-30\%$ for $L < 10^{10}\,L_\odot$ to $\geq 90\%$ for $L > 10^{12}\,L_\odot$. These results are very reminiscent of what is observed for optically selected starbursts and AGNs (cf. Gallimore & Keel 1993). (3) The most luminous ULIGs appear to be preferentially associated with evolved mergers rather than distant tidal encounters (e.g. Joseph & Wright 1985). The typical morphology of the ULIG hosts is illustrated in Fig. 38, which shows contour maps made from red CCD images of five BGS ULIGs from Sanders et al. (1988).

Further evidence for the role of evolved mergers in triggering ULIGs comes from sub-arcsecond near-infrared imaging, which reveals that double nuclei are a common feature in these objects (Carico et al. 1990; Mazzarella et al. 1992; Majewski et al. 1993). In all, six of the ten BGS ULIGs observed between these three studies show double cores. Further observations with the Keck telescope and the NICMOS camera on HST will increase these types of observations dramatically.

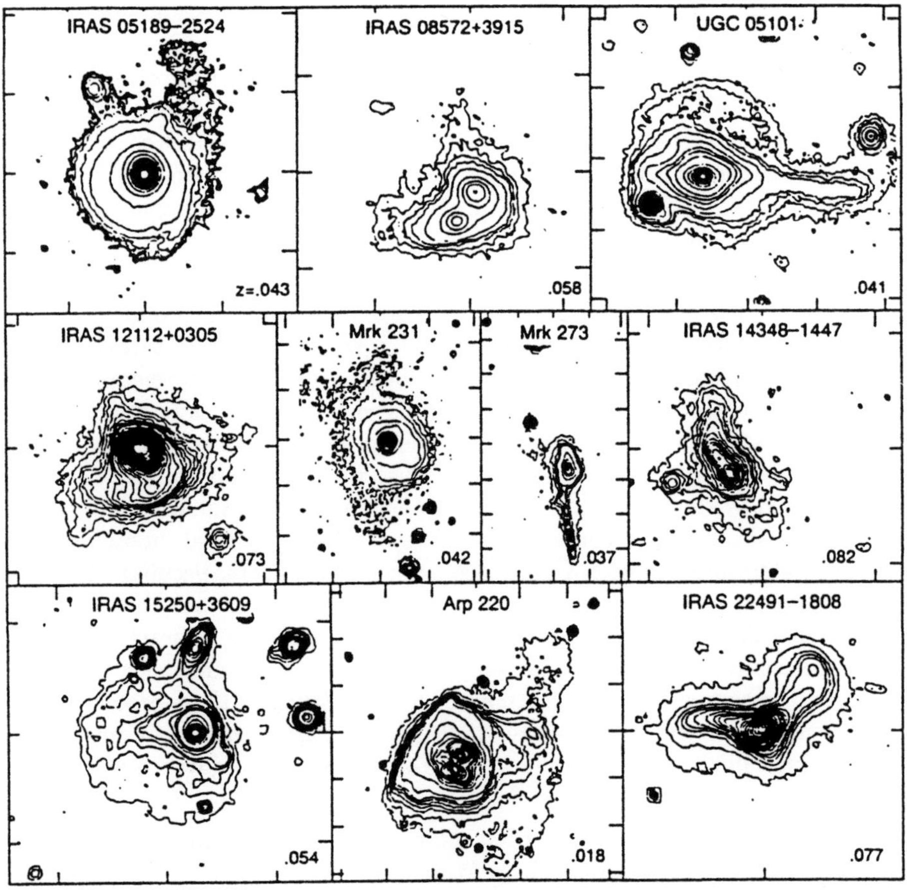

Fig. 38. Red isophotal contours of 5 ultraluminous infrared galaxies, from Sanders et al. (1988).

5.9 The Power Source: Starbursts or Dusty AGNs?

One of the enduring debates in this field centers on the nature of the primary power source for these extraordinary objects. The only other objects known to have comparably high bolometric luminosities are quasars, and this fact, coupled with the AGN-like spectra and compact structure of many ULIGs, leads one naturally to speculate that the ULIGs are simply dust-enshrouded quasars. On the other hand, many of the properties of LIGs show clear signatures of starbursts, especially the enormous molecular disks, blue continua and stellar spectra, and there are strong parallels in the association of both types of objects with interactions and mergers. So it is tempting to attribute the luminosities of the LIGs and ULIGs to dust-enshrouded starbursts. Harwit et al. (1987) have pointed out that emission from the dissipation of the colliding molecular media can be an important source of infrared luminosity, and this may be important in objects such as NGC 6240 (Keel 1990).

In any case we are dealing with extraordinary, unique objects, either the local analogs to protogalaxies, or the evolutionary precursors to quasars, or both. This makes the reward for understanding the origins of the IRAS galaxies a very rich one. The issue is unsettled, and my own view is that we are seeing a mixture of mega-starbursts and dusty AGNs; indeed it may be that both processes are closely associated in the most luminous objects. Here I summarize the evidence for each of these processes, and point out cases where both starbursts and AGNs are almost certainly present. For an excellent discussion of the various arguments see Heckman (1991).

Evidence for Starbursts. Direct evidence for the importance of starbursts comes from $0.2 - 3\,\mu$m spectra, which show reddened blue continua and strong absorption features of young stars in most LIGs. The critical question is how much of the total bolometric luminosity can be associated with this star formation, since we know that the central infrared core remains largely hidden at visible wavelengths. The fraction of the dust heating from this visible star formation can be estimated by scaling the observed SFR by the extinction, as estimated from the Balmer decrement of the ionized gas or from the reddening of the visible spectrum. This will tend to underestimate the true SFR, because the optical spectrum is heavily weighted toward regions of lowest extinction. Armus et al. (1989) carried out such an analysis for a sample of LIGs with strong starburst signatures, and concluded that the visible star formation accounts for a significant fraction (typically several tens of percent or higher) of the total infrared luminosity.

Regardless of the question of how much star formation takes place in the infrared cores of the LIGs, the ubiquitous presence of young stellar absorption features and HII region-like spectra in off-nuclear spectra provides convincing evidence that circumnuclear starbursts are present in most of the IRAS galaxies. Thus the debate centers not on whether starbursts power the infrared activity but rather whether starbursts power most or all of the activity.

More circumstantial support for a central starburst comes from the presence of massive molecular disks in the LIGs. The sizes and masses of the disks are generally consistent with the sizes and luminosities of the infrared emitting regions. When such circumnuclear disks are found in nearby galaxies they are inevitably accompanied by intense star formation. This suggests strongly that star formation is an important if not dominant process in the LIGs. However the presence of molecular gas in itself is not direct proof of a dominant starburst, since the gas could conceivably be feeding a massive central black hole instead. However it is difficult to imagine that such dense and massive disks could survive for long without forming copious numbers of stars.

The X-ray properties of LIGs potentially offer a powerful discriminant between central starbursts and AGNs, as discussed by Heckman (1991). Hard X-rays (> few keV) should be detectable into the cores of the FIR emitting regions, and AGNs —broad-line objects in particular— are very luminous in the $2-10$ keV region (e.g., Lawrence & Elvis 1982). The observed X-ray properties of LIGs are very heterogeneous. Objects with AGN-like spectra are often detected (e.g., David et al. 1992; Boller et al. 1993), but the X-ray/infrared ratios in most LIGs are 10 – 1000 times lower than in quasars and Seyfert 1 nuclei (Rieke 1988). This is consistent with starbursts being the predominant power source for the FIR emission. However Seyfert 2 nuclei are also weak X-ray emitters, and hence the AGN hypothesis cannot be ruled out on the basis of X-ray properties alone.

As discussed earlier most LIGs tend to lie on the infrared vs radio correlation observed for starburst galaxies, and this has been cited as circumstantial evidence for compact starbursts being the primary power source (e.g., Condon et al. 1991). Radio-loud AGNs are characterized by much lower FIR/radio ratios, but many Seyfert galaxies, especially Seyfert 2 systems, lie near (but slightly displaced from) the correlation (Bicay et al. 1995). Whether this implies the presence of star formation in Seyfert 2 systems or Seyfert activity in many LIGs (or both!) is unclear.

Evidence for AGNs. The case for AGN-powered ULIGs is summarized by Sanders et al. (1988). The Seyfert-like spectra observed in some LIGs provides direct evidence for a buried AGN. The evidence is the strongest for objects with broad-line nuclei (e.g., Mrk 231). These represent a minority of the LIGs but their fraction increases among the most luminous ULIGs. Objects with dominant Seyfert 2 spectra are also candidates for buried AGNs, but as mentioned earlier those with LINER-like spectra are more ambiguous, as the emission can be produced by an AGN or by other mechanisms. It is also possible that some Seyfert 2 AGN spectra are produced by concentrated starbursts instead (Terlevich & Boyle 1993).

The compact structures in some LIGs may offer the strongest evidence for a significant AGN component. The luminosities of the VLBI components are far stronger than can be accounted for by individual supernovae (Lonsdale et al. 1993), and the association of these sources with objects containing AGN-like optical spectra points strongly toward a central compact object. The presence of VLBI jets in some objects is even stronger evidence (e.g., Mazzarella et al. 1991). The presence of very small mid-IR sources in Arp 220 and other ULIGs is also consistent with the AGN interpretation, though a very compact starburst cannot be ruled out.

A potentially powerful test for AGNs is the detection of X-ray variability. Boller, Fink, & Schaeidt (1994) have reported variability in IRAS 15564+6359. The object shows factor-of-two variations on 1500 sec timescales and order-of-magnitude variations on timescales of a day. The X-ray spectrum shows both starburst and AGN-like components, yet the optical spectrum is HII region-like. This is convincing evidence for a buried AGN in at least this object, and it would be interesting to place variability limits on larger samples of LIGs.

Conclusions. These data illustrated the diversity in the population of infrared-luminous galaxies, and suggest that both processes —enormous starbursts and luminous AGNs— are important in subclasses of the LIGs and ULIGs. When one examines the observed properties of the LIGs taken together, two distinct classes of objects seem to emerge in the extremes, one with Seyfert spectra, compact radio structure, and strong hard X-ray emission, and another with HII region like optical spectra, strong stellar continua, extended radio structure, and weak X-ray emission. Some properties, such as massive central molecular disks and evidence of tidal interactions appear to characterize both types of objects. This suggests strongly that we are observing the effects of buried AGNs and starbursts in different subclasses of the LIGs and ULIGs, and probably a composite of both processes in a large percentage of the LIGs. The statistics also point to an increasing role of AGN-like activities in the most luminous objects. Whether this reflects a real physical transition or simply the masking effects of starbursts at lower luminosities is not entirely clear.

Although the discussion here has been restricted to infrared-selected luminous objects, it is interesting that many of the evidences of composite starburst/AGN structure are now seen in more traditional samples of AGNs as well. For example, massive molecular disks are seen in the centers of many Seyfert galaxies, and many quasars appear to be surrounded by circumnuclear starburst regions (see Heckman 1991, 1994, and references therein). We have already seen that the interaction environments of luminous quasars and infrared galaxies are similar as well. This suggests an obvious continuity between the ULIGs and quasars, in which central "engines" and starbursts both play important roles.

An interesting case study, and a possible "smoking gun" of the links between ULIGs and Seyfert galaxies or quasars, is Markarian 463, as studied by Mazzarella et al. (1991). It is an ultraluminous galaxy ($L_{\rm IR} > 10^{12}\,{\rm L}_\odot$) in a double-nucleus Markarian galaxy, with the nuclei separated by 4″, or 6 kpc; hence it is likely to represent an intermediate stage of a merger. Both nuclei exhibit Seyfert 2 spectra, and one shows an underlying broad-line nucleus in polarized light, as well as subarcsecond radio structure and a compact radio jet. The galaxy also contains a massive central CO disk with an inferred mass of $10^{10}\,{\rm M}_\odot$. As pointed out by these authors, we are probably observing a system that is in the midst of a transition from an ULIG to a radio-loud quasar or radio galaxy. Another example of a composite object is in the interacting system Mrk 266 = NGC 5256 (Wang et al. 1997).

The question remains as to the global role of starbursts vs AGNs across the spectrum of IRAS galaxies, and the related question of how the two processes may be physically related to one another. A detailed discussion of these issues is beyond the scope of this lecture. Rowan-Robinson (1995) has attempted to address these questions by analyzing the luminosity functions of quasars and IRAS galaxies and by comparing their spectral energy distributions at all wavelengths. He concludes that starburst galaxies are an order of magnitude more numerous than quasars, but that the importance of the latter increases, being comparable for $L_{\rm bol} > 10^{13}\,{\rm L}_\odot$. He also concludes, consistent with the evidence reviewed above, that starbursts inevitably accompany a Seyfert or quasar event, but that not all starbursts need form an AGN. This is not the only attempt to unify these phenomena (see Sanders & Mirabel 1996 for references to others), but it proves a good illustration of how the disparate observations described here can be understood within a common observational picture.

6 Spectral Diagnostics of Starbursts

Most of the preceding discussion has focussed on observations of interacting galaxies, and comparisons of their properties with those of normal galaxies. In order to understand the physical processes that trigger and regulate the star formation in these systems, we need to translate these observations into quantitative measures of the starburst properties. At the top of our wish list would be measures of the basic temporal properties of the starburst, the current star formation rate (SFR), the total mass of stars formed in the burst, and the duration and age of the starburst. Beyond that we would like to measure such quantities as the star formation efficiency, the shape and limiting masses of the stellar initial mass function (IMF), the chemical composition, and the mechanical energy outputs of the burst population.

Fulfilling this wish list is a daunting task, because no single spectral diagnostic provides complete information about these properties, and it is often difficult to disentangle the various parameters in the observed spectrum. The task is more difficult yet in the luminous infrared starbursts, due to the heavy extinction and contamination from non-stellar radiation. Fortunately, several properties of the interacting galaxies simplify the modelling task. The spectra of many objects are dominated by a single massive young population, and the strong emission of this starburst population across the spectrum allows us to combine information from several spectral indicators. For example, the hydrogen emission-line fluxes (e.g. Hα) provide direct information on the current SFR in the bursts, while the FIR and visible continua are sensitive to the SFR integrated over the history of the bursts. By making use of the entire spectrum we can impose physically meaningful constraints on the range of star formation properties in these systems, and begin to understand the processes that drive the starbursts (Lectures 7–8).

This lecture describes the methods that are used to constrain the physical properties of starbursts, and presents several examples of applications to interacting galaxies. The methods used to analyze optically visible and dust-obscurred starbursts are similar in principle but quite different in practice, so I discuss each separately. I conclude by addressing the question of whether infrared-luminous starburst galaxies are powered by a stellar population with an anomalous IMF.

6.1 Evolutionary Synthesis Models

Two different approaches have been used to model the integrated spectra of composite stellar populations. The "spectral synthesis" technique reconstructs the integrated spectrum from an empirical library of stellar spectra (or star clusters), which spans the range of types expected to be present in the population. This method has been widely applied to elliptical galaxies and spiral bulges, where the range of stellar ages and compositions is reasonably well constrained. For active star forming galaxies, an "evolutionary synthesis" technique is often used, where theoretical stellar evolution and spectral models are combined to synthesize the integrated spectra for an arbitrary star formation history. A grid of such models is then compared with an observed spectrum to constrain the properties of the system. Most studies of interacting galaxies and starbursts have been based on evolutionary synthesis models, so I will restrict my discussion to this method. For a full discussion of spectral modelling, I refer the reader to the proceedings of a recent conference (Leitherer, Fritze-v. Alvensleben, & Huchra 1996).

The basic steps used in constructing an evolutionary synthesis model were described in Section 2.1. A grid of stellar evolution models covering the full range of stellar masses provides effective temperatures and bolometric luminosities as a function of time. This is combined with stellar atmosphere models to convert the temperatures and luminosities into a grid of theoretical spectra for each stellar mass and time step. The latter are then combined, weighted by the IMF, to produce the synthetic spectra of model star clusters as a function of age, often referred to as "single-burst models". Single-burst spectra for different ages are then combined to synthesize the integrated spectrum of a population with an arbitrary star formation history. For normal galaxies a smooth age distribution is assumed, usually approximated by an exponential function (see Section 2.1), while for interacting systems one usually superimposes a young burst population on top of an underlying older population. A variant on this technique combines published theoretical isochrones for different age populations, but the end product is similar.

The result of this exercise is a set of model spectra for a range of IMFs and star formation histories. Several synthesis programs and model grids are available in the literature, with probably the most widely used version being that of Bruzual & Charlot (1993). An extensive grid designed especially for starburst systems has been published by Olaffson (1989). Many other examples are given in Leitherer et al. (1996). The same approach can be used to synthesize the broadband colors. The luminosity Hα (or other recombination lines) is derived by first integrating the spectrum shortward of the Lyman limit at 912 Å to derive the ionizing photon flux, then using recombination theory to convert these to nebular line luminosities. The effects of reddening, extinction, and density bounding can be incorporated into the calculation as well. The far-infrared (FIR) flux can be derived likewise by integrating the spectrum over the ultraviolet and visible, weighted by the dust optical depth.

Evolutionary synthesis models of this sort provide excellent fits to the integrated spectra of normal galaxies. This is illustrated in Fig. 39, which shows model fits (Bruzual & Charlot 1993) to the integrated spectra of four elliptical and spiral galaxies (Kennicutt 1992b). The models adopt a fixed chemical composition (solar) and fixed IMF, with the only free parameters being the age and the e-folding time of the SFR, denoted as τ in Fig. 39. Although the best fits are obtained by varying the galaxy age, reasonable fits can be obtained with a fixed age in the range of $10 - 15\,\mathrm{Gyr}$.

The straightforward approach of evolutionary synthesis modelling should not convey the false sense that applying these models is easy! The computation of the synthesized spectrum is a tedious but relatively straightforward task; interpret-

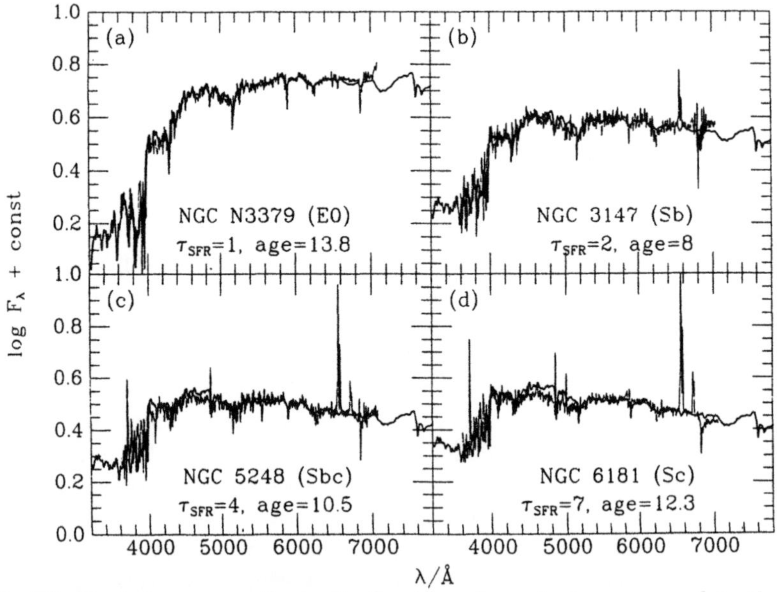

Fig. 39. Evolutionary synthesis fits for 4 nearby galaxies compared to the observed integrated spectra, from Bruzual & Charlot (1993).

ing the results is much more difficult. The main problem is that the integrated spectrum is determined by several independent parameters (star formation history, IMF, age, reddening, abundance), and many of these parameters produce similar effects on the integrated spectrum. Breaking these degeneracies is problematic, when dealing with normal galaxies or starbursts. Below I list some of the most serious complications.

Age-Metallicity Degeneracy: For most populations it is very difficult to distinguish a change in metal abundance from a change in the stellar age distribution; a galaxy with weak metal lines may simply be metal-poor, or the metal lines from a relatively metal-rich population (mainly from cool stars) may be diluted by the continuum of younger, hotter stars. This age-metallicity degeneracy can be partly removed by using UV, visible, or infrared colors to independently constrain the age, or by using spectral indices that are specially designed to distinguish between age and metallicity (Rose 1985). To complicate matters, most published synthesis models are computed for a single chemical composition (usually solar), but of course most real galaxies differ in mean composition and contain a wide range of abundances. Synthesis models for non-solar compositions are available, but their calibration uncertainties increase rapidly the farther one goes from solar abundances.

Reddening: Interstellar reddening and extinction will affect the continuum colors and emission-line fluxes, mimicking the signature of an older population (see Fig. 40). Dereddening the composite spectrum of an active star forming galaxy is problematic, because the extinction along a given line of sight is highly variable, different stellar components are reddened by different amounts, and the

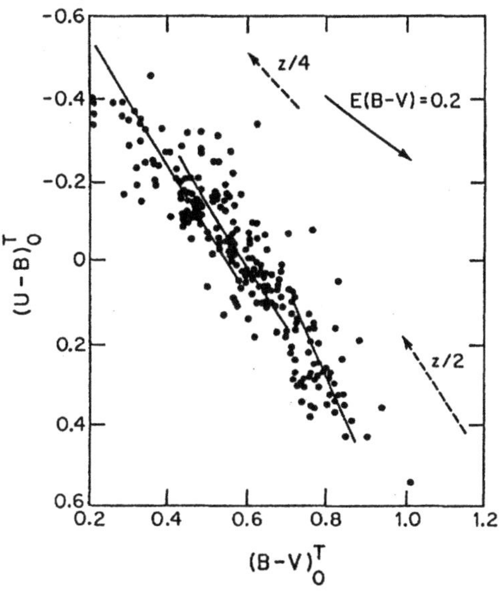

Fig. 40. A series of models for the *UBV* colors of star forming galaxies, showing the effects of varying the IMF, reddening, and metallicity, as described in the text. Figure taken from Kennicutt (1986).

integrated spectrum is heavily weighted toward the least reddened regions. The emission-line spectrum provides some independent constraints on the reddening, but in very dusty regions (galactic nuclei, LIGs) this breaks down as well. One can avoid this problem by synthesizing spectral line indices only, or by observing in the infrared.

IMF: Most synthesis models assume a particular form for the IMF, usually the solar neighborhood IMF as parameterized by Salpeter (1955) or Scalo (1986). Unfortunately, the visible spectrum is relatively insensitive to the IMF shape, and changes in the IMF are nearly degenerate with changes in the star formation history. Moreover, visible and near-infrared spectra provide little useful information about the upper and lower ends of the IMF, because stars in those mass ranges contribute little to the visible continuum. This problem can be alleviated by using other observables that are sensitive to the extremes of the IMF, for example emission lines to constrain the upper IMF and the dynamical mass of the population to constrain the low-mass component.

Age: For an evolved stellar population, changing the assumed age of the galaxy shifts the model SEDs and spectra in ways that are nearly degenerate with star formation history, IMF, metal abundance, etc. This problem is less severe for a dominant young population, where age has a very strong effect on the spectrum. However for post-burst populations there is a strong degeneracy between the inferred age and strength of the starburst (below).

Pre-Existing Stellar Populations: Although some starbursts are so strong that they dominate the spectrum of the host galaxy at all wavelengths, to date there is not a single conclusive case of a galaxy with a purely young population; there is almost always a pre-existing stellar population, with its own star formation history, that contributes significantly to the integrated spectrum. In many instances it is difficult to separate the inferred properties of the star formation burst from the properties of this underlying population (Bica et al. 1990).

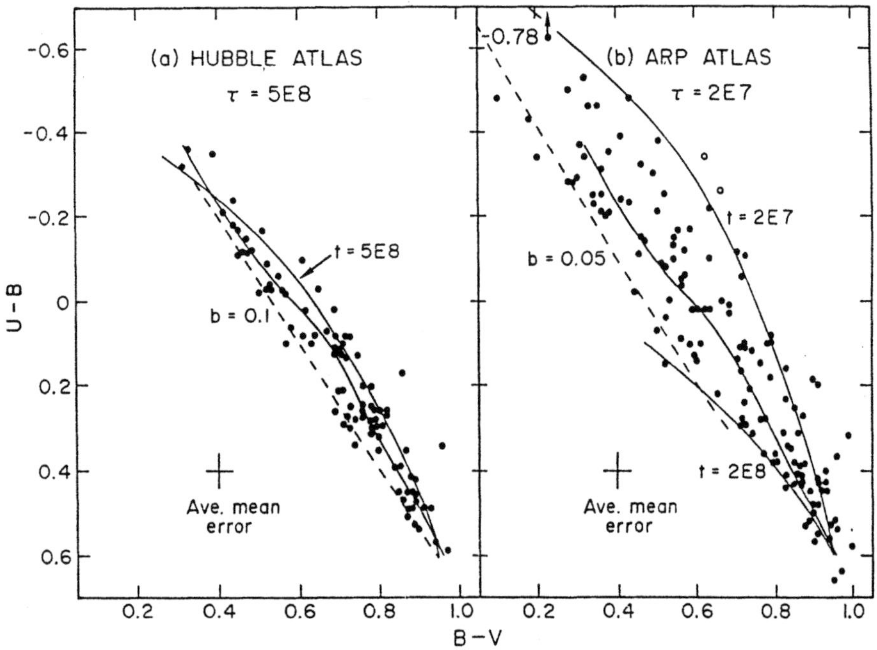

Fig. 41. *UBV* colors of normal and peculiar galaxies, with evolutionary synthesis models, from Larson & Tinsley (1978). See text for a description of the models.

An important manifestation of this problem for interacting galaxies is the degeneracy between burst age and strength. The spectra of post-starburst galaxies are dominated by strong Balmer absorption lines (e.g., Fig. 14) superimposed on an older underlying spectrum, the so-called "E+A" spectrum. Since the luminosity of a starburst population declines rapidly with age, such a composite spectrum can be reproduced by a relatively small but young burst, or alternatively by an older but larger population. Varying the assumed spectrum of the underlying pre-burst population allows for even more freedom in the inferred age and strength of the starburst. For this reason, most analyses of the ages and strengths of evolved starbursts quote a range of properties that are consistent with the observed spectrum. In some cases the age of the burst can be constrained independently (e.g., via kinematic dating or via mass consumption arguments). A promising technique based on the Ca II H+Hϵ absorption feature has been described recently by Leonardi & Rose (1996).

The effects of several of these degeneracies on the integrated colors of galaxies are illustrated in Fig. 40, taken from Kennicutt (1986). The lines shows the loci of colors for 3 different model sequences where the birthrate parameter b is held fixed (one value for each line), while the IMF is varied. The arrows show the effects of changing the reddening or metallicity of the models, as computed by Larson & Tinsley (1978). Compare this with the locus of constant IMF, variable birthrate history models in Fig. 41. Note that varying any of these parameters shifts the models in directions parallel to the main 2-color sequence of galaxy colors, so disentangling the various parameters from colors alone is difficult.

I hope that this discussion will not discourage the reader from applying synthesis models, but I do hope that I have conveyed the care that must be taken when interpreting the models. It is ironic that the subtleties that Larson & Tinsley described 20 years ago are being rediscovered by a new generation of modellers, as the same techniques are applied to starburst galaxies and high-redshift objects. Despite these difficulties, the spectra of interacting galaxies reveal some unmistakable properties that allow us to define the parameter space of starburst populations in interactions and mergers.

6.2 Applications: Integrated Colors of Interacting Galaxies

The most famous application of synthesis modelling to interacting galaxies is the study of UBV colors by Larson & Tinsley (1978; LT). Most of their conclusions are still applicable today, and the simplicity of the multi-color approach makes it an excellent case study for illustrating the general features of spectral modelling. For an updated application of this technique, see Fritze-v. Alvensleben & Gerhard (1994a).

Figure 41 shows the famous LT comparison of colors of interacting and normal galaxies, but this time with several sets of models superimposed. The lower solid curve (left panel) shows a grid of model populations with fixed IMF, fixed abundance, fixed age, and exponentially declining star formation histories; these models provide an excellent fit to the colors of normal galaxies, with the dispersion in colors around this line consistent with observational errors (cf. Fig. 3).

These steady-state evolutionary models cannot reproduce the blue envelope and color spread seen in the interacting and peculiar galaxies (right panel). To reproduce their colors, LT computed a grid of models in which star formation bursts were added to the steady state models. Their grid encompassed starbursts of duration $20 - 500\,\mathrm{Myr}$, and strengths (b) ranging from 0.001 to 1. The "strength" in this case is defined as the fraction of stars in the galaxy produced during the burst. For any particular galaxy, the integrated colors could be reproduced with several combinations of burst strength, duration, and age, due to the degeneracies described earlier, and the fact that the color of the galaxy prior to the burst is not known. However it is possible to define a parameter space of burst strengths, durations, and ages that encompasses the locus of observed colors. LT concluded that the range of galaxy colors could be reproduced by bursts with duration $20\,\mathrm{Myr}$ or longer (upper limit $500\,\mathrm{Myr}$) and strengths of less than 5% of the total stellar mass. Applying these relative numbers to an L^* galaxy yields peak SFRs in the range $20 - 200\,\mathrm{M}_\odot\,\mathrm{yr}^{-1}$ and total masses formed of up to $5 \times 10^9\,\mathrm{M}_\odot$. These results still provide an excellent characterization of the starburst properties of the Arp atlas interacting galaxies.

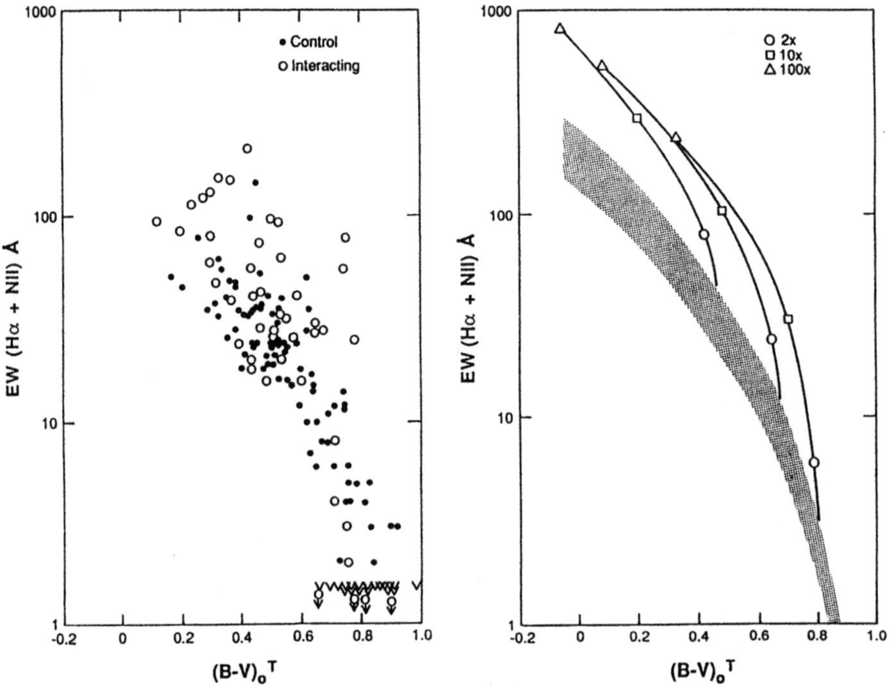

Fig. 42. Hα + [NII] emission-line equivalent widths and $B - V$ colors of normal (control) and interacting galaxies galaxies (left), and burst model calculations (right), from Kennicutt et al. (1987).

6.3 Applications: Hα Emission, Colors, and SFRs

Kennicutt et al. (1987) extended this approach to incorporate modelling of the Hα emission. The advantage of measuring the Hα fluxes and EWs is that they provide much more direct constraints on the instantaneous SFRs and upper IMF properties than the colors alone. This is illustrated in Fig. 42, which shows a two-color diagram of Hα emission EWs plotted against $B - V$ color.

The left panel of Fig. 42 shows that the interacting galaxies are characterized by stronger Hα emission and bluer colors (Lecture 3), but there is excess Hα emission even at a fixed color. This result provides unambiguous evidence that the blue colors are produced by shortlived starbursts. The right panel shows the expected loci of Hα EWs and colors for models similar to those computed by LT. The shaded region shows the expected locus for steady-state star formation with a roughly Salpeter IMF (shading indicating the effects of $0 - 1$ mag of extinction at Hα). Once again the normal galaxies (solid points) are reproduced with steady star formation. However, most of the interacting galaxies lie above the shaded region, and starbursts are required to reproduce their EWs. The lines in Figure 42 show a grid of models with burst amplitude varied from $0 - 100$, superimposed on 3 examples of underlying stellar populations. In this case the amplitude is defined as the ratio of the burst SFR to the SFR prior to the burst.

The properties of the interacting galaxies require relatively short bursts ($\leq$ 50 Myr), and absolute SFRs of $\sim 1-60\,M_\odot\,yr^{-1}$, assuming an average extinction of 1 mag at Hα. The total mass of stars formed in the largest bursts is of order $10^9 - 10^{10}\,M_\odot$, similar to the LT results, if a normal IMF is assumed. Note however that the EWs of the galaxies with the strongest Hα emission cannot be reproduced by the models in Fig. 42, even with infinitely strong bursts. Kennicutt et al. (1987) suggested that these objects require a top-heavy IMF. I return to this point in Section 6.5.

6.4 Applications: Spectral Modelling of Evolved Starbursts

The techniques described above break down for populations which are observed well after the peak star formation. As shown in Lecture 3 (Fig. 14), many evolved mergers exhibit post-starburst spectra, with little evidence of current star formation, but with the Balmer-dominated spectra of a massive evolved starburst. This result should not be surprising. We have already learned from the modelling of *UBV* colors and Hα emission that the period of peak star formation is typically a few times 10^7 yr (LT; Kennicutt et al. 1987), considerably shorter than the dynamical timescale of most interactions, which must be of order 10^8 yr. Indeed the discordance in these timescales provides an interesting challenge for modelling the star formation in these systems (Lecture 8). In any case, we can expect a considerable fraction of any sample of interacting galaxies to be observed after the peak episode of star formation, and in order to measure the properties of those starbursts a full spectral synthesis is required.

Detailed modelling of the optical spectra of starburst galaxies has been carried out by Bernlöhr (1992; 1993a, b), Fritze-v. Alvensleben & Gerhard (1994b), and Leonardi & Rose (1996). Figures 43 and 44 show fits to two famous mergers, NGC 520 (Bernlöhr 1993a) and NGC 7252 (Fritze-v. Alvensleben & Gerhard 1994b), and demonstrate the current state of the art in this area. The fits are subject to uncertainties introduced by the degeneracies described earlier, especially the burst age vs strength degeneracy, but it is still possible to constrain the burst properties to very interesting ranges. For example the optically visible (NW) component of NGC 520 (Fig. 15) is best fitted by a starburst that ended $2-3 \times 10^8$ yr ago (the other component of NGC 520 contains a dust-obscurred starburst that is still active). The total mass of stars produced in the bursts is not well constrained, but Bernlöhr (1993a) estimates an upper limit of 20% for the fraction of the galaxy's stars that has formed in the merger. NGC 7252 is a more evolved merger, and in this case the mass of stars formed in the interaction is better constrained. The synthesis modelling indicates a mean burst age of 1.3 – 2 Gyr, a burst duration of $1 - 5 \times 10^8$ yr, and a total burst mass comprising 20 – 50% of the total stellar mass of the galaxy! As will be discussed in Lecture 7, the transformation of 20 – 50% of the mass of NGC 7252 from gas to stars in a dynamical timescale (compared to $\sim 10-50$ dynamical times for a normal spiral galaxy) presents a challenge to our theoretical understanding of the large-scale star formation in galaxies.

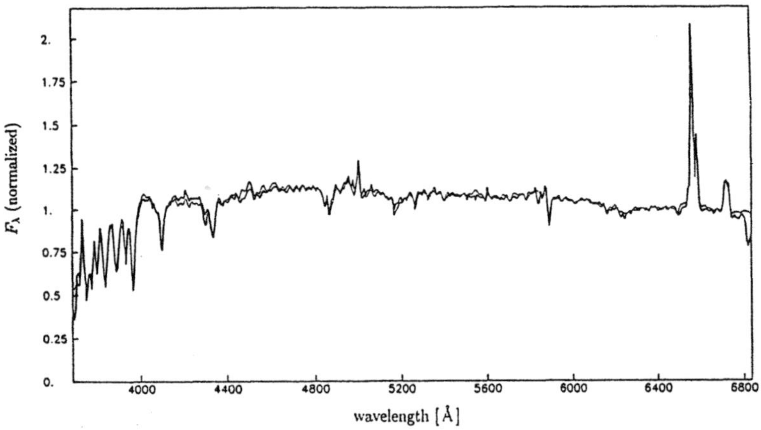

Fig. 43. Observed spectrum and synthesis model of NGC 520, from Berhlöhr (1993a).

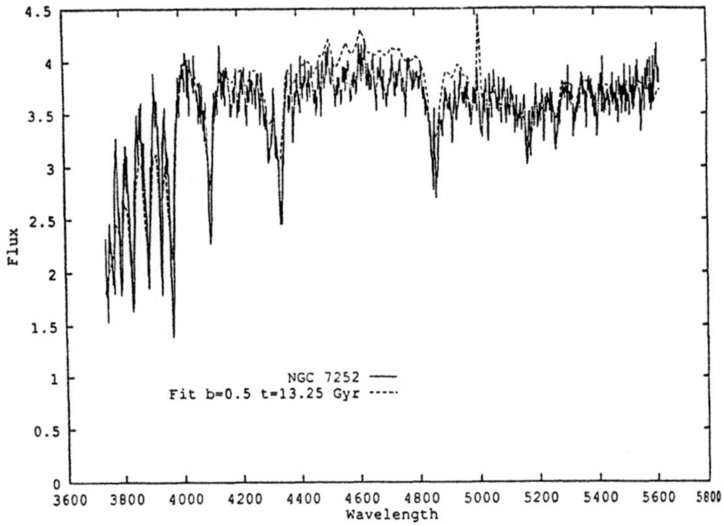

Fig. 44. Observed spectrum and synthesis model of NGC 7252, from Fritze-v. Alvensleben & Gerhard (1994b).

Modelling of a larger sample of starbursts by Bernlöhr (1993b) suggests that NGC 7252 is by no means a unique system. He fitted long-slit spectra for 30 interacting and merging systems, in order to estimate the typical range of burst properties, much in the spirit of the LT study. He concluded that most of the bursts are relatively shortlived, with periods of peak star formation lasting 15 – 50 Myr (average ~ 20 Myr). In addition there is evidence of delays between bursts in individual pair members of up to 200 Myr. This is reminiscent of the LT and Kennicutt et al. (1987) results, which suggest that the typical burst lifetimes are several times shorter than the dynamical timescale of the interaction.

The masses of stars produced in the bursts are difficult to constrain accurately, due to the strength vs age degeneracy and the dependence of the inferred burst masses on the assumed IMF. In several cases of strong bursts (e.g., NGC 520, NGC 7714, NGC 3690) application of a standard IMF would yield masses, integrated over the lifetime of the burst, of 20 – 100% of stellar masses of the galaxies, and larger than the likely mass available in gas prior to the interactions. Bernlöhr concludes from this that the IMF in the mergers is likely to be significantly deficient in low-mass stars. I discuss the issue of the IMF in starbursts further in the next section. Regardless of the IMF, it is apparent that the integrated masses of stars formed in the most extreme mergers are prodigious, probably consuming a large fraction of the available interstellar gas in the galaxies, and forming of order 10% of the total stellar mass or more.

6.5 Infrared Starbursts and the IMF in Starbursts

The most extreme examples of interaction induced starbursts are located in the dusty cores of LIGs, and the spectral diagnostics discussed up to now are of limited use. However, similar techniques can be employed using infrared diagnostics of the stellar population. For details I refer the reader to reviews by Rieke (1991) and Joseph (1991), as well as the references cited below.

The principles of the spectral modelling are similar to those described already, and in some respects the underlying assumptions are simpler. For example, in most LIGs one can safely assume that the starburst dominates the bolometric luminosity of the object, which is directly measurable from the FIR luminosity (assuming no AGN contribution). Further constraints on the luminosity of the stellar component come from the near-infrared colors and spectra, especially the CO bandhead at $2.29\,\mu$m. The ionization rate and massive SFR can be inferred from near-infrared recombination lines (e.g. Brγ) or from higher-order recombination lines in the submillimeter. Other important spectral diagnostics include the near-IR and mid-IR forbidden line spectrum, which constrains the effective temperatures and masses of the ionizing stellar population, and the stellar velocity dispersion of the starburst, as measured for example from the CO bandhead, which can provide an important constraint on the dynamical mass of the population.

Detailed modelling of several infrared luminous starbursts has been carried out by Rieke et al. (1980; 1993); Bernlöhr (1992), Doane & Matthews (1993); Doyon, Joseph, & Wright (1994); Prestwich, Joseph, & Wright (1994), and references therein. To simplify the discussion I will concentrate on M 82, which is the best studied case to date (Rieke et al. 1980, 1993; Bernlöhr 1992). The ionizing luminosity for M 82 is $\sim 10^{54}$ photons sec^{-1}, a prodigious luminosity given the small size of the galaxy. The corresponding SFR for a local IMF (stellar mass limits $0.1 - 100\,M_\odot$) is $\sim 10 - 30\,M_\odot\,yr^{-1}$, depending on whether one uses a Salpeter or Scalo (1986) IMF. Comparable SFRs are implied by the FIR luminosity of M 82 (for the same IMF). However, there are indications from the infrared forbidden line spectrum that stars more massive than $\sim 30\,M_\odot$ are absent, either because they never formed or because we are observing M 82 after the star formation peak. In this case, the SFR for an IMF truncated at 30 $M_\odot$ (but still extending to the bottom end of the local IMF) would be $\sim 25-70\,M_\odot\,yr^{-1}$. By any measure M 82 contains a prodigious starburst, with a massive SFR comparable to or larger than the most luminous normal galaxies (e.g., M 101, M 51).

Several straightforward arguments can be used to show that SFRs of several tens of solar masses per year are unrealistic. Photometric modelling shows that the age of the burst is at least 10^7 yr, and hence the SFRs given above would imply a stellar mass of up to $10^9\,M_\odot$ formed over the duration of the M 82 starburst. This is inconsistent with the dynamical mass estimate for the region ($< 2.5 \times 10^8\,M_\odot$) and the K-band luminosity ($M_K \sim -22.5$). It would also imply an efficiency of star formation of nearly 100% in the region. Rieke et al. (1993) explored a range of parameters, adjusting the ionizing luminosities, infrared luminosities, and upper mass limits over the allowable ranges, and concluded that most models with a solar neighborhood IMF and upper mass limit of $\sim 30\,M_\odot$ overpredict the mass of the starburst by a factor of ~ 5 on average. This discrepancy was first pointed out by Rieke et al. (1980), who proposed that the problem could be solved if the M 82 starburst did not form stars with masses below about $3\,M_\odot$. Similar conclusions were reached (though different in quantitative detail) for M 82 and other luminous starbursts by Joseph (1991), Doane & Mathews (1993), Rieke et al. (1993), and Prestwich et al. (1994).

Similar hints of a top-heavy IMF have come from modelling of optically-selected starbursts. Kennicutt et al. (1987) found that the strongest Hα-emitting starbursts could only be understood if they were all very young (unlikely) or if the IMF were enriched in massive stars relative to the solar neighborhood IMF. Likewise, Bernlöhr (1992; 1993a,b) found that applying a standard IMF to his sample of starburst spectra resulted in unrealistically large mass fractions and star formation efficiencies. In all of these cases the discrepancies could be relieved if the ratio of high to low-mass stars were skewed by factors of a few in the starbursts. There is very indirect evidence that a similar skewing of the IMF may be required in supergiant HII regions such as 30 Doradus (Kennicutt & Chu 1988).

Given the number of studies and the wide range of empirical approaches used, there now seems to be a strong case for a systematic difference between the IMFs in luminous starbursts –whether observed in the infrared or optical– and the time averaged IMF in the solar neighborhood, in the sense that stars of a few solar masses or less are under-represented by factors of a few in the starbursts. However I believe that the precise nature of this IMF difference is much less understood; while the observations of M 82 and other LIGs can be fitted with an IMF truncated below a few solar masses, less drastic alternatives have not been excluded. I suspect that at least part of the IMF discrepancy between the starbursts and the solar neighborhood IMF is due to an under-representation of massive stars in the Scalo (1986) IMF that is often used as the reference for the solar neighborhood. Modelling of the integrated colors and Hα emission of nearby normal spirals shows that the Scalo IMF underpredicts the number of massive stars by a factor of $2-3$ (Kennicutt 1983; Kennicutt et al. 1994), and direct evidence for a shallower IMF comes from color-magnitude diagrams of star forming regions in nearby galaxies (e.g., Hunter et al. 1995). Another concern with the truncated IMF model is the strong dependence of the inferred low-mass cutoff on the high-mass limit; adjusting the upper mass limit from $20 - 30\,M_\odot$ to $60 - 100\,M_\odot$ lowers the total inferred SFRs, and relieves the mass deficit considerably. Although some of the evidence for a upper mass limit is strong, other analyses based for instance on the strength of the He I $2.06\,\mu$m line are weaker (see Shields 1993). Even with these uncertainties, the case for a skewed IMF in starbursts seems to be quite strong, but differences involving a shallower slope for all masses, rather than a truncated lower mass limit, may be able to fit most of the data equally well (e.g., Bernlöhr 1993b; Doyon et al. 1994). Whatever the interpretation, these data offer exciting evidence for a systematic difference in not only the absolute SFRs, but also in the physical character of the star formation in massive starbursts.

7 Triggering and Regulation of Star Formation: The Problem

One of the main challenges in this field is understanding the physical processes that trigger and regulate the star formation bursts in interacting galaxies. This is a subject at the frontier of our understanding of galactic evolution. There is no comprehensive theory of what determines the SFR in a merger, a starburst, or in any galaxy for that matter. Indeed, a realistic physical description of star formation is one of the major missing ingredients in theoretical models of galaxy evolution in general. Interacting galaxies are important in this context, because they provide us with local analogs to protogalaxies over a wide range of physical scales and star formation environments.

The ultimate solution of this problem probably rests with the generation of young scientists that is reading this book. With that optimistic view in mind, I will begin by describing the physical nature of the problem we confront (this lecture), then I review some of the ideas for how starbursts may be triggered and maintained (Lecture 8).

7.1 Introduction: Star Formation as a Nonlinear Process

The first lesson we can learn from observations of interacting galaxies is that the processes driving the global SFR are highly nonlinear in nature. To understand this, consider a simple model in which the total SFR scales simply in proportion to the total mass of gas, the linear form of the Schmidt (1959) law. Such a law produces an exponentially declining SFR (R):

$$R \propto \rho_{\mathrm{g}} \propto \frac{\mathrm{d}\rho_{\mathrm{g}}}{\mathrm{d}t}\,, \tag{3}$$

$$\rho_{\mathrm{g}}(t) \propto \rho(0)\,\mathrm{e}^{-t/\tau}\,. \tag{4}$$

This simple model provides a reasonable description to the star formation histories of many normal galaxies (see Lecture 2), and indeed the exponential approximation is often adopted in evolutionary synthesis models. However, the model fails to account for the star formation bursts that are observed so frequently in interacting systems. For a linear Schmidt law the total gas mass of the system determines the global SFR, and since the interaction can only deplete gas (via ejection or star formation), the total SFR can never increase with time. The very existence of starbursts is *bona fide* evidence for a very nonlinear dependence of the SFR on gas density or other physical property of the system.

To quantity this argument, consider a more general form of the Schmidt law, in which the SFR scales as an arbitrary power n of the gas density:

$$R \propto \frac{\mathrm{d}\rho_{\mathrm{g}}}{\mathrm{d}t} \propto \rho_{\mathrm{g}}^{n}\,. \tag{5}$$

For a nonlinear Schmidt law ($n > 1$), gas compression during the interaction will produce an enhanced SFR, with the enhancement in SFR determined by the

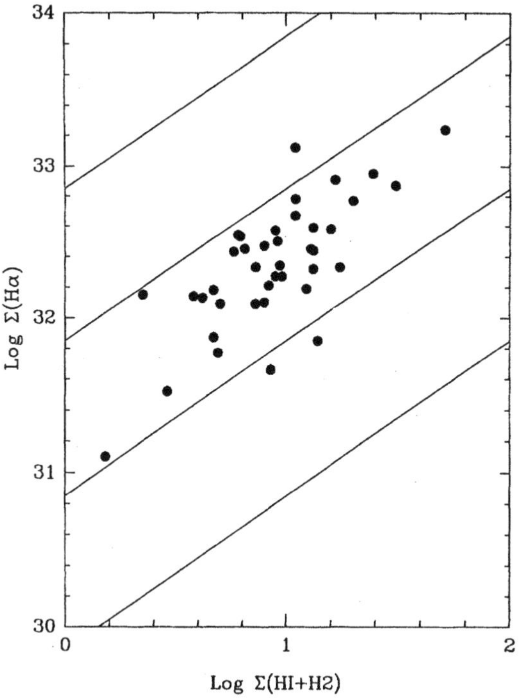

Fig. 45. Correlation between average Hα surface brightness of optical disks and the average surface density of atomic plus molecular gas. Lines indicate constant efficiencies or timescales for gas consumption, as described in the text.

compression factor and the index n of the star formation law. As an example, for a quadratic Schmidt law ($n = 2$), and a typical compression factor of four (what might be expected from strong tidal shocks), the SFR per unit volume would increase 16-fold, while the total SFR would increase by a factor of four (16/4 to account for the decreased gas volume). In general, for a Schmidt law of index n defined in this way, a gas compression by factor A will result in an increase in SFR by factor A^{n-1}.

This simple scaling argument tells us something very important about the physical processes that trigger the star formation in the strongest starbursts. As discussed in previous lectures, many interacting galaxies are characterized by starbursts with very high amplitudes, as high as $10 - 100$ times the SFR prior to the burst. Such systems require either a very nonlinear star formation law ($n > 4$ for average compression $A = 4$), or a nonlinear law *and* a very high gas compression factor, for example $A \geq 10 - 100$ for $n = 2$.

7.2 The Schmidt Law in Normal Galaxies

Several observations of normal galaxies suggest that the SFR on global scales is well represented by a Schmidt law with index $n \sim 1.5$ (Kennicutt 1989, 1997a, and references therein). As an illustration, Fig. 45 shows the correlation between average disk surface brightness in Hα, proportional to the SFR per unit area, and average gas surface density. The plot is an updated version of one shown in Kennicutt (1989), and is based on a compilation of galaxies with spatially resolved Hα, HI, and CO distributions, so that the various surface densities can be measured in a self-consistent manner. Note that volume densities are rarely observable, so we have recast the Schmidt law in terms of projected surface densities; I will use the notation N for the power law index, to distinguish it from the 3-dimensional index n.

The lines shown in Fig. 45 correspond to a slope $N = 1$, and correspond to global star formation efficiencies of 1%, 10%, and 100% per 10^8 years, using the SFR calibration and extinction corrections in Kennicutt (1983). The observed relation is clearly nonlinear, with a best-fitting slope $N \simeq 1.5$. Similar results have been derived by other authors using a variety of star formation tracers (e.g., Donas et al. 1987; Buat et al. 1989; Buat 1992; Boselli 1994; Deharveng et al. 1994). The average zeropoint of the relation in Fig. 45 corresponds to a SFR of $\sim 4 - 5\%$ per 10^8 yr; in other words the average star forming spiral converts $4 - 5\%$ of its interstellar gas to stars every 0.1 Gyr (within the optical radius). The reciprocal of this rate is the average gas depletion timescale; in the absence of stellar recycling, the timescale is $\sim 2 - 2.5$ Gyr. Recycling actually increases these times by a factor of $1.5 - 2$ (Kennicutt et al. 1994). The form of the local star formation law is discussed separately in Lecture 8.

The relatively shallow slope of the global Schmidt law in normal disks emphasizes the extraordinary character of the star formation in interacting starburst galaxies. Applying the results described earlier for an $N = 1.5$ Schmidt law, average density compression factors of $100 - 10000$ are required to increase the global SFR by factors of $10 - 100$. Clearly the star formation law in such regimes is different; either some physical mechanism acts to produce a much more nonlinear Schmidt law, or the efficiency of star formation at a given density is much higher (or both).

7.3 Star Formation Regimes in Interacting Galaxies

This parameterization of the star formation in terms of an effective Schmidt law provides a very useful operational framework for identifying different physical regimes of star formation in interacting galaxies. Although there is a continuous range in absolute SFRs among these systems, the physical environments subdivide nicely into three distinct regimes, with a very different physical mechanism responsible for triggering the star formation in each case. I shall refer to these as: a "linear regime", where the observed star formation properties can be understood by simply extrapolating the normal Schmidt law; a "nonlinear regime", where a much steeper star formation law (threshold) is required; and a "catastrophic regime", uniquely associated with the central starbursts in violent mergers, where the SFRs approach their limiting values and the main triggering mechanism is the violent radial transport of the ISM in a merging galaxy.

The Linear Regime. The results presented in Lectures 3 and 4 showed that the majority of tidal interactions produce only modest enhancements in the global SFR, factors of a few or less. The intermediate SFRs in these galaxies can be readily explained within the framework of the normal Schmidt law for spiral disks. Wallin (1990) finds that typical gas compression factors of < 5 in the disks of interacting spirals, corresponding to SFR enhancements of factors of $2-5$ for a Schmidt law with $N = 1.5-2$. Somewhat larger enhancements might be produced if the interactions transport modest amounts of interstellar gas from the outer HI envelopes into the denser star forming disks. Other mechanisms such as orbit crossing and cloud collisions may increase the SFR by modest factors over what would be expected from the normal Schmidt law as well (e.g., Olson & Kwan 1990; Elmegreen, Kaufman, & Thomasson 1993; see Lecture 8). Since no special mechanisms are required to account for these galaxies, I will not discuss this regime further. The reader should bear in mind that a considerable fraction of all interacting systems, perhaps a majority, fall into this regime and can be understood without invoking different star formation triggering mechanisms.

The Nonlinear Regime. Some interacting systems show evidence for much stronger enhancements in global star formation. Several types of systems fall into this category. Many early-type, gas-poor interacting systems show increases of an order of magnitude or more in their global SFRs (Lecture 3), much larger than can be accounted for with a shallow Schmidt law. Other systems, such as M 51, show more modest increases in their integrated SFRs, but exhibit strongly localized star formation, often in spectacular grand-design spiral patterns. The arm/interarm SFR contrasts in these galaxies tend to be much stronger than the respective contrasts in gas density, requiring a nonlinear triggering mechanism. The post-burst mergers (Lecture 6) are yet another subclass of interactions that require a nonlinear star formation law. Synthesis modeling of these objects shows that $\sim 10-50\%$ of the stellar mass of the galaxy formed on timescales of few times 10^7 to 10^8 years. This represents a increase of $10-30$ in the SFR per unit gas mass over normal disks, and requires either enormous gas compression factors or order-of-magnitude increases in the effective star formation efficiency at a given gas density, if not both. None of these systems can be understood in the context of the shallow $N \simeq 1.5$ law that characterizes normal disks.

Several physical mechanisms have been proposed to account for this nonlinear behavior, and given the wide range of galaxy types within this category, more than one mechanism is likely to be important. I discuss these in more detail in the next lecture. Gravitational stability thresholds for cloud growth and star formation offer an especially attractive mechanism for producing rapid star formation in initially quiescent systems, and they also can account for a wide range of properties of normal galaxies (Kennicutt 1989, 1997a). As mentioned earlier, cloud collisions are probably important in many interacting systems. Radial transport of gas, directly or via the formation of a strong bar, is clearly important for fueling the nuclear starbursts observed in many systems. These subjects are discussed in detail in the next lecture.

The Catastrophic Regime. The most dramatic starburst regime, and one quite distinct the others, occurs in the central regions of the ultraluminous starburst galaxies. These regions are characterized by extraordinarily high SFRs ($10 - 1000\,M_\odot\,yr^{-1}$, enormous central gas concentrations ($10^8 - 10^{11}\,M_\odot$ in the central kiloparsec), and star formation efficiencies that are typically an order of magnitude higher than in normal disks or Galactic molecular clouds. This regime is unique to the nuclear starbursts; comparable regions are not observed in normal galaxies.

The extreme properties of the ULIGs are all the more remarkable when one appreciates that these systems could not be forming stars any more rapidly! Consider, for example, a galaxy comparable in mass and scale to the Milky Way, which contains $10^{11}\,M_\odot$ in gas. What is the maximum SFR that could be sustained in such a system? Even ignoring the limiting effects of feedback from stellar winds and supernovae, the minimum time for assembling the gas in the center is the dynamical (free-fall) timescale, which for this example is roughly $1 - 2 \times 10^8$ yr. The maximum SFR of such a system, assuming that the gas could be converted to stars with 100% efficiency over that timescale, is then $\sim 500 - 1000\,M_\odot\,yr^{-1}$. Likewise, the maximum bolometric luminosity that could be produced by stellar processing from such a population is of order $10^{-4}\,mc^2$ for a normal IMF (e.g., Salpeter), up to $10^{-3}\,mc^2$ for an IMF comprised entirely of massive stars. The former corresponds to a bolometric luminosity of $\sim 10^{13}\,L_\odot$ for a $10^{11}\,M_\odot$ galaxy, similar to the limit observed for ultraluminous starburst galaxies. The ULIGs —if they are powered in substantial part by star formation— must be forming stars at close to the physical limit for a self-gravitating system.

A variation on this calculation was presented by Heckman (1994), who estimated the limiting SFRs in terms of the internal velocity dispersion of the system, using the virial theorem to relate the dynamical timescale to the velocity. Fig. 46, taken from this paper, shows lines of maximum bolometric luminosity (following the assumptions described above) along with the observed luminosities and linewidths of the IRAS BGS sample (points) and normal galaxies (the enclosed region in the bottom of the plot). The lines closest to the upper envelope of the observed distribution correspond to the maximum SFR for a Salpeter IMF and 100% star formation efficiency, or for an IMF composed of massive stars only and 10% star formation efficiency. The coincidence of this upper envelope with the maximum SFR models suggests that many of the ULIGs are forming stars near their physically limiting rates.

This chain of argument breaks down if most of the infrared luminosity comes from a buried quasar, because the efficiency of energy production from a quasar is ~ 100 times larger, and the mass consumption arguments are less interesting. However, the evidence presented in Lecture 6 indicates that star formation powers at least a substantial fraction of the luminosity in all but the most powerful ULIGs, so it is likely that these objects are forming stars as rapidly as allowed by fundamental dynamical considerations. It is interesting that these physical arguments can be applied to show that candidate "hyperluminous" IRAS galaxies such as FSC 10214+4724, with an apparent bolometric luminosity of order $10^{14}\,L_\odot$, could not possibly be powered by star formation; instead it appears to be a gravitationally lensed AGN (Eisenhardt et al. 1996).

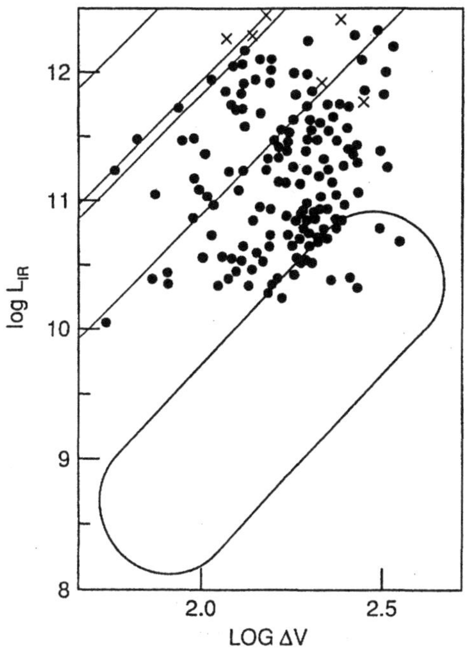

Fig. 46. Relationship between the internal velocity dispersion and infrared luminosity for a sample of IRAS-selected galaxies (points), quasar-like objects (crosses), and optically-selected galaxies (enclosed diagonal region). Lines show upper bounds for limiting SFRs as described in the text. Figure from Heckman (1994).

If we assume that at least a substantial fraction of the bolometric luminosity of most ULIGs is produced by star formation, then the physical implications are dramatic. ULIGs contain central gas disks comparable in mass to the global interstellar gases of galaxies, and are forming stars near the maximum rates consistent with the timescale for forming the disks. The mass of stars produced over the life of the burst are comparable to those of the central bulges and nuclei of giant galaxies.

We can now understand why such extraordinary starbursts are uniquely associated with violent mergers. The requisite bolometric luminosities, SFRs, and gas consumption rates can *only* be fueled by literally transporting most of the entire ISM of a galaxy into its center over a timescale of $\sim 10^8$ yr or less. This inflow, in turn, can only be driven by a rapid evolution of the host galaxy's gravitational potential (Lin, Pringle, & Rees 1988; Barnes & Hernquist 1991), either in a violent merger or in the initial collapse of a protogalaxy. Galaxies cannot simply self-destruct in this way without such an external perturbation.

Based on these arguments, it appears that the primary triggering agent for the LIGs and ULIGs is the fueling of enormous masses of gas to the centers of the merging galaxies, a subject discussed at length in Lecture 8 by Josh Barnes. Converting such large masses of gas into stars on timescales of $10^7 - 10^8$ yr, with nearly 100% efficiency, may involve other mechanisms that are unique to this environment (next lecture), but from a dynamical perspective there is no serious impediment; the collapse time for such massive central disks are a small fraction of the starburst lifetime.

7.4 Summary

The characteristic SFRs, burst lifetimes, and gas masses in different types of interacting galaxies provide valuable insights into the physical conditions in the star forming regions, and clues to the physical mechanisms that trigger and drive the star formation. While it is clear that the enormous diversity in the properties of the starburst galaxies requires more than one dominant triggering mechanism, I hope that I have made a convincing case that asymptotic physical regimes can be identified, and that at least the basic physical nature of the dominant fueling and triggering mechanisms can be isolated. I discuss possible physical mechanisms for producing these different behaviors in the next lecture.

8 Triggering and Regulation of Starbursts: Theoretical Ideas

As discussed in the previous lecture, the diversity of star forming environments in interacting galaxies suggests that several distinct physical processes are probably responsible for triggering starbursts in different environments. The modest SFR enhancements observed in weak interactions can be understood in terms of local gas compression combined with the normal Schmidt law observed in quiescent disks. The physics underlying this Schmidt law may be more complicated (and is beyond the scope of this lecture), but probably involves increased cloud collision rates and/or shorter collapse timescales at higher density (e.g., Larson 1987; 1991). In any case, no new physics is needed to account for the star formation in these mildly interacting galaxies.

More highly nonlinear evolution in the SFR is observed in galaxies with strong global starbursts, and in early-type interacting galaxies. This requires some other physical mechanism to alter the linearity of the star formation law (i.e., the slope of the Schmidt law), and/or the efficiency of star formation at fixed density (the zeropoint of the Schmidt law). Finally, the most luminous nuclear starbursts appear to be triggered by the rapid radial transport of gas during a strong tidal encounter or merger. The physics relevant to this process is detailed in Josh Barnes' lectures, and I will only comment briefly on those aspects that are most relevant to the issue of starburst triggering. The problem of AGN triggering is a subject of its own, and I will not address it here.

8.1 Gravitational Star Formation Thresholds

Highly nonlinear star formation features are observed in a wide range of normal and interacting galaxies. The most common feature observed is an abrupt outer edge to the star forming disks of late-type spirals, well inside the gaseous HI disk (Fig. 48). Similar thresholds are observed in Magellanic irregular galaxies (e.g., Davies, Elliott, & Meaburn 1976; Hunter & Gallagher 1986; Skillman 1987; Taylor et al. 1994). Although star formation is not completely suppressed outside the threshold region, there is an abrupt transition radius in most late-type disks. Other thresholds are observed in gas-rich S0 galaxies (e.g., Eder 1990) and in low surface brightness galaxies (e.g., van der Hulst et al. 1993), where galaxies with gas masses of up to 10^{11} $M_\odot$ possess abnormally low SFRs, far below the extrapolation of a linear Schmidt law. Another nonlinearity is often observed in the disks of grand-design spiral galaxies, where relatively small arm vs interarm contrasts in gas column density induce order-of-magnitude contrasts in the SFR (e.g., Knapen et al. 1992; Rand 1993.).

The common feature in these observations is the presence of a star formation threshold, where relatively large changes in local SFRs are observed over a much smaller range of gas surface densities. A quantitative illustration of the thresholds is shown in Fig. 47, which plots the relationship between radially averaged Hα surface brightness (proportional to the SFR per unit area) and the azimuthally

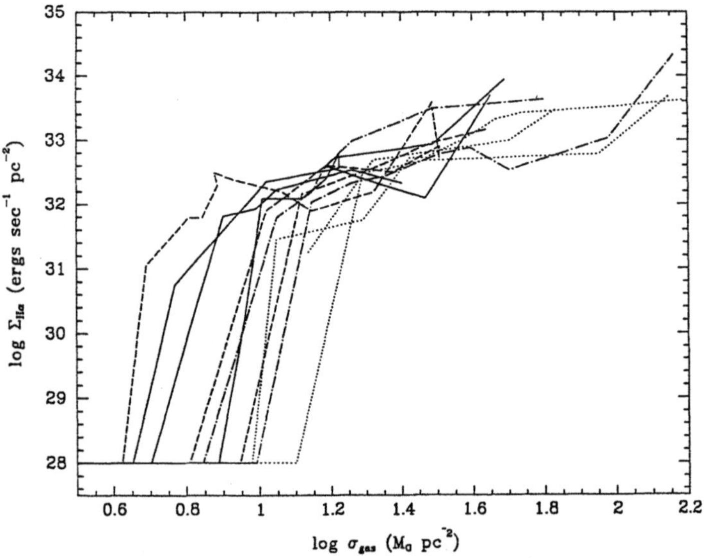

Fig. 47. Correlation of azimuthally averaged Hα surface brightness with average gas surface density for nearby star forming spirals, from Kennicutt (1997b).

averaged gas density, for a sample of active star forming galaxies studied by Kennicutt (1997b). At high gas densities, the star formation law is reasonably well represented by a Schmidt law, with slope and zeropoint similar to that observed for disks as a whole (see Fig. 45). However, below a critical threshold density the SFR drops abruptly. The critical column density varies between about 10^{21} and 10^{22} H cm^{-2} ($\sim 1 - 20\,M_\odot\,pc^{-2}$), and there is evidence for systematic variation in this threshold density between different types of galaxies and within the same galaxy, as discussed below. A visual illustration of a typical outer disk threshold is shown in Fig. 48.

Kennicutt (1989) proposed that these observed thresholds could be physically associated with a large-scale gravitational threshold for the growth of density perturbations in the gas disk (cf. Quirk 1972). He estimated the critical densities for a set of star forming spirals in terms of the Toomre (1964) criterion for the growth of large-scale perturbations in a self-gravitating gas disk:

$$\Sigma_c = \alpha \frac{\kappa c}{\pi G}\,, \tag{6}$$

$$Q \equiv \frac{\Sigma_c}{\Sigma_g}\,, \tag{7}$$

where the critical surface density Σ_c is dictated by the epicyclic frequency of rotation κ, the gas velocity dispersion c, and a dimensionless constant α which takes into account the deviation of real disks from the idealized Toomre model. The critical parameter for determining whether cloud growth and star formation is suppressed is the famous Toomre Q parameter, the ratio of the critical density

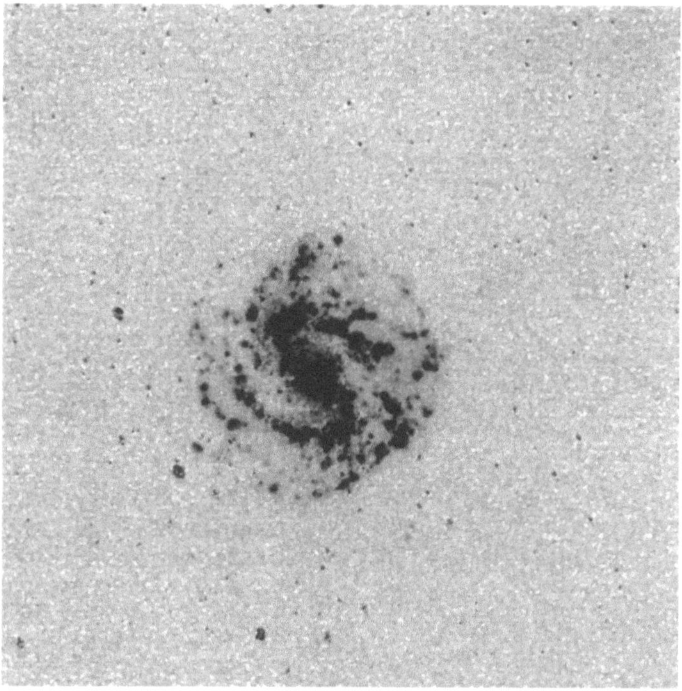

Fig. 48. A deep Hα image of M83, showing the sharp threshold in the local SFR.

at a given radius to the actual gas density. In this picture, star formation should be strongly suppressed in regions with $Q \gg 1$, and it should proceed unimpeded in regions with $Q \ll 1$. In intermediate regimes, where $Q \sim 1$, the disk would be globally stable against cloud growth but could be easily perturbed locally. Since star formation tends to deplete the disk until Q approaches unity, one would expect evolved disks to hover in the metastable regime near $Q \sim 1$. For typical values of κ and c the critical density is of order 10% of the total disk surface density. This potential supply of fresh gas for star formation is the key to understanding the fueling of starbursts in interacting early-type galaxies.

This admittedly oversimplistic gravitational model is surprisingly successful at reproducing many of the observed star formation properties of normal galaxies, as summarized in Kennicutt (1989, 1997). Figure 49, taken from Kennicutt (1989), shows the typical radial variation in Q for samples of active star forming spirals (top) and relatively quiescent early-type systems (bottom). It is interesting that even though the absolute gas density drops by $1-2$ orders of magnitude over the radius of a typical disk, the stability index Q varies by factors of a few or less, suggesting that the thresholds partly regulate the radial behavior of the gas distribution. The thresholds also predict with surprising precision the threshold radii and column densities of the star forming disks; see Fig. 49 and Kennicutt (1989) for details. The same criterion successfully accounts for the

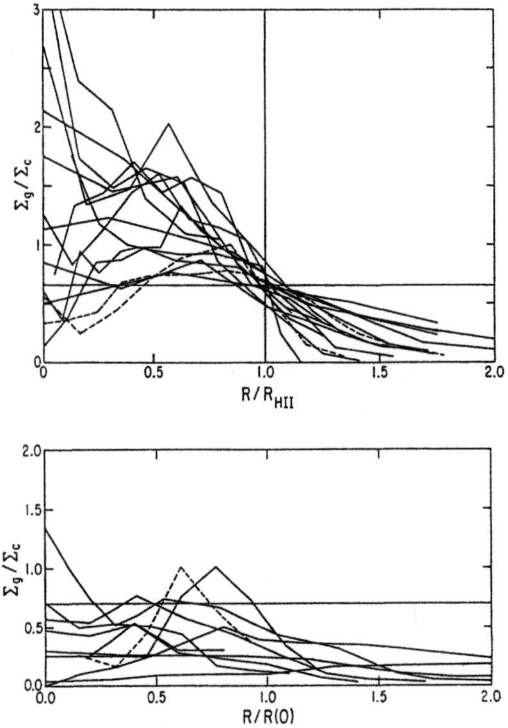

Fig. 49. Radial behavior of the Toomre stability parameter Q for active star forming disks (top) and early-type disks with low SFRs (bottom). Plots taken from Kennicutt (1989).

observed star formation properties of low surface brightness galaxies (van der Hulst et al. 1993), gas-rich S0 galaxies (Kennicutt 1989; Eder 1990), and star forming irregular galaxies (Taylor et al. 1994).

I suspect that the thresholds can also account for the rapid increases in SFRs observed among relatively gas-poor interacting galaxies (Lecture 3). Such disks in isolation will lie below the density threshold for efficient cloud formation and star formation (see Figure 49), but when perturbed in a tidal interaction, local compression in density waves may drive the gas locally above the threshold. This appears to be happening today in M 51, where the radially-averaged density lies near or below the threshold, but the modest tidal compression in the spiral arms is sufficient to drive efficient star formation (Rand 1993). Another application to interacting galaxies was made by Charmandaris, Appleton, & Struck-Marcell (1993), who attribute the strong star formation in the ring galaxy to a threshold (see also Struck-Marcell 1991). Further tests of this model for interacting galaxies in general is limited by the dearth of 2-dimensional maps of the atomic and molecular hydrogen in the galaxies, and it remains an important area of investigation for the future.

The problem of gravitational thresholds in the central regions of normal and barred galaxies has been explored by Kenney, Carlstrom, & Young (1993) and Elmegreen (1993). They consider alternatives to the simplistic Q treatment of Kennicutt (1989) and consider the effects of an inner Lindblad resonance on the critical density and subsequent star formation. Elmegreen finds that very efficient conversion of gas to stars is expected, consistent with the observed properties of many nuclear starburst rings.

8.2 Cloud Collision Models

Star formation thresholds offer an attractive mechanism for fueling star formation in an otherwise quiescent disk, thereby inducing a highly nonlinear response in the SFR. However this mechanism can only remove a gravitational barrier to star formation; by itself it will not increase the SFR above the $\sim 5\%$ per 10^8 yr value observed in gas-rich normal disks. Another mechanism is needed to account for the much higher global star formation efficiencies that are often observed in strongly interacting systems. As will be discussed later, no special mechanisms may be needed to account for the very large SFRs in nuclear starbursts, because the relevant dynamical timescales in those regions are already an order of magnitude or more shorter than in typical spiral disks. However other mechanisms may be needed to account for the high SFRs observed over larger scales, especially in some merger remnants (e.g., NGC 7252).

As discussed by Keel (1991), cloud collisions are expected to be commonplace in tidal encounters, induced for instance by orbit crossing, a tidally-induced bar, or a strong spiral density wave. Simulations of these effects have been carried out by Olson & Kwan (1990), Elmegreen et al. (1994), and Hernquist & Mihos (1995 and references therein). An important feature in both of these simulations is an increase in the cloud velocity dispersion in the encounter, which increases the frequency and speed of cloud collisions and raises the gravitational threshold density. This may favor the formation of unusually massive bound clouds which can form stars with higher efficiency.

Jog & Solomon (1992) and Jog & Das (1992) have proposed that high-velocity cloud collisions in tidal interactions will enhance the star formation efficiency through the formation of a high pressure intercloud medium. Their calculations suggest that dense molecular clouds will not collide while the larger diffuse HI clouds will collide at high velocity and form a high-pressure intercloud medium. The enhanced star formation efficiency is produced by shock compression of the giant molecular clouds, either in the colliding disks of the merging galaxies (Jog & Solomon 1992) or by giant clouds falling into the center of a starburst galaxy (Jog & Das 1992). As discussed in Lecture 5, several observational tracers of the ISM in LIGs are consistent with the presence of an unusually high pressure ISM.

8.3 Radial Transport of Gas: Clues from Barred Galaxies

As discussed in Lecture 7, the most critical prerequisite for an ultraluminous starburst appears to be the radial transport of large masses of interstellar gas — up to $10^{10} - 10^{11}\,M_\odot$— into the center of a merger remnant. A detailed discussion of the relevant theory and examples of simulations are given in Lecture 8 by Josh Barnes. In this section, I briefly discuss direct evidence for smaller-scale feeding in normal barred galaxies, and simulations of central feeding and star formation in mergers.

Nearby barred galaxies provide valuable clues to the processes that may transport the gas to the centers of interacting systems. Reviews of the relevant theory and observations can be found in Athanassoula (1994) and Kennicutt (1994), respectively, as well as in the proceedings of a recent IAU meeting devoted to barred galaxies (Buta, Elmegreen, & Crocker 1996). Bar fueling in particular is reviewed by Friedli & Martinet (1997).

Observations of star formation in barred galaxies reveal two distinct types of objects. The first are predominantly early-type "strong bar" systems, in which there is evidence of a central bulge, strong shocking of gas, and a dearth of star formation within the bar, and often evidences of rings of star formation outside the bar. The other type of object, "weak bar" systems, is characterized by small or absent bulges, and star formation within the bar, often extending continuously from the galactic center to the end of the bar. For our purposes perhaps the most interesting features observed in the strong bar systems are very high central concentrations of gas and star formation, often concentrated in a nuclear ring of dense young star clusters, "hotspots", or "twin peaks" (Kenney 1994). The circumnuclear structures appear to be located at or near the inner Lindblad resonance, and often surround a central AGN.

Hydrodynamical simulations of gas flows in barred potentials reproduce most of the observed features in the strong and weak-bar systems (e.g., Friedli & Benz 1993; Athanassoula 1994; Combes 1994; Piner, Stone, & Teuben 1995). The results suggest that the circumnuclear activity is fueled by a modest radial inflow of gas that is driven in turn by the interaction of the bar with the gas disk. The mass inflow rates are often quite modest, of order $1\,M_\odot\,yr^{-1}$or less (Kenney & Lord 1991), so such bars are not likely to produce a very luminous starburst. However this process may be very important for driving the lower level nuclear activities seen in many interacting and isolated spirals. Ho, Filippenko, & Sargent (1996) find evidence for enhanced nuclear star formation in early-type barred galaxies (mainly strong bar type) but no evidence of enhanced nuclear AGN activity.

8.4 Simulations of Starbursts in Merging Galaxies

An extensive series of hydrodynamical simulations of star formation in merging galaxies has been carried out by Mihos, Hernquist, and Barnes (Barnes & Hernquist 1991; Mihos, Richstone, & Bothun 1992; Mihos & Hernquist 1994; Hernquist & Mihos 1995; and references therein). A more detailed discussion and critique is presented in Barnes's lectures, and I only briefly summarize the relevant highlights. A general result of the simulations is that it is relatively easy to stimulate modest to large enhancements in the SFR even in minor mergers. However, the star formation in most of the simulations tends to take place over the dynamical lifetime of the encounter, whereas observations suggest that the period of peak star formation is considerably shorter, the "timescale problem" alluded to earlier.

Mihos & Hernquist (1994) suggest that a critical element in determining the duration and timing of the central starburst is the presence of a central bulge.

In their simulations the bulge acts to stabilize the galaxies against an early bar-driven inflow, and when the final inflow does take place, it fuels a more rapid burst, lasting of order 50 Myr for realistic input parameters. If these results are shown to apply in general, they will weaken the physical analogy that is often drawn between present-day ULIGs and collapsing protogalaxies. In these simulations the existence of a central mass concentration is a prerequisite for fueling an luminous central starburst, and hence a previous episode of nuclear star formation or merging may be required to produce these objects.

9 The Cosmological Role of Interactions and Starbursts

In the previous lectures I have described how interactions can alter —often profoundly— the star formation properties and the evolution of individual galaxies. The question remains: How important are interactions within the larger context of star formation and galaxy formation in the universe? How different would the universe look today if there had been no interactions or mergers of galaxies? We have seen ample evidence that an interaction can literally reshape or rebuild the galaxies involved, but we have also seen that an unbiased census of interacting systems shows a wide range of responses, with modest enhancements in star formation vastly outnumbering the ultraluminous starbursts that tend to attract our attention.

The kinds of questions we shall address in this last lecture are as follows: What fraction of the stars forming today are formed in interactions or mergers? Was this fraction different at earlier cosmological epochs? What fraction of the mass of galaxy disks was assembled from mergers? What are the other large-scale effects of starbursts on the universe?

I begin this lecture by briefly describing the central role that mergers play in contemporary hierarchical cosmological models. I then review systematic studies of interaction and merger rates —today and at previous epochs— and the fraction of stars and metals formed in these interactions. I conclude by discussing how the kinematics of disks can be used to constrain the merger history of the universe and the hierarchical cosmological models. Some of these subjects are also addressed in Lecture 9 by François Schweizer, and I encourage the reader to compare our perspectives on these important problems.

9.1 Interactions in Hierarchical Cosmology

Interactions are a central element in the formation and evolution of galaxies in current hierarchical cosmological models, and before addressing the observational constraints, it is useful to understand the basic features and predictions of this theoretical picture. For more detailed reviews see White (1990, 1994).

The most popular and successful variant of this theory is the Cold Dark Matter (CDM) model. In this picture, most of the mass density of the universe is contained in nonbaryonic "cold" form, and structure in the universe grows from a spectrum of primordial density fluctuations. The power spectrum of these perturbations, determined by the simplest form of cosmological inflation theory, peaks on small linear scales, so the first nonlinear structures to grow are low-mass protogalactic condensations. Larger structures, including large galaxies, groups, and clusters of galaxies, grow through the gravitational aggregation of these smaller objects, in a "bottom-up" cosmology. Since the dark matter dominates the total mass density, the growth of this structure is dominated by the mergers of dark halos, with the baryonic matter (gas) carried with the dark matter. Eventually the gas cools, condenses within the dark matter halos, and forms stars. The observable properties of a galaxy are determined by its condensation and merger history, in which the dark matter plays the dominant role, and by

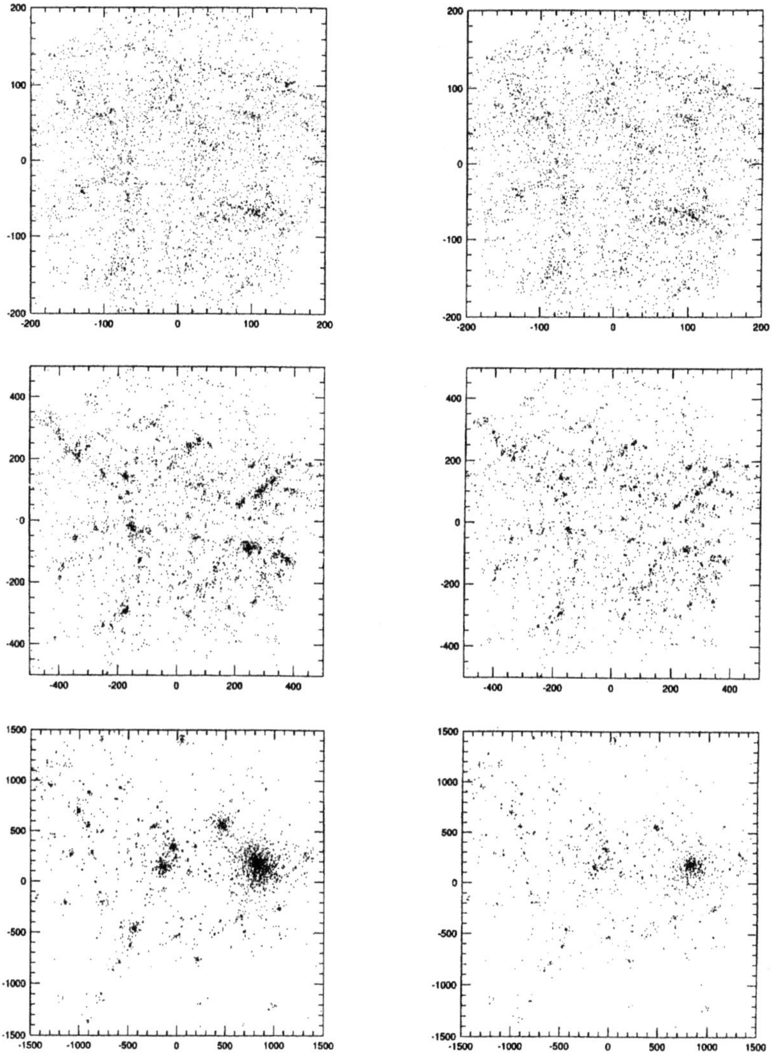

Fig. 50. A numerical simulation of the evolution of structure in a CDM universe. The panels on the left follow the evolution of dark matter particles, while those on the right follow the evolution of the gas. The time steps correspond to redshifts of 6.4 (top), 1.9 (middle), and 0 (bottom). Figure taken from Navarro & Benz (1991).

the ratio of the cooling time of the gas to the dynamical (collapse) timescale (e.g., Dekel & Silk 1986; White & Rees 1978).

The central role that mergers play in this process is illustrated in Fig. 50, which shows an example of a CDM simulation from Navarro & Benz (1991). From top to bottom the simulations correspond to redshifts of 6.4, 1.9, and 0, with linear scales in units of kiloparsecs. The plots on the left show the evolution of the dark matter, while those on the right show the evolution of the gas. This example

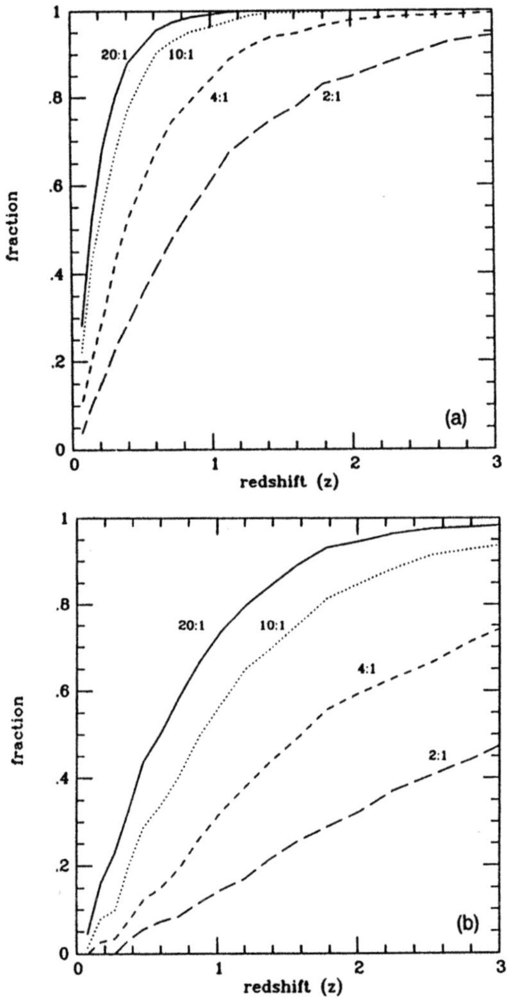

Fig. 51. Cumulative fractions of galaxy halos with circular velocity $220\,\mathrm{km\,s^{-1}}$ that undergo mergers by redshift z, for a CDM model with cosmological density $\Omega = 1$ (top) and $\Omega = 0.2$ (bottom). Each curve shows the merger fractions for companions of different masses relative to the primary galaxy. Figure taken from Kauffmann & White (1993).

shows several generic properties of the CDM models. Much of the structure forms only at late epochs; galaxy formation is a protracted process. Another feature that is especially relevant here is the pervasive evidence of merging; most large objects do not form in a single collapse event, but form instead from the successive mergers of several objects. This is quantified in Fig. 51, which shows the results of another CDM simulation by Kauffmann & White (1993). Plotted are the fraction of present-day halos which have undergone merging events since redshift z, for models of $\sim L^*$ galaxies with cosmological mass densities $\Omega = 1$ (upper) and $\Omega = 0.2$ (bottom). Note that the merger history is a strong function of Ω, with most of the merging occurring at relatively recent epochs in a high-density universe.

It is worth pointing out that the very definition of galaxy formation becomes much less well defined in this picture, relative to traditional models. In the single-collapse picture of galaxy formation (Eggen, Lynden-Bell, & Sandage 1962; ELS), all of the physical processes that one might associate with galaxy formation —the growth of a gravitational perturbation, the assembly of the mass, the collapse of the gas, the formation of the first stars, the formation of the first metals, the formation of the present-day halo, disk, and bulge— occur almost simultaneously. There is no ambiguity in defining the epoch of galaxy formation or the appearance of a protogalaxy. But how do we define the epoch of galaxy formation in the hierarchical picture? When the mass in the present day halo was first assembled? When the gas first condensed within this halo? When the first generation of stars formed? When the bulk of the current mass was assembled in one place? When most of the stars were formed? Depending on which definition one adopts, one can define the "formation" epoch of most galaxies to range anywhere from within a billion years after the Big Bang ($z > 3$) to the present epoch, when many are still merging and forming stars!

9.2 Interaction-Induced Star Formation Today

Several investigators have used measures of SFRs in interacting galaxies to estimate the global fraction of star formation that is triggered by interactions. Larson & Tinsley (1978) estimated that roughly 5% of the stellar mass of a typical Arp atlas interacting galaxy could be formed in the starburst. Given that such galaxies represent a small fraction of the local population, this implies that interactions were responsible for a tiny fraction, a few percent or less, of the present-day star formation in the universe. On the other hand, Condon (1982) estimated that $\sim$ 50% of the present-day star formation in the universe could be associated with tidal interactions or "cloud collisions", based on a radio continuum survey of nearby galaxies. Part of the order-of-magnitude difference in these results could be traced to differences in definition of interactions, but most reflected the lack of hard data on global SFRs of galaxies and local interaction rates.

Kennicutt et al. (1987) attempted to measure directly the fraction of local interaction-induced star formation, by combining the SFRs measured in their samples of interacting and noninteracting galaxies with the selection statistics of these samples. Their complete sample of galaxy pairs was drawn from a well-defined statistical algorithm, so it was straightforward to estimate the fraction of all galaxies that were in pairs. They concluded that $6 \pm 3\%$ of current star formation was triggered by interactions. Strictly speaking this value was an upper limit, because the statistics were based on a catalog of paired (but not necessarily interacting) galaxies. On the other hand, their survey tended to underestimate the SFRs of heavily obscured nuclear starbursts, thus their 6% estimate is probably accurate to within the 50% error quoted.

A more recent calculation of the local merger rate was carried out by Keel & Wu (1995). They compiled a sample of nearby merger candidates and derived a more robust estimate of the local merger fraction using a self-consistent treatment of the local luminosity function. They derived a present-day merger rate of 4.2 and 0.33 per Hubble time for paired spirals and all spirals, respectively (pairs as defined comprise 7.9% of all spirals). If a typical merger forms $10-30\%$ of the stellar mass in the galaxies, which is a reasonable estimate based on the results from Lecture 6, then the Keel & Wu results would imply that mergers are responsible for $\sim 3-10\%$ of all present-day star formation.

9.3 Interaction-Induced Star Formation in the Past

Deep imaging of distant clusters and field galaxies with HST provides striking qualitative evidence that galaxy interactions and mergers were much more frequent in the past (e.g., Burkey et al. 1994; Dressler et al. 1994; Colless et al. 1994; Glazebrook et al. 1995; Abraham et al. 1996). Several groups have used data of this kind to make quantitative estimates of the cosmic evolution in the merger rate. These are usually parameterized in terms of the evolution of the pair fraction or merger rate with redshift, though the reader should be warned that most analyses are based on measures of the pairing/merger rate as a function of apparent magnitude, with the redshift dependence inferred indirectly.

An early attempt to measure the evolution of the merger rate from ground-based data was by Zepf & Koo (1989). They noticed a tendency for faint galaxies to have an excess of companions within 3″. Comparing the pairing fraction at $B < 22$ with that observed in local samples yielded a pair frequency that increased as $(1+z)^{4.0\pm2.5}$. However, their result was based on a small number of pairs, and the excess itself was only significant at the 90% confidence level.

The availability of high-resolution images from HST and CFHT has stimulated several recent studies of the interaction rates at high redshift. These results are detailed in Lecture 9 by François Schweizer, so I only summarize the main results here. Most workers have reported a sharp increase in the number of apparent galaxy pairs at faint magnitudes, confirming the Zepf & Koo results. For example Burkey et al. (1994) derive a pair fraction of 34% for $I = 18-22$, as compared to a local fraction of $\sim 7\%$. When expressed in terms of the pair fraction, $n(z) \propto (1+z)^m$, various groups have derived values $m = 3.5 \pm 0.5$ (Burkey et al. 1994), 2.4 ± 1.0 (Carlberg, Pritchett, & Infante 1994), and 4.0 ± 1.5 (Yee & Ellingson 1995). The depth the surveys vary but the rates apply out to redshifts of $0.7-1$ in most cases. There is considerable disagreement between these authors in how to convert these pair evolution rates into merger evolution rates; I refer the reader to the respective papers. In any case, if these indices apply out to high redshifts, then the fraction of all galaxies that undergo a major merger over their lifetimes is very high. Recall that Keel & Wu (1995) derived a present-day merger rate of 0.33 per Hubble time for an average spiral, thus with the evolution suggested above, these rates exceed unity by $z \sim 1$.

This interpretation has been challenged by Woods, Fahlman, & Richer (1995), who analyzed a deep CFHT field and found no excess of physical pairs over the local fraction; most of the pairs they do observe are attributed to random superpositions. It is not clear to me how the Woods et al. results can be reconciled with the other studies cited above. If the steep evolution indicated by the HST studies extends to higher redshifts, however, the preponderance of physical pairs and mergers should be unmistakable, and one can expect this difference in interpretations to be clarified in the near future.

9.4 Disk Kinematics and the Merger Rate

The results in Section 8.2 suggest that the global role of interaction-induced star formation in the universe today is modest, amounting to perhaps a few percent of the total star formation at $z = 0$. Independent confirmation of this picture comes from the kinematics of galaxy disks, which also suggest a relatively low merger rate, at least among disk-dominated systems.

The small scale heights and low velocity dispersions of disks make them dynamically fragile, and excellent "seismometers" of recent merger activity. This argument was applied quantitatively to the Galactic disk by Tóth & Ostriker (1992), who estimated that no more than 4% of the mass of the disk within the solar circle could have been accreted in the last 5 Gyr. Numerical simulations by Quinn, Hernquist, & Fullagar (1993) supported this conclusion, though it is possible that a somewhat larger fraction of the *halo mass* could be accreted without destabilizing the disk (Navarro, Frenk, & White 1995).

The discovery of several examples of nearby spirals with counter-rotating gaseous and/or stellar disks demonstrates that disks can undergo mergers or episodic formation and form stable cold dynamical systems (see Lecture 5 by Schweizer). For example, if the bulk of the material accreted is gaseous, it may be able to dissipate most of its orbital energy without disrupting the existing stellar disk. At first glimpse, these examples would appear to contradict the conclusion above that mergers play a minor role in spirals. However, recent surveys suggest that the fraction of the total disk mass contained in these counter-rotating components is quite small. H. Rix and K. Kuijken (private communications) have each undertaken systematic surveys for disk counter-rotation in a total of $\sim 30 - 40$ S0 and early-type spiral galaxies, and have failed to detect counter-rotation in any at the level of a few percent. Gaseous counter-rotating disks are more common (Bertola, Buson, & Zeilinger 1992; Kennicutt 1996), but the typical masses involved are of order $10^4 - 10^8\ M_\odot$, again a few percent or less of the total mass. Likewise, a statistical analysis of polar ring galaxies indicates that only about 5% of S0 galaxies have possessed a polar ring over their lifetimes (Whitmore et al. 1990).

These results are very interesting in that they provide direct evidence that some S0 and spiral galaxies have experienced a merger with a galaxy or gas cloud over the past several Gyr, and nevertheless survived as normal disk galaxies. However, when one averages over all of the types of disks with peculiar kinematics, the total fraction of disk mass involved appears to be consistently 5% or less, similar to the Tóth & Ostriker (1992) estimate, and remarkably close to the estimates of merger-induced star formation in Section 9.2. Even allowing a factor of two for prograde accretions (which would leave no observable kinematic signature in the disk), it would appear that mergers are responsible for $< 10\%$ of disk formation over the past several Gyr.

How do these results compare with the predictions of the CDM models given in Fig. 51? If we naively multiply the 5% upper limit on counter-rotating disk material by a factor of two for prograde accretions, and say another factor of two to allow for mergers that have destroyed spirals since $z \sim 1$, we derive an upper limit of $\sim$ 20% on the fraction of "disks" that have formed by mergers in recent epochs. This is perhaps consistent with the CDM predictions for an $\Omega = 0.2$ universe (bottom plot) but appears to be flatly inconsistent with the predictions for an $\Omega = 1$ universe. As pointed out by Navarro et al. (1995), however, the predicted number of disk mergers is considerably lower than the halo merger rates plotted in Fig. 51; in their $\Omega = 1$ simulation fewer than 30% of spiral disks have grown by more than 10% in the past 5 Gyr. Hence it may be possible to reconcile the observations of disk kinematics with a high-density CDM universe. In any case, this example serves to illustrate how observations of galaxy interaction rates can have important implications for the largest issues in contemporary cosmology.

9.5 Global Effects of Starbursts and Superwinds

Although this lecture has emphasized the cosmological implication of the interaction and merger rates themselves, the starbursts which are triggered by mergers may have other important consequences for the formation and evolution of galaxies. Heckman et al. (1990) and Heckman (1994) have estimated the global rates of star formation, metal production, and energy injection into the intergalactic medium (IGM) from starbursts and explored their consequences.

As discussed in previous sections, an ultraluminous starburst galaxy can produce the equivalent lifetime output of an L^* galaxy over a timescale of order 10^8 yr. In absolute terms, an ULIG may inject a few times 10^{61} erg in radiant energy, $10^8 - 10^9\,M_\odot$ of metals, and $10^{59} - 10^{60}$ erg in mechanical energy from stellar winds and supernovae. The associated superwind will form a bubble with ram pressure of order $10^6\,\mathrm{K\,cm^{-3}}$ extending over Galactic scales, removing much of the galaxy's ISM, and potentially obliterating the ISM of any galaxy in the vicinity. The superwind cavity will continue to grow and may reach dimensions of 1 Mpc or more (Heckman 1994). These are very rough estimates, but they illustrate the impact that the most luminous starbursts may have on their surroundings, especially at early cosmological epochs when they may have been much more common that at present.

10 Concluding Remarks

I hope that these lectures have conveyed the enormous richness of star formation processes that are revealed in interacting galaxies. The past decade has seen enormous progress in the systematic measurement and physical understanding of star formation in galaxies, and much of this rapid progress has been driven by interest in the extraordinary properties of interacting galaxies, the extreme starburst galaxies in particular.

What can we expect in the next decade? It would take another entire lecture to discuss the most important outstanding problems, so instead I will highlight a few areas that I expect will be especially fruitful. On the observational side, it is important to probe the limits of physical conditions that are present in star forming galaxies, particularly in the ultraluminous galaxies. As discussed in Lecture 5, most of the extraordinary properties of the ULIGs can be understood as being powered by starbursts that are close to the physical limits set by dynamical timescales and the IMF. However the possibility remains that much or most of the bolometric luminosity of the most extreme systems comes not from star formation but rather from compact accretion-driven processes. It is important to clearly separate the luminosity of the star forming regions in these objects; once that is accomplished we can move on to tackle the physical couplings of the starburst and AGN processes, the IMF in the starbursts, and the physical triggering of these extraordinary events. The higher resolution capabilities offered by ISO and SIRTF should help to disentangle this problem, and add enormously to our inventory of high-luminosity infrared galaxies. Similarly, much remains to be learned about the local star formation properties of galaxies spanning the full range of SFRs and interstellar environments. Although we tend to parameterize and study the SFRs of galaxies on global scales, it is quite possible that much of the physics of the large-scale SFR is driven on local scales, and we can learn much more about the nature of star forming regions —their numbers, mass spectra, clustering properties, and IMFs— as well as the physical properties of their associated interstellar media. Last but certainly not least, the revolutionary opening of the high-redshift universe to ground-based and space-based observations opens a new era, in which it will be possible to apply our powerful star formation diagnostic techniques as a function of cosmological lookback time, to map out the star formation history of the universe, and to directly address the role of interactions and mergers in this history. This undoubtedly will be one of the most exciting fields in all of astronomy over the next 10 years.

With the anticipated commissioning of several large ground-based telescopes, a major infrared observatory, millimeter arrays, and possibly even a Next Generation Space Telescope, one has every reason to be optimistic that we will go far toward solving the problems outlined above in the coming decade. With any luck we will put into place the major pieces of a coherent picture of the evolution of galaxies, and the role of interactions in this big picture. However I am less optimistic that in that period we will have constructed a similarly complete physical understanding of the processes that drive that evolution. As demonstrated amply in the lectures by Josh Barnes, a combination of numerical simulations and semi-analytical modelling has produced major breakthroughs in our understanding of the behavior of gas and stars in interactions and mergers, and the role of these processes in galaxy formation and structure formation in a larger context. However it is interesting (and ironic) that future progress in this field rests in considerable part in improving our understanding of large-scale star formation and feedback effects, so that we can provide improved numerical "recipes" to the modellers. Although considerable progress has been made in isolating the primary mechanisms that influence the SFR on large scales (Lectures 7 and 8), the simple 35-year old Schmidt law still defines the state of the art in most models, and the recipes for feedback are even more *ad hoc.* I suspect that the real prize in solving the problem of galactic evolution —either in interacting or normal galaxies— rests with constructing this physical picture. In this respect interacting galaxies will play a crucial role, as the laboratories for conducting perturbation experiments and revealing the dynamical processes that regulate, maintain, and trigger star formation. May the next decade prove as exciting and productive as the last!

References

Aalto, S., Booth, R.S., Black, J.S., & Johansson, L.E.B. 1991, A&A 247, 291

Aalto, S., Booth, R.S., Black, J.S., & Johansson, L.E.B. 1995, A&A 300, 369

Abraham, R.G., Tanvir, N.R., Santiago, B.X., Ellis, R.S., Glazebrook, K., & van den Bergh, S. 1996, MNRAS 279, L47

Armus, L., Heckman, T.M., & Miley, G.K. 1989, ApJ 347, 727

Armus, L., Heckman, T.M., & Miley, G.K. 1990, ApJ 364, 471

Armus, L., Neugebauer, G., Soifer, B.T., & Matthews, K. 1995, AJ 110, 2610

Arp, H. 1966, Atlas of Peculiar Galaxies (Pasadena: Caltech); also ApJS 14, 1

Athanassoula, E. 1994, in Mass-Transfer Induced Activity in Galaxies, ed. I. Shlosman (Cambridge: Cambridge Univ. Press), 143

Augarde, R., & Lequeux, J. 1985, A&A 147, 273

Baan, W.A. 1993, in Astrophysical Masers, ed. A.W. Clegg & G.E. Nedoluha (New York: Springer), 73

Bahcall, J.N., Kirhakos, S., Schneider, D.P. 1995, ApJ 450, 486

Baldwin, J., Phillips, M., & Terlevich, R. 1981, PASP 93, 5

Balzano, V.A. 1983, ApJ 268, 602

Barnes, J.E., & Hernquist, L.E. 1991, ApJ 370, L65

Bergvall, N., & Johansson, L. 1995, A&AS 113, 499

Bernlöhr, K. 1992, A&A 263, 54

Bernlöhr, K. 1993a, A&A 268, 25

Bernlöhr, K. 1993b, A&A 270, 20

Bertola, F., Buson, L.M., & Zeilinger, W.W. 1992, ApJ 401, L79

Bica, E., Alloin, D., & Schmidt, A. 1990, MNRAS 242, 241

Bicay, M.D., Kojoian, G., Seal, J., Dickinson, D.F., & Malkan, M.A. 1995, ApJS 98, 369

Bohlin, R.C., Cornett, R.H., Hill, J.K., O'Connell, R.W., & Stecher, T.P. 1990, ApJ 352, 44

Boller, T., Fink, H., & Schaeidt, S. 1994, A&A 291, 403

Boller, T., Meurs, E.J.A., Brinkmann, W., Fink, H., Zimmermann, U., & Adorf, H.-M. 1992, A&A 261, 57

Boroson, T.A., Persson, S.E., & Oke, J.B. 1985, ApJ 293, 120

Boselli, A. 1994, A&A 292, 1

Bothun, G.D., Halpern, J.P., Lonsdale, C.J., Impey, C., & Schmitz, M. 1989, ApJS 70, 271

Bregman, J.N., Schulman, E., & Tomisaka, K. 1995, ApJ 439, 155

Bruzual, G., & Charlot, S. 1993, ApJ 405, 538

Buat, V. 1992, A&A 264, 444

Buat, V., Deharveng, J.M., & Donas, J. 1989, A&A 223, 42

Buat, V., Donas, J., & Deharveng, J.M. 1987, A&A 185, 33

Buat, V., & Xu, C. 1996, A&A 306. 61

Burkey, J.M., Keel, W.C., Windhorst, R.A., & Franklin, B.E. 1994, ApJ 429, L13

Bushouse, H.A. 1986, AJ 91, 255

Bushouse, H.A. 1987, ApJ 320, 29

Bushouse, H.A., & Gallagher, J.S. 1984, PASP 96, 273

Bushouse, H.A., Lamb, S.A., & Werner, M.W. 1988, ApJ 335, 74

Buta, R., Elmegreen, B.G., & Crocker, D.A., Barred Galaxies (San Francisco, ASP Conference Series Vol. 91)

Caldwell, N., Kennicutt, R., Phillips, A.C., & Schommer, R.A. 1991, ApJ 370, 526

Caldwell, N., Kennicutt, R., & Schommer, R. 1994, AJ 108, 1186
Calzetti, D., Kinney, A.L., & Storchi-Bergmann, T. 1994, ApJ 429, 582
Carico, D.P., Graham, J.R., Matthews, K., Wilson, T.D., Soifer, B.T., Neugebauer, G., & Sanders, D.B. 1990, ApJ 349, L39
Carlberg, R.G., Pritchet, C.J., & Infante, L. 1994, ApJ 435, 540
Carral, P., Hollenbach, D.J., Lord, S.D., Colgan, S.W.J., Haas, M.R., Rubin, R.H., & Erickson, E.F. 1994, ApJ 423, 223
Casini, C., & Heidmann, J. 1975, A&A 39, 127
Charmandaris, V., Appleton, P.N., & Marston, A.P. 1993, ApJ 414, 154
Colless, M., Schade, D., Broadhurst, T.J., & Ellis, R.S. 1994, MNRAS 267, 1108
Combes, F. 1994, in Mass-Transfer Induced Activity in Galaxies, ed. I. Shlosman (Cambridge: Cambridge Univ. Press), 170
Combes, F. 1995, Highlights of Astronomy, 10, 521
Comte, G., Augarde, R., Chalabaev, A., Kunth, D., & Maehara, H. 1994, A&A 285, 1
Condon, J.J. 1983, ApJS 53, 459
Condon, J.J., Condon, M.A., Gisler, G., & Puschell, J.J. 1982, ApJ 252, 102
Condon, J.J., Helou, G., Sanders, D.B., & Soifer, B.T. 1990, ApJS 73, 359
Condon, J.J., Huang, Z.-P., Yin, Q.F., & Thuan, T.X. 1991, ApJ 378, 65
Cottrell, G.A. 1978, MNRAS 184, 259
Courvoisier, T.J.L., & Mayor, M. 1990, Active Galactic Nuclei, Saas-Fee Twentieth Advanced Course of the SSAA, (Berlin: Springer)
Cox, P.N., Krugel, E., & Mezger, P.G. 1986, A&A 155, 380
Cutri, R.M., & McAlary, C.W. 1985, ApJ 296, 90
Dahari, O. 1984, AJ 89, 966
Dahari, O. 1985, ApJS 57, 643
David, L.P., Jones, C., & Forman, W. 1992, ApJ 388, 82
Davies, R.D., Elliott, K.H., & Meaburn, J. 1976, Mem RAS 81, 89
Dekel, A., & Silk, J. 1986, ApJ 303, 39
Deharveng, J.M., Sasseen, T.P., Buat, V., Bowyer, S., Lampton, M., & Wu, X. 1994, A&A 289, 715
de Vaucouleurs, G., de Vaucouleurs, A., & Corwin, H.G. 1976, Second Reference Catalog of Bright Galaxies (Austin: University of Texas Press)
Devereux, N.A., & Hameed, S. 1997, AJ 113, 599
Devereux, N.A., & Young, J.S. 1990, ApJ 350, L25
Devereux, N.A., & Young, J.S. 1991, ApJ 371, 515
Djorgovski, S. 1994, in Mass-Transfer Induced Activity in Galaxies, ed. I. Shlosman (Cambridge: Cambridge Univ. Press), 452
Doane, J.S., & Mathews, W.G. 1993, ApJ 419, 573
Donas, J., Deharveng, J.M., Laget, M., Milliard, B., & Huguenin, D. 1987, A&A 180, 12
Doyon, R., Joseph, R.D., & Wright, G.S. 1994, ApJ 421, 101
Dressler, A., & Gunn, J.E. 1983, ApJ 270, 7
Dressler, A., Oemler, A., Sparks, W.B., & Lucas, R.A. 1994, ApJ 435, L23
Eder, J.A. 1990, Ph.D. thesis, Yale University
Eggen, O.J., Lynden-Bell, D., & Sandage, A. 1962, ApJ 136, 748 (ELS)
Eisenhardt, P.R., Armus, L., Hogg, D.W., Soifer, B.T., Neugebauer, G., & Werner, M.W. 1996, ApJ 461, 72
Elmegreen, B.G. 1993, Proc. 4th EIPC Conference, 335
Elmegreen, B.G. 1994, ApJ 425, L73
Elmegreen, B.G., Kaufman, M., & Thomasson, M. 1993, ApJ 412, 90

Ferland, G.J., & Netzer, H. 1983, ApJ 264, 105
Friedli, D., & Benz, W. 1993, A&A 268, 65
Friedli, D., & Martinet, L. 1997, in Starburst Activity in Galaxies, ed. J. Franco et al., RevMexAA (Serie de Conferencias) 6, 177
Fritze-v. Alvensleben, U., & Gerhard, O.E. 1994a, A&A 285, 751
Fritze-v. Alvensleben, U., & Gerhard, O.E. 1994b, A&A 285, 775
Fuentes-Williams, T., & Stocke, J.T. 1988, AJ 96, 1235
Gallagher, J.S., Hunter, D.A., & Tutukov, A.V. 1984, ApJ 284, 544.
Gallimore, J.F., & Keel, W.C. 1993, AJ 106, 1337
Gehrz, R.D., Sramek, R.A., & Weedman, D.W. 1983, ApJ 267, 551
Glazebrook, K., Ellis, R., Santiago, B., & Griffiths, R. 1995, MNRAS 275, L19
Harwit, M., Houck, J., Soifer, B.T., & Palumbo, G. 1987, ApJ 315, 28
Heckman, T.M. 1990, in Paired and Interacting Galaxies, ed. J.W. Sulectic, W.C. Keel, & C. Telesco (Washington: NASA Conference Publication CP-3098), 359
Heckman, T.M. 1991, in Massive Stars in Starburst Galaxies, ed. C. Leitherer, T. Heckman, C. Norman, & N. Walborn (Cambridge: Cambridge Univ. Press), 289
Heckman, T.M. 1994, in Mass-Transfer Induced Activity in Galaxies, ed. I. Shlosman (Cambridge: Cambridge Univ. Press), 234
Heckman, T.M. 1996, in The Interplay Between Massive Star Formation, the ISM, and Galaxy Evolution, ed. D. Kunth, B. Guiderdoni, M. Heydari-Melayari, & T.X. Thuan (Paris: Editions Frontieres), 159
Heckman, T.M., Armus, L, & Miley, G.K. 1990, ApJS 74, 833
Heckman, T.M., Balick, B., & Crane, P.C. 1980, A&AS 40, 295
Heckman, T.M., Bothun, G.D., Balick, B., & Smith, E.P. 1984, AJ 89, 958
Heckman, T.M., Dahlem, M., Eales, S.E., Fabbiano, G., & Weaver, K. 1996, ApJ 457, 616
Heidmann, J., & Kalloghlian, A.T. 1973, Afz 9, 71
Hernquist, L., & Mihos, J.C. 1995, ApJ 448, 41
Hibbard, J.E. 1995, Ph.D. thesis, Columbia University
Hibbard, J.E., & van Gorkom, J.H. 1996, AJ 111, 655
Hill, G.J., Becklin, E.E., & Wynn-Williams, C.G. 1988, ApJ 316, L11
Hintzen, P., Menon, T., Palumbo, G., & Persic, M. 1989, ApJ 341, 679
Ho, L.C., Filippenko, A.V., & Sargent, W.L.W. 1995, ApJS 98, 477
Ho, L.C., Filippenko, A.V., & Sargent, W.L.W. 1996, ApJ 462, 183
Holmberg, E. 1958, Medd Lunds Astr Obs, 2, No. 136
Houck, J.R. et al. 1984, ApJ 278, L63
Huchra, J.P. 1977, ApJ 217, 928
Hummel, E. 1981, A&A 96, 111
Hummel, E., van der Hulst, J.M., Kennicutt, R.C., & Keel, W.C. 1990, A&A 236, 333
Hunter, D.A., & Gallagher, J.S. 1985, ApJS 58, 533
Hunter, D.A., & Gallagher, J.S. 1986, PASP 98, 5
Hunter, D.A., Shaya, E.J., Holtzman, J.A., Light, R.M., O'Neill, E.J., & Lynds, R. 1995, ApJ 448, 179
Hutchings, J.B., Crampton, D., Campbell, B., Gower, A.C., & Morris, S.C. 1982, ApJ 262, 48
Impey, C., & Bothun, G. 1989, ApJ 341, 89
Israel, F.P., & van der Hulst, J.M. 1983, AJ 88, 1736
Jog, C.J., & Das, M. 1992, ApJ 400, 476
Jog, C.J., & Solomon, P.M. 1992, ApJ 387, 152

Joseph, R.D. 1991, in Massive Stars in Starbursts, ed. C. Leitherer, T. Heckman, C. Norman, & N. Walborn (Cambridge: Cambridge Univ. Press), 259
Joseph, R.D., Meikle, W.P.S., Robertson, N.A., & Wright, G.S. 1984, MNRAS 209, 111
Joseph, R.D., & Wright, G.S. 1985, MNRAS 214, 87
Joy, M. & Lester, D.F. 1988, ApJ 331, 145
Joy, M. et al. 1989, ApJ 339, 100
Karachentsev, I.D., & Karachentseva, V.E. 1974, Astr Zh 51, 724
Kauffmann, G., & White, S.D.M. 1993, MNRAS 261, 921
Kazarian, M.A., & Kazarian, E.S. 1987, Afz 28, 487
Keel, W.C. 1990, AJ 100, 356
Keel, W.C. 1991, in Dynamics of Galaxies and Their Molecular Cloud Distribution, ed. F. Combes & F. Casoli (Dordrecht: Kluwer), 243
Keel, W.C. 1993, AJ 106, 1771
Keel, W.C. 1996, AJ 111, 696
Keel, W.C., Kennicutt, R.C., Hummel, E., & van der Hulst, J.M. 1985, AJ 90, 708
Keel, W.C., & van Soest, E.T.M. 1992, A&AS 94, 553
Keel, W.C., & Wu, W. 1995, AJ 110, 129
Kenney, J.D.P. 1994, in Mass-Transfer Induced Activity in Galaxies, ed. I. Shlosman (Cambridge: Cambridge Univ. Press), 78
Kenney, J.D.P., Carlstrom, J.E., & Young, J.S. 1993, ApJ 418, 687
Kenney, J.D.P., & Lord, S.D. 1991, ApJ 381, 118
Kennicutt, R.C. 1983, ApJ 272, 54
Kennicutt, R.C. 1986, in Stellar Populations in Galaxies, ed. C. Norman, A. Renzini, & M. Tosi (Cambridge: Cambridge Univ. Press), 125
Kennicutt, R.C. 1988, ApJ 334, 144
Kennicutt, R.C. 1989, ApJ 344, 685
Kennicutt, R.C. 1990a, in The Interstellar Medium in Galaxies, ed. H.A. Thronson & J.M. Shull, (Dordrecht: Kluwer), 405
Kennicutt, R.C. 1990b, in Paired and Interacting Galaxies, ed. J.W. Sulenctic & W.C. Keel (Washington: NASA Conference Publication CP-3098), 269
Kennicutt, R.C. 1992a, ApJ 388, 310
Kennicutt, R.C. 1992b, ApJS 79, 255
Kennicutt, R.C. 1992c, in Star Formation in Stellar Systems, ed. G. Tenorio-Tagle, M. Prieto, & F. Sánchez (Cambridge: Cambridge Univ. Press), 191
Kennicutt, R.C. 1994, in Mass-Transfer Induced Activity in Galaxies, ed. I. Shlosman (Cambridge: Cambridge Univ. Press), 131
Kennicutt, R.C. 1996, in New Light on Galaxy Evolution, ed. R. Bender & R.L. Davies (Dordrecht: Kluwer), 11
Kennicutt, R.C. 1997a, in The Interstellar Medium in Galaxies, ed. J.M. van der Hulst (Dordrecht: Kluwer), in press
Kennicutt, R.C. 1997b, ApJ, submitted
Kennicutt, R.C., & Chu, Y.-H. 1988, AJ 95, 720
Kennicutt, R.C., Edgar, B.K., & Hodge, P.W. 1989, ApJ 337, 761
Kennicutt, R.C., & Keel, W.C. 1984, ApJ 279, L5
Kennicutt, R.C., Keel, W.C., & Blaha, C.A. 1989, AJ 97, 1022
Kennicutt, R.C., Keel, W.C., van der Hulst, J.M., Hummel, E., & Roettiger, K.A. 1987, AJ 93, 1011 (K87)
Kennicutt, R.C., & Kent, S.M. 1983, AJ 88, 1094
Kennicutt, R.C., Tamblyn, P., & Congdon, C.W. 1994, ApJ 435, 22

Kim, D.-C., Sanders, D.B., Veilleux, S., Mazzarella, J.M., & Soifer, B.T. 1995, ApJS 98, 129
Knapen, J.H., Beckman, J.E., Cepa, J., van der Hulst, J.M., & Rand, R.J. 1992, ApJ 385, L37
Larson, R.B. 1987, in Starbursts and Galaxy Evolution, ed. T.X. Thuan, T. Montmerle, & J. Tran Tranh Van, (Paris: Editions Frontierières), 467
Larson, R.B. 1991, in Star Formation in Stellar Systems, ed. G. Tenorio-Tagle, M. Prieto, & F. Sánchez (Cambridge: Cambridge Univ. Press), 127
Larson, R.B., & Tinsley, B.M. 1978, ApJ 219, 46 (LT)
Laurikainen, E., & Salo, H. 1995, A&A 293, 683
Lawrence, A., & Elvis, M. 1982, ApJ 256, 410
Lawrence, A., Rowan-Robinson, M., Leech, K., Jones, D.H.P., & Wall, J.V. 1989, MNRAS 240, 329
Leech, K.J., Rowan-Robinson, M., Lawrence, A., & Hughes, J.D. 1994, MNRAS 267, 253
Leitherer, C. 1990, ApJS 73, 1
Leitherer, C., Fritze-v. Alvensleben, U., & Huchra, J. 1996, From Stars to Galaxies (San Francisco: ASP Conference Series Vol. 98)
Leitherer, C., & Heckman, T.M. 1995, ApJS 96, 9
Leitherer, C., Vacca, W.D., Conti, P.S., Filippenko, A.V., Robert, C., & Sargent, W.L.W. 1996, ApJ 465, 717
Leonardi, A.J., & Rose, J.A. 1996, AJ 111, 182
Lester, D.F., & Gaffney, N.I. 1994, ApJ 431, L13
Lin, D.N.C., Pringle, J.E., & Rees, M.J. 1988, ApJ 328, 103
Lisenfeld, U., Völk, H.J., & Xu, C. 1996, A&A 306, 677
Liu, C.T., & Kennicutt, R.C. 1995a, ApJ 450, 547
Liu, C.T., & Kennicutt, R.C. 1995b, ApJS 100, 325
Lonsdale, C.J., & Helou, G. 1987, ApJ 314, 513
Lonsdale, C.J., Persson, S.E., & Mathews, K. 1984, ApJ 287, 95
Lonsdale, C.J., Smith, H.E., & Lonsdale, C.J. 1993, ApJ 405, L9
MacKenty, W.J. 1989, ApJ 343, 125
Maiolino, R., Ruiz, M., Rieke, G.H., & Papadopoulos, P. 1997, ApJ, 485 552
Majewski, S.R., Hereld, M., Koo, D.C., Illingworth, G.D., & Heckman, T.M. 1993, ApJ 402, 125
Maoz, D., Filippenko, A.V., Ho, L.C., Macchetto, D., Rix, H.-W., & Schneider, D.P. 1996, ApJS 107, 215
Markarian, B.M. 1967, Afz 3, 55
Markarian, B.M., Lipovetskii, V.A., Stepanian, J.A., Erastova, L.K., & Shapovalova, A.I. 1989, Soobsch. Spets. Astrof. Obs. 62, 5
Martin, C.L. 1996, Ph.D. thesis, University of Arizona
Matthews, K., Neugebauer, G., McGill, J., & Soifer, B.T. 1987, AJ 94, 297
Mazzarella, J.M. 1989, Ph.D. thesis, University of Michigan
Mazzarella, J.M., Gaume, R.A., Soifer, B.T., Graham, J.R., Neugebauer, G., & Matthews, K. 1991, AJ 102, 1241
Mazzarella, J.M., Soifer, B.T., Graham, J.R., Hafer, C.I., Neugebauer, G., & Matthews, K. 1992, AJ 103, 413
McCarthy, P. 1993, ARA&A 31, 639
McCarthy, P., Heckman, T., & van Breugel, W. 1987, AJ 93, 264
McCleod, K., & Rieke, G.H. 1995, ApJ 441, 96
McQuade, K., Calzetti, D., & Kinney, A.L. 1995, ApJS 97, 331

Melnick, J., & Mirabel, I.F. 1990, A&A 231, L19

Menon, T.K. 1995, MNRAS 274, 845

Meurer, G.R., Heckman, T.M., Leitherer, C., Kinney, A., Robert, C., & Garnett, D.R. 1995, AJ 110, 2665

Mihos, J.C., Richstone, D.O., & Bothun, G.D. 1992, ApJ 400, 153

Mihos, J.C., & Hernquist, L. 1994, ApJ 431, L9

Mirabel, I.F. 1992, in Star Formation in Stellar Systems, ed. G. Tenorio-Tagle, M. Prieto, & F. Sánchez (Cambridge: Cambridge Univ. Press), 479

Mirabel, I.F., & Sanders, D.B. 1988, ApJ 335, 104

Moles, M., del Olmo, A., Perea, J., Masegosa, J., Márquez, I., & Costa, V. 1994, A&A 285, 404

Moles, M., Márquez, I., & Pérez, E. 1995, ApJ 438, 604

Monaco, P., Giuricin, G. Mardirossian, F., & Mezzetti, M. 1994, ApJ 436, 576

Morgan, W.W., & Mayall, N.U. 1957, PASP 69, 291

Moshir, M. et al. 1992, Explanatory Supplement to the IRAS Faint Source Survey, Version 2, JPL D-10015 8/92 (Pasadena: JPL)

Navarro, J.F., & Benz, W. 1991, ApJ 380, 320

Navarro, J.F., Frenk, C.S., & White, S.D.M. 1995, MNRAS 275, 56

Norris, R.P., Allen, D.A., Sramek, R.A., Kesteven, M.J., & Troup, E.R. 1990, ApJ 359, 291

Olofsson, K. 1989, A&AS 80, 317

Olson, K.M., & Kwan, J. 1990, ApJ 349, 480

Okamura, S.K., Kawabe, R., Ishiguro, M., Kasuga, T., Morita, K.I., & Ishizuki, S. 1991, in Dynamics of Galaxies and Their Molecular Cloud Distributions, ed. F. Combes & F. Casoli (Dordrecht: Kluwer), 425

Osterbrock, D.E. 1989, Astrophysics of Gaseous Nebulae and Active Galactic Nuclei (Mill Valley: University Science Books)

Petrosian, A.R. 1982, Afz 18, 548

Petrosian, A.R., & Turatto, M. 1995, A&A 297, 49

Piner, B.G., Stone, J.M., & Teuben, P.J. 1995, ApJ 449, 508

Pogge, R.W., & Eskridge, P.B. 1987, AJ 93, 291

Prestwich, A.H., Joseph, R.D., & Wright, G.S. 1994, ApJ 422, 73

Quinn, P.J., Hernquist, L., & Fullagar, J.P. 1993, ApJ 403, 74

Quirk, W.J. 1972, ApJ 176, L9

Rafanelli, P., Violato, M., & Baruffolo, A. 1995, AJ 109, 1546

Rand, R.J. 1993, ApJ 410, 68

Rand, R.J. 1994, A&A 285, 833

Rieke, G.H. 1988, ApJ 331, L5

Rieke, G.H. 1991, in Massive Stars in Starbursts, ed. C. Leitherer, T. Heckman, C. Norman, & N. Walborn (Cambridge: Cambridge Univ. Press), 205

Rieke, G.H., Lebofsky, M.J., Thompson, R.I., Low, F.J., & Tokunaga, A.T. 1980, ApJ 238, 24

Rieke, G.H., Loken, K., Rieke, M.J., & Tamblyn, P. 1993, ApJ 412, 99

Rieke, G.H., & Low, F.J. 1972, ApJ 176, L95

Roberts, M.S. 1963, ARA&A 1, 149

Romanishin, W. 1990, AJ 100, 373

Rose, J.A. 1985, AJ 90, 1927

Rowan-Robinson, M. 1995, MNRAS 272, 737

Rowan-Robinson, M. et al. 1984, ApJ 278, L7

Salpeter, E.E. 1955, ApJ 121, 161

Sandage, A. 1961, The Hubble Atlas of Galaxies (Washington: Carnegie Institution of Washington)

Sandage, A. 1986, A&A 161, 89

Sanders, D.B. 1992, in Relationships Between Active Galactic Nuclei and Starburst Galaxies, ed. A. Filippenko (San Francisco, ASP Conference Series Vol. 31), 303

Sanders, D.B., Scoville, N.Z., & Soifer, B.T. 1991, ApJ 370, 158

Sanders, D.B., & Mirabel, I.F. 1996, ARA&A 34, 749

Sanders, D.B., Soifer, B.T., Elias, J.H., Madore, B.F., Matthews, K., Neugebauer, G., & Scoville, N.Z. 1988, ApJ 325, 74

Saunders, W., Rowan-Robinson, M., Lawrence, A., Efstathiou, G., Kaiser, N., Ellis, R.S., & Frenk, C.S. 1990, MNRAS 242, 318

Scalo, J.M. 1986, Fund Cos Phys 11, 1

Schade, D., Lilly, S.J., Crampton, D., Hammer, F., Le Fevre, O., & Tresse, L. 1995, ApJ 451, L1

Schmidt, M. 1959, ApJ 129, 243

Schombert, J.M., Wallin, J.F., & Struck-Marcell, C. 1990, AJ 99, 497

Scoville, N.Z., Hibbard, J.E., Yun, M.S., & van Gorkom, J.H. 1994, in Mass-Transfer Induced Activity in Galaxies, ed. I. Shlosman (Cambridge: Cambridge Univ. Press), 191

Searle, L., Sargent, W.L.W., & Bagnuolo, W.G. 1973, ApJ 179, 427

Sekiguchi, K., & Anderson, K.S. 1987, AJ 94, 644

Sekiguchi, K., & Wolstencroft, R.D. 1992, MNRAS 255, 581

Sekiguchi, K., & Wolstencroft, R.D. 1993, MNRAS 263, 349

Shields, J.C. 1993, ApJ 419, 181

Shlosman, I. 1994, Mass-Transfer Induced Activity in Galaxies (Cambridge: Cambridge Univ. Press)

Simkin, S.M., Su, H.J., & Schwarz, M.P. 1980, ApJ 237, 404

Skillman, E.D. 1987, in Star Formation in Galaxies, ed. C.J. Lonsdale Persson, (Washington: NASA Conf Publ 2466), 263

Smirnov, M.A., & Tsvetkov, D.Yu 1981, Pis'ma Astr. Zh. 7, 154

Smith, E.P., & Hintzen, P. 1991, AJ 101, 410

Smith, E.P., & Kassim, N.E. 1993, AJ 105, 46

Soifer, B.T., Houck, J.R., & Neugebauer, G. 1987, ARA&A 25, 187

Soifer, B.T., & Neugebaurer, G. 1991, AJ 101, 354

Soifer, B.T. et al. 1984, ApJ 278, L71

Steidel, C.C., Giavalisco, M., Pettini, M., Dickinson, M., & Adelberger, K.L. 1996, ApJ 462, L17

Stocke, J.T. 1978, AJ 83, 348

Stockton, A. 1978, ApJ 223, 747

Storchi-Bergmann, T., Kinney, A.L., & Challis, P. 1995, ApJS 98, 103

Struck-Marcell, C. 1991, ApJ 368, 348

Sulentic, J.W. 1976, ApJS 32, 171

Taylor, C.L., Brinks, E., Grashuis, R.M., & Skillman, E.D. 1995, ApJS 99, 427

Taylor, C.L., Brinks, E., Pogge, R.W., & Skillman, E.D. 1994, AJ 107, 971

Telesco, C.M. 1988, ARA&A 26, 343

Telesco, C.M., Dressel, L.L., & Wolstencroft, R.D. 1993, ApJ 414, 120

Telesco, C.M., Wolstencroft, R.D., & Done, C. 1988, ApJ 329, 174

Telles, E., & Terlevich, R. 1995, MNRAS 275, 1

Terlevich, R., & Boyle, B. 1993, MNRAS 262, 491

Tinsley, B.M. 1972, A&A 20, 383

Tinsley, B.M., & Danly, L. 1980, ApJ 242, 435
Tomita, A., Tomita, Y, & Saito, M. 1996, PASJ 48, 285
Toomre, A. 1964, ApJ 139, 1217
Toomre, A. 1977, in The Evolution of Galaxies and Stellar Populations, ed. B.M. Tinsley & R.B. Larson (New Haven: Yale Univ. Observatory), 401
Toomre, A., & Toomre, J. 1972, ApJ 178, 623
Tóth, G., & Ostriker, J.P. 1992, ApJ 389, 5
Turner, A. 1997, Ph.D. thesis, University of Arizona
van den Broek, A.C. 1993, A&A 269, 96
van der Hulst, J.M. 1979, A&A 71, 131
van der Hulst, J.M., Kennicutt, R.C., Crane, P.C., & Rots, A.H. 1988, A&A 195, 38
van der Hulst, J.M., Skillman, E.D., Smith, T.R., Bothun, G.D., McGaugh, S.S., & de Blok, W.J.G. 1993, AJ 106, 548
Veilleux, S., Kim, D.-C., Sanders, D.B., Mazzarella, J.M., & Soifer, B.T. 1995, ApJS 98, 171
Veilleux, S., & Osterbrock, D.E. 1987, ApJS 63, 295
Vorontsov-Velyaminov, B.A. 1959, Atlas and Catalog of Interacting Galaxies (Moscow: Sternberg Astronomical Institute)
Vorontsov-Velyaminov, B.A. 1977, A&AS 28, 1
Wallin, J.F. 1990, AJ 100, 1477
Wang, J., Heckman, T.M., Weaver, K.A., & Armus, L. 1997, ApJ 474, 659
Walterbos, R.A.M., & Greenawalt, B. 1996, ApJ 460, 696
Walterbos, R.A.M., & Schwering, P.B.W. 1987, A&A 180, 27
Weedman, D.W., Feldman, F.R., Balzano, V.A., Ramsey, L.W., Sramek, R. A., & Wu, C.-C. 1981, ApJ 248, 105
Weedman, D.W., & Khachikian, E.Ye. 1969, Astrofisica 5, 113
White, S.D.M. 1990, in The Interstellar Medium in Galaxies, ed. H.A. Thronson & J.M. Shull (Dordrecht: Kluwer), 371
White, S.D.M. 1994, in Mass-Transfer Induced Activity in Galaxies, ed. I. Shlosman (Cambridge: Cambridge Univ. Press), 464
White, S.D.M., & Rees, M.J. 1978, MNRAS 183, 341
Whitmore, B.C., Lucas, R.A., McElroy, D.B., Steiman-Cameron, T.Y., Sackett, P.D., & Olling, R.P. 1990, AJ 100, 1489
Whitmore, B.C., & Schweizer, F. 1995, AJ 109, 960
Woods, D., Fahlman, G.G., & Richer, H.B. 1995, ApJ 454, 32
Xu, C., & Sulentic, J.W. 1991, ApJ 374, 407
Yee, H.K.C., & Ellingson, E. 1995, ApJ 445, 33
Yun, M.S., Ho, P.T.P., & Lo, K.Y. 1993, ApJ 411, L17
Zepf, S.E., & Koo, D.C. 1989, ApJ 337, 34.

Observational Evidence for Interactions and Mergers

François Schweizer

Carnegie Institution of Washington, Department of Terrestrial Magnetism,
5241 Broad Branch Road, N.W., Washington, DC 20015-1305, USA

1 Interactions in Our Time

During the past four decades, galaxy interactions and induced star formation have slowly but steadily gained astronomers' attention. First, there seemed to be just a few "peculiar galaxies" with deformed bodies and filamentary extensions. Then, *gravitational* interactions between closely approaching galaxies were shown to explain some of these deformations and filaments as *tides*. The hypothesis was advanced that such interactions lead to galactic *mergers*, and a few candidate merging galaxies were identified and studied. Recently, the few have become many. As ground- and space-based observations have opened up new spectral windows, and as computers have become tools to simulate complex interactions, we have come to realize that galaxies grew from initial density perturbations through episodic accretions and full-fledged mergers, some of which continue to the present time. Gas plays a crucial role in this process of *prolonged galaxy building*. Because of its dissipative nature, gas tends to get crunched into molecular form during collisions and mergers, turning into fuel for intense *bursts of star formation*. Thus, galaxy interactions and induced star formation seem now topics of central interest in our efforts to understand galaxy formation and evolution.

This first chapter of my written lectures expands on the above topics. The goal is to introduce not only my assigned subject of *Observational Evidence for Interactions and Mergers*, but also the subjects of the *Dynamics of Galaxy Interactions*, covered by Dr. Barnes in parallel lectures, and of *Induced Star Formation*, covered by Dr. Kennicutt in his lectures. Both these lecturers do, of course, present more detailed introductions in their own first chapters. But all three of us wish to emphasize how intertwined the subjects of galaxy interactions and induced star formation have become, and how observations and theory mutually stimulate each other. Because this first chapter presents an overview, its level is more elementary than that of the following chapters. Personally, I have always found it easier to absorb and integrate scientific knowledge in the context of the history of the subject. Therefore, I wish to begin by retracing the development of our understanding of galaxy interactions and of the role they play in galaxy evolution. Even in the more technical chapters that follow, I shall occasionally revert to this historical approach. An understanding of past successes and failures seems quintessential for any scientist trying to maximize the returns of his or her research while avoiding pitfalls along the way.

Those of us fortunate enough to have been able to participate in the 26th "Saas-Fee" Advanced Course held in Les Diablerets during the last week of March 1996 owe a great debt to its three organizers, Drs. Louis Martinet, Daniel Friedli, and Daniel Pfenniger of the Observatoire de Genève. They, their various helpers, and the Swiss Society of Astrophysics and Astronomy made it possible for us all to enjoy an enriching week of scientific exchange and learning. We thank them very much. I also gratefully acknowledge the support of the Carnegie Institution of Washington and the National Science Foundation[1] not only during the preparation and write-up of my lectures, but also during the years of research leading to results partially described in these lectures. Finally, I thank the many friends, fellow astronomers, and organizations who kindly granted me permission to reproduce their figures and photographs.

Fig. 1. Edwin Hubble (1889–1953) inspecting a photographic plate. Photograph courtesy of the Carnegie Institution of Washington.

1.1 Hubble's Morphological Sequence

During the period 1922–1936, Edwin Hubble (Fig. 1) made at least three fundamental discoveries. The first two brought him instant fame, while the third turned into a major puzzle that we have begun to solve only during the past 25

[1] Through Grants AST–92 21423 and AST–95 29263.

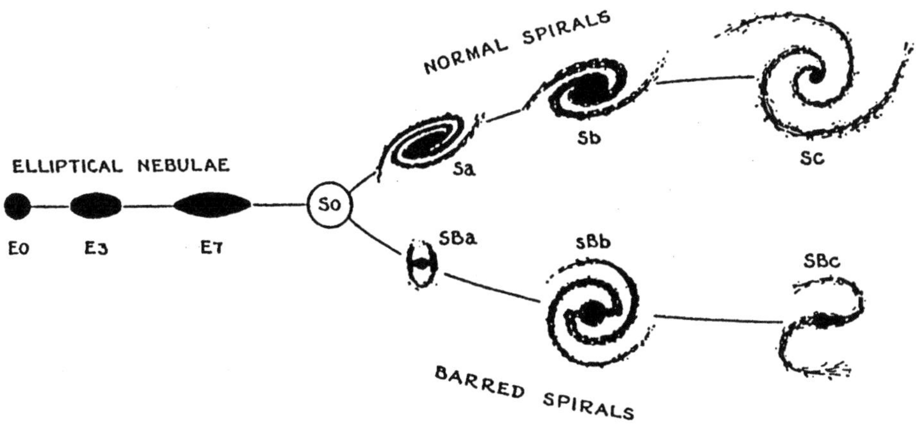

Fig. 2. Hubble's morphological sequence of extragalactic nebulae, from Hubble (1936).

years. Hubble's first famous discovery was that our own Milky Way galaxy is just one among millions of similar galaxies populating the universe. His second, even more famous discovery was that this universe itself is expanding. His third fundamental discovery, however, is less well known to the general public. After years of studying the morphology of thousands of galaxies, Hubble concluded that galaxies could be ordered in a sequence that is defined by two extreme shapes (Hubble 1926, 1936).

At the left end of Hubble's morphological sequence are the blobby, featureless galaxies which he named "Elliptical Nebulae" (Fig. 2). At the right ends of the branching sequence, which astronomers quickly nicknamed the "tuning-fork diagram", are the magnificent spiral galaxies of type Sc. And in between these extremes lie mixed-type galaxies that feature an elliptical-like bulge at their center, but a disk – often with spiral arms – farther out.

Most such disks share two characteristic properties: They are relatively thin, with vertical thicknesses typically less than 10% of their diameters, and stars and gas in them rotate around the disk center on nearly circular orbits. Compared to the mean rotational velocities, the random velocities of individual stars are small. The fraction of the disk mass in cold gas (mostly atomic and molecular hydrogen) is of order 10%, varying systematically along the Hubble sequence from near zero for S0 galaxies to 20%–30% for some Sc and Sd spirals. As this gas rotates around the center, it gets crunched in the spiral arms and forms new, young stars that give the spiral arms their splendid appearance and bluish color. In contrast to these disk galaxies, elliptical galaxies appear blobby, no matter from which direction they are seen. Stars in them tend to move with large random velocities, each following some space-filling, three-dimensional orbit. The shapes of ellipticals range from nearly spherical to American-football like to triaxial (i.e., three unequal axes). However, none of these ellipticals are nearly as flat as spiral disks. Also, because most ellipticals have no cold gas left to form young stars, they contain only relatively old stars that tend to give them a reddish

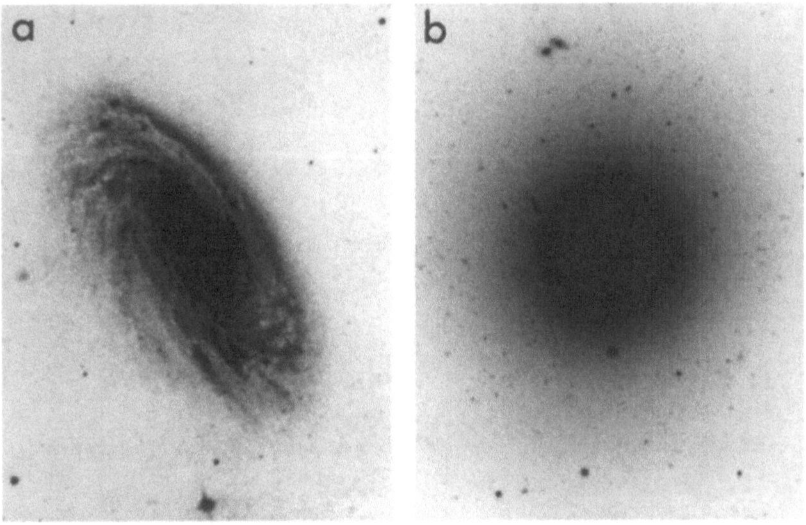

Fig. 3. Two fundamentally different types of galaxies: (a) Disk galaxy with spiral arms and (b) elliptical galaxy M87 (from Schweizer 1986).

hue. As Fig. 3 shows, even on black-and-white photographs the different shapes of disk galaxies and ellipticals are striking.

Why, then, is there this *dichotomy of shapes* among galaxies? This is the puzzle that Edwin Hubble left us with. By the early 1960s, astronomers thought they had found the answer. By then, Sandage (1961) had published *The Hubble Atlas of Galaxies*, emphasizing regular, well behaved, one could even say beautiful galaxies. Also, much progress had been made in disentangling the kinematics of disk and halo stars in our own Galaxy. In a very influential paper, Eggen, Lynden-Bell, & Sandage (1962) summarized the evidence for plunging orbits of halo stars and deduced that the stellar halo of the Milky Way must have formed rapidly from free-falling gas clouds, while the stellar disk must have formed more slowly from gas accumulating in a pancake-shaped disk. Theoretical considerations suggested to them that the collapse of our Galaxy and formation of its halo had occurred within a short period of a few 10^8 yr. As happens so often in science, astronomers generalized from the Milky Way to other galaxies, and soon the view spread that *all* galaxies had formed from huge protogalactic gas clouds shortly after the big bang. In most cases, these gas clouds had supposedly collapsed into gaseous disks, in which stars then formed slowly and continuously. Such disks became the spiral galaxies that make up about 80% of all major galaxies. In other cases, stars had formed much faster during the free-fall phase of the protogalactic collapse, leading to the blobby galaxies called ellipticals. Therefore, ellipticals were thought to be the oldest and least organized galaxies. There was only one hitch in this generalized view of galaxy formation: Nobody could ever explain why the stars of ellipticals should have formed so fast and early during the collapse.

Fig. 4. Fritz Zwicky (1898–1974) at Caltech. Photograph courtesy of the California Institute of Technology.

1.2 Peculiar Galaxies: Zwicky and Arp

The solution to the puzzle posed by the dichotomy of shapes, and to the problem of vigorous star formation during collapse, was to come from a completely unexpected direction. Astronomers had long noticed that a few percent of all galaxies show strong distortions, most often when two disk galaxies appear to be colliding. A famous example is that of NGC 4038/4039, a close pair of disk galaxies from which protrude two long filaments reminiscent of the antennae of some giant bug (see Fig. 9 below). This interesting pair must have caught already Hubble's attention, since it is at his instigation that Duncan (1923) photographed and described it.

The first to start photographing such interacting galaxies systematically was the Swiss astrophysicist Fritz Zwicky (Fig. 4). He was a genius who thought of, among other things, supernovae, neutron stars, and gravitational lenses decades ahead of most astronomers. Shortly after his arrival as a young professor at the California Institute of Technology in Pasadena, he had an 18-inch Schmidt camera built and installed at Palomar.[2] From photographs of peculiar galaxies taken with this camera, Zwicky concluded early on that many of the narrow filaments must be tides stemming from gravitational interactions. He may also have read Holmberg's (1941) famous paper, which reached a similar conclusion (see first chapter by Dr. Barnes). Anyway, in a 1953 issue of *Physics Today*, whose cover is graced by his sketch of the interacting galaxies in M51, Zwicky (1953) described the filaments as *tides* and *countertides*, guessed correctly that

[2] This "small" Schmidt camera was the first built in the United States and was so successful that it lead to the design and building of a second, 48-inch Schmidt camera for Palomar.

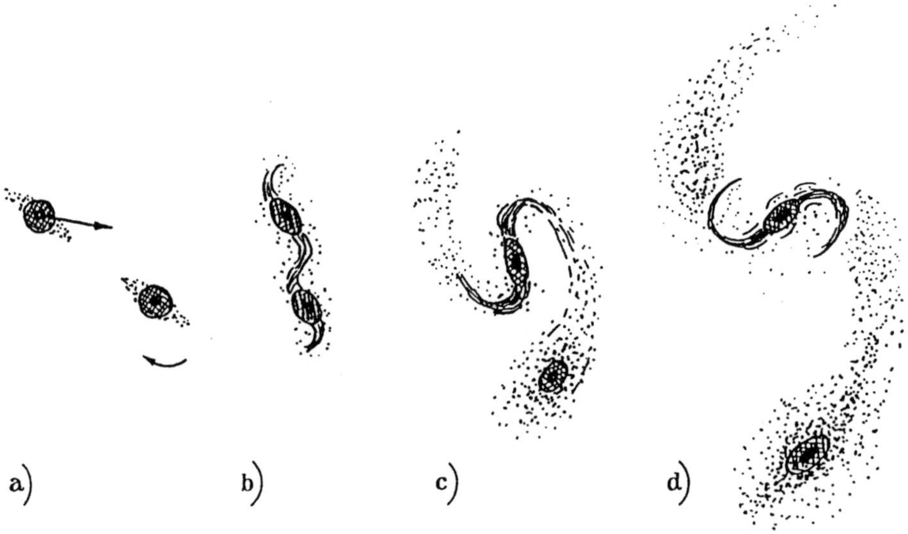

Fig. 5. Schematic drawing by Zwicky (1956) of possible bridge formation between two interacting galaxies, resulting in a configuration resembling the pair Arp 96. Lower galaxy in frames **a** and **b** becomes upper galaxy in frames **c** and **d**.

the narrowest of them may be broad sheets of matter seen nearly edge-on, and emphasized that most of the visible matter in galactic *bridges* and *tails* must consist of stars. Figure 5 reproduces his sketch of "the possible formation of an intergalactic bridge between two galaxies passing each other" (Zwicky 1956; see also Zwicky 1959).

Yet another Pasadena astronomer not quite happy with the tranquil world of galaxies projected by the *Hubble Atlas* was Halton Arp. He was struck not only by Zwicky's photographs of interacting galaxies, but also by the many faint images of peculiar, distorted galaxies that he and others (e.g., Vorontsov-Velyaminov 1959) could typically see on deep plates obtained with the Palomar 48-inch Schmidt camera. Sensing that "the peculiarities ... represent perturbations, deformations, and interactions which should enable us to analyze the nature of the real galaxies which we observe and which are too remote to experiment on directly", Arp spent four years photographing many such galaxies with the Palomar 200-inch telescope. In 1966, he published the *Atlas of Peculiar Galaxies*, which contains high-resolution images of 338 interesting systems (Arp 1966). In arranging the photographic prints six to a page for publication, Arp displayed an uncanny knack for grouping together galaxies of similar interaction types, unbeknown as the exact nature of at least some of these interactions may have been then. Figure 6 shows Plate 41 reproduced from his *Atlas*. As will become clear during the course of these lectures, Arp's Atlas turned out to be an invaluable source of material during the subsequent development of the subject of galactic interactions and mergers.

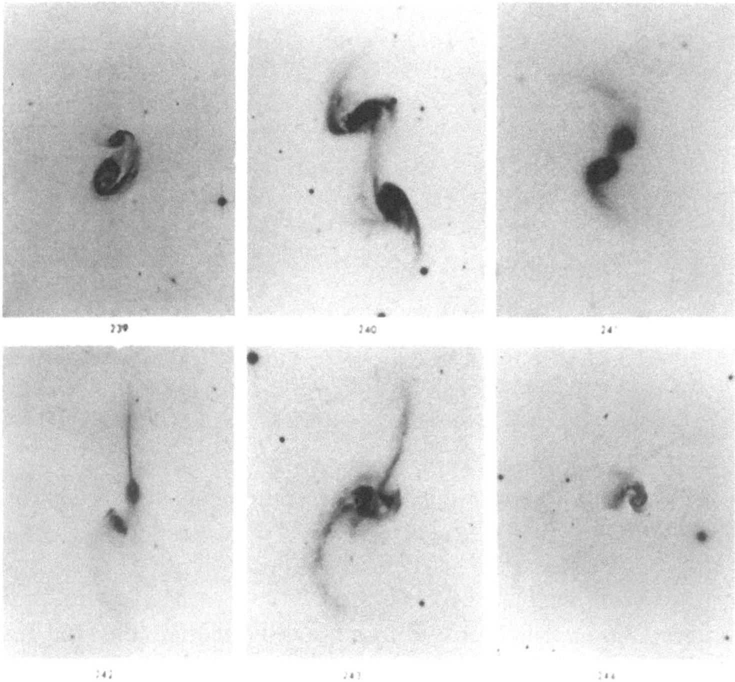

Fig. 6. Six pairs of gravitationally interacting galaxies, presented as Plate 41 in Arp's (1966) Atlas. Reproduced with kind permission of Dr. H. Arp.

1.3 "TT": Galactic Bridges and Tails

Just about the time that Zwicky and Arp published their observational studies, computers available to scientists became powerful enough to make possible some first numerical model simulations of gravitationally interacting galaxies. The very first models, representing disk galaxies as central point masses surrounded by test particles within the plane of a perturber's orbit, were already successful at reproducing the development of tidal arms (Pfleiderer & Siedentopf 1961; Pfleiderer 1963). A whole slew of tidal-interaction models followed in the early 1970s. These models still used point masses surrounded by test particles to represent interacting galaxies, but allowed for inclined disks and different mass ratios between the simulated galaxies (Yabushita 1971; Wright 1972; Clutton-Brock 1972; Toomre & Toomre 1972; Eneev, Kozlov, & Sunyaev 1973).

Among these various papers, the one written by the brothers Alar and Juri Toomre (1972) stands out. It is so well-known that it is often simply referred to as "TT". While still a graduate student working on his Ph. D. thesis, I was given a preprint of this paper by Ivan King. Upon reading it, I knew instantly that after completion of my thesis I would want to begin observing interacting and merging galaxies because TT offered, for the first time, a key to galaxy evolution. Besides being extremely well written and systematic, this paper had two significant advantages over all others from that period. First, it presented not only

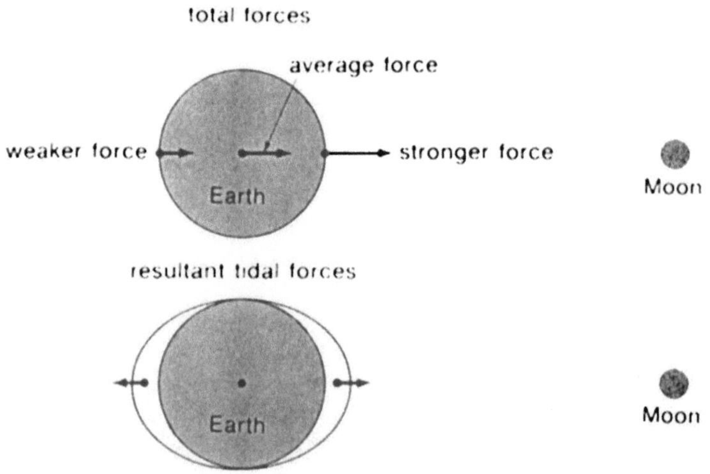

Fig. 7. Sketch illustrating how the Moon generates ocean tides on Earth. Diagram from King (1976).

generic interaction models, but also four specific models aimed at reproducing all available observations of four real pairs of interacting galaxies, including M51 and the Antennae. And second, it made two fascinating predictions concerning the eventual fate of strongly interacting disk galaxies: Such galaxies would tend to merge rather quickly, and their merged remains would masquerade as elliptical galaxies.

That gravitational tides can be so fierce as to induce galactic mergers is not immediately clear to a casual observer of the relatively mild ocean tides on Earth. As we all know and Fig. 7 illustrates, the Earth's oceans get stretched by the differential gravitational attraction exerted on them by our Moon. Earth's side nearer to the Moon gets attracted more than the center, and Earth's far side less, resulting in a two-sided bulge of water that we observe as twice-daily tides during Earth's rotation around its axis. The *differential*, tidal forces between the near and far side depend on the *third* power of the Moon's distance from Earth. If our Moon were not 30 Earth diameters away as in reality, but only 2.5 Earth diameters as sketched in King's (1976) figure, ocean tides would be more than 1700 times as high as they are, and there would be no Amsterdam, London, or New York!

It is such fierce tides between closely approaching disk galaxies that TT modeled. Figure 8 shows their model of M51. In this model, a companion galaxy with a central point mass one quarter as massive as that of the main disk has flown by on the sketched relative orbit. The initial disks of test particles surrounding the two point masses have been deformed by the interaction and now sport asymmetric tides, i.e., one "bridge" and one "tail" for each model galaxy.

As TT demonstrated convincingly, *long tails* take a long time to develop, and the longest tails are the product of strong gravitational interactions between

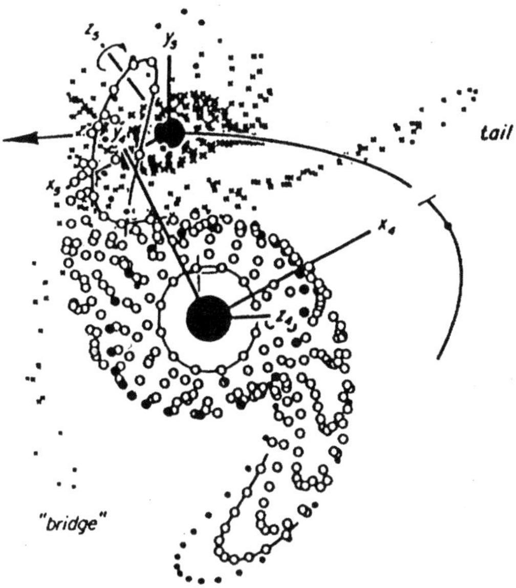

Fig. 8. Model of the recent encounter between NGC 5194 and NGC 5195, a galaxy pair known as M51. Diagram from Toomre & Toomre (1972).

corotating disks of nearly equal mass.[3] Figure 9 compares the TT model for "The Antennae" with a photograph of these galaxies. The model galaxies have equal central point masses and reached perigalacticon slightly more than half an orbital period ago. Hence, the tidal tails ejected near perigalacticon have had time to grow long.[4] This tidal-interaction model successfully reproduces the apparently *crossed* tails of NGC 4038/4039, which in 1970 were a challenge to explain. Note that in such equal-mass encounters the tidal bridges tend to be much less prominent than the tails. Hence, *two* long filaments emerging from some relatively isolated and perhaps chaotic-looking galactic system are likely to be two tidal tails, rather than a bridge and a tail. Such paired tails are a strong indicator that two disk galaxies of comparable mass are, or were, involved in a collision (i.e., a strong tidal interaction).

At least as important as these successful numerical simulations of four well-known pairs of interacting galaxies was the Toomres' emphasis on three fundamental concepts associated with strong tidal interactions: tidal friction, orbit decay, and violent relaxation.

Tidal friction is a natural consequence of the energy exchange occurring when two N-body systems interact gravitationally. Even though the detailed forms of energy exchange cannot easily be guessed without the help of N-body simulations, it is intuitively clear that the generation of such major tides as those observed in The Antennae must cost energy. As Holmberg (1941) guessed and

[3] The disk rotations must also be prograde, i.e., in the same direction as the orbital motion, as explained in detail in the chapters by Dr. Barnes.

[4] For a still very instructive demonstration of tail growth, see the beautiful sequence of models shown by Toomre & Toomre (1973), where the two model galaxies are colored differently and displayed orbiting and deforming each other.

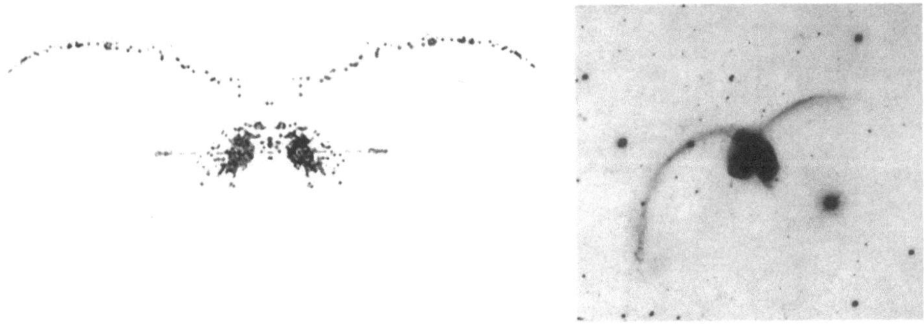

Fig. 9. TT model of an equal-mass disk encounter leading to long, apparently crossed tidal tails, compared with a photograph of NGC 4038/4039 ("The Antennae"). Diagram from Toomre & Toomre (1972), photograph by author.

Alladin (1965) estimated even before TT, a main source of such energy is the orbital motion of the interacting galaxies. Hence, energy transferred from the orbits to tidal deformations leads to braking and *orbit decay*, and the galaxies must eventually merge. During such mergers the gravitational field tends to fluctuate rapidly, leading to *violent relaxation* (Lynden-Bell 1967) of the involved matter. As Lynden-Bell argued for a single collapsing galaxy, violent relaxation tends to drive the stellar distribution toward an end state resembling that observed in elliptical galaxies. Perhaps with such results in mind, Toomre & Toomre (1972) pointed out that the end stage of a disk–disk merger should be "a single three-dimensional pile of stars" resembling an elliptical galaxy. Besides pointing to nearly a dozen candidate mergers mostly in Arp's Atlas (Fig. 10), TT and Toomre (1977a) emphasized that there should be *at least* one-third as many remnants stemming from such mergers during the past Hubble time as there are elliptical galaxies. In fact, the number of merger remnants may well be considerably higher since interactions and mergers must have been more frequent in the younger, denser universe. Or, as Toomre (1977a) stated so eloquently, referring to galaxies near enough to be included in the NGC: "If several hundred merged remnants of former disks do not now constitute many of the ellipticals, where else have they possibly gone?"

Fig. 10. Sequence of "eleven NGC prospects for ongoing mergers" sketched by Toomre (1977a); from left to right, NGC 4038/39, 4676, 7592, 7764A, 6621/22, 3509, 520, 2623, 3256, 3921, and 7252.

1.4 Objections to the Toomres' Hypothesis

The tidal-interaction models published by TT and others immediately generated several objections from astronomers who had bet on non-gravitational forces being responsible for the observed deformations. Some of these objections were relatively minor. After giving talks during the mid-1970s, I would typically hear objections like "The restricted three-body approach is too simple!" or "The Toomre models of M51 and The Antennae have too many free parameters!" Perhaps the most widely advanced objection was that "Angular-momentum conservation prevents mergers!" This simplistic objection is easily refuted if one points out that – indeed – angular momentum must be conserved, but that only in two-body systems are the consequences transparent. Already with three bodies, matters get much more complicated. And two interacting galaxies consisting of billions of stars each have many degrees of freedom to transfer angular momentum between, say, their orbits and internal motions.[5]

More serious objections were raised and published especially after Alar Toomre (1977a) forcefully reiterated the merger hypothesis of elliptical formation (e.g., Ostriker 1980; Tremaine 1981; van den Bergh 1982; Gunn 1987). Although these objections have mostly been refuted by now, it is still instructive to list them since we need to know past pitfalls to help us navigate around future ones. Among about a dozen published objections, perhaps the five most challenging were as follows:

(1) Elliptical galaxies obey color–luminosity and metallicity–luminosity relations. How could such relations hold if ellipticals were assembled from spirals of various metallicities?
(2) Measured surface brightnesses and velocity dispersions suggest that the central densities in ellipticals are orders of magnitude higher than in spirals. How could spiral–spiral mergers, which scatter stars, increase the central densities of the remnants, when in fact they should decrease them?
(3) Scale lengths for the disk components of spirals are typically 2–5 kpc, while those for the cores of ellipticals are generally $\lesssim$1 kpc. Again, what process in mergers could possibly lead to such a central contraction?
(4) The specific frequency of globular clusters in ellipticals (i.e., the number of clusters per unit luminosity of the host galaxy) is about 5–10× higher than in spirals. How could ellipticals then be the sum of two merged spirals?
(5) Most giant ellipticals occur in rich clusters of galaxies, yet the high velocity dispersions in these clusters produce few slow, merger-inducing collisions.

[5] One memorable, though only half-serious, objection stems from Gérard de Vaucouleurs (1974) at the IAU Symposium No. 58. Perhaps mildly annoyed by all the attention that interacting galaxies were receiving, de Vaucouleurs exclaimed: "Interacting or colliding systems form an interesting subject of study, but are not new types of galaxies. After a collision a car is a wreck, not a new type of car!" Had he been fully aware of the Toomres' hypothesis of elliptical formation through disk mergers, he might have had to consider that, in the strange world of galaxies, two cars can indeed collide and coalesce into one limousine.

If disk mergers formed ellipticals, shouldn't ellipticals be rare, rather than plentiful, in rich clusters?

Objection (5) ignores the fact, known already to Roos & Aarseth (1982), that in an expanding universe most mergers among cluster galaxies occur predominantly during the early expansion and subsequent collapse of the cluster. And Objections (1)–(4) have all come to fall because they ignored the simple fact that *most mergers involve gas.*[6] The following section briefly describes how astronomers have come to realize the fundamental role played in mergers by gas and star formation.

1.5 Mergers with Gas: Galaxy Building

Already Zwicky (1956) and Arp (1969) knew that interacting galaxies often show signs of unusually vigorous star formation. Infrared observations during the 1970s confirmed that some of these galaxies seem to undergo major bursts of star formation which – Larson & Tinsley (1978) estimated from *UBV* colors – may convert up to ~5% of the total mass from gas into stars. Yet, it was the 1983 sky survey by the Infrared Astronomical Satellite (*IRAS*) that drove home the point just how enormously vigorous star formation can be in mergers. Among the thousands of new infrared sources discovered during this survey are many relatively distant, optically inconspicuous galaxies. Ground-based follow-up observations including optical imaging and measurements of redshifts have shown that many of these galaxies emit most of their luminosity, in some cases up to 99% of it, in the infrared ($\lambda\lambda \approx 8$–$120\ \mu$m). Interestingly, among the galaxies of highest bolometric luminosity ($L > 10^{12}\ L_\odot$), nearly all seem to feature tails and deformations indicative of major mergers (Sanders et al. 1988b; Sanders 1990). Their rates of star formation exceed that of the Milky Way by factors

[6] Given that roughly 80% of all galaxies are spirals, one might think that the potential importance of gas in mergers had long been recognized. Yet, well into the 1980s this simple fact was widely ignored, and most merger skeptics thought and wrote only in terms of purely stellar, dissipationless galaxy mergers. Thus, they ignored not only much evidence of exceptional star formation in interacting galaxies stemming from the 1970s, but also the marvelous little section in TT entitled *"Stoking the Furnace?"*

Even three years after the launch of the infrared satellite *IRAS* and the ensuing ground-based observations of "super starbursts", no less an astrophysicist than Jim Gunn strongly believed in the validity of Objection (2). Summarizing the 1986 Santa Cruz Summer Workshop on Nearly Normal Galaxies and, specifically, what had been said about "Spheroids, Mergers, Disks, and all That", Gunn exclaimed: *"I do not think you can make rocks by merging clouds"*. By 'rocks' he meant the dense centers of ellipticals, by 'clouds' the relatively fluffy centers of spirals. I sat in the audience incredulous, having earlier presented to the Workshop the observational evidence for merger-induced starbursts and the extraordinarily high molecular-gas densities measured at the centers of mergers by people like Judith Young and David Sanders (see Sect. 1.5). If such observational facts could not refute the fallacy of Objection (2), what else could?

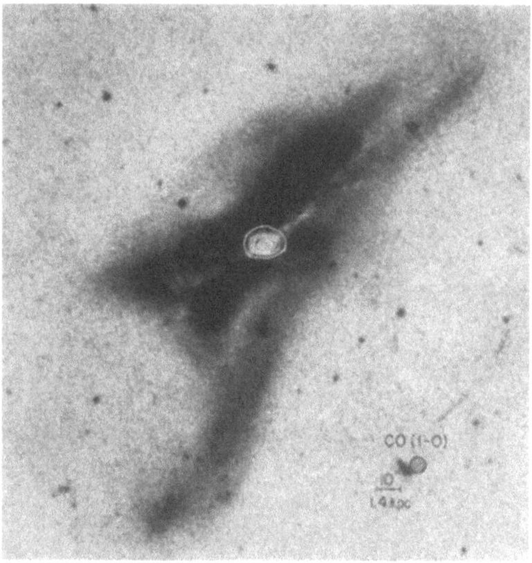

Fig. 11. Map of CO emission in merging pair of galaxies NGC 520. Notice the enormous central concentration of the molecular gas. Figure from Sanders et al. (1988a).

of typically 10–100. Since intense star formation depends on ample supplies of dense gas for fuel, the high star-formation rates observed in these ultraluminous galaxies imply that major mergers of spirals involve much gaseous dissipation (Joseph & Wright 1985; Barnes & Hernquist 1991; Barnes 1994).

It is a happy coincidence that just around the time *IRAS* carried out its survey, the detectors for millimeter-wave radio telescopes had become sensitive enough to detect CO emission from galaxies beyond the Local Group. Observations of ultraluminous infrared galaxies revealed large amounts of molecular gas, typically $4 \times 10^9 - 4 \times 10^{10}\,\mathcal{M}_\odot$ or 2–20 times the molecular-gas content of the entire Milky Way, in them (Young et al. 1984, 1989). But the biggest surprise was just how concentrated toward the center some of this molecular gas is. Figure 11 shows an isophotal map of CO emission in the advanced merger NGC 520, obtained with the Owens Valley millimeter-wave interferometer by Sanders et al. (1988a). About 1/3 of the total CO emission stems from a small central region of only 1.5 kpc diameter. The molecular-gas mass within this region is $\sim 2 \times 10^9\,\mathcal{M}_\odot$. In more luminous mergers, molecular-gas masses in excess of $10^{10}\,\mathcal{M}_\odot$ have been observed to be concentrated in similarly small regions. Some of the central *gas* densities clearly reach into the range of central mass densities measured in ellipticals, suggesting a process by which such *stellar* mass densities in Es may be formed. Hence, these CO observations emphasize that merger-induced gaseous dissipation may play a significant role in forming some of the high mass densities observed at the centers of galactic spheroids.

More dilute gas in the outer parts of merging galaxies also seems to play a larger role than one might have guessed only a few years ago (Hibbard 1995). Plate 1 shows the distributions of neutral hydrogen (H I) in five merger galaxies observed with the Very Large Array by Hibbard & van Gorkom (1996). These five galaxies were selected from the merger sequence by Toomre (1977a) and range in separation from a wide post-collision pair to two coalesced remnants. Surprisingly large amounts of H I (several $10^9\,\mathcal{M}_\odot$) are associated with the tidal tails of the more advanced mergers and are observed to now be *falling back* into the remnants, just as Barnes (1988) predicted. Though decreasing with time, such inflow of gas temporarily stored at large radii may nourish central star formation and perhaps even nuclear activity for prolonged periods of order 1 Gyr or more.

Further evidence for ongoing and past starbursts in interacting galaxies stems from optical spectra. Interestingly, the starbursts in the two galaxies of an interacting pair may proceed at significantly different rates or peak at different times. As the example of NGC 7714/15 shows, one galaxy may be intensely star forming and highly infrared luminous while the other displays a spectrum typical of an aging (~45 Myr old) and inactive starburst (Bernlöhr 1993). Since the two galaxies presumably experienced the strongest tidal forces at the same time of perigalacticon, there must be other factors (e.g., gas distribution, disk orientation, sense of rotation) that influence the process of stimulated star formation. Clearly, unraveling the episodic star-formation histories of interacting galaxies is a challenging task. In several of his chapters, Dr. Kennicutt describes in detail the methods that have been developed to tackle this task and the interesting results obtained so far.

One might think that unraveling the past star-formation history of a merger remnant is especially difficult. Yet, the dramatic increase in spatial resolution recently achieved by the *Hubble Space Telescope* (*HST*) promises significant help. Specifically, the discovery of systems of *young globular clusters* apparently formed during major mergers (e.g., Holtzman et al. 1992; Whitmore et al. 1993) is providing us with a new opportunity to age-date simple, coeval stellar populations created during the merger-induced starbursts. These freshly minted star clusters add yet another piece of evidence that mergers and their associated starbursts may play a significant role not only in shaping certain galaxies, but also in building up their stellar and cluster populations. The subject of merger-induced cluster formation looks so promising that I devote a whole chapter to it (Chap. 7).

The unprecedented high-resolution imaging capability of *HST* is also leading to rapid progress in studying galaxy interactions and evolution at higher redshifts. Beyond $z \approx 0.3$ we are beginning to see galaxies that are both significantly *younger* and *more gas rich* than nearby galaxies. Objects at $z = 0.5$ are only about 1/2 as old as the oldest local objects, and objects at $z = 3$ are even only ~12% as old. With the repaired *HST* now resolving galaxies at $z \approx 1-2$, we can begin to directly study the evolution of the gas contents, star-formation rates, and morphologies of various kind of galaxies as a function of epoch. Already, *HST* images of galaxy clusters at $z = 0.4-0.6$ have begun to clarify the causes of the Butcher-Oemler effect. Butcher & Oemler (1978) had discovered that clusters in that redshift range possess systematically higher fractions of blue galaxies than nearby clusters do, and follow-up studies of these blue galaxies revealed that many show spectroscopic signatures of major, aging starbursts. The

HST images now suggest that increased rates of interaction and less mature, more gaseous disk galaxies may both contribute to the Butcher-Oemler effect (see Chap. 9 and chapters by Dr. Kennicutt).

Even the spectacular fireworks of quasars may be related to mergers of gas-rich galaxies. In the late 1970s, Alan Stockton found that at least some quasars seem to be associated with close companion galaxies and may be gravitationally interacting with them. A break-through was the discovery that the quasar 3CR249.1 features two long tails of fluorescent gas and perhaps even a double body, signatures reminiscent of merger galaxies like The Antennae (Stockton & MacKenty 1983). Several more, similarly shaped quasars have since been found (Chap. 9). If – as is now widely believed – black holes fueled by gas are the central engines of such quasars, the main challenge from Stockton's observations is to understand how gravitational interactions or mergers can bring fresh gas not just to the central kiloparsec of these galaxies, but to within a few astronomical units from the presumed black hole. As Dr. Barnes details in his chapters, there are gravitational and dissipative mechanisms that send gas in merging galaxies plunging toward the center, at least down to scales of 10–100 pc. How the gas proceeds from there to the very center, bridging another factor of $\sim 10^5$ in radius, is the subject of much current research and speculation.

The main lessons we have learned from *IRAS* and follow-up observations are that (1) most galaxy interactions and mergers involve gas, (2) the stirring of such gas leads to intense bursts of star formation, and (3) such episodes of vigorous star formation may be an essential part of "galaxy building" (e.g., Larson 1990; Kennicutt 1990).

1.6 *N*-body Simulations: From $N = 2$ to 2×10^6

The past decade has also brought tremendous progress in our theoretical understanding of galactic mergers and of the gaseous processes involved. Much of this progress has been achieved through numerical N-body simulations of interacting galaxies. Computers of ever greater speed and storage capacity have made it possible to dramatically increase the number N of mass points that represent two model galaxies and to also include gas with its different physics.

The model simulations of the early 1970s all represented two interacting galaxies by two point masses, each surrounded by a disk of test particles. These simulations were remarkably successful at reproducing some of the observed tidal deformations (e.g., bridges and tails) and were used into the 1980s to explain the then new phenomenon of "ripples" or "shells" in elliptical galaxies (Toomre 1981b; Quinn 1984). However, since two point masses move on their elliptical or parabolic orbits forever, these early $N = 2$ simulations could not reproduce the effects of tidal friction and orbit decay. Hence, simulated models of *mergers* had to await increased computing power and the development of efficient N-body simulation codes.

First N-body simulations of merging galaxies were run with typically $N \approx$ 200–500 mass particles. Figure 12 shows a simulation by White (1979) in which two rotating spherical model galaxies, each represented by 250 mass particles,

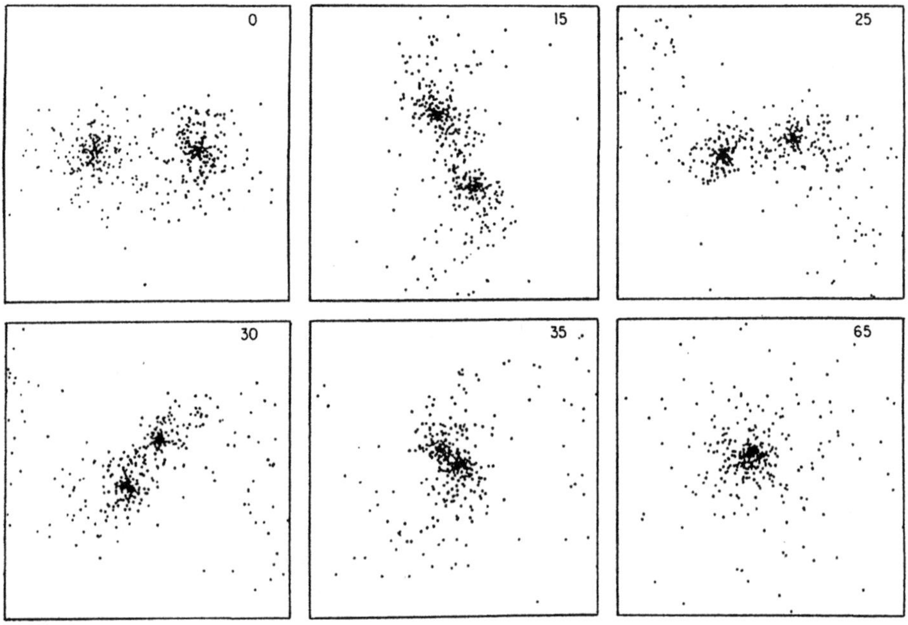

Fig. 12. Simulated merger of two rotating spherical model galaxies, each with 250 mass particles. View is onto plane of parabolic orbits, and the galaxies orbit counterclockwise. Diagram from White (1979).

are followed from their parabolic collision through their final merger. These and similar simulations involving disk galaxies (e.g., Gerhard 1981) showed that prograde encounters, in which the galaxy and orbital spins are roughly aligned, lead to faster merging than retrograde encounters. They also yielded triaxial, slowly rotating model remnants with density profiles not unlike those inferred from observations of elliptical galaxies (Gerhard 1983a).

Yet, it is the modern N-body simulations with $N \approx 10^5$ that have helped elucidate the main dynamical processes that occur during galactic mergers (Barnes 1988, 1992). In his chapters, Dr. Barnes explains the development of the necessary codes, his and others' efforts at including dark matter and gas in the models, and all the interesting effects and results found with such models. The ability to play with model galaxies, some with gas and others without, has increased our understanding of mergers in ways that observations alone never could have. Not only can we now explicitly determine the influence of dark halos and of gaseous dissipation, but we can also study the orbital characteristics of stars in merger remnants, distinguishing stars of one galaxy from those of the other. These are results that observers can only dream of at present. Therefore, the simulations of galactic mergers seem to have caught up with the observations and are now often leading the way toward an increased understanding.

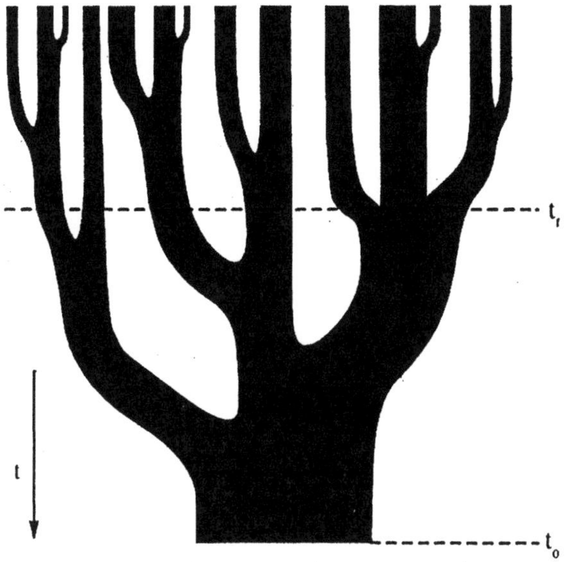

Fig. 13. "Merger tree" schematically depicting the growth of a galaxy halo as the result of a series of mergers; time increases from top to bottom. Diagram from Lacey & Cole (1993).

As if such successes were not enough, similar computational techniques can now also be applied to studies of mergers in the cosmological context of an expanding universe. Such simulations are often run with $N = 1-2 \times 10^6$ mass particles and aim at reproducing the growth of galaxies from small primordial density fluctuations through hierarchical mergers. Figure 13 shows a schematic representation of a "merger tree" depicting the growth of a galaxy halo as a function of time (Lacey & Cole 1993). The price for such ambitious simulations of cosmological evolution, of course, is a decreased resolution for individual galaxies, where details of the merger process itself are often lost. However, these simulations are providing valuable new insights into the possible merger histories of average galaxies. Thus, they complement the more detailed N-body simulations of individual encounters. By synthesizing the results of both kinds of simulations with observations to ever higher redshifts we can now hope to eventually understand the growth of galaxies and to solve the puzzle of Hubble's morphological sequence.

1.7 Some Key Questions

In lieu of a summary, let me write down some of the main questions that research on galaxy interactions and induced star formation is trying to answer:

- What is the meaning of Hubble's morphological sequence?
- What is the relative importance of Nature vs. Nurture in galaxy evolution?
- What are the main dynamical processes driving galaxy evolution?
- What is the relative importance of mere interactions, minor mergers, and major mergers in shaping galaxies of various morphological types?

- What are the star formation histories of different types of galaxies?
- What are the relative fractions of stars formed quiescently vs. in violent episodes?
- How fast do galaxies assemble?
- How has the mix of galaxy types evolved during the history of the universe?

Obviously, several of the above questions are closely interrelated. In the following chapters of the present book, most of these questions are addressed by at least one of the three lecturers, and some by two or even all three lecturers. As always in science, answers to these questions remain often tentative and incomplete. Yet, I sincerely believe that during the memorable week of lectures in late March 1996, we all learned from each other, and I hope that the seeds for new ideas and more complete answers were then planted.

1.8 General References

The literature on normal, interacting, and merging galaxies is vast. Although the following chapters cite many of the most important papers and sources, the books and articles listed below might help a reader broaden his background knowledge, gain first-hand evidence, or improve her understanding of the historical development of the subject.

For inspiration and reality checks on galaxies both undisturbed and interacting, the following three atlases present invaluable collections of photographs obtained with large ground-based telescopes:

The Hubble Atlas of Galaxies, by A. Sandage (1961). Carnegie Institution of Washington, Washington, D.C.

The Carnegie Atlas of Galaxies, by A. Sandage and J. Bedke (1994). Carnegie Institution of Washington, Washington, D.C.

Atlas of Peculiar Galaxies, by H. Arp (1966). California Institute of Technology, Pasadena; also published in: ApJS 14, 1

For a thorough introduction to galaxy dynamics, the seminal paper in the field of galaxy interactions and mergers, and a modern review with very complete references, consult and read:

Galactic Dynamics, by J. Binney and S. Tremaine (1987). Princeton University Press, Princeton (esp. Chap. 7 on "Collisions and Encounters of Stellar Systems").

Galactic Bridges and Tails, by A. Toomre and J. Toomre (1972). ApJ 178, 623.

Dynamics of Interacting Galaxies, by J.E. Barnes and L. Hernquist (1992). ARA&A 30, 705.

Finally, the following five books contain conference proceedings and course lectures from the period 1977–1994 that I have found useful in preparing the present lectures. Each volume presents a synthesis of views on galaxy structure and evolution representative for its time. Browsing through these volumes will allow the reader to assess the development of ideas and the rate of progress achieved in the field of galaxy dynamics and interactions. These volumes are:

The Evolution of Galaxies and Stellar Populations (1977). Proceedings of Yale Conference (eds. B.M. Tinsley & R.B. Larson). Yale University Observatory, New Haven.

Morphology and Dynamics of Galaxies, by J. Binney, J. Kormendy, and S.D.M. White (1982). Twelfth Advanced Course of Astrophysics at Saas-Fee (eds. L. Martinet & M. Mayor). Geneva Observatory, Sauverny.

Nearly Normal Galaxies (1987). Proceedings of Eighth Santa Cruz Summer Workshop in Astronomy and Astrophysics (ed. S.M. Faber). Springer Verlag, New York.

Dynamics and Interactions of Galaxies (1990). Proceedings of Heidelberg Conference (ed. R. Wielen). Springer Verlag, Berlin.

The Formation and Evolution of Galaxies (1994). Fifth Canary Islands Winter School of Astrophysics at Tenerife (eds. C. Muñoz-Tuñón & G. Sanchez). Cambridge University Press, Cambridge.

2 Tidal Signatures

Galactic tides are striking signatures that contain valuable information about the structure and dynamics of galaxies. For example, the lengths of tidal tails can be used to estimate the masses of the dark-matter halos that surround disk galaxies (Dubinski et al. 1996). But above all, the tides' morphology and kinematics yield strong clues about the past history and dynamics of gravitational interactions.

Before describing the various tidal signatures, a brief review of some of the terminology seems appropriate. The term *interacting galaxies* is applied mostly to *gravitationally* interacting pairs or multiplets of galaxies, although non-gravitational interactions occur also (e.g., plasma-jet-induced star formation in a neighbor galaxy, or gas stripping due to a galaxy's motion through intracluster gas). Gravitational interactions are often called "weak" or "strong" depending on the strength of the visible deformations, the inferred distance of closest approach, or the inferred relative space velocity of the interacting galaxies. The term *colliding galaxies* is normally reserved for strong gravitational interactions involving either a direct hit or a very close flyby.

The term *merging galaxies* is often used loosely ("merger"), but should really be reserved for systems showing considerable evidence of ongoing or imminent coalescence. For example, M51 is *not* a merger by this definition, since the time scale for coalescence is poorly known and coalescence may not occur until 10 Gyr in the future. The much more specific term *merger remnant* indicates that two galaxies merged into one. It implies that the merging is complete in the central region of the remnant galaxy, even though tidal signatures of various sorts may survive in the outskirts.

In the following, §§2.1–2.4 discuss the observed morphology, stellar and gaseous contents, and kinematics of galactic bridges and tails; the formation of dwarf galaxies in tails; the tidal nature of beautiful two-armed spirals; and evidence that some ring galaxies are products of galactic collisions.

2.1 Bridges and Tails: Morphology, Contents, & Kinematics

The various numerical-simulation papers of the early 1970s, and especially TT, did much to popularize the notion that galactic bridges and tails are excellent signatures of strong tidal interactions in the recent past. Observers have mostly used the striking morphology of such features as an indicator of tidal interactions. Until recently, relatively few studies were dedicated to the stellar and gaseous contents of bridges and tails. However, with the increased sensitivity of the *Very Large Array* (VLA) radio studies of the H I content of bridges and tails have become feasible and are leading to rapid advances in our understanding of both the gaseous contents and the kinematics of these tidal signatures.

Morphology

Of the two tides generated in a disk galaxy during the close passage of a companion, the one extending toward the companion is called the "bridge", while the

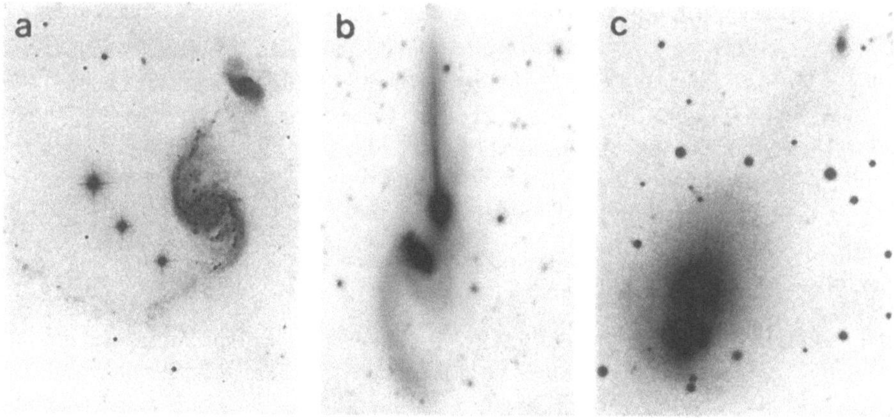

Fig. 14. Three pairs of gravitationally interacting galaxies: (a) NGC 2535/2536, (b) NGC 4676, and (c) NGC 750/751. The disk galaxies in the first two pairs produce spectacular tidal bridges and tails, while due to their higher internal velocity dispersions the two ellipticals in the third pair produce broad tidal fans. Reproduced from Arp (1966) and Hibbard (1995).

one extending in the opposite direction is called the "tail". Given the frequent display of M51 in advertisements, the tidal extension leading from its eastern spiral arm northward to NGC 5195 must be the most often seen galactic bridge, though few people would know it by this technical term (see Fig. 8).

Figure 14 shows three pairs of galaxies with bridges and tails. The two long filaments emerging from the bodies of the "Mice" (= NGC 4676) vividly justify the use of the word "tails". Notice the importance of the orientation of each tail for its appearance: The northern tail emerges from a disk seen nearly edge-on and appears straight and narrow, while the southeastern tail emerges from a less inclined disk and appears curved, broader, and of lower surface brightness. As Zwicky (1953) realized already, these aspect-dependent differences suggest that bridges and tails are relatively thin, curved sheets of tidally extracted disk material. Bridges and tails can take many forms, depending on not only the viewing geometry, but also the mass ratio of the two interacting galaxies and, especially, the internal velocity dispersion of their stars. When this velocity dispersion is high, as in the case of the pair of interacting ellipticals NGC 750/751 (Fig. 14), the tides take the form of broad, fuzzy fans of luminous matter that are hardly reminiscent of bridges and tails. Much interesting information has been extracted from these broad tidal features (see, e.g., Borne et al. 1994; Colina & Borne 1995; Combes et al. 1995), but is not discussed here for lack of space.

In exceptional cases, the morphology of bridges or tails can yield extra information beyond that discussed already. For example, in NGC 3808 (= Arp 87) the bridge from the main spiral appears wrapped around the edge-on companion at least 1.5 times, thus providing direct evidence for the occurrence of *mass transfer* from one galaxy to the other. This evidence is unambiguous because on each of its two crossings in front of the companion galaxy, the wrapped bridge material absorbs light from the companion. Hence, this material must contain dust. Another case of obvious mass transfer via a bridge occurs in AM 0327−285 (de Mello et al. 1995). Such mass transfers were already apparent in TT's simple interaction models and are now thought to be relatively frequent. When they involve significant amounts of gas, these mass transfers tend to induce intense bursts of star formation in the accreting galaxy.[7]

Stellar and Gaseous Content

The colors of bridges and tails tend to be bluer than, or similar to, the colors of the disks of the associated galaxies. This fact was well known to Zwicky and Arp, who exploited it by photographing interacting galaxies with blue-sensitive plates. Many modern studies with CCD detectors have confirmed such bluish colors and have concluded that tidal bridges and tails contain more young stars than does the average galactic disk (e.g., Schombert et al. 1990). Such color studies tend to reinvent the wheel since long-available spectroscopic observations of H II regions along bridges and tails provide stronger, direct evidence for the presence of young O and B stars there (see below). The main value of CCD photometry of bridges and tails is that it yields good estimates of the percentage of light in these features. Averaged over many interacting systems, about 25% of the total light in V seems to come from tidal extensions (Schombert et al. 1990). In the "Mice", about 1/3 of the total emission in R stems from the two long tails (Hibbard & van Gorkom 1996). Given the often low surface brightness of tidal features, such large light fractions may come as a surprise. However, these same low-surface brightness features often extend over large areas, thus explaining their significant contribution to the total light. Since in red passbands the light is a relatively good estimator of the visible mass, these observations suggest that strong gravitational interactions eject of the order of 10%−30% of the visible-disk mass in the form of tides.

Optical spectra of bridges and tails have long supported the view that most of the optical emission stems from stars (Zwicky 1958; Zwicky & Humason 1961). Such spectra mostly show continuum emission and, in rare cases of high surface-brightness features, also stellar absorption lines. In addition, these spectra often also show knotty [O II] $\lambda3727$ and/or Hα line emission indicative of giant H II regions. The "Mice" are particularly well studied and interesting: On spectra obtained with the slit aligned along the northern tail, the above two emission lines are seen along the whole tail length (Theys et al. 1972; Stockton 1974;

[7] As Arp (1969) noted, companions at the ends of "spiral arms" (read: galactic bridges) tend to have unusually high surface brightness and often show evidence for major starbursts. Although Arp speculated that these companions might be young and ejected from their associated galaxies, Gary Grasdalen – in 1972 a fellow student at the University of California in Berkeley – pointed out to me that inflows of gas into the deep potential wells of early-type companions may be the cause of these starbursts. Much recent work (see Shlosman 1994) shows that his guess was right on the mark.

Mihos et al. 1993). Moreover, strong Balmer *absorption* lines are visible in this tail, in the main body of component A, and in part of the body of component B (Stockton 1974). Such Balmer absorption-line spectra are typical of aging starbursts. Thus, the combined spectroscopic evidence suggests that the generation of tides in NGC 4676 was associated with major starbursts in each galaxy and that some residual star formation persists in H II regions along the northern tail.

In an effort to test the hypothesis that the formation of galactic bridges and tails may be associated with magnetic fields, Arp (1962) found tentative, photographic evidence for polarized light in the long bridge connecting the two spirals of Arp 295. However, an attempt in 1974 to confirm and measure this polarization with a photoelectric photometer yielded a null result to within 2%–3% (Borra & Arp 1974). This absence of polarization is consistent with the notion that stars are the dominant source of optical emission in that bridge.

Much interesting new information about the gas content of bridges and tails has recently become available because of the increased sensitivity and resolution of the VLA. The advantage of using this interferometer to observe H I in interacting galaxies is that it yields not only the gas distribution, but also the detailed gas kinematics wherever H I is found. In a sample of five interacting systems, the H I seems to extend along most of the optical bridges and tails and, occasionally, out to twice the distance of their apparent ends (Hibbard & van Gorkom 1996). Plate 1 shows the H I distributions (*blue*) of these systems superposed on optical images (*yellow–green–white*). The masses of H I found in the bridges and tails are surprisingly large, often $2\text{–}10\times10^9\,\mathcal{M}_\odot$ (for $H_0 = 75$). Presumably this abundance of gas reflects the fact that a fair fraction of the total H I in spiral galaxies resides in the outskirts of the disks, from where material is flung into tidal tails during the interaction. This gas abundance also explains why many bridges and tails form stars and appear bluish.

The five interacting systems studied by Hibbard & van Gorkom (1996) are a subset of Toomre's (1977a) proposed merger sequence and represent progressive stages of merging. In Arp 295, the two interacting galaxies are still widely separated, while in NGC 3921 and NGC 7252 they have completely merged. Interestingly, the HI distributions seem to vary systematically along this sequence. In Arp 295 the majority of the H I resides in the interacting disks, while in the two merger remnants most of the H I resides in the long tails, and the remnants' bodies are devoid of H I. These different gas distributions imply either a transformation or a redistribution of HI during the merger process (see Chap. 3).

One puzzle posed by these high-resolution H I observations remains to be solved. In several interacting systems, the optical-light and H I distributions along the tidal features appear shifted against each other by up to 1–2 kpc in projection, a highly significant amount (Hibbard & van Gorkom 1996). Apparently, the tidally extracted stars and gas have separated from each other. The cause or causes of this separation (e.g., resistance from an intergalactic or intragroup medium?) are presently not understood at all.

Kinematics

One of the advantages of observing interacting galaxies with the VLA in the 21-cm line is that H I is present along most optical bridges and tails and often extends beyond the optical limits. Therefore, the velocity fields of tidal features can be measured with better spatial coverage than was possible through optical spectroscopy of a few H II regions. Plate 2 displays color-coded ***velocity maps*** in five interacting galaxy systems observed by Hibbard & van Gorkom (1996; cf. with Plate 1). The kinematical information contained in such detailed maps yields invaluable constraints when one tries to model these interacting systems through N-body simulations. An example of a comparison between observed H I velocities and model velocities is presented in Chap. 3.

Progress in determining velocity fields of interacting galaxies has also been achieved through optical Fabry-Pérot spectroscopy. Velocity maps obtained in this manner for NGC 7753/7752 give evidence of mass transfer between the main spiral galaxy and its companion, though in which direction this transfer occurs is unclear (Marcelin et al. 1987). Similarly, new velocity maps of "The Antennae" contain many more data points than previously available (Amram et al. 1992). The advantage of such optically obtained maps is their high spatial resolution. However, their coverage along bridges and tails is not nearly as complete as that now obtainable via H I observations.

Some common characteristics to emerge from kinematical studies of many interacting galaxies are as follows. Radial velocities of tails relative to the parent galaxies are typically 100–400 km s^{-1}. Tail lengths divided by tail velocities yield estimates of the tail ejection times. In making such estimates, one has to keep in mind that the tail lengths are measured in projection against the plane of the sky, while the tail radial velocities represent the velocity component perpendicular to that plane. But in a statistical sense, dividing the projected tail lengths by the radial velocities yields a first estimate good to better than a factor of two. For example, the southern tail of the "The Antennae" (see Fig. 15) has a projected length of $15' \approx 80\,\mathrm{kpc}$ (for $H_0 = 75$) and moves with a radial velocity of about $+100\,\mathrm{km\ s^{-1}}$ relative to the main body, whence – if deceleration can be neglected – it was ejected of order 800 Myr ago.[8] Since for several similar tails the relative radial velocities are more typically 200–300 km s^{-1}, interaction ages counted since the last pericenter passage are typically somewhat shorter, of order 200–600 Myr.

One old objection to TT models of interacting galaxies was that both the simulated and observed tails appeared too thin. The argument was that, if indeed gravity takes as long to produce these tails as the TT models imply, then the velocity dispersion of the former disk stars should have widened the tails. In 1976, I checked on this criticism by taking deep photographs of "The Antennae" with the Cerro Tololo 4-m telescope (Schweizer 1978). Figure 15 displays, side by side, prints of a 50-min exposure reproduced at normal contrast and of two coadded plates totaling 3.5 hours of exposure. As these photographs show, the

[8] This estimate makes use of the well-known conversion relation: $1\,\mathrm{km\ s^{-1}} = 1.02\,\mathrm{pc\ Myr^{-1}} = 1.02\,\mathrm{kpc\ Gyr^{-1}}$.

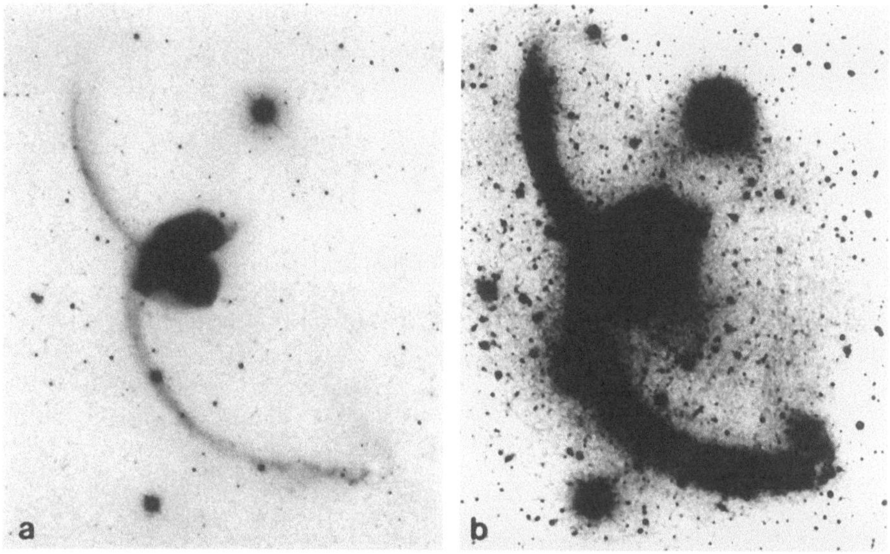

Fig. 15. NGC 4038/4039 (The Antennae) photographed with CTIO 4-m telescope. (a) Exposure of 50 minutes; and (b) superposition print of two plates totaling 3.5 hours of exposure. On the deep print, note the increased width of the tails and the dwarf stellar system near the tip of the southern tail. Photographs from Schweizer (1978).

tails do indeed appear about 3× wider on the deep image than they do on the regular print. On that deep image, the tails appear ~1/10th as wide as they are long, which implies a mean stellar velocity dispersion of about 10–20 km s^{-1}. Such a velocity dispersion is typical for disk stars, whence the criticism was defused. This question should be readdressed and could now be answered more quantitatively through CCD imaging and photometry.

2.2 Dwarf-Galaxy Formation in Tails

The fate of the tidally ejected tail material is of great interest. Zwicky (1956) was the first to propose that some of this material might survive in the form of self-gravitating dwarf galaxies. Presumably, he was struck by the knotty structure of some tails, and perhaps he even saw the low-surface-brightness patch near the tip of the southern tail of "The Antennae", shown in Fig. 15 here (cf. with his own Fig. 3 sketch).

On plates taken through an Hα interference filter, one can see four HII regions near the tip of this tail (Schweizer 1976b). Their radial velocities, measured from optical spectra, are 1690–1711 km s^{-1}, in good agreement with the 1710 km s^{-1} velocity of H I measured at the position of the optical patch (van der Hulst 1979a). This agreement suggests a physical association between this patch, which presumably is a dwarf galaxy, and the tail (Schweizer 1978). Figure 16 shows an enlarged photograph of this region, plus segments of my old spectra of two of

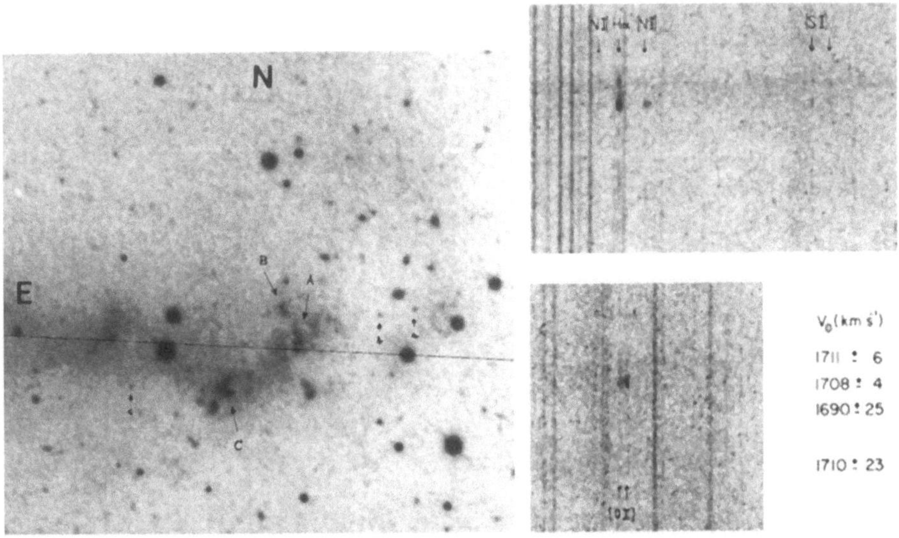

Fig. 16. (*left*) Region covering tip of tidal tail of NGC 4038 and nearby optical dwarf (enlargement from CTIO 4-m plate); capital letters A, B, and C identify giant H II regions. (Line marks slit position of spectrum not shown here.) (*right*) Image-tube spectra of H II regions A and C, around Hα at the top and around [O II]λ3727 at the bottom, slit was at PA = 124°.

the H II regions. For comparison, Fig. 17 displays a modern spectrum of the H II region C obtained by Mirabel et al. (1992). This spectrum is typical of a high-excitation H II region of relatively low O/H abundance, as often found in dwarf galaxies. The Hα luminosity of this H II region exceeds that of the entire Orion nebula by a factor of ~300. Whereas on the old photographs the optical dwarf appears separated from the tip of the tidal tail (Fig. 15), a VLA map of the H I distribution shows unmistakably that this distribution contains both the tip and the optical low-surface-brightness dwarf (Mahoney et al. 1987). This map is shown in Fig. 18. Note that the H II regions coincide with one of the two peaks of the extended H I distribution. It is this whole region that is now considered a likely dwarf galaxy in formation. With an absolute luminosity of $M_V = -14.4$, diameter of 10 kpc, H I mass of $4 \times 10^8\ \mathcal{M}_\odot$ (for $H_0 = 75$), and an oxygen abundance of $12+\log[\mathrm{O/H}] = 8.4$, this region does indeed have properties typical of dwarf galaxies of similar luminosity (Mirabel et al. 1992). Therefore, it seems not unlikely that this system may remain gravitationally bound and may, in the future, orbit the remnant of the merged NGC 4038/4039 disks as a new dwarf galaxy. Alternatively, this system may form a future *pair* of orbiting dwarfs, since the two H I peaks – and the corresponding separation between the tip of the optical tail and the low-surface-brightness patch to its Northwest – may indicate that *two* self-gravitating subsystems have formed already.

The record-holding galaxy pair in terms of tail lengths is IRAS 19254−7245, nicknamed "The Superantennae" by Mirabel et al. (1991). From tip to tip, the

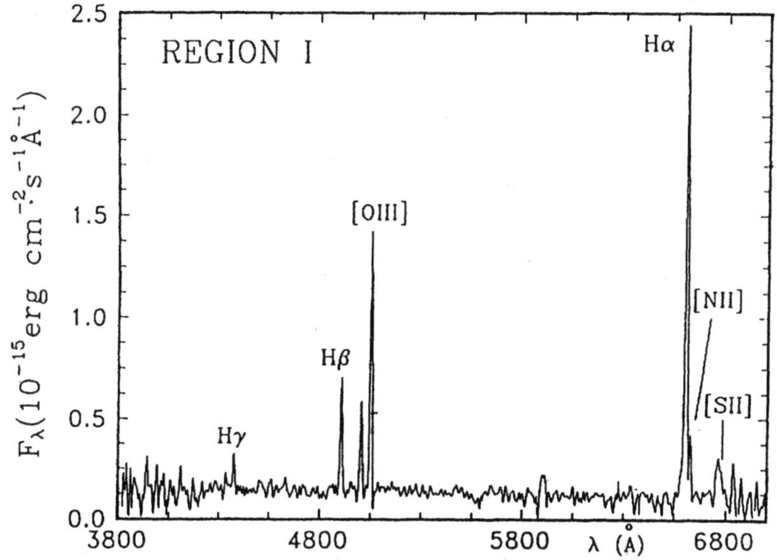

Fig. 17. Optical spectrum of H II region C in tail of NGC 4038, obtained with ESO 3.6-m telescope. Note relatively low O/H abundance typical of dwarf galaxies. Diagram from Mirabel et al. (1992), who call this H II region Region I.

tidal tails of this gigantic system extend over a projected distance of 350 kpc ($H_0 = 75$). The two interacting disk galaxies are separated by only 10 kpc in projection. One harbors a Seyfert nucleus, while the other features a starburst. Mirabel et al. list nine individual condensations in the tails. Of these, the one at the tip of the southern tail is very blue, indicating strong ongoing star formation at 200 kpc projected distance from the parent galaxy. It, too, may be a future dwarf galaxy. Further examples of candidate dwarfs at the ends of tidal tails are described by Duc & Mirabel (1994) and Duc (1995).

Recently, enthusiasm for the formation of dwarf galaxies from tidal debris has run high and may be in danger of going overboard. In an imaging study of 42 compact groups of galaxies, Hunsberger et al. (1996) identify 47 knots in 15 tails of interacting group members as candidate dwarf galaxies. They point out that *if* the majority of these "dwarfs" are gravitationally bound, then up to half of the total dwarf population in compact groups may be the product of interactions among giant parent galaxies.

However, it seems nowhere certain that most knots in tidal tails are destined to become self-gravitating dwarf galaxies. Clearly, tail fragments will orbit the remnants of mergers for a long time. The question is whether such fragments will remain cohesive enough to be recognized as dwarf galaxies, and whether their properties (e.g., dark-matter content) will resemble the properties of dwarf galaxies. Perhaps the most positive evidence so far in favor of tidal dwarf formation is the observation that the velocity dispersion of the gas tends to increase significantly in H I condensations along tidal tails (Hibbard et al. 1994; Hibbard

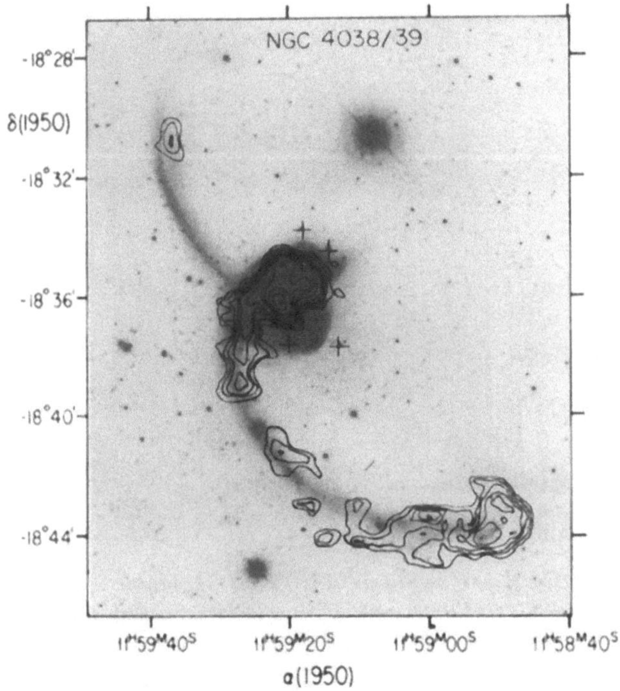

Fig. 18. Distribution of H I in "The Antennae" (contour lines), superposed on optical photograph. Note extensive mass of H I covering both the tip of the southern tail and the optical low-surface-brightness patch. Figure by Mahoney et al. (1987), kindly provided by Dr. J.M. van der Hulst.

& van Gorkom 1996). A particularly interesting aspect of tidal dwarf formation are the observed H II regions and the implied formation of young stars at large galactocentric distances. These stars form from gas not nearly as metal poor as one might expect from its remote location, since the gas originally resided in the outer disk of a major spiral galaxy. Thus, when in our own Galactic neighborhood we observe dwarf spheroidals with stellar populations of mixed age and mixed metallicity, we should keep in mind that some of these companions may have originated closer to home than their present distance would indicate.

2.3 Beautiful Two-Armed Spirals

Under the title "Certain Exceptional Spirals?" Toomre & Toomre (1972) pointed out that (1) several of the most beautiful two-armed spiral galaxies feature major nearby companions, and (2) their apparently tidal bridges and tails seem to connect smoothly onto their inner spiral structures. Since then, a large body of work has demonstrated that *swing amplification* of tidal deformations can lead to very strong, if relatively short-lived, spiral density waves (Toomre 1977c, 1981a; see Combes 1994 for a review). Therefore, even some of the prototypical Sb- and Sc-type spirals may owe their grand designs to recent tidal interactions.

Figure 19 shows three examples of galaxies with tidally excited spiral structure. NGC 7753/7752 (=Arp 86) is a close kin of M51, with kinematical evidence of tidal interaction (Marcelin et al. 1987) and a beautiful spiral structure

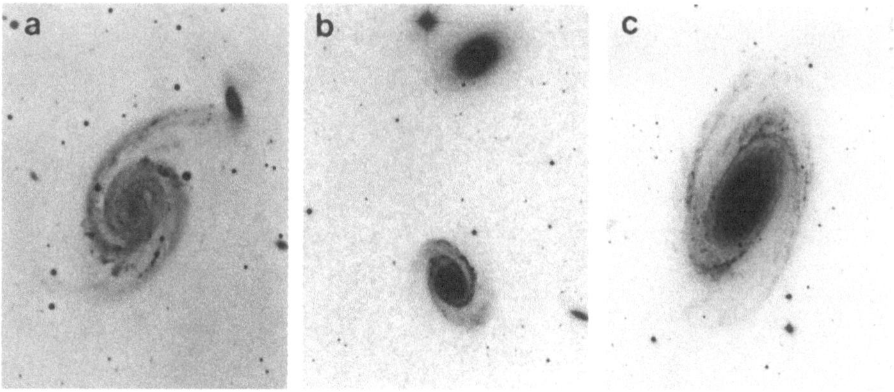

Fig. 19. Three beautiful two-armed spirals likely to have been excited by tidal interactions: (a) NGC 7753 with companion NGC 7752, (b) NGC 5364 (Sc) with NGC 5363 (E pec), and (c) M81. Photographs from Arp (1966), author, and Palomar Sky Survey.

winding from the bridge and tail inward nearly into the center. NGC 5364, described by Sandage (1961) as "one of the most regular galaxies in the sky", is an Sc galaxy with a giant elliptical companion of similar luminosity (NGC 5363) about 2.5 diameters away. The velocity difference between the Sc and E is only $+102\,\mathrm{km\,s^{-1}}$, lending credence to the Toomres' long-standing guess that the two form a physical pair and NGC 5364 owes its beautiful spiral structure to a recent close passage. Finally, M81 is yet another well-known *Hubble Atlas* spiral (type Sb) suspected by TT to have interacted with a neighbor galaxy, famous M82. Its strong spiral pattern has been studied optically (Schweizer 1976a; Elmegreen et al. 1989; Kaufman et al. 1989) and in H I (Rots 1975; van der Hulst 1979b), and has been interpreted dynamically in terms of spiral density waves (Visser 1980; Lowe et al. 1994, and refs. therein). The evidence for past tidal interactions is strong: A magnificent map of the H I distribution, pieced together from VLA observations at 12 different positions and reproduced in color on the cover of *Nature* Magazine (Yun et al. 1994), shows a long H I filament extending from M81 to NGC 3077, where it bends sharply and continues toward M82. Also, an H I bridge extends from M81 nearly to M82, containing in it the Magellanic-type dwarf galaxy Holmberg IX. Here, then, is evidence for a likely *triple* interaction between M81, M82, and NGC 3077 and for the tidal formation of both a dwarf galaxy and beautiful spiral structure in M81!

The subject of spiral-density-wave generation is complex and mostly beyond the scope of these lectures. Observationally, some of the most interesting questions are statistical in nature: (1) How frequent are such spiral-structure generating interactions (Kormendy & Norman 1979)? And (2), how much is the star-formation efficiency enhanced during these interactions? These questions are addressed at least in part in the chapters by Drs. Barnes and Kennicutt. There is one specific question which, I believe, deserves observational attention and might lend itself for a Ph.D. thesis: How does tidally excited spiral struc-

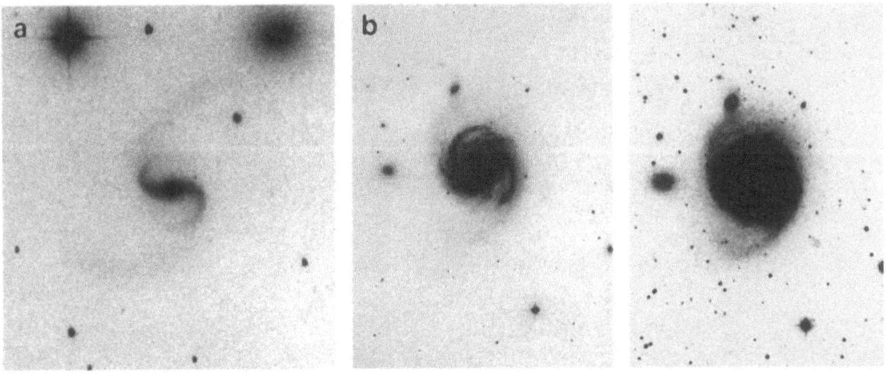

Fig. 20. Two galaxies with inner spiral structures that do not smoothly join onto outer tidal structures. (a) Arp 96, a disk galaxy that has interacted with a massive-looking elliptical. (b) Two views of M100 (=NGC 4321), an Sc galaxy in Virgo with tightly wound inner spiral arms and a more open outer spiral structure of low surface brightness. Photographs from Arp (1966) and by author.

ture evolve with time? There are galaxies where the tidal bridges & tails and the induced spiral structure can be seen co-existing, yet not joining smoothly. Such galaxies may yield observational clues about the swing-amplification process and the disk parameters that control it. Figure 20 shows two examples: Arp 96 and M100. In the outskirts of the disk galaxy of Arp 96, the bright spiral arms have a shallower pitch angle than the faint bridge and tail. One gets the impression that the disk's spiral structure has begun to decouple from the tidal structure. Similarly, the Sc galaxy M100 features relatively tightly wound inner spiral arms superposed on a more open, faint outer spiral structure. The latter appears tidal and could be due to any of the three companion galaxies or – given M100's location in the Virgo cluster – perhaps to a more massive, though more distant cluster member.

2.4 Collisional Ring Galaxies

Ring galaxies formed through nearly "bull's-eye" collisions are a rare phenomenon and may, therefore, be of minor importance in our attempts to understand the evolution of an average galaxy. Yet, they are striking signatures of tidal interaction and as such belong among the gems that Nature once in a while provides us with. At the very least, they are delightful curiosities.

Morphology

Ring galaxies have been known since Zwicky's (1941) discovery of the now famous "Cartwheel" galaxy. Vorontsov-Velyaminov (1959, 1962) and others found several more, so that Arp (1966) was able to include images of half a dozen ring galaxies in his Atlas. Figure 21 shows some examples.

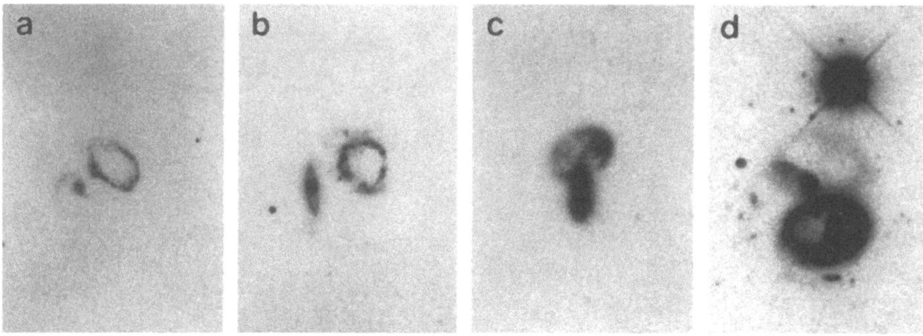

Fig. 21. Four classical ring galaxies and their companions: (a) Arp 146, (b) Arp 147, (c) Arp 148, and (d) II Hz 4. Notice the companions' rings in Arp 147 and II Hz 4. Photographs courtesy of Arp (1966) and Lynds & Toomre (1976).

There are three different kinds of galactic ring structures that have often been confused with each other. The first kind are the collisional rings to be discussed here, often called classical ring galaxies or ring galaxies for short. Some of these rings appear empty, while others feature some sort of off-centered nucleus in or near them. The second kind of ring structures occur in normal and barred spiral galaxies. These structures – often described as nuclear, inner, or outer rings – are thought to be associated with Lindblad and other resonances and are not directly related to galactic collisions. Finally, the third kind of ring structures are the "polar rings" around some disk and especially S0 galaxies. These polar rings are discussed in Chap. 5.

For several decades, ring galaxies were a complete puzzle. Any self-gravitating ring-shaped system of stars should form clumps and quickly lose its ring symmetry. A first helpful hint emerged when Cannon, Lloyd, & Penston (1970) noticed that many ring galaxies have nearby companions and often look as if their nucleus and disk had separated. But it was a graduate student at Columbia University, John Theys (1973), who first pointed out a crucial detail: Nearly without exception, when there are companions near a ring galaxy, at least one of these companions lies within 25° of the apparent minor axis of the ring. It is this fact that gave Alar Toomre and Theys & Spiegel (1976) the idea that companions may be lying near the *true* ring axis, though out of the ring plane, and may have punched through a former disk galaxy, thus producing the ring.

Kinematics and Dynamics

Spectrograms of ring galaxies obtained with the spectrograph slit oriented at different position angles across the ring show that these galaxies do not only rotate, but also either contract or expand (Theys & Spiegel 1976, 1977; Fosbury & Hawarden 1977). The distinction between contraction and expansion is difficult observationally since it is rarely clear which side of the ring lies nearer

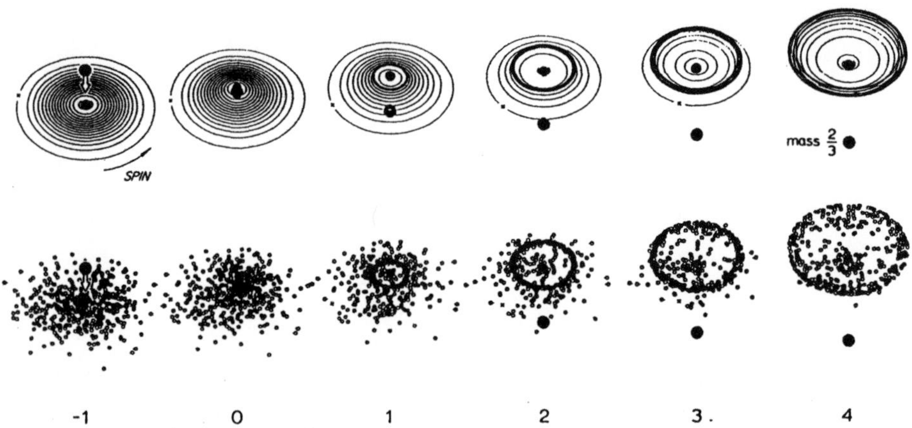

Fig. 22. Model simulation of ring-galaxy formation. A Gaussian-disk galaxy suffers an axial penetration by a more pointlike companion of 2/3 its mass. Top row shows evolution of a disk model consisting of a central point mass and 16 rings, while bottom row shows evolution of an equivalent model with mass particles. Diagram from Lynds & Toomre (1976).

to us. Figure 22 shows a model simulation of ring-galaxy formation that reproduces both the observed rotation and expansion, if expansion it is (Lynds & Toomre 1976). In this simulation by Toomre, a Gaussian-disk galaxy experiences an axial, "bull's-eye" collision with a compact companion of two thirds its own mass. The companion free-falls along the axis, punches through the main disk, and continues its path on the other side. For a brief period of time, the central force felt by the disk is nearly doubled because of the intruder's added mass. As this force diminishes after the central passage, the disk material that had been pulled inward rebounds with a vengeance, forming an outward-running density wave. According to this model, ring galaxies represent tidal signatures generated during strongly interpenetrating collisions.

The clinching piece of evidence validating this proposed formation mechanism stems not from any observed kinematics, but from a discovery made by Roger Lynds. On very deep exposures of the ring galaxy II Hz 4, Lynds found that the apparently compact companion near the minor axis of the bright ring is itself surrounded by a faint, off-centered ring (see Fig. 21). Obviously, if *two disk galaxies* interpenetrated each other along a nearly common disk axis, then *each* should develop a ring-like density wave in response. Hence the binary ring system! This discovery clinched the argument in favor of Toomre's model, despite the lack of measured expansion for either of the two rings. Perhaps because of this 1976 model's obvious success, relatively little theoretical work has been done on ring galaxies since. The puzzle has, in essence, been solved.

In retrospect, one can see that the companion galaxy of Arp 147 also sports a ring of its own (Fig. 21). However, because this second ring is seen nearly edge-on its true nature was not recognized immediately.

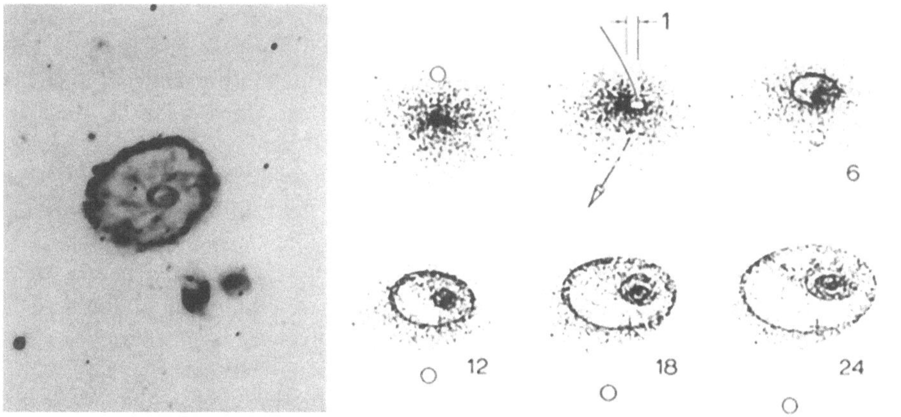

Fig. 23. "Cartwheel" galaxy with off-centered inner ring and spokes, compared to a collisional model by Toomre (1978). The inner ring represents disk material on its second rebound after one of the companion galaxies punched through the former disk galaxy. Photograph courtesy of Dr. V.M. Blanco.

Stellar and Gaseous Contents

With the realization that some ring galaxies feature major starbursts along their perimeter, interest in their stellar and gaseous content has recently revived. Perhaps the best-studied such starburst occurs in the well-known "Cartwheel" galaxy, first discussed by Zwicky (1941).

Figure 23 shows a deep photograph of the "Cartwheel" obtained by Victor Blanco and an off-center collisional model by Toomre. The unusual features of the "Cartwheel" are a second, inner ring and a set of spiral-shaped "spokes" that seem to link this ring to the outer main ring. The latter has a diameter of 44 kpc ($H_0 = 75$), rotates at ~250 km s^{-1}, and expands with ~90 km s^{-1}(Fosbury & Hawarden 1977). The diameter and expansion velocity date the ring-producing collision to about 300 Myr ago, although which of three companion galaxies was the culprit remains unclear (see Higdon 1996). The ionized gas in the main ring is metal deficient by a factor of 3–20 when compared with the Orion nebula, suggesting that the victim's disk gas was relatively unprocessed. As Fosbury & Hawarden point out, it takes a starburst with about 10^6 OB stars to produce the observed amount of ionization in this gas. Note that Toomre's (1978) simple model is impressive in its ability to reproduce both the inner ring and its off-center location relative to the main ring. In this model, a half-mass companion has passed parabolically at ~15% from the center through the disk of a galaxy of unit radius. The inner ring consists of disturbed disk material on its second rebound.

As new model calculations and observations demonstrate, the Cartwheel and related ring galaxies still hold some surprises for us in store. A modern model of this galaxy, based on N-body simulations and including gas and hydrodynamics, explains the mysterious spiral "spokes" in terms of instabilities in the gas behind the expanding density wave (Hernquist & Weil 1993). The spirality is due to shearing of the rotating disk material. Surprisingly, images taken with a ground-based telescope through an interference filter show that all the Hα emission stems from the outer ring and none from the spokes or inner ring (Higdon 1995). This suggests that the inner ring consists only of stars. But why do the supposedly gaseous spokes emit mainly continuum radiation? High-resolution *HST* images of the "Cartwheel" may yield an answer (Borne et al. 1996). Besides providing evidence of ongoing star and cluster formation in the spectacular starburst of the ring, these images suggest the existence of a strong *radial age gradient.* The spokes appear to consist of expanding stellar associations and feature several small rings or bubbles indicative of *past* star formation and/or supernova explosions. The inner ring does indeed seem to consist of a smoother stellar population, which may be dominated by slightly older stars now ascending the giant branch. This ring also shows dust lanes and comet-shaped structures that my trace gaseous shocks due to gas falling back at supersonic velocities of order 100 km s^{-1} (Struck et al. 1996). Finally, the very center of the "Cartwheel" seems to consist of a lens-like core with a nucleus surrounded by bluish clusters.

Similar age progressions may have been detected in several other ring galaxies from the ground, though none with the clarity afforded by the *HST* observations of the "Cartwheel". In the most youthful rings, star formation rates are typically an order of magnitude higher than in normal spiral galaxies of the same mass (Appleton & Struck-Marcell 1987; Marston & Appleton 1995). In the "Cartwheel" itself, the inferred star-formation rate is an extraordinary 67 $\mathcal{M}_{\odot}$ yr^{-1} (Higdon 1995). Ample molecular gas to fuel such starbursts has been detected in over a dozen ring galaxies (Higdon et al. 1995; Horellou et al. 1995). Presumably, some of this molecular gas formed recently from H I through compression in the density wave.

Seyfert nuclei have been discovered in several ring galaxies, some embedded in the rings themselves (e.g., de Vaucouleurs & de Vaucouleurs 1975; Wakamatsu & Nishida 1987). Such active galactic nuclei suggest that the tidally induced commotion in the gas of these former disk galaxies can deliver some of the necessary fuel to the nuclei at relatively high rates. For a detailed review of the above and other aspects of ring galaxies, see Appleton & Struck-Marcell (1996).

Ring galaxies still present at least one major puzzle: What are these severely impacted galaxies eventually turning into? Do they evolve into an earlier, more gas-poor Hubble type, moving from, say, a pre-collision Sc to an Sb, Sa, or even S0? Or do some of them eventually absorb their intruder and become peculiar, taking for example the form of polar rings around central S0 galaxies (see Chap. 5)?

2.5 Summary

The main results of this chapter are as follows:

- Galaxies involved in strong tidal interactions tend to form bridges and tails, the morphology and kinematics of which can yield valuable clues about the galaxies' structure and encounter geometry. In interacting *disk* galaxies, these tides take the form of relatively thin, curved sheets of extracted disk material.
- Galactic bridges and tails contain stars and often surprisingly large amounts of H I gas. Star formation in them reveals itself through giant H II regions and bluish *UBVR* colors.
- In tails rich in H I, the gas tends to clump and may occasionally lead to the formation of some self-gravitating dwarf galaxies.
- Beautiful two-armed spirals often show signs of having been excited tidally. Examples are M51, M81, M100, NGC 5364, and NGC 7753.
- Rare, ring-shaped galaxies appear to have suffered recent, nearly bull's-eye collisions with relatively compact companions. These collisional ring galaxies can be explained as expanding density waves in former disks. Their stellar populations are often dominated by starbursts and can display strong age gradients across the rings.
- Among puzzles remaining to be solved are (1) the cause of the observed displacements between gas and stars in some bridges and tails, and (2) the unknown nature of the descendants of collisional ring galaxies.

3 Recent Merger Remnants

Theory suggests that tidal friction leads to orbit decay, which in turn leads to merging and violent relaxation. Hence, the challenge for observers has been to identify galaxies that are merger remnants. As TT realized first, tidal tails are signatures uniquely suited to identify the remnants of mergers involving major disk galaxies.

In preparation of a search for such remnants, simple considerations suggest that remnants likely to be identifiable as such should be of the order of 10^9 yr old. This is because the typical rotation period for a major disk galaxy is ~250 Myr, or call it one galactic year, and final orbit decay must occur on a time scale of a few rotation periods. If only 10^8 yr or ~0.4 galactic years have elapsed since the encounter of two disk galaxies initially following parabolic orbits, any possible merger must clearly still be in progress. Thus, detailed measurements of, say, the velocity field in such a system yield at best a snapshot of a confused and rapidly evolving configuration. After 10^9 yr or ~4 galactic years, on the other hand, any imminent merger should have progressed to near completion, yet tidal signatures will hopefully still be strong and recognizable. Hence, one might expect such merger remnants to be relatively easy to find and profitable to study. Finally, though many galaxies may have experienced mergers of bits and pieces early in their history, after 10^{10} yr or ~40 galactic years the most easily recognizable tidal signatures are likely to have decayed. Hence, identifying merger remnants of that age unambiguously must be quite difficult.

The present chapter discusses two recent merger remnants, NGC 7252 and NGC 3921, in considerable detail. Both remnants have many properties in common with elliptical galaxies and are likely 0.5–1 Gyr old protoellipticals. Thus, they form a crucial link in the chain of evidence supporting the formation of ellipticals through major mergers. The chapter ends with a brief description of some properties of even younger merger remnants.

3.1 A Rosetta Stone: NGC 7252

The galaxy NGC 7252 is now widely considered to be a prototypical merger remnant. Because of its chaotic looking body and two long tails, TT had included it among five galaxies likely to be "at a fairly advanced stage of a merger", and based in part on my early observational results Toomre (1977a) made it the last of his famous merger sequence. Since then, various observational studies of this remnant have revealed some of the complex processes occurring during galactic mergers: the violent relaxation of the stellar body, a galaxy-wide aging starburst, the building of a counterrotating core at the center, the prolonged return of tidally ejected tail material, and the formation of a new subsystem of globular clusters. Hence, NGC 7252 has played the role of a Rosetta stone in helping us decipher the many connections between disk mergers and elliptical galaxies.

The present section §3.1 describes the merger signatures of this galaxy, the properties of its main body, and the evidence for gas falling back from the tidal

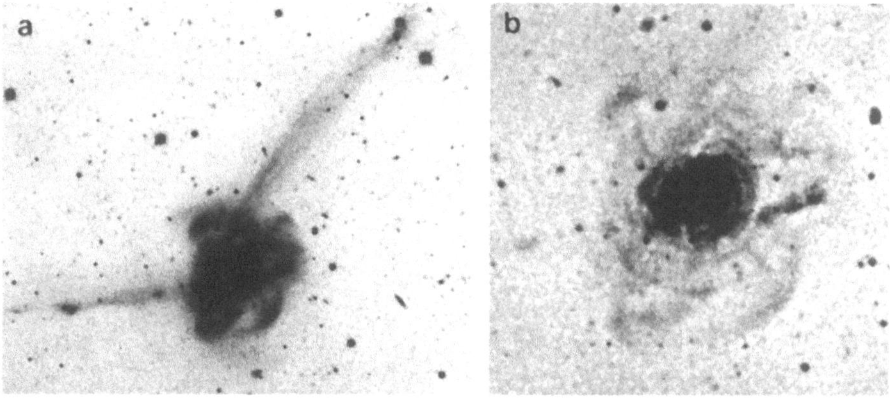

Fig. 24. (a) Blue-light photograph of NGC 7252 and (b) digitally processed and masked image of the main body. Note the two long tidal tails and the rich fine structure within the main body. The NW tail connects to the SSE loop and the E tail probably to the W loop. Photograph and image from Schweizer (1982, 1990).

tails. The connections between the fine structure of NGC 7252 and that of many ellipticals is discussed in Chap. 6, and the genesis of young globular clusters within this galaxy in Chap. 7.

Signatures of a Merger Remnant

What characteristics should one look for when trying to identify remnants of recently merged pairs of disk galaxies?

The following five characteristics may be somewhat too restrictive, but make – if all present in one single galaxy – a strong case in favor of a recent merger remnant (Schweizer 1978): (1) A *pair* of long tidal tails is the safest indicator of two participants. (2) Ideally, the candidate remnant should be a relatively isolated galaxy to exclude the possibility of tidal damage from neighbors. (3) In a merger remnant, as opposed to an ongoing merger, there should be a single nucleus. (4) Relative to this nucleus, the two tails should move in opposite directions since tidal-interaction models tell us that tails originate from the far sides of the two merging disks (TT). And (5), motions in the main body might still be rather chaotic if the merger took place recently.

NGC 7252 is regarded as a prototypical merger remnant because it displays all five of these characteristics (Schweizer 1982). First, its two tails shown in Fig. 24 are long indeed: their *projected* lengths amount to $280'' \approx 130\,\mathrm{kpc}$ for the northwest tail and $170'' \approx 80\,\mathrm{kpc}$ for the eastern tail ($H_0 = 50$). These tails and the associated loops suggest that not so long ago there were two disk galaxies that interacted tidally. Second, NGC 7252 is also unusually isolated. There is no physical companion galaxy of comparable brightness within a surrounding sphere of $\gtrsim 2.4\,\mathrm{Mpc}$ radius. Thus, the chance that the two "tails" could be a tidal bridge and tail created by the recent close passage of a companion galaxy

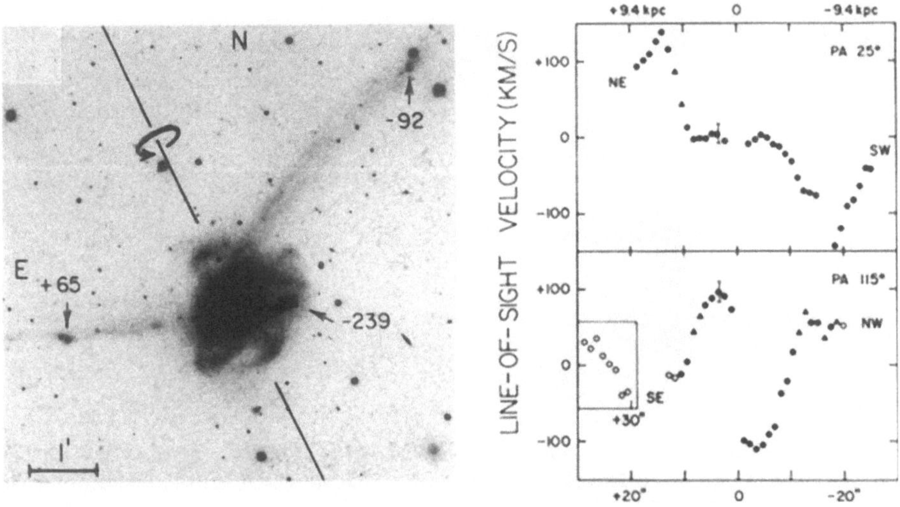

Fig. 25. Observed line-of-sight velocities of ionized gas in NGC 7252, in km s^{-1} relative to center. (*left*) Rotation axis of central ionized-gas disk and velocities of giant H II regions near the tips of the tails and in western loop. (*right*) Velocities measured with slit aligned along major axis (*bottom right*) and minor axis (*top right*) of central gas disk. Note the velocity reversals and near discontinuities. Figures from Schweizer (1982).

is essentially zero.[9] Third, NGC 7252 indeed features a single main body with a light distribution centered on a single nucleus (Fig. 24), whence the central regions of the two component galaxies must have merged completely.

Fourth, Fig. 25 illustrates that the ends of the two tidal tails do indeed move in opposite directions, as predicted by TT for crossed tails. Relative to the center of the galaxy, the giant H II region near the tip of the NW tail approaches us with $-92\,\mathrm{km\ s^{-1}}$, while that near the tip of the E tail recedes from us with $+65\,\mathrm{km\ s^{-1}}$. Division of the projected tail lengths by these relative velocities yields estimated kinematic ages of 1.0 Gyr and 0.9 Gyr, respectively, since tail ejection began ($H_0 = 50$). Because of projection uncertainties, these ages are uncertain by about a factor of two, leading to an estimated merger age of $1.0^{+1.0}_{-0.5}$ Gyr. This merger age is reckoned since the last close encounter that led to both the formation of the tidal tails and the merger.

And fifth, the motions of at least the ionized gas in the main body of NGC 7252 are indeed surprisingly complex. Near the center, there is a disk of ionized gas that shows a very regular rotation pattern. This disk extends out to a radius of $8'' \approx 4\,\mathrm{kpc}$ and rotates around the axis sketched in the left part of Fig. 25. However, beyond this central disk, the ionized gas shows dramatically

[9] As Hibbard et al. (1994) point out, the fact that the two tails and their connecting loops appear to lie in quite different planes also argues against their representing tides from a single, disturbed disk.

different, chaotic velocities. When one aligns the spectrograph slit with the apparent *major* axis of the gas disk, the measured radial velocities reverse direction beyond the disk edge (Fig. 25, bottom right panel). When one aligns it with the apparent minor axis, the velocities suddenly increase from near zero within the disk to roughly $\pm 140\,\mathrm{km\ s^{-1}}$ beyond its edge (Fig. 25, top right panel). Thus, the central gas disk appears to *counterrotate* relative to the surrounding ionized gas, which itself moves in a more chaotic manner and around a different axis. This kinematic signature, when discovered in 1976, was quite unexpected and points strongly toward two surviving motion systems. In hindsight, of course, it is easy to think that one should have expected to find two motion systems in the remnant of two recently merged disk galaxies. At present, the motions of the outer ionized gas have still not been fully mapped and are poorly understood. They may indicate either mainly rotation or perhaps the presence of major shock regions, where gas falling back from the tails hits gas distributed throughout the main body of the remnant.

Uncertain details notwithstanding, the above five characteristics of NGC 7252 leave little room for doubt that this galaxy is the remnant of two former disk galaxies that merged beginning about 1 Gyr ago.

Properties of the Main Body

On photographic plates and CCD images of NGC 7252, the main body appears as a relatively smooth and symmetric light distribution with faint superposed loops and arcs. The rich fine structure can be rendered more visible through various enhancement techniques. On some published photographs (e.g., Arp 1966; Schweizer 1982; and Fig. 24b here), the enhancement has been achieved through electronic or manual dodging during the enlargement process. In images processed by computer, a similar enhancement can be achieved by subtracting a certain percentage of the mean light distribution. This technique, which imitates photographic unsharp masking (Mees & James 1966; Malin 1977), reduces large-scale contrast while preserving small-scale contrast. The image shown in the right panel of Fig. 24 was produced by digitizing three IIIa-J plates obtained with the CTIO 4-m telescope, coadding them in the computer, and subtracting 90% of the background density after smoothing the image with running median and Gaussian filters (Schweizer & Ford 1986). Besides the loops protruding from the main body, this masked image also shows various fine ripples and the NW tail crossing in front of the remnant before connecting to the SSE loop. Given the orbital motions of the loop and ripple material, it seems likely that this fine structure will mix with the general light distribution and become nearly invisible over the next few Gyr.

Despite this surviving fine structure, the mean light distribution of NGC 7252 already resembles that of elliptical galaxies to a surprising degree. Figure 26 shows radial brightness profiles at low and high spatial resolution. The low-resolution profile to the left was obtained from multi-aperture photometry by differentiation and shows that the mean surface brightness in V closely follows a $r^{1/4}$ law typical of E galaxies (de Vaucouleurs 1953; Kormendy 1977). This law is

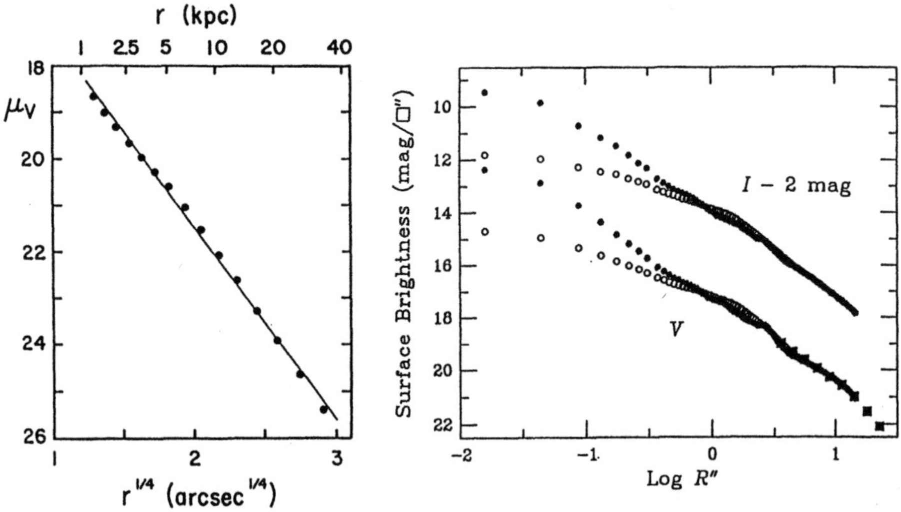

Fig. 26. Radial brightness profiles of NGC 7252. (*left*) The mean V-surface-brightness profile observed from the ground closely follows an $r^{1/4}$-law over 7 mag in brightness. (*right*) The central V- and I-brightness profiles measured with *HST* follow a power law of index $\gamma \approx -1.3$ down to the resolution limit at $r \approx 0\farcs05$ (*open circles:* observed, *filled dots:* deconvolved). Figures from Schweizer (1982) and Whitmore et al. (1993).

characteristic of stellar systems that have *violently relaxed* (Lynden-Bell 1967), which include remnants of merged galaxies of near-equal mass (Barnes 1988, 1992; Hernquist 1992). In NGC 7252, the $r^{1/4}$ law seems to hold approximately over a range of 7 mag, or a factor of 600 in intensity, and has been confirmed through CCD photometry in the R passband (Hibbard et al. 1994). The effective or half-light radius of this light distribution is $r_e = 14'' \approx 7\,\mathrm{kpc}$ ($H_0 = 50$).

Near the center, the light distribution does not flatten out, but keeps rising right into the nucleus. The high-resolution brightness profile displayed to the right in Fig. 26 was obtained with the *Hubble Space Telescope* and shows an unresolved, single-peaked light distribution down to the resolution limit of $\sim 0.05''$ (Whitmore et al. 1993). This projected light distribution is well described by a power law of index $\gamma \approx -1.3$ and resembles the light distributions observed at the centers of many giant ellipticals. Within the central unresolved region ($r < 25\,\mathrm{pc}$), the surface brightness reaches a value of at least $V_0 = 12.37$ mag arcsec^{-2}, implying a mean volume luminosity density in excess of $4.3 \times 10^3\,L_\odot$ pc^{-3}. Again, this central luminosity density falls well within the range of 10^1–$10^7\,L_\odot\,\mathrm{pc}^{-3}$ observed among ellipticals.

Interestingly, the light emitted by the main body of NGC 7252 is dominated by *young*, A-type main-sequence stars. Figure 27 shows two spectra obtained with small apertures centered on the nucleus and at a radius of $14''$($\sim$7 kpc). Both spectra display strong Balmer absorption lines, indicating that the young stars occur not only at the center of the galaxy, but also at a distance of 7 kpc

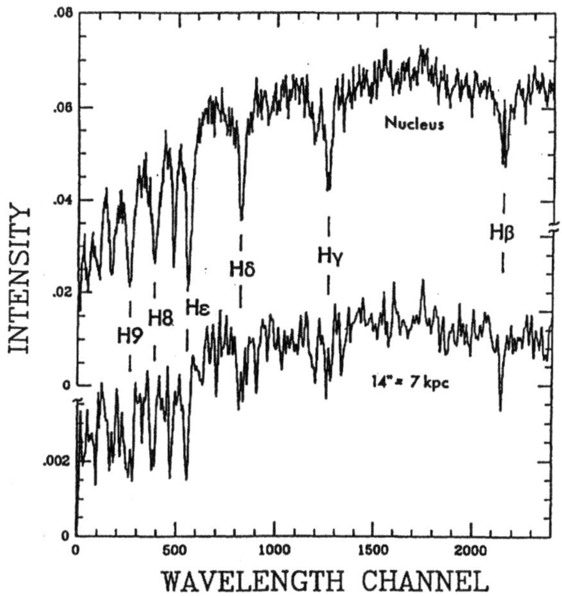

Fig. 27. Spectra of NGC 7252 at the nucleus and at $r = 14''$ (7 kpc), obtained with Las Campanas 2.5-m telescope. Spectral range covers rest wavelengths $\lambda\lambda$ 3720–5010. Note strong Balmer absorption lines indicative of young main-sequence stars. Mean spectral type is about A7 V. Diagram from Schweizer (1990).

from it in the main body. The mean spectral type is A7 V, suggesting that the main-sequence turnoff of the young population occurs around that spectral type. Interpreting the nuclear spectrum in terms of a merger-induced starburst, Fritze-von Alvensleben & Gerhard (1994) find that (1) two pretty massive, gas-rich Sc galaxies must have been involved, (2) the burst began about 1.3–2 Gyr ago, and (3) this burst formed about 20%–50% (by mass) of the present stars. Given that the gas fraction in Sc galaxies typically is $\lesssim$30% of the luminous mass, the upper limit of the quoted mass range seems high. Yet, there can be little question that a very major burst of star formation took place during the merger that led to NGC 7252. The fact that the young stars appear to be well mixed throughout the main body suggests that much of this starburst occurred in the two precursor galaxies early during the merger, whence the freshly formed stars participated in the ensuing violent relaxation.

Within the central gas disk, star formation continues to the present time. In addition to ionized gas this disk contains about $8 \times 10^9 \mathcal{M}_\odot$ of *molecular* gas (Dupraz et al. 1990; converted to $H_0 = 50$), an ample reservoir of fuel for continued star formation. Spatial and velocity maps of the CO emission obtained with the Owens Valley Radio Interferometer show that the CO disk is cospatial with the HII disk and rotates like it (Wang et al. 1992). Hence, the molecular gas, too, rotates in the opposite direction from the outer ionized gas. Like the ionized-gas disk, the CO disk seems to feature a small hole at its center ($r \approx 300$ pc).

Hubble Space Telescope images of the region covered by this gas disk reveal beautiful spiral structure, highlighted by OB associations and young star clusters (Whitmore et al. 1993; Miller et al. 1997). Figure 72 in Chap. 7 shows an image of this spiral structure, which extends out to 3–4 kpc radius. The presence of a "minispiral" at the center of a merger remnant supposed to turn into an elliptical galaxy poses a challenge: How can spiral structure be excited that deep in the potential well, when gas disks are known to be highly dissipative? A source of continued excitation seems necessary. As Toomre (1990) pointed out for normal spiral galaxies, one good way of keeping spiral structure excited in shearing gas disks is through prolonged gaseous infall. Interestingly, just such a prolonged infall of gas has been discovered in NGC 7252, as discussed below. The gas infall rate seems to be comparable to the star-formation rate, suggesting that it is tail gas that keeps fueling star formation at the center. Therefore, in NGC 7252 we seem to be witnessing the merger-induced building of an oddly rotating *stellar* core similar to those observed in giant ellipticals (Schweizer 1990).

Tail Gas: What Goes Up Must Come Down

The VLA observations of the H I distribution in NGC 7252 yield several interesting results (Hibbard et al. 1994).

First, all the observed H I is associated with the tidal tails and the W loop. As Plate 1 shows, H I is virtually absent from the relaxed main body, creating the impression of a huge central "hole". The present atomic gas content within this region is $<2 \times 10^8 \mathcal{M}_\odot$ (for $H_0 = 50$), a value typical of upper limits established for ellipticals (Knapp et al. 1985). The "gap" between the H I in the tidal outskirts and the central H_2 disk is filled with ionized gas. Two possible sources of ionization in this extensive "gap" region are the intense radiation field due to the young stars (especially those within the central gas disk) and shocks that colliding gas clouds on intersecting orbits must experience.

Second, the large amount of $8 \times 10^9 \mathcal{M}_\odot$ of H I in the tails suggests that two gas-rich galaxies merged. In fact, when taking into account that the present tail material represents less than half of the total material originally ejected into the tails (see below), one can infer that the two merging galaxies must both have been major Sc spirals. Hence, the merger type of NGC 7252 was Sc–Sc.

Third, the tail gas shows several lumps and clumps, especially near the tips of the optical tails. There, major H I clumps are associated with the giant H II regions. These clumps show a distinct increase in the internal velocity dispersion, are likely self-gravitating, and provide yet another example of the possible formation of dwarf galaxies in tidal tails (see §2.2).

Fourth, the H I velocities measured along the tails clearly show that toward the center *tail gas is falling back into the main body.* The upper part of Fig. 28 displays the distribution of HI with velocity contours superposed, while the lower part of the figure displays a profile of radial velocity vs. projected position along an east–west line. The upper graph illustrates the superiority of H I velocity mapping over optical measurements (cf. Fig. 25): detailed radial velocities are now available along the whole length of both tails. These velocities show that over most of its length the NW tail is swinging toward us, while near the center – where this tail clearly lies between us and the main body – it is moving *away* from us and is, therefore, falling into the remnant. Similarly, the E tail generally swings away from us, but – if indeed it connects behind the main body

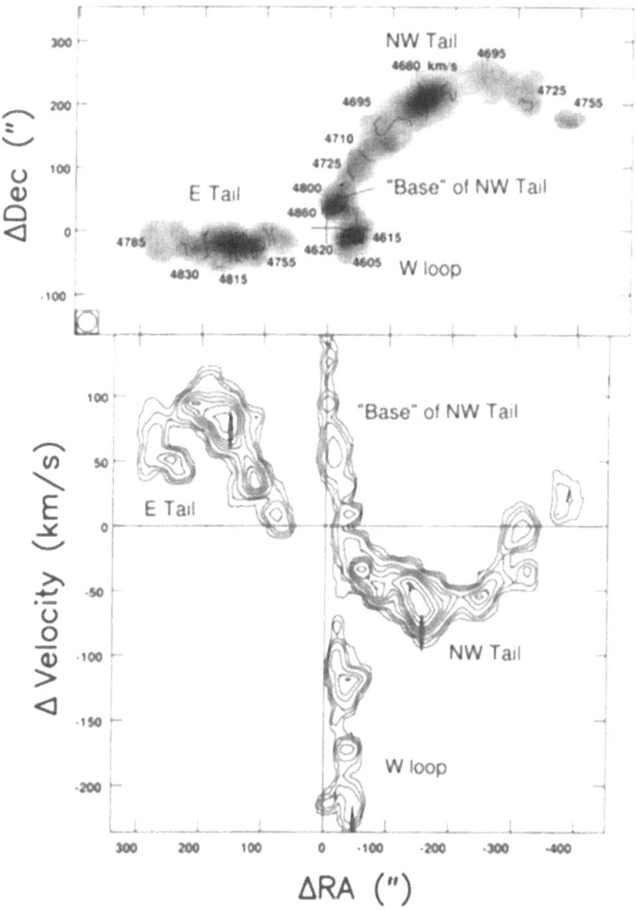

Fig. 28. Kinematics of H I in NGC 7252. (*top*) Map of H I distribution with superposed velocity contours, and (*bottom*) radial velocities of H I plotted vs. right ascension. Note velocity reversal near the base of the NW tail, indicative of gas falling in. Diagrams from Hibbard et al. (1994).

to the W loop – it reverses direction in its central part, moves toward us, and again is seen falling into the remnant. This falling back of tail material was predicted by Barnes (1988, 1992) based on his model simulations of disk–disk mergers that include massive dark halos. According to these simulations (see also chapters by Dr. Barnes), there is at any given time a "turn-around radius" within which tidally ejected material should be falling back into the remnant. As time progresses, this turn-around radius moves outward. The above H I observations provide brilliant observational confirmation of Barnes's model prediction.

Detailed velocity maps along tidal features yield powerful constraints for model simulations of past mergers. For example, the above H I velocities in NGC 7252 invalidate a first merger model proposed for this galaxy by Borne & Richstone (1991). In that model, the velocity reversal at the edge of the central gas disk was reproduced by postulating that the two disk galaxies rotated retrograde relative to their orbits. Whereas the model reproduces both the observed morphology and the central velocity reversal reasonably well, it fails to reproduce the newly observed tail velocities. This example demonstrates that the

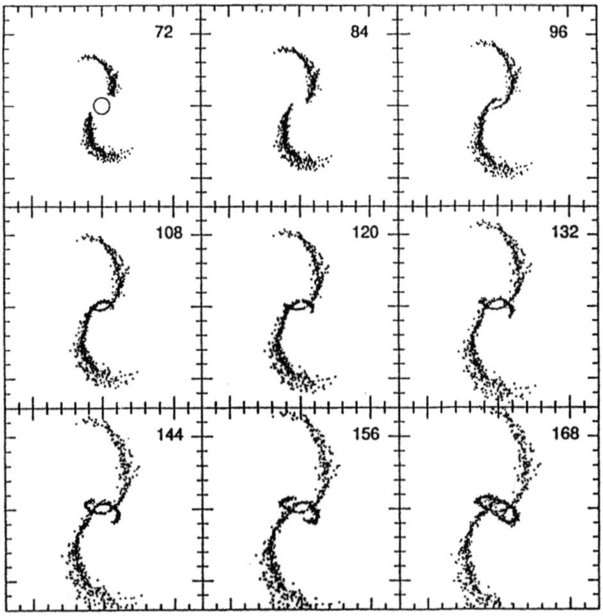

Fig. 29. Model of loop formation in NGC 7252. Tail particles are plotted as seen from above orbital plane at nine different epochs, ranging from moment of merger to ~1.5 Gyr later. Note how returning tail particles form loops. Figure courtesy of Hibbard & Mihos (1995).

morphology alone provides insufficient constraints for making a reliable estimate of the merger geometry that leads to a specific remnant.

A second, more successful merger model of NGC 7252 has been computed by Hibbard & Mihos (1995) through a series of N-body simulations ($N \lesssim 64,000$). Experiments with different relative disk orientations demonstrate that both disks have to rotate in a *prograde* manner to reproduce the observed tail motions. The final model reproduces these motions and the infall of material in considerable detail. In addition, this model also yields interesting estimates of the return rate of H I as a function of time. At present, this return rate is about 4–8 $\mathcal{M}_{\odot}\,\mathrm{yr}^{-1}$ ($H_0 = 50$), while in the past it was several times larger. The model also predicts that about one half of the remaining HI will return to within $5\,r_e$ during the next 5 Gyr. Figure 29 illustrates how this might happen. Tail particles are plotted as seen from above the orbital plane at 9 different times, beginning with the time of the central merger and ending about 1.5 Gyr later. The small circles in the first and last frames have a radius of $4\,r_e$. Note how tail particles falling back toward the center form loops as time progresses. These loops resemble the two main loops observed in NGC 7252. Thus, loop formation seems to be a direct manifestation of cold material returning from the tidal tails to the main body. As discussed in Chap. 4, the same mechanism may explain the formation of ripples ("shells") in elliptical galaxies.

3.2 The Dynamically Young Remnant NGC 3921

The galaxy NGC 3921 (= Arp 224 = Mrk 430) is another candidate merger proposed by TT because of its two long tails. In the mid-1970s, I took image-tube plates of it in the near infrared and found a single nucleus, whence it became a good candidate for being a recent merger remnant. However, two things then made life difficult. First, outside its nucleus NGC 3921 showed barely any emission lines, making it difficult to measure and map velocities. And second, this candidate remnant appeared to be in a small group of galaxies and failed, therefore, to fulfill the isolation criterion mentioned in §3.1. I especially worried about an apparent companion not far from the end of the northern tail. In retrospect I need not have worried: That apparent companion is now known to be the dominant member of the galaxy cluster Abell 1400 in the distant background.

Merger Remnant in a Group

Based on measured radial velocities, at least three of the galaxies surrounding NGC 3921 are likely physical companions (Schweizer 1996). Hence, NGC 3921 lives in a small tight group not unlike a Hickson group, where astronomers have long suspected that mergers are relatively frequent. Figure 30 shows two images of the galaxy. One is a deep CCD exposure emphasizing the two long, crossed tidal tails, while the other is a masked enlargement emphasizing the rich fine structure of the main body. The deep exposure shows the nearest of the three likely physical companions, an anonymous galaxy, to the west of the remnant. Given our improved understanding of mergers based on N-body simulations (Barnes & Hernquist 1992), the relatively crowded group environment of NGC 3921 can hardly cast any serious doubt anymore on the fact that with its two crossed tails, but single main body, this galaxy must be the remnant of a merger between two disk galaxies of nearly equal mass.

The main body of NGC 3921 shows much interesting fine structure. Masked images like the one displayed in Fig. 30b reveal loops of material, various ripples in the light distribution, and a number of faint plumes. Two peculiarities stand out. First, the main body appears quite asymmetric. Especially the nucleus is strikingly offset from the apparent center of the general light distribution. This offset was already noted by Arp (1966) and Zwicky & Herzog (1966), who called the nucleus "eccentric".[10] And second, there appears to be a faint "crown" of material protruding from the luminous envelope of the galaxy to the north. This "crown" consists of a half dozen plumes or streamers that appear to diverge radially away from a strong ripple in the light distribution. Presumably, these plumes represent some splatter of stars torn from one or both of the tidally interacting galaxies. This at present mysterious signature extends over ~17 kpc in the east–west direction and should be of use in future attempts at modeling this past merger through N-body simulations.

[10] Hence, it is tempting to call NGC 3921 an eccentric in a group...

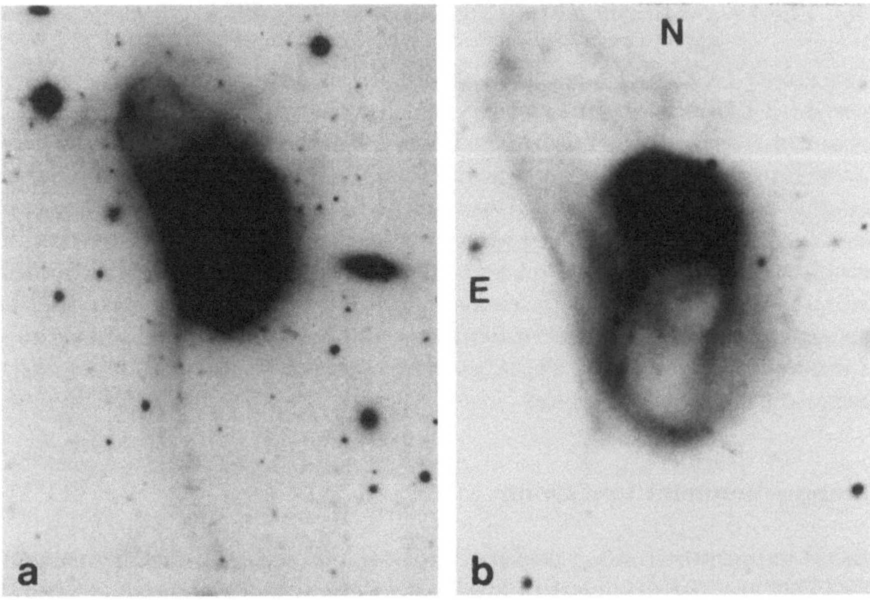

Fig. 30. Images of merger remnant NGC 3921. (a) Deep image obtained with Palomar 5-m telescope, showing the two crossed tails. (b) Same image enlarged and digitally masked to show loops and ripples in main body; north is up (from Schweizer 1996).

Signs of Youth: Sloshing Isophotes, Aging Starburst

The eccentric position of the nucleus of NGC 3921 is only the most visible sign of a strangely distorted main body. Figure 31 shows, side by side, an isophotal map of this body and a graph tracing the wandering of the centers of successive isophotes on the plane of the sky. Note the "sloshing" behavior of successive isophotes, whose centers first shift to the southeast of the nucleus and then, in a major way, to the north, west, and finally south. Note that the scale of the right panel is 5× enlarged relative to that of the left panel. In their wandering over the sky, the isophote centers shift by >2 kpc from the nucleus. The maximum offset of an isophote center is ~23% of the isophotal radius for an isophote that contains about 45% of the total light. Clearly, the luminous matter making up the main body of NGC 3921 is still in disequilibrium, indicating that this remnant is dynamically young.

Interestingly, despite this obvious disequilibrium the azimuthally averaged, *mean* radial light distribution is already well represented by a $r^{1/4}$ law (Stanford & Bushouse 1991; Schweizer 1996, esp. Fig. 10); there is no trace of an exponential component left. Hence, in NGC 3921 as in NGC 7252 the violent relaxation – though obviously incomplete – has been effective in redistributing the luminous matter of the two former disks. The mean light distribution has an effective radius of $r_e = 14'' \approx 8\,\text{kpc}$, which falls right in the middle of effective radii measured in ellipticals of the same luminosity. As in many ellipticals and in NGC 7252, the central light distribution remains unresolved even with *HST*.

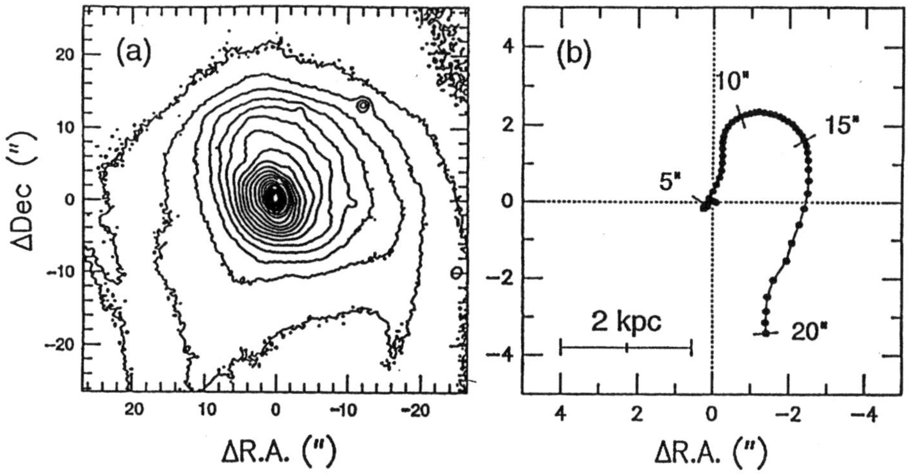

Fig. 31. (*left*) Central isophotes of NGC 3921 in blue-green light, and (*right*) wandering of isophote centers relative to position of nucleus. Crossbars mark center location for isophotes of semi-major axes $a = 5''$, $10''$, $15''$, and $20''$. Note "sloshing" isophotes and isophote-center excursions of up to $\sim$2 kpc from nucleus (from Schweizer 1996).

The blue-light spectrum of the central region ($r \lesssim 4''$) features the strong Balmer absorption lines typical of aging starbursts (see chapters by Dr. Kennicutt). Although one can use sophisticated modeling to analyze such spectra (e.g., Fritze-von Alvensleben & Gerhard 1994), I wish to emphasize that at least in the near ultraviolet such spectra are *not* very composite. Some people mistake the presence of the K line of Ca II in addition to the Balmer absorption lines for a sign of spectral compositeness. Yet, the K line appears in the spectra of all A stars, where the ratio between it and the flanking lines H8 and Hϵ+H is used as a criterion to determine the spectral subtype of the stars. In the case of NGC 3921, this line ratio points toward A3 V–A5 V stars being the dominant source of near-UV light. These stars presumably populate the main-sequence turnoff of the aging starburst population. Stars of this spectral type have masses of about 2.0–2.5 $\mathcal{M}_\odot$ and main-sequence lifetimes of 0.5–1.0 Gyr (Schaller et al. 1992). This suggests that the merger-induced starburst peaked about 0.5–1.0 Gyr ago, in rough agreement with estimates of the merger age based on the main-body asymmetry and on dynamical time scales.

Due to this starburst population the mean color of NGC 3921 [$(B–V)_e = 0.75$] is bluer than the typical color of ellipticals. Yet, the radial color *gradients* measured in *UBVI* within the main body fall well within the range of such gradients observed in ellipticals (e.g., Peletier et al. 1990; Goudfrooij et al. 1994). Thus, *if* the stellar populations within the main body are well mixed and feature no strong radial age gradient, the colors of the main body will grow redder without significantly changing the color gradients, and within a few Gyr NGC 3921 will resemble a normal elliptical in both its colors and radial color gradients.

Tail Gas: An Sa–Sc Merger?

The H I distribution of NGC 3921 displays yet another new and striking phenomenon (Hibbard 1995; Hibbard & van Gorkom 1996). As Plate 1 shows, only one of the two tidal tails has H I associated with it: the S tail is very H I rich, while the NE tail shows no detected H I. The tentative interpretation of this dichotomy is that a gas-rich galaxy collided and merged with a gas-poor galaxy. According to this interpretation, NGC 3921 is a Sa–Sc, or perhaps even a S0–Sc, merger. If this interpretation is correct, the eccentric and mildly active nucleus may represent the center of the former bulge of the Sa or S0 galaxy. This interpretation would agree with models suggesting that bulges in merging disk galaxies retard the onset of star formation, leading to late but strong starbursts (Mihos & Hernquist 1994, 1996). However, the same interpretation also creates a puzzle. As *HST* observations of NGC 3921 show, the gas-free NE tail contains some young stellar associations and clusters (Schweizer et al. 1996). How could these clusters have formed without the presence of considerable amounts of gas?

As in the tails of NGC 7252, the radial velocities of HI measured in the S tail of NGC 3921 show a strong reversal near the "base" of the tail. However, this velocity reversal cannot be interpreted unambiguously in terms of gaseous infall near the center, because at present we do not know whether the S tail passes in front or behind the main body. If the gas does fall in near the tail base, then the sudden drop in H I surface density there suggests that atomic H I gets rapidly converted into another, either ionized or perhaps molecular, gas phase through some agent (e.g., radiation field, shocks).

3.3 NGC 3921 & NGC 7252 as Likely Protoellipticals

Except for being demonstrably younger than most low-redshift giant ellipticals, NGC 3921 and NGC 7252 share many properties with these ellipticals and among themselves. Both merger remnants feature a luminous main body ($-22.5 \gtrsim M_V \gtrsim -23$) with a single nucleus. Over the next 5–10 Gyr, these main bodies will fade by about 1–1.5 mag and will have absolute magnitudes comparable to those of high-luminosity ellipticals. Their mean radial light distributions are well approximated by a $r^{1/4}$ law over $\sim$7 mag, testifying to the efficiency with which violent relaxation redistributes luminous matter during mergers. Of course, the remnant bodies share this form of light distribution with all ellipticals. These bodies' *central* light distributions resemble those of ellipticals as well. They have power-law shapes and reach central luminosity densities in excess of $10^3 L_\odot \text{ pc}^{-3}$. The radial *UBVI* color gradients within the main bodies are similar to those observed in ellipticals, even though the colors themselves are bluer. Finally, the bodies of both NGC 3921 and NGC 7252 show rich fine structure reminiscent of the weaker fine structure observed in many ellipticals (see Chap. 4). In these two remnants, the mechanism generating much of the fine structure is known: streams of former disk material are falling back from the tidal tails at a diminishing rate and provide a long-lasting source of dynamically cold matter

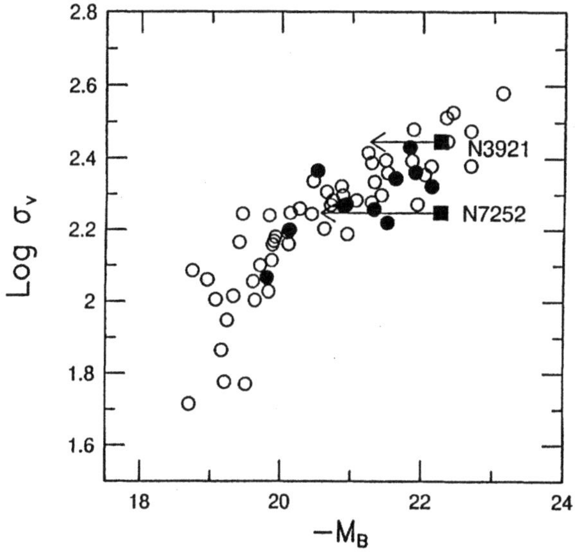

Fig. 32. Luminosity–velocity-dispersion relation for 13 merger remnants (*filled dots* and *squares*) and 53 elliptical galaxies (*open circles*). Fading vectors are drawn for NGC 3921 and NGC 7252. Diagram after Lake & Dressler (1986).

to form sharp-edged features like ripples and shells. In short, NGC 3921 and NGC 7252 very much look like ~1 Gyr old protoellipticals.

Little is presently known about the detailed *stellar* kinematics of these two merger remnants. However, their central velocity dispersions have been measured and have removed one of the main objections raised early on by critics of the merger hypothesis of E formation. The objection was that the stellar velocity dispersion in merger remnants would not agree with that in ellipticals of similar luminosity. Simple theoretical estimates based on the assumption of homologous mergers seemed to suggest that a doubling in luminosity should *not* be accompanied by any significant increase in the velocity dispersion σ_v (Ostriker 1980). If so, remnants of merged disk galaxies would fall off the L–σ_v relation (Faber & Jackson 1976) valid for elliptical galaxies. Yet, Lake & Dressler (1986) checked this simplistic prediction observationally and found it to be wrong. The velocity dispersions of candidate merger remnants do, in fact, fall right onto the L–σ_v relation observed for elliptical galaxies.

Figure 32 shows the velocity dispersions of merger remnants and ellipticals measured by Lake & Dressler and plotted versus blue absolute magnitude M_B. Among the 13 measured merger remnants, two are from the Toomre (1977a) sequence: NGC 3921 and NGC 7252. The figure suggests that the merger remnants, especially when corrected for the predicted fading due to their aging starbursts, fall on the same L–σ_v relation as the elliptical galaxies. Hence, at least in their central stellar kinematics, merger remnants in general – and NGC 3921 and NGC 7252 specifically – do mimic giant elliptical galaxies. It remains for future studies to test whether the stellar kinematics further out in the bodies of merger remnants agrees with that in ellipticals as well. As a first step, Lake & Dressler (1986) measured stellar rotation in three of their remnants and found one slow, one moderate, and one fast rotator.

Table 1. Differences Between NGC 3921 and NGC 7252

	NGC 3921	NGC 7252
Merger age	0.7 ± 0.3 Gyr	$1.0^{+1.0}_{-0.5}$ Gyr
Hubble types of components	Sa–Sc (or S0–Sc)	Sc–Sc
Isophotes of main body	"Sloshing"	Symmetric
Nucleus	LINER	Quiescent
Brightest globular clusters	$M_V = -12$ to -14	$M_V = -14$ to -17

So far, we have emphasized properties common to NGC 3921, NGC 7252, and giant ellipticals. However, *differences* between individual merger remnants may eventually teach us something about differences among elliptical galaxies. Parameters that could influence the structure and stellar-population content of remnants are the merger age, the merger geometry, and the structure and gas content of the component galaxies. Table 1 summarizes the most important differences found so far between NGC 3921 and NGC 7252. The obvious asymmetries of the remnant body of NGC 3921 may be due to a younger merger age. The LINER-type nucleus of this remnant (Stauffer 1982a, b) may also owe its mild activity to a younger merger age, or perhaps to the bulge of the presumed Sa (or S0) component having delayed the onset of the starburst, as explained in §3.2. Finally, the lesser luminosity of the most luminous young globular clusters in NGC 3921, when compared with those in NGC 7252, may reflect a less violent starburst due to the presence of less gas (see §7.3). Presumably, a Sa–Sc merger cannot quite match the fireworks of a Sc–Sc merger. To improve on such speculation, models detailing the merger geometry of NGC 3921 and including gas dynamics for both NGC 7252 and NGC 3921 are needed. But the message emerging already from these observations is that in addition to the merger geometry the *Hubble types* and *gas contents* of merging galaxies are likely to imprint the structure and stellar populations of the remnants.

3.4 Even Younger Merger Remnants

The sky survey carried out by the *IRAS* satellite (launched in 1983) has brought to the fore a class of galaxies that seem to be in transition between ongoing mergers and young merger remnants. Optically, these galaxies are often quite inconspicuous, yet in the infrared they shine nearly as bright as quasars. Many of them emit >90% of their bolometric luminosity, and some up to 99%, in the far infrared ($\lambda = 8$–1000μm).

Perhaps the two most famous such galaxies are Arp 220 and NGC 6240. Neither was included in Toomre's (1977a) sequence, yet both seem to clearly be pairs of merging galaxies. Their infrared luminosities are enormous ($7 \times 10^{11}\ L_\odot$ and $1.5 \times 10^{12}\ L_\odot$, respectively), giving these galaxies "superstar" status. The

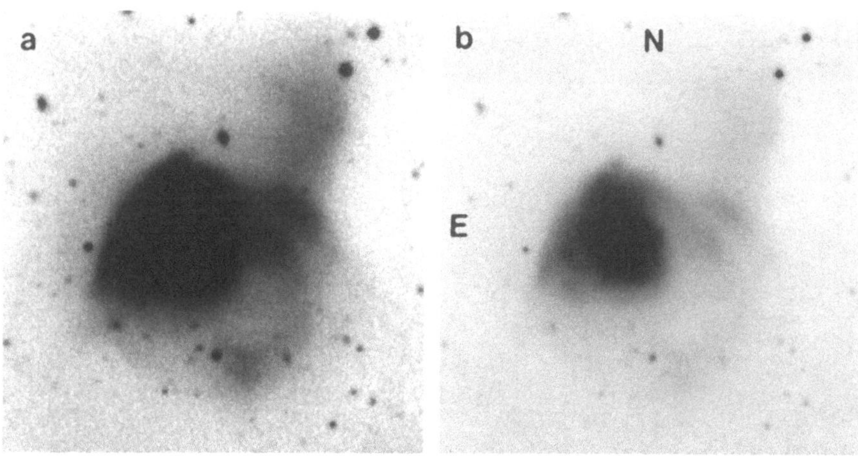

Fig. 33. Image of Arp 220 obtained by Dr. P. Seitzer with Kitt Peak 4-m telescope; (a) high-contrast and (b) low-contrast display.

sources of much of this luminosity seem to be dust-enshrouded starbursts and, in the case of Arp 220, perhaps also an active nucleus (Sanders et al. 1991). For a discussion of the fascinating phenomena that occur in the gas of these and related galaxies, consult the chapters written by Dr. Kennicutt. Here, I wish only to give a few details relating to the merger status of these galaxies, and especially to the kinematics of their *stars*. I find two things most remarkable: First, although dozens of detailed studies have been published for each object we still do not know what types of galaxies collided with each other in either case. And second, despite this embarrassing ignorance, both objects show signs of turning into protoellipticals.

Figure 33 shows a CCD image of Arp 220 reproduced at high and low contrast. On both photographs (and on the one in Arp's Atlas), the sharp boundary of the light distribution to the northeast is striking and reminiscent of tidally limited material (like the SW arm of M51) or of a bent disk seen nearly edge-on. On the high-contrast photograph (Fig. 33a) various appendices are visible, of which the one protruding to the northwest is most prominent. But, are these appendices tidal tails or are they just some splatter of stars? If there are any tails, is there just one or are there two? As these unanswered questions illustrate, we do not know at present what kinds of galaxies collided in Arp 220 and what the merger geometry was.

Figure 34 displays two photographs of NGC 6240 published by Fosbury & Wall (1979), both made from the same plate. The masked photograph of the main body (Fig. 34b) shows messy wisps, some of which are dominated by line emission. The deep print (Fig. 34a) seems to show two faint tails, though there is disagreement on whether these luminous filaments are of tidal origin. So far, no "tail" velocities have been measured. Hence, although NGC 6240 does look like the possible remnant of a disk–disk collision, we will not know for certain

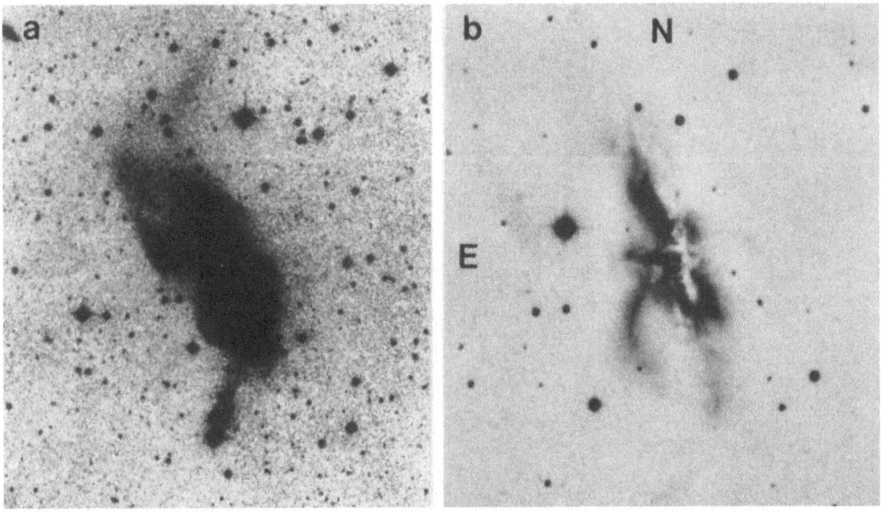

Fig. 34. Photographs of NGC 6240. (a) Deep print of IIIa-J plate obtained with ESO 3.6-m telescope, showing two faint outer filaments (tidal tails?). (b) Masked photograph of main body showing dust lanes and luminous wisps (enlarged 2×). Photographs from Fosbury & Wall (1979).

until somebody manages to measure velocities in the tails, presumably from H I observed with the VLA.

Modern infrared-array detectors are now permitting studies of Arp 220 and NGC 6240 that pierce through the optically obscuring dust. Table 2 summarizes some of the results. Very close double nuclei have been found in both objects (Arp 220: Norris 1988, Graham et al. 1990; NGC 6240: Fried & Schulz 1983), suggesting that these are former pairs of galaxies whose central regions are just about to complete merging. In Arp 220 the velocity difference between the two

Table 2. Some Properties of Arp 220 and NGC 6240

	Arp 220	NGC 6240
Redshift $cz_{\rm hel}$ (km s^{-1})	5453 ± 30	7473 ± 30
Projected separation of nuclei[a]	$0.''95 = 330$ pc	$1.''8 = 850$ pc
Δv between nuclei (km s^{-1})	~200	$\lesssim 100$
Central σ_v (km s^{-1})	150 ± 21	360 ± 20
$r^{1/4}$-type light distribution	Yes	Yes

[a] For $H_0 = 75$.

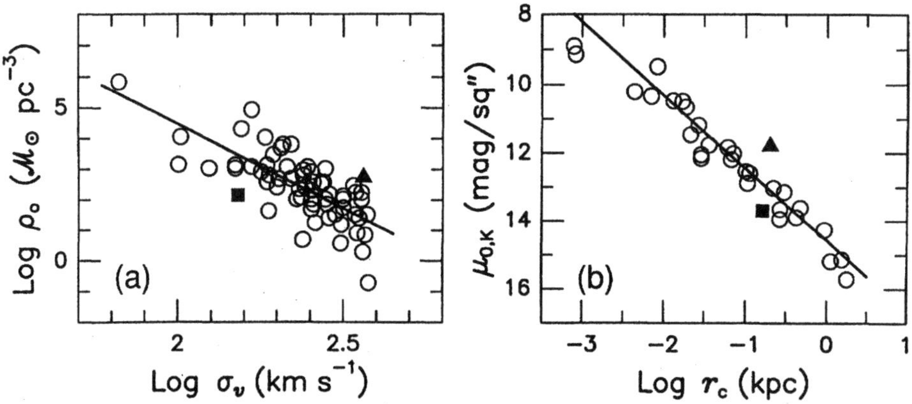

Fig. 35. Positions of Arp 220 (*square*) and NGC 6240 (*triangle*) in (a) mass-density vs. velocity-dispersion diagram and (b) central K-band surface-brightness vs. core-radius diagram for ellipticals and bulges (*circles*). Diagrams after Doyon et al. (1994).

nuclei seems to be of the order of 200 km s^{-1}, though the measurement is based on the Paβ and Brγ emission lines and may, therefore, not accurately reflect the relative motions of the stars (Larkin et al. 1995). In NGC 6240 the velocity difference between the two nuclei seems to be less than 100 km s^{-1} (Lester & Gaffney 1994). In both galaxies, the low relative velocities of the double nuclei agree with the notion of very late-stage mergers. Mean central velocity dispersions of the stars have been measured from the CO band at 2.3 μm due to red giants or supergiants (Doyon et al. 1994; Lester & Gaffney 1994). These velocity dispersions, in combination with infrared surface photometry (Wright et al. 1990), suggest that both Arp 220 and NGC 6240 have central mass densities, velocity dispersions, and surface brightnesses consistent with their lying close to the fundamental plane of elliptical galaxies (see Fig. 35). Hence, violent relaxation and gaseous dissipation seem to have accomplished most of their work even in these late-stage mergers or nascent merger remnants.

3.5 Summary

The main results of this chapter are as follows:

- Major mergers of nearly equal-mass disk galaxies do occur at the present time and produce elliptical-like remnants (e.g., NGC 3921 and NGC 7252).
- When involving at least one galaxy with copious gas, such mergers tend to trigger major bursts of star formation (>5%–10% by mass).
- Besides the merger geometry, the Hubble types and gas contents of the merging galaxies may determine the structure of the remnants and may imprint their stellar populations.
- During a period of several Gyr after a major merger, the prolonged return of stars and gas from the tidal tails may generate fine structure such as loops and ripples ("shells") and help build an oddly rotating core.

4 Early-Type Galaxies: Isophotal Shapes and Fine Structure

This is the first of three chapters devoted to the structure of E and S0 galaxies and to observational clues concerning their origins. The present chapter describes what has been learned about the structure and possible merger formation of these galaxies from analyses of their isophotal shapes and fine structure. The following Chap. 5 then deals with some of the kinematic signatures found in both early-type galaxies and merger remnants, and Chap. 6 presents an attempt to synthesize the observations into an evolutionary scenario.

Until the mid 1970s, elliptical galaxies were thought to be uniformly old, rotationally supported, and structurally simple. The discovery of their slow rotation (Bertola & Capaccioli 1975; Illingworth 1977) made a first serious dent in this picture. Following this discovery, studies with new photographic emulsions, CCD detectors, and increasingly sophisticated image-analysis software have demonstrated that E and S0 galaxies are, in fact, quite complex. It is their structural and kinematic complexities that offer us hope to unravel their formation history.

4.1 Structure of Ellipticals from Isophotal Analysis: Triaxiality

Figure 36 displays an isophotal map of the elliptical galaxy NGC 1549 and several profiles derived from isophotal analysis (Jedrzejewski 1987b). The isophotal map of this elliptical is representative for the whole class. It shows that the isophotes are (1) close to being mathematical ellipses, (2) nearly perfectly concentric, and (3) slightly twisted against each other, i.e., the position angle of the apparent major axis changes as a function of radius. From these three properties, one can immediately derive important insights into the structure of elliptical galaxies. First, for the isophotes to appear elliptical in projection, the bodies of ellipticals must be ellipsoidal (oblate, prolate, or triaxial). Second, the high degree of isophote concentricity indicates that, no matter through what process these galaxies formed (mergers or not), the central equilibration must have been fast or all ellipticals are very old. In either case, there seem to be no long-lived vibrational modes of any major amplitude.[11] And third, the major-axis twists indicate triaxiality, as we shall see in a moment.

Before describing the method of isophotal analysis, let us make a brief visit to its hall of fame. The idea of using Fourier analysis to study isophotal shapes goes back at least to Kibblewhite (1969). The first to apply the Fourier technique extensively to elliptical galaxies was Carter (1978). Based on this technique, Kent (1983), Lauer (1983, 1985a), and Jedrzejewski (1987a) wrote software packages that have been used by many astronomers since. Lauer (1985b) recognized that the isophotes of many ellipticals show small, but systematic deviations from perfect ellipses and coined the terms "boxy" and "disky" to describe their shapes.

[11] Think of the strongly eccentric nucleus of the recent merger remnant NGC 3921 and how fast it may, or may not, reach center position.

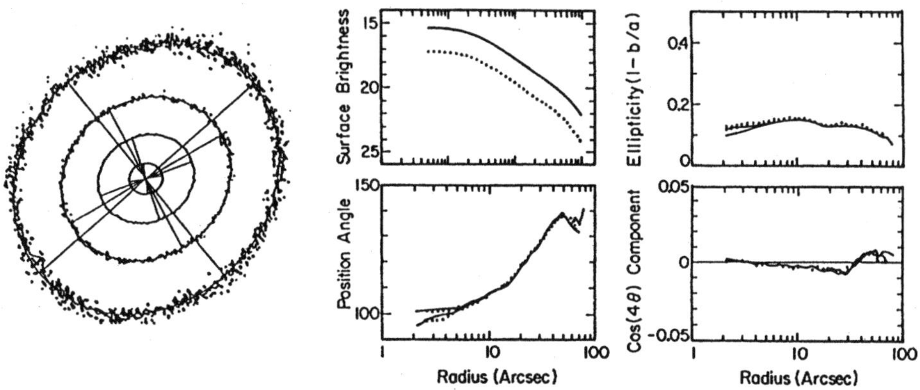

Fig. 36. Isophotes of NGC 1549 (type E2) and radial profiles of surface brightness, major-axis position angle, ellipticity, and cos(4θ) coefficient derived from isophotal analysis. Long and short lines superposed on isophotes represent major and minor axes. Diagrams from Jedrzejewski (1987b).

Among the many dozens of studies done since this early period are several important Ph.D. theses, including those by Franx (1988), Peletier (1989), Jørgensen (1993), and Goudfrooij (1994). An outstanding review of the main results obtained by isophotal analysis until the mid 1980s is given by Jedrzejewski (1987b).

In essence, the method of isophotal analysis via Fourier series works as follows. Given a set of isophotes, one typically first fits to each isophote an ellipse of the form

$$I(\phi) = I_0 + C_1 \cos\phi + S_1 \sin\phi + C_2 \cos 2\phi + S_2 \sin 2\phi\,, \qquad (1)$$

where I is the surface brightness and ϕ is an azimuthal angle measured at the center of the galaxy. In practice, an isophote is generated from a CCD image of an elliptical galaxy by taking a thin slice in surface brightness: one selects all picture elements ("pixels") that have counts within a narrow range. As the above-referenced papers by Kent, Lauer, and Jedrzejewski explain in more detail, an iterative procedure is then used to obtain the best fit. The Fourier coefficients C_1,S_1 yield corrections ΔX_c,ΔY_c to the adopted center coordinates X_c,Y_c of a preliminary ellipse, while the coefficients C_2 and S_2 yield corrections $\Delta\epsilon$ to its ellipticity ϵ and ΔP to its position angle P, respectively. After iterating a few times, this procedure yields the surface brightness $I_0(a)$, center coordinates $X_c(a)$ and $Y_c(a)$, ellipticity $\epsilon(a)$, and position angle $P(a)$ of the best-fitting ellipses as functions of the semi-major axis a.[12] Because the isophotes of most ellipticals are nearly exactly concentric, the procedure is often simplified by determining the center of the light distribution first and then keeping the centers

[12] Note that, since the components of Fourier series are orthogonal, errors in the various measured quantities are independent to first order. This means that the measured surface brightnesses are independent of remaining small errors in the center coordinates, ellipticities, and position angles of the isophotes.

of the fitted ellipses fixed at this location. The first three framed plots in Fig. 36 show the profiles $I_0(a)$, $\epsilon(a)$, and $P(a)$ obtained in this manner for NGC 1549.[13]

Once this first step of the isophotal analysis has been completed, higher-order Fourier coefficients are computed in a second step. The complete Fourier series for the surface brightness along a certain path is

$$I(\phi) = I_0 + \sum_{i=1}^{n} \{C_i \cos(i\phi) + S_i \sin(i\phi)\}\,, \tag{2}$$

where values of $n = 4$ to $n = 6$ are commonly used. The paths chosen to compute the coefficients C_3,S_3, C_4,S_4, ... are the best-fit ellipses obtained in the first step. Although this procedure may appear a bit complicated to the novice, in practice the various software packages available for the task save the user from having to deal with details. They take him or her by the hand, ask that a pointer be placed on the center of the light distribution, compute the centroid and the Fourier coefficients as described, and then print out a list of $I_0(a)$, $X_c(a)$, $Y_c(a)$, $\epsilon(a)$, $P(a)$, and $C_i(a)$,$S_i(a)$ $(3 \le i \le n)$ as functions of a. The fourth framed plot in Fig. 36 shows the radial profile of the coefficient $C_4(a)$ obtained in this manner for NGC 1549.

Isophotal analyses of hundreds of ellipticals have yielded many interesting results. These results are here presented according to the highest-order Fourier coefficients necessary for their derivation. The remainder of this section discusses results obtained by simply fitting ellipses (i.e., derived from C_i,S_i for $i \le 2$), while the next section (§4.2) deals with results obtained from the third- and fourth-order coefficients.

First and foremost, isophotal analyses confirm the visual impression from maps that the isophotes of E galaxies are very well approximated by ellipses (e.g., Fig. 36). The deviations of the isophotes from mathematical ellipses are typically smaller than 1% in radius. As stated already, this isophotal ellipticity implies that E galaxies have ellipsoidal bodies, but does not by itself discriminate between spheroidal and triaxial shapes.

Second, in nearly all ellipticals the isophotes are indeed concentric to a very high degree. The centers of individual isophotes wander from the mean center by typically less than 1% of the isophotal semi-major axis, especially after effects of superposed fine structure (e.g., ripples) are taken into account. This implies either fast central equilibration or a great age for all ellipticals.

Third, isophotal analyses confirm that a de Vaucouleurs (1953) law of the form $\log I_0(a) \propto a^{1/4}$ is a good first approximation to most elliptical light distributions over typically 5–7 mag in surface brightness.[14] As Lynden-Bell (1967)

---

[13] As is common practice in isophotal analysis, the intensity coefficient C_4 has been transformed into a *spatial* coefficient measuring the radial deviations of the isophotes from a perfect ellipse and expressed as a fraction of the semimajor axis; for details, see Jedrzejewski (1987a) or Lauer (1985b).

[14] The approximation seems best for ellipticals of $M_V \approx -22$. Less and more luminous ellipticals are better fitted by laws with exponents deviating from 1/4, often written as $1/n$ (Schombert 1987; Kormendy & Djorgovski 1989; Caon et al. 1993).

argued, this kind of light distribution is symptomatic of *violent relaxation*. However, this does not necessarily imply a merger formation of all elliptical galaxies if other mechanisms can be found that induce violently fluctuating gravitational fields (e.g., early collapses of many gaseous clumps). Perhaps equally interesting is the fact that the radial surface brightness profiles show *small, but significant and often localized deviations* from $r^{1/n}$-laws ($\lesssim$0.2 mag). When measured with high precision ($\lesssim$0.02 mag), the surface-brightness profiles of no two ellipticals are alike. Presumably, such deviations carry information about *incomplete* relaxation, but they have not been studied in any systematic manner so far.

Fourth, in most E galaxies the isophotal ellipticity changes as a function of radius. It can increase, decrease, or vary irregularly with increasing radius (Liller 1961; King 1978; Jedrzejewski 1987a; Franx et al. 1989a; and many since). Unfortunately, the easily measured ellipticity profiles do not, by themselves, convey much useful information concerning the three-dimensional shape of ellipticals. Only in connection with detailed velocity information or in large statistical analyses are they of use.

Fifth, many E galaxies show significant *isophotal twists* (Evans 1952; Williams & Schwarzschild 1979). When present, these twists indicate that an elliptical cannot be a spheroid (i.e., oblate or prolate), but must either be a triaxial ellipsoid (Lindblad 1956) or have *intrinsic* axis twists. Such intrinsic twists are seen in remnants of simulated disk–disk mergers and can be long-lived (>15 mean crossing times, see Gerhard 1983b). It is not quite easy to visualize how triaxial ellipsoids without intrinsic axis twists can produce isophotal twists in projection. A marvelous color picture by Bertola (1981a) helps: When one looks at the triaxial ellipsoid along any of its three axes, the projected light distribution does not – of course – show any isophotal twists. However, if the axis ratios of isodensity surfaces change as a function of radius and one looks at an angle to all three axes, then isophotal twists appear and can get very large (10–90°) when the ellipticity $\epsilon(a)$ approaches zero. The discovery of isophotal twists by Williams & Schwarzschild (1977, 1979) came at a time when triaxiality had been invoked by Binney (1976, 1978a,b) to explain the surprisingly slow rotation of some ellipticals (Bertola & Capaccioli 1975; Illingworth 1977). Such slow rotation suggests that the structure of ellipticals must be supported at least in part by large velocity dispersions, in which case these galaxies seem likely to be triaxial. The discovery of isophotal twists strongly supported this hypothesis.

A second important morphological indicator of triaxiality is the observed paucity of nearly round, E0-type galaxies. Figure 37 shows the distribution of *mean* ellipticities ϵ for 217 E galaxies studied by Franx, Illingworth, & de Zeeuw (1991). All ellipticities were measured from CCD images through isophotal analyses. The resulting distribution (shown hatched) differs dramatically from the distribution of ellipticities computed from the apparent major and minor axes measured by de Vaucouleurs et al. (1976) and shown as a thick line. Whereas the latter distribution peaks sharply at $\epsilon = 0$, the new distribution based on CCD data contains much fewer E0 galaxies. Apparently, visual measurers of axis ratios find it difficult to distinguish between round images (type E0) and images with

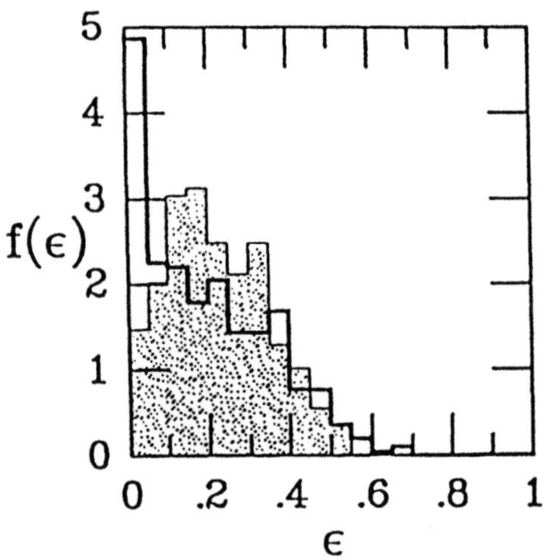

Fig. 37. Distribution of mean apparent ellipticity ϵ in E galaxies, determined from CCD images through isophotal analysis (*shaded*) and from photographic plates through visual measurements of major and minor axes (*thick line*). The paucity of objectively measured E0 galaxies indicates triaxiality. Diagram from Franx et al. (1991).

small ellipticities (E1 and E2) and bias their estimates toward E0. The newly established paucity of E0 types indicates that *few ellipticals feature two equal axes*; hence the majority must be triaxial (de Zeeuw 1994).

Triaxiality of E galaxies is also indicated by their kinematics. In triaxial systems observed at random angles, rotational motions should be seen not only along the major axis, but also along the minor axis (Contopoulos 1956; Binney 1985). Indeed, minor-axis rotation has been observed in about one quarter of all measured ellipticals and exceeds major-axis rotation in several of them (Franx et al. 1989b). In order to pin down the axis ratios $a:b:c$ as a function of radius in any given galaxy, one needs both good surface photometry and accurate two-dimensional velocity fields. Only a few ellipticals have been observed in such detail so far. Figure 38 shows a velocity map of NGC 5982, constructed from spectroscopic observations with the slit at four position angles across the nucleus (Wagner 1990). The velocity field is strongly distorted despite the fact that this E3 galaxy shows no isophotal twist, and the minor-axis rotation is obvious. A similar velocity map has recently been obtained by Statler et al. (1996) for NGC 1700. A detailed interpretation of their data suggests that within $r \lesssim 35'' = 2.5\,r_e$ the body of NGC 1700 is nearly oblate, while beyond $2.5\,r_e$ the twisting of both the morphological and kinematic axes indicates increasing triaxiality, an intrinsic axis twist, or both. The strong rotation observed at large radii, when paired with the near oblateness within $2.5\,r_e$, suggests that "NGC 1700 owes its present form to a merger of three or more stellar systems 2–4 h^{-1} Gyr ago" (where $h = H_0/100$, Statler et al. 1996).

In summary, surface photometry and isophotal analyses of elliptical galaxies strongly suggest that these galaxies are often triaxial, may have intrinsic axis twists, and have experienced violent relaxation.

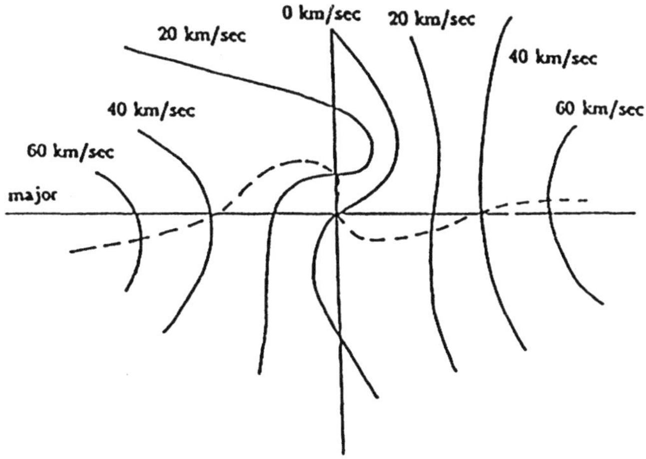

Fig. 38. Velocity map for NGC 5982, an E3 galaxy without any significant isophotal twist. Note the distorted velocity field and minor-axis rotation. Diagram from Wagner (1990).

4.2 Embedded Disks and Boxiness in E and S0 Galaxies

The higher-order Fourier coefficients C_3,S_3, ..., C_n,S_n yielded by isophotal analyses of early-type galaxies contain much interesting information. This section concentrates on results derived from the 3rd- and 4th-order coefficients and, especially, from C_4.

Figure 39 illustrates the shapes of model isophotes with strongly non-zero coefficients C_3, S_4, and C_4 (Peletier 1989). As the figure shows, a positive C_3 indicates egg-shapedness of the isophotes. The C_3 terms in the isophotes of early-type galaxies are usually quite weak, and *all* higher-order terms are rarely as strong as the 5% deviations from perfect ellipticity shown in the figure. When non-zero 3rd-order terms are detected, they nearly always reflect isophotal perturbations caused by patches of absorbing dust. Hence, the coefficients of such terms are wavelength dependent and are strongest on blue images of ellipticals and weakest on infrared images (Peletier et al. 1990; Goudfrooij et al. 1994).

As Fig. 39b shows, a positive S_4 coefficient indicates rhomboidal-shaped isophotes. However, in early-type galaxies this coefficient is usually near zero. In contrast, the coefficient C_4 is often significantly positive or negative and contains valuable information about the internal structure of the galaxy. If $C_4 > 0$ (Fig. 39c), the isophotes protrude beyond a best-fit ellipse near both the major and minor axes and are said to be *disky* (Lauer 1985b). As Fig. 5 in Jedrzejewski (1987b) nicely illustrates, such isophotes can be produced by adding the light of a highly inclined disk to that of an ellipsoidal body along the latter's apparent major axis. On the other hand, if $C_4 < 0$ (Fig. 39d) the isophotes appear pushed in and flattened near the major and minor axes and are said to be *boxy*. A powerful technique for displaying the presence of significant 4th-order terms consists of subtracting from the image of an elliptical galaxy a best-fit model

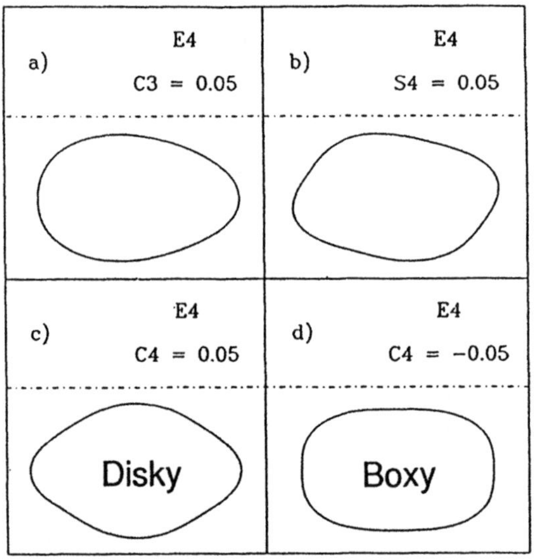

Fig. 39. Model isophotes with higher-order Fourier coefficients C_3, S_4, and C_4 that strongly deviate from zero. Positive values of C_4 indicate "disky" isophotes, while negative values indicate "boxy" isophotes. Diagram after Peletier (1989).

with strictly elliptical isophotes and forming a map of the residuals. In galaxies that are disky at all radii, the residual map shows excess light along the major and minor axes, while in uniformly boxy galaxies the residual map shows missing light along the principal axes and excess light along two diagonal lines (e.g., Jedrzejewski 1987b).

In most ellipticals the degree of diskiness or boxiness varies markedly with distance from the center, and some ellipticals contain both boxy and disky regions at different radii. Figure 40 shows radial profiles of the coefficient $C_4(a)$ in seven ellipticals observed by Jedrzejewski (1987a). Whereas NGC 3379 is perfectly elliptical with no significant 4th-order terms at any radius, NGC 4697 is strongly disky throughout, NGC 4387 is boxy throughout, and NGC 3818 and NGC 7029 both are disky inside but boxy further out. An extreme example of coexisting diskiness and boxiness (NGC 3610, not shown in Fig. 40) is discussed in more detail below.

Isophotal analyses that include 4th-order terms have been performed for well over 100 early-type galaxies and lead to the following main results: (1) Many ellipticals feature weak embedded disks. (2) The distribution of C_4 coefficients is compatible with the hypothesis that *all* ellipticals possess weak embedded disks, those without any apparent diskiness having disks seen nearly face-on. (3) There is a continuum from very weak-disk ($\lesssim$1% of total light) ellipticals through strong-disk (10%–20% of light) ellipticals to S0 galaxies (e.g., Michard 1985; Capaccioli et al. 1990; Capaccioli & Longo 1994). (4) Strongly disky ellipticals tend to rotate relatively fast (as measured by $v_{\rm m}/\sigma_v$, the ratio between the mean rotational velocity and the velocity dispersion), while strongly boxy ellipticals show a large spread in rotational velocities (Carter 1987; Bender 1988a; Bender et al. 1989). And (5), diskiness and boxiness – even in extreme forms – can coexist (e.g., Bender 1990; Seitzer & Schweizer 1990; Schweizer & Seitzer 1992).

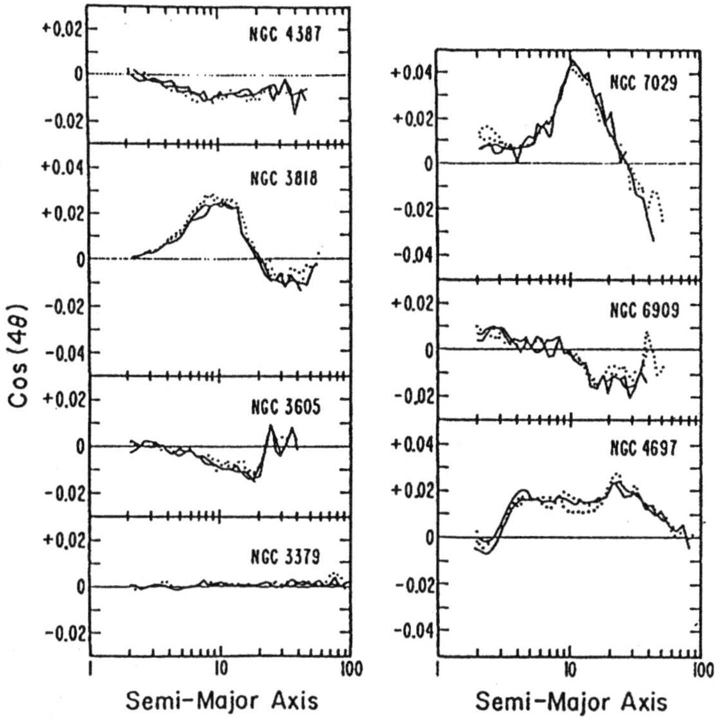

Fig. 40. Radial profiles of Fourier coefficient C_4 in seven elliptical galaxies. *Solid* lines show profiles from B images, *dotted* lines from R images. Diagram from Jedrzejewski (1987a).

The discovery that boxy ellipticals tend to be radio-loud and have high X-ray luminosities L_X, while disky ellipticals tend to be radio-quiet and have low L_X, led to the suggestion that boxy ellipticals may preferentially be merger remnants (Bender et al. 1989), while disky ellipticals may have formed in a different manner (Bender 1990; Scorza & Bender 1995). Although this suggestion has been widely accepted, I seriously doubt that it is correct, as explained in more detail in Chap. 6. Here, I simply wish to emphasize that ellipticals exist that combine strong inner disks with strongly boxy exteriors. These ellipticals invariably show clear signs of having experienced major mergers. The galaxy NGC 3610 is one of them.

NGC 3610: A Boxy E5 with Central Disk and Much Fine Structure

This galaxy is classified as an E5. On short exposures it indeed appears rather flattened. However, on deep exposures NGC 3610 appears virtually square, a case of extreme outer boxiness (see Fig. 1 in Seitzer & Schweizer 1990).

Figure 41 shows an isophotal map of this galaxy, plus a strongly masked CCD image that displays deviations from perfect ellipticity. The isophotes illustrate

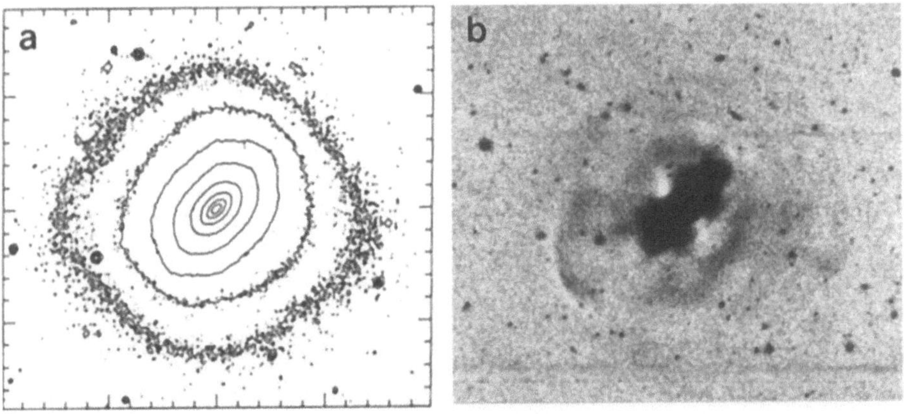

Fig. 41. Elliptical galaxy NGC 3610: (a) Isophotal map covering $200'' \times 180''$, and (b) masked image (scale 30% smaller). Note isophotes changing from E5 near center to nearly square in the outskirts, and ripples and plumes indicative of a recent merger.

the dramatic change from an E5 near the center to a nearly square body further out. Within the central E5 body, there is a high-surface-brightness *disk* seen approximately edge-on and extending out to at least $10'' \approx 1.5\,\mathrm{kpc}$ (Scorza & Bender 1990). The disk-to-spheroid luminosity ratio is 0.07 in the V passband. When spectroscopically separated from the spheroid, this inner disk appears dynamically very cold, with $v_{\mathrm{m}}/\sigma_v = 4.5$ (Rix & White 1992). Concerning any possible merger origin of NGC 3610, this cold inner disk only tells us that it cannot have formed before the last major merger. Instead, if there was such a merger the embedded disk must have formed within the remnant, perhaps in a fashion similar to the disk currently forming in NGC 7252 (Chap. 3). However, since many disk galaxies show no evidence of past mergers and disks are long-lived structures, the inner disk of NGC 3610 by itself tells us neither that there was a merger nor when any presumed merger might have occurred.

The boxiness of the outer isophotes of NGC 3610 is extreme. Like disks, boxiness is thought to be long lived since it presumably reflects the preferential population of certain box orbits by a significant fraction of all stars. Therefore, even though such preferential population is likely due to a major merger (Binney & Petrou 1985; Barnes 1992; Heyl et al. 1994), it still does not tell us *when* this merger occurred. Because the inner part of the boxy spheroid of NGC 3610 rotates fast and in the same direction as the embedded disk, Scorza & Bender (1995) have suggested that this elliptical cannot be the remnant of a major merger. Thus, we seem to face a dilemma: Is NGC 3610 the remnant of a major merger or is it not, and if it is, when did the merger occur?

The answer lies in NGC 3610's fine structure. As the masked image in Fig. 41 displays only partially, NGC 3610 features various ripples, luminous debris, a central "X-structure" (see §4.5), and various plumes at an angle to the central disk. This makes NGC 3610 the galaxy with the richest fine structure among 36 ellipticals imaged in a similar manner (Seitzer & Schweizer 1990). Both the ripples ("shells") and the luminous plumes are unmistakable signatures of a *recent* merger involving disk material, since such structures are relatively short-lived (e.g., Hernquist & Quinn 1988; Barnes 1988, 1992; Hernquist & Spergel 1992). In agreement with these morphological signatures of a recent merger, NGC 3610 also shows spectral signatures of such a merger: In the *UBV* passbands this galaxy appears too blue to be an old elliptical (Schweizer & Seitzer 1992), and its spectrum features enhanced Hβ absorption indicative of an aging starburst (Schweizer et al. 1990). Overall, the evidence is strong that a *major* merger involving gas-rich disk galaxies either formed, or at least occurred in, NGC 3610 sometime during the past 5–7 Gyr.

Interestingly, two other ellipticals with extremely boxy outer isophotes also feature ripples: IC 3370 and NGC 3640. Most likely these two galaxies are relatively recent merger remnants as well. Of these two ellipticals, IC 3370 features a strong inner disk, while NGC 3640 does not. Modern N-body simulations of disk–disk mergers show that such mergers can produce remnants with boxy, disky, or boxy *and* disky isophotes (Heyl et al. 1994). The detailed appearance of the remnants depends on the encounter geometry, structure of the progenitor galaxies, and orientation of the remnant to the line of sight. As Dr. Barnes explains in his parallel lectures, the remnants of merged disk galaxies with nearly aligned disk and orbital spins can be quite fast rotating themselves. Thus, the apparent dilemma concerning NGC 3610 and similar ellipticals is solved: even relatively fast rotating ellipticals with marked inner disks can be produced through major disk–disk mergers.

4.3 Ripples ("Shells") in Ellipticals

Ripples in elliptical galaxies have helped convince many skeptics that mergers do occur in these galaxies and may even be associated with their formation.

When Malin (1979) reported the discovery of these arc-like, sharp-edged features in M89 (= NGC 4552) and called them "shells", his announcement created quite a stir (see also Malin & Carter 1980, 1983). Ellipticals had just been found to rotate slowly and display isophotal twists, fostering the view that they are dynamically hot systems supported by relatively high, anisotropic velocity dispersions. The presence of sharp-edged features in them introduced a complication: Given the velocity dispersions of 100–300 km s^{-1} $\approx$ 100–300 pc Myr^{-1}, any such features consisting of stars should diffuse in $\ll$1 Gyr. So, what do ripples consist of, and what is their origin?

Figure 42 shows two ellipticals with strong ripples: Arp 230 and NGC 5018.[15] The ripples of Arp 230 show a striking and rare geometry: their distance from

[15] As it turns out, Arp (1966) had already included five systems similar to Malin's M89 in the *Atlas of Peculiar Galaxies* and had used the term "shells" to describe the ripples in Arp 230, while for the other systems he used the terms "rings" and "arcs". My own preference for the term "ripple" is that, like "arc", it is descriptive without implying any specific spatial topology, whereas the terms "shell" or "ring" do. However, for most purposes the words "ripple" and "shell" can be used interchangeably.

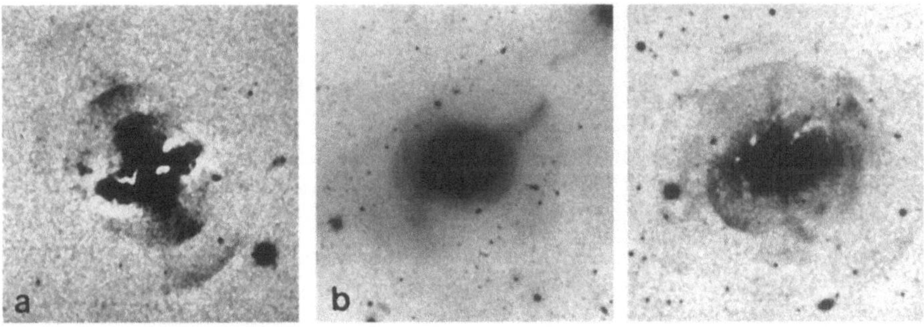

Fig. 42. Two E-type galaxies with ripples: (a) Arp 230 with "interleaving" ripples, and (b) NGC 5018 with luminous plumes or tails (deep image) and irregular ripples (enlarged masked image).

the center increases in such a fashion that if the galaxy were mirror-imaged onto itself, the ripples on one side of the nucleus would interleave in radius with those on the other side. The ripples of NGC 5018, on the other hand, are placed irregularly, but are more representative of ripple systems typically observed in elliptical galaxies.

The frequent association of ripples with tails, both in ellipticals and especially in recent merger remnants (e.g., NGC 7252 and NGC 3921), suggested early on that ripples may consist of disk debris and may form during mergers (Schweizer 1980, 1983; Quinn 1982). Models aimed at reproducing interleaving-ripple systems have emphasized radial accretions of small companion galaxies (Quinn 1984; Dupraz & Combes 1986, 1987; Hernquist & Quinn 1988), while modern N-body simulations show that ripples are also a natural by-product of major disk–disk mergers (see chapters by Dr. Barnes). So far, the observed properties of ripple systems do not allow us to distinguish unambiguously between minor and major mergers for their origin, but overall the evidence increasingly seems to favor major mergers.

The main properties of ripples in elliptical galaxies are as follows (see reviews by Athanassoula & Bosma 1985, Prieur 1990, and Schweizer 1993): Over half of all field ellipticals show ripples. Typically, the number of ripples per galaxy is about 3–10, with an observed range of 1–20^+ ripples. As Prieur emphasizes, most ripple systems in elliptical galaxies are neither of the interleaving-in-radius kind nor very symmetric. There seems to be a preference for ripples to occur along the apparent major axis. The innermost ripples are often located so near to the nucleus that *dynamical friction* must be invoked to explain how matter from a companion galaxy ended up so deep in the potential well of the elliptical (Dupraz & Combes 1986). In such cases, it seems unlikely that the ripples formed merely from mass transferred during the close flyby of a disk companion, and full-fledged accretions of companions, or even major mergers of near-equal galaxies, seem necessary.

The combined luminosity of ripples is surprisingly high: values corresponding to 5% – 25% of the total luminosity of the underlying elliptical are typically measured in B or V (Fort et al. 1986; Schombert & Wallin 1987; Prieur 1988; Schweizer, unpublished). Though ripples tend to have a low surface brightness, they often cover large areas. This explains their relatively high fractional luminosity. In fact, the measured luminosities tend to represent lower limits since, when measuring ripple luminosities, it is easy to miss flux from extended low-luminosity parts that blend with the background. The *colors* of ripples are typically similar to those of the host galaxy or slightly bluer, with mean color differences of $\langle \Delta(B-R) \rangle \approx -0.1\,\mathrm{mag}$ to $-0.2\,\mathrm{mag}$ (above refs. and McGaugh & Bothun 1990). Hence, it seems plausible that the mass-to-luminosity ratio of ripple material does not differ much from that of the luminous matter in the underlying galaxy. If so, the relatively high fractional luminosities of ripple systems indicate that the accreted galaxies must have been relatively massive.

Spectra of ripples show continuum light but no emission lines, indicating that ripples consist mostly of stars and feature little or no ionized gas (Schweizer 1980; Quinn 1982; Bosma et al. 1985; Pence 1986; Fort et al. 1986). This rules out all early explanations in terms of gaseous and/or explosive phenomena. Ripple *velocities* would be invaluable in trying to determine the kinematics and dynamics of these features, but they are extremely difficult to measure. Quinn (1982) and Bosma et al. (1985) detected glitches of about 10–30 km s^{-1} in stellar absorption-line velocities measured across some relatively bright ripples. However, there have been no confirming modern measurements of such velocity glitches, and there is a serious need for new absorption-line measurements.

Neutral hydrogen has been detected in the ripple galaxies NGC 5128 and NGC 2865 (Schiminovich et al. 1994, 1995). Figure 43 shows contours of the H I distribution in NGC 2865 superposed on an optical image of this elliptical (E4). The H I appears in the form of fragments typically near, but not coincident with, ripples and shows high, apparently rotational velocities (100–300 km s^{-1}) relative to the galaxy center. Though this gas may well form some fragmentary outer disk, I have serious doubts about Schiminovich's assertion that it correlates with the optical ripples and thus measures their velocities. Instead, the H I may be tidally ejected gas returning from some former tails and may be associated with ripples only through chance superposition. This interpretation would favor the view that NGC 2865 and NGC 5128 are remnants of major mergers involving at least one gas-rich disk galaxy. In either case, the velocities of the ripples need to be measured optically to allow a direct comparison with the velocities of nearby H I fragments.

Are ripples in elliptical galaxies signatures mostly of minor accretions or of major mergers? Throughout the 1980s, the view prevailed that ripples indicate the "infall of a gas-rich companion" into a pre-existing elliptical galaxy. However, there are two serious problems with this view: (1) The fractional luminosities of ripple systems are too high, as stated already; and (2) there seem to be insufficient numbers of dwarf galaxies around field ellipticals to explain the high accretion rate inferred from ripple statistics (see Sandage's objection

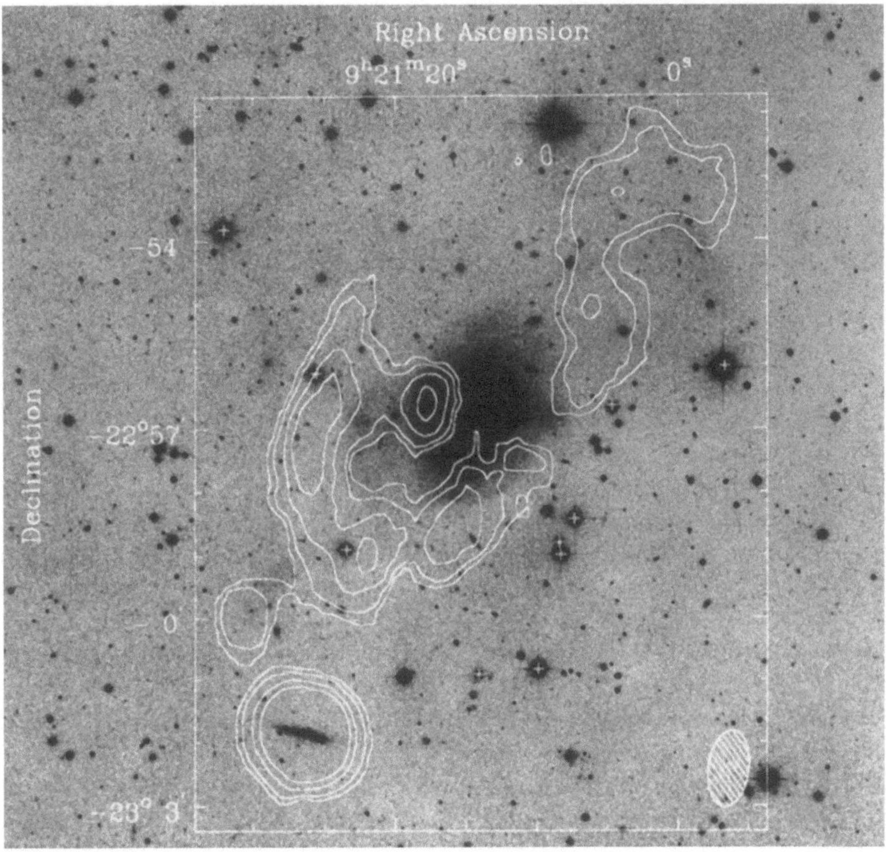

Fig. 43. Contour map of neutral hydrogen in NGC 2865 (E4) overlaid on a deep optical image. Illustration from Schiminovich et al. (1995).

after Schweizer's 1983 review). Recently, the emphasis has increasingly shifted toward major disk–disk mergers, which seem able to explain better many of the observed ripple properties: the relatively high fractional luminosity, the presence of ripples deep in the potential well, the frequent association of ripples and tails in merger remnants and ellipticals, and the observed return of matter from the tidal tails (see Chap. 3, esp. Fig. 29). It is these returning disk debris that often seem to provide a plentiful and long-lasting source of material for ripple formation (Barnes 1988; Hernquist & Spergel 1992; Hibbard & Mihos 1995). However, especially in galaxy groups minor mergers and mass transfers must occur and may contribute to ripple formation as well.

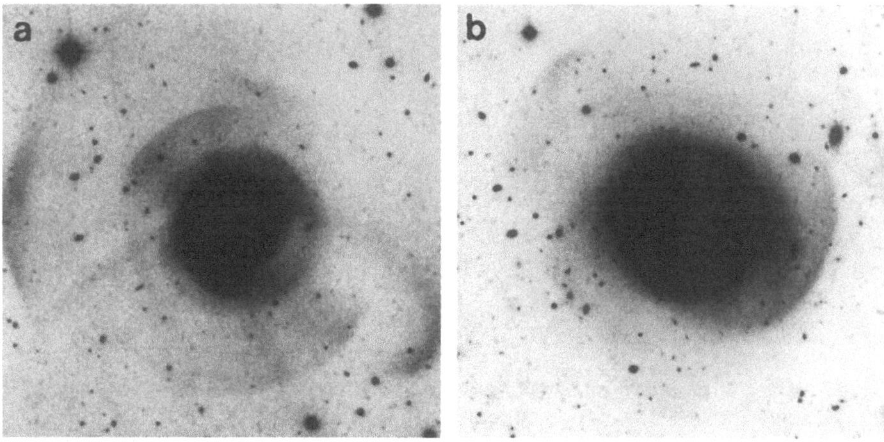

Fig. 44. Ripples in two early-type disk galaxies: (a) NGC 474 (=Arp 227, type S0/a) and (b) NGC 3619 (Sa). Partially masked CCD images from Schweizer & Seitzer (1988).

4.4 Ripples in S0–Sb Galaxies

Ripples have been found not only in ellipticals, but also in early-type disk galaxies of Hubble types S0 to Sa, and perhaps even in one Sb galaxy (Schweizer & Seitzer 1988). Figure 44 shows two examples: NGC 474 (S0/a) and NGC 3619 (Sa). In disk galaxies there is, of course, an abundance of dynamically cold matter, whence one has to worry about transient spiral waves or other forced vibrations that may have been induced in the existing disk by the close passage of some companion. There are two main pieces of evidence that at least some of the observed ripples consist of extraneous matter. First, rather than displaying central symmetry the detected ripples often interleave in radius on opposite sides of the center, a striking signature of matter accretion and phase-wrapping (e.g., Quinn 1984). And second, a significant fraction of the ripples have pitch angles either of opposite signs within one galaxy or of opposite sign to the pitch angle of existing spiral arms. Hence, at least these ripples seem unlikely to be density perturbations trailing in a rotating disk. Since they would have to be *leading* density perturbations, an external origin seems more likely.

The one galaxy of type Sb that seems to feature ripples is NGC 3310. Besides these ripples, it also sports a radially aligned filament of luminous matter that may be the remnant tail of an accreted companion galaxy.[16] Further evidence for a recent accretion is provided by the strong density wave associated with the inner spiral structure (van der Kruit 1976), a clear offset between the nucleus and the rotation center (Walker & Chincarini 1967), and an unusual central reversal of the abundance gradient (Balick & Heckman 1981). Interestingly, the disk of NGC 3310 seems to have survived this accretion relatively intact, perhaps because it is gas rich and dissipates energy efficiently.

[16] Walker & Chincarini (1967) called this radial filament and the bright western arc perpendicular to it the "bow-and-arrow" structure of NGC 3310.

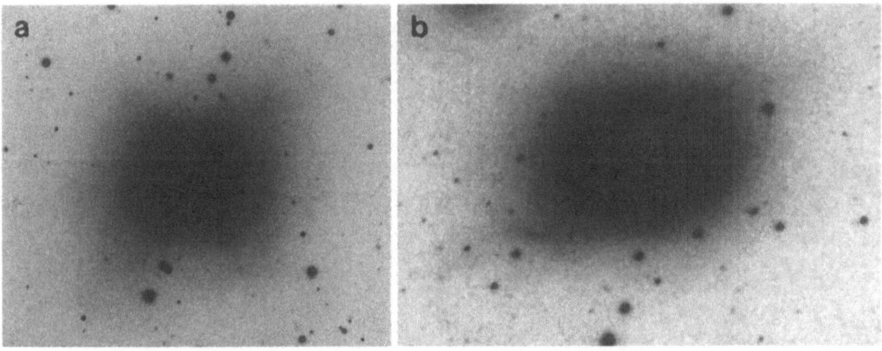

Fig. 45. Rarer forms of fine structure in elliptical galaxies. (a) X-structure in IC 3370 (E2–3), and (b) two filamentary arms or tails in NGC 1700 (E3).

The fraction of early-type disk galaxies with evidence for ripples of an external origin drops rapidly from Hubble type S0 (~30%) to type Sb (<1%). Since the bulge-to-disk ratio decreases markedly along the same type sequence, both the ripples and the bulges may have formed due to accretions. One can speculate that accretions of relatively massive companions (10% – 50% as massive as the main disk galaxy) may form S0 galaxies with relatively massive bulges, many ripples, and often thickened disks, while accretions of relatively minor companions ($\lesssim$10%) may form Sa and Sb galaxies with their less massive bulges, fewer ripples, and thinner disks. If so, Hubble's morphological sequence may represent mostly a sequence of decreasing merger damage.

4.5 X-Structure and Other Fine Structure

Besides ripples, early-type galaxies display various other kinds of fine structure. In the case of NGC 3610, we have already mentioned boxiness, "X-structure", and luminous plumes and filaments. Figure 45 illustrates two relatively rare kinds of fine structure: X-structure in IC 3370, an E2–3 galaxy with extremely strong outer boxiness, and two centrally symmetric filamentary arms or tails in NGC 1700, an E3 galaxy with rhomboidal outer isophotes. Note that to better show the X-structure, the image of IC 3370 has been masked by subtracting a purely elliptical model light distribution.

IC 3370 is yet another elliptical with very boxy outer isophotes, but a strong inner disk. Even mere visual inspection of the SRC Sky Survey film or of a direct CCD image shows immediately that there is extra light forming some sort of an "X" along the two diagonals of the boxy isophotes. It is this extra light that has come to be called X-structure (Schweizer & Ford 1985; Whitmore & Bell 1988). It shows especially well after an elliptical light distribution has been subtracted. The relation between this X-structure and boxiness is not very clear. It appears that X-structure may occur only in extremely boxy ellipticals and bulges, but that not all very boxy galaxies show X-structure.

There has been some controversy about how boxiness is generated. The two main mechanisms invoked have been *mergers* injecting stars into box orbits (Binney & Petrou 1985; Barnes 1992) and *bars* inducing vertical instabilities in disks (Combes et al. 1990; Pfenniger & Friedli 1991). Since IC 3370 has a strong inner disk and a spheroid rotating fast and in cylindrical fashion (Jarvis 1987), the question arises as to which mechanism formed its extreme boxiness and X-structure. In this case, there can be no doubt that a merger was involved. The galaxy features not only irregular outer ripples, but also 3–4 faint inner ripples that occur in the northeast sector of the "X" and are *exactly bounded by the X-structure.* Interestingly, this northeast sector and its counterpart to the southwest appear filled in with luminous matter, while the two other sectors appear empty. Isophotal analysis shows that about 10% – 20% of the total light is involved in this remarkable X-structure, whence the merger must have involved a relatively major disk galaxy. If a bar did play a significant role, it may have formed as a transient during the presumed merger (Mihos et al. 1995).

An alternative way to populate box orbits is to have a small companion galaxy fall into a flattened potential at an angle to the plane of symmetry (Hernquist & Quinn 1989). Whereas such an accretion puts mostly material from the *companion* into these box orbits, the transient bar in the newer Mihos et al. model puts mostly material from a preexisting *disk* into such orbits.

X-structure occurs not only in ellipticals, but also in S0 galaxies with strong disks. The galaxy IC 4767 represents an extreme case: about 1/3rd of the total blue luminosity stems from the bulge, another third from the disk, and the last third from the X-structure (Whitmore & Bell 1988). Since many S0 galaxies with strongly box-shaped bulges have companions and IC 4767 itself lies near the center of a cluster, Whitmore & Bell suggest that either accretion of a minor companion or mass transfer may have played a role. Of course, a bar instability induced by a close passage of one of the visible companions might also be responsible. In this case, there are no ripples to indicate a recent merger or accretion.

Finally, luminous plumes, tails, or "jets" of stars occur in about 20% of all E and S0 galaxies (Schweizer 1993). The nearly symmetric placement of two filaments observed in NGC 1700 (Fig. 45) is very rare. More often, plumes and jets occur at random orientations and locations, presumably being luminous debris left over from some merged disks. Rather than showing yet another elliptical with such linear features, I refer you to the remnants of simulated disk–disk mergers depicted by Barnes elsewhere in this volume. These remnants resemble real ellipticals to a remarkable degree, even in the plumes and jets that they sport. In fact, three of the four remnants of merged disks shown in Fig. 9 of Barnes (1992) feature crossed jets reminiscent of those observed, e.g., in NGC 4125 (Bertola 1981b; Bertola et al. 1984). Since *major* disk mergers produce these features so easily in N-body simulations, why should they not also produce them in real ellipticals?

4.6 Summary

The observations and methods of analysis discussed in this chapter yield valuable information concerning the fine structure and possible merger origin of early-type galaxies. Yet, they leave us with a problem that defines the main theoretical needs and observational desiderata, as follows.

- **Main result:** There is much circumstantial evidence connecting ripples, tails, extreme boxiness, and X-structures in early-type galaxies to past mergers and/or accretions. Low-velocity-dispersion material is needed to form the observed sharp-edged ripples. The cold, former-disk matter falling back from tidal tails may be a frequent source of ripple formation in early-type galaxies, marking these galaxies as likely remnants of major disk mergers. Occasionally, cold matter may also be supplied through the accretion of small companions or through mass transfers from nearby galaxies.
- **Problem:** So far, all models of ripple formation have been only generic, rather than aimed at reproducing specific, observed ripple systems.
- **Proposed solution:** There is a serious need for N-body model simulations capable of reproducing in detail the observed fine structure of individual E and S0 galaxies. Achieving this goal would represent a success similar to TT's modeling of galactic bridges and tails in M51 and three other systems. As galaxies to be modeled, I propose Arp 230, NGC 1316, NGC 5128, and NGC 3610.
- **Observational desiderata:** (1) Measure radial velocities of ripples through absorption-line spectroscopy; (2) establish the relation between optical ripples and HI features apparently associated with them; (3) perform large-field CCD photometry of E–S0–Sa galaxies to quantify any correlations between ripples, plumes, boxiness, X-structure, etc; and (4) study fine structure in early-type galaxies as a function of their environment (field vs. groups vs. clusters).

5 Kinematic Signatures of Past Mergers

Mergers affect not only the morphology of galaxies, but also their kinematics and dynamics. Whereas many of the morphological signatures of past mergers are relatively short-lived (Chaps. 3 & 4), the kinematic signatures that I am about to describe tend to be long-lived. To the best of our present knowledge, they last for periods of the order of a Hubble time or longer. Therefore, these signatures identify galaxies where mergers presumably took place, but due to their longevity they rarely permit the dating of these mergers.

Kinematic peculiarities such as misaligned gas disks or counterrotating stars have been found in various kinds of galaxies. The present chapter discusses three categories of such objects: (1) S0 galaxies with polar rings; (2) ellipticals with misaligned gas disks and/or oddly rotating stellar cores; and (3) early-type disk galaxies (S0–Sb) that feature counterrotating disks of gas or, occasionally, even stars. Especially the recent discoveries of counterrotating stars in disks and bulges have led to much excitement and to the realization that disk galaxies may be less fragile than previously thought. If so, mergers may reshape such galaxies without necessarily destroying their disks.

5.1 Galaxies with Polar Rings

Historically, it is the S0 galaxies with polar rings that first impressed upon astronomers the power of kinematic signatures for revealing past mergers. Hence, we discuss these objects first.

A Second Event!

Figure 46 shows three examples of S0 galaxies with polar rings. The first of these, NGC 2685, was recognized early on by Sandage (1961) as being "perhaps the most unusual galaxy in the Shapley-Ames catalogue". He noted that the central amorphous body appears like a spindle surrounded by helical filaments, which quickly led astronomers to nickname this curious object the "Helix Galaxy". But what is the true shape of the apparent spindle? Is this object shaped like a cigar or more like a pancake?

The peculiar nature of the helical filaments was reemphasized when Marie-Helene Ulrich (1975) measured line-of-sight velocities from the [O II] doublet in them and found them to be rotating around the apparent *major* axis of the "spindle". This result was unexpected, and I remember long and heated discussions among Pasadena astronomers whether this could really be true. It seemed to imply that the filaments form a ring rotating around a *prolate* galactic body. Not long thereafter, Schechter & Gunn (1978) settled the question by measuring both emission- and absorption-line velocities along the apparent major and minor axes. Their measurements clearly confirmed Ulrich's claim about the filament rotation and demonstrated that the "spindle" is, in fact, a *rotating stellar disk* viewed nearly edge-on. The crux of the demonstration is that the stellar line-of-sight velocity dispersion σ_v is significantly lower than the mean rotation

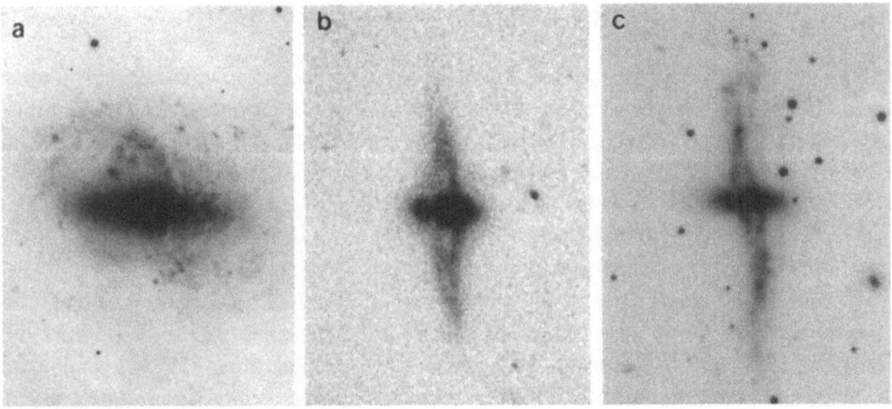

Fig. 46. Three S0 galaxies with polar rings: (a) NGC 2685 ("Helix Galaxy", from plate by Hubble); (b) anonymous galaxy A0136−0801; and (c) NGC 4650A.

velocity of $v_{\rm rot} = 115\,{\rm km\,s}^{-1}$ along the major axis: $v_{\rm rot}/\sigma_v > 1$. Therefore, the stars form a dynamically cold disk, and the ionized gas in the helical filaments rotates over the poles of this disk. The two nearly orthogonal rotation systems also show up in H I, where two nested disks are observed. The main H I disk is aligned with the optical "spindle" and extends far beyond its boundary, while the small central H I disk coincides with the helical filaments, suggesting that some HI cloud or gas-rich companion galaxy fell in (Shane 1980; Toomre 1977b).

Other S0 galaxies sport more extended polar rings than NGC 2685. Figure 46 shows two such systems: A0136−0801 (Schweizer et al. 1983) and NGC 4650A (Sérsic & Agüero 1972; Laustsen & West 1980). Further examples can be found in the atlas by Whitmore et al. (1990). In the about half dozen cases where the kinematics has been investigated in some detail, the polar ring invariably rotates at nearly right angles to the apparent "spindle", which itself shows the kinematic signature of a relatively cold stellar disk. (A few ellipticals with polar rings are also known.) The polar rings look morphologically similar to collisional ring galaxies and have, therefore, occasionally been confused with the latter. However, there is a crucial difference between the two kinds of rings. In all measured polar-ring galaxies, the "spindle" sits at the center of the ring and is at rest with respect to the ring, while in collisional ring galaxies the intruder lies well away from the center and moves relative to the ring. Therefore, while collisional ring galaxies are transient configurations, polar rings clearly represent relatively long-lived, near-equilibrium structures.

Since the collapse of a single protogalactic gas cloud could hardly lead to two orthogonal rotation systems within one galaxy, the main message from the about 30 known S0 galaxies with polar rings is that their formation history must have included a *second event.* Such events may not be as rare as might appear at first sight. Statistical studies show that a few percent of all S0 galaxies feature polar rings (Schweizer et al. 1983; Whitmore et al. 1990). If mergers and accretions occur from random directions, secondary gas disks may form at random angles to preexisting disks, but may survive for long periods only when located near the poles, where differential precession and gaseous settling are slow (Christodoulou et al. 1992; de Zeeuw 1994). According to this *statistical selection* hypothesis, the few percent of S0 galaxies with polar rings represent merely the visible tip of an iceberg, and *most* S0 galaxies are likely to have experienced a second event.

What Kinds of Interactions or Mergers? The "Failed Bulge" Hypothesis

Although two distinct spin systems in one galaxy strongly suggest that some accretion or merger took place, the details of this process are unclear at present.

In a few lucky cases of S0 galaxies with polar rings, there is direct evidence for a past merger. For example, deep CCD images of the relatively isolated galaxy MCG −5-7-1 (= ESO 415-G26) reveal several ripples indicating that not just a gas cloud, but some former companion galaxy with stars was involved in forming this S0's extremely narrow ring (Whitmore et al. 1987; van Gorkom et al. 1987). In several other cases, kinematic perturbations in the S0 galaxies themselves seem to provide evidence for past mergers (Reshetnikov & Combes 1994). Until recently, it has generally been assumed that these mergers involved gas-rich companions falling in nearly over the poles of a disk galaxy.

Polar rings may occasionally also form during close flybys of companions through mass transfer from the disk of one galaxy to the other. This mechanism seems to be at work in NGC 3808 (= Arp 87), where a bridge extends from the main spiral towards the companion and wraps around its poles at least 1.5 times. Model simulations confirm the ability of mass transfers to create rings of material around the accreting galaxy (Hernquist & Quinn 1988). Since the cross section for close flybys is much larger than that for direct hits leading to mergers, I used to think that perhaps the majority of polar rings around S0 galaxies were formed through flybys and mass transfers.

However, new observations seem to increasingly favor a slightly different formation mechanism: At least in galaxies with *large* polar rings it appears that the S0 galaxy itself may be the secondary structure, rather than the polar ring. I call this the *failed-bulge hypothesis.* Observations of both the H I contents of polar rings and the spectra of central S0 galaxies seem to support this hypothesis.

Single-dish and VLA observations of S0 galaxies with polar rings show that these systems are surprisingly rich in HI. The ratio of HI mass to blue luminosity, $\mathcal{M}(\mathrm{H\,I})/L_B$, is exceptionally high when compared to that of S0 galaxies, but matches the ratio observed in late-type spirals (Schechter et al. 1984a). The mean H I mass for about 20 systems is $\langle\mathcal{M}(\mathrm{H\,I})\rangle = 1.2\times10^{10}\,\mathcal{M}_\odot$ (for $H_0 = 50$; see van Gorkom et al. 1987; Richter et al. 1994). This high amount of H I is typical of full-grown spirals and can hardly have been delivered by some dwarf galaxy or through a mere mass transfer. Moreover, the H I distributions are aligned with the polar rings rather than with the underlying S0 galaxies, and the H I generally displays well ordered disk rotation (Fig. 47). Therefore, it very

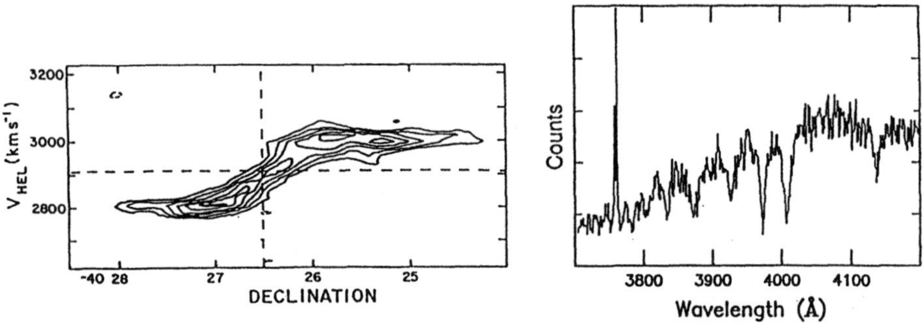

Fig. 47. (*left*) H I rotation curve of the polar ring and (*right*) post-starburst spectrum of the central S0 galaxy of NGC 4650A. Diagrams after van Gorkom et al. (1987) and Schechter et al. (1984b).

much looks as if many polar rings were normal, late-type disk galaxies before experiencing their second event (see, e.g., Arnaboldi et al. 1997).

The spectra of the S0 components often show signs of past starbursts, supporting the notion that at least in some cases it is these central components that are the secondary structures. Figure 47 shows the spectrum of the "spindle" in NGC 4650A. The Balmer absorption lines are extremely strong, indicating an aging starburst population (Schechter et al. 1984b). Weaker, but still enhanced Balmer absorption lines have been detected in other S0 galaxies with polar rings, including in A0136−0801 (Schweizer et al. 1983). Together with the vast amounts of H I in the polar rings, these post-starburst spectra suggest that the second event may have been the near-central infall of a smaller galaxy into a larger late-type spiral, from over the latter's poles. It is unclear at present whether such an infall created all stars of the S0 galaxy from the combined gas of the intruder and the main disk, or whether it left a remnant of the intruder itself stuck at the center.[17] In either case, it seems fair to call the central S0 galaxy a "failed bulge" since, had the accretion not occurred almost perpendicularly to the larger disk galaxy but at a shallower angle, the result might have been a co- or counterrotating bulge.

Clearly, this failed-bulge hypothesis for the formation of S0 galaxies with large polar rings needs to be tested through N-body simulations that include gas and various mass ratios for the accreted companions. It may well be that in some cases the secondary structure is the S0 galaxy (e.g., NGC 4650A), while in others it is the polar ring (e.g., NGC 2685). Hopefully, such simulations will also help us understand why some polar rings appear extremely narrow (e.g., MCG −5-7-1).

[17] A way to distinguish between the two possibilities may be through absorption-line gradients in the central S0 galaxies. Equally enhanced Balmer lines at all radii would seem to favor a wholesale formation of the S0 galaxy, while only centrally enhanced Balmer lines might favor a stuck intruder that experienced a central starburst fueled by gas from the large disk galaxy.

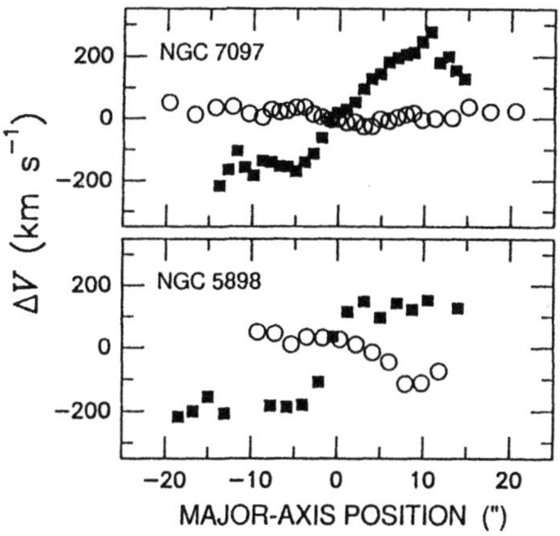

Fig. 48. Rotation curves of ionized gas (*filled squares*) and stars (*open circles*) in two elliptical galaxies. Note that the gas *counter*rotates to the stars. Diagrams after Caldwell et al. (1986) and Bertola et al. (1990).

5.2 Inner and Outer Gas Disks in Ellipticals

Although elliptical galaxies used to be thought of as being nearly devoid of gas, many are now known to possess not only extensive atmospheres of X-ray emitting, 10^6–10^7 K gas, but also disks of $\sim 10^4$ K ionized gas and, occasionally, even of cold gas and dust. Many of these disks are strong signatures of past accretions and mergers because they are severely warped and/or rotate in a sense opposite to the stars.

Figure 48 compares the rotation curves of the ionized gas in two ellipticals with those of the stars. The galaxy NGC 7097 (E4) is the first elliptical in which a gas disk *counter*rotating to the stars was discovered (Caldwell et al. 1986). Note that the mean rotation velocity of the stars is very slow, while the gas rotates with $\sim$200 km s^{-1}. The other galaxy, NGC 5898 (E0), is one of many studied by Bertola et al. (1990). It, too, shows pronounced counterrotation of the ionized gas. Among 18 ellipticals with dust lanes, Bertola et al. find two with corotating gas disks, three with counterrotating gas disks, and nine with gas disks rotating either orthogonally or at a skewed angle to the stars. This is strong evidence indeed that the gas and dust are not shed by the stars of these galaxies, but are remnants of accretions or mergers.

The case of the elliptical IC 2006 (E1) is particularly interesting. The ionized gas near the center is very turbulent and rotates in the opposite sense from the stars. In addition to this inner gas disk there is an outer ring of H I which counterrotates as well (Schweizer et al. 1989; Franx et al. 1994). The fact that gas masses at 3 kpc and 19 kpc radius share the same sense of rotation suggests that perhaps here, as in the S0 galaxies with large polar rings, a large disk galaxy with H I may have been struck by a major companion to, in this case, form an elliptical. The stellar velocity fields of IC 2006 and similar ellipticals should be mapped out to several effective radii since they may yield interesting new clues about how exactly these galaxies formed.

The disks of gas and dust in ellipticals are often markedly warped. Even before the discovery of counterrotation in some of them, astronomers of the early

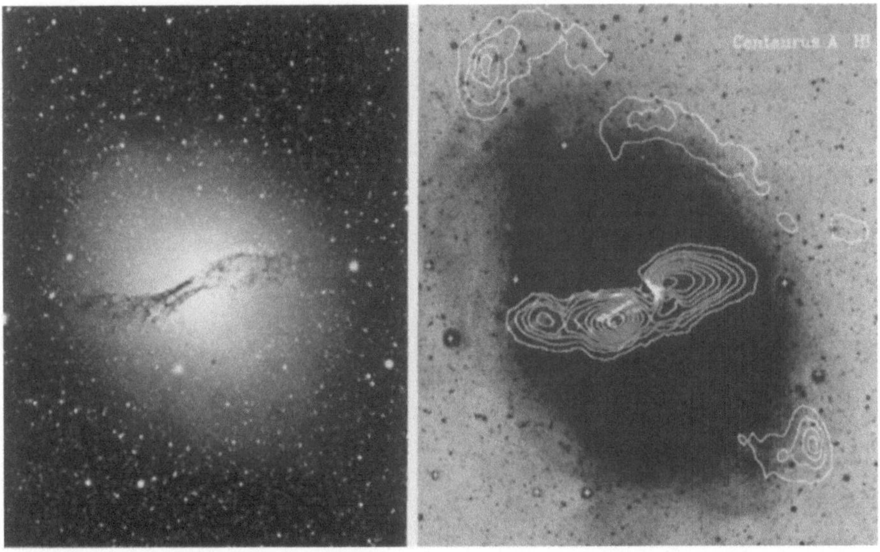

Fig. 49. (*left*) Blue-light photograph of NGC 5128 and (*right*) H I contours superposed on a high-contrast photograph. Note the warped disk of dust and gas, the outer H I fragments, and the faint optical ripples. Photographs courtesy of Graham (1979) and Schiminovich et al. (1994).

1980s were wondering about how to explain the warps. One favorite hypothesis was that warps arise when gas orbits in triaxial galaxies that are tumbling. Various equilibrium configurations are then possible, some of which lead to warped disks resembling the observed ones (Merrit & de Zeeuw 1983). When the sense of rotation for the gas is known, the shape of the warp allows one to predict which way the elliptical galaxy should tumble. It turns out that the most famous of all elliptical galaxies with warped gas disks, NGC 5128, rotates or tumbles in the *opposite* sense from what was then predicted.

Figure 49 shows a photograph of NGC 5128 and a contour map of the H I distribution. The warped dust lane bounding the disk of HII regions is prominent in the masked photograph (Graham 1979) and agrees in shape with the inner H I disk (Schiminovich et al. 1994). Given the rotation curve measured from the H II regions, the stellar body of NGC 5128 would have to tumble with its northeast side receding from us to explain the observed warp as an equilibrium configuration (van Albada et al. 1982), yet the stars rotate such that the northeast side approaches with $\sim$40 km s^{-1} (Bertola et al. 1985; Wilkinson et al. 1986). This observation strongly supports the notion that the central gas disk is in disequilibrium and still settling (Tubbs 1980; Simonson 1982). By inference, there must have been a recent merger. Direct evidence for such a merger is present in the form of many ripples ("shells") in the main body of the galaxy (Malin et al. 1983). But was this event really the accretion of a gas-rich dwarf, as many have believed, or was it a major merger between two galaxies of comparable mass?

The distribution and kinematics of the H I in NGC 5128 both favor a major merger. In addition to the central HI disk, there are a number of HI fragments at ∼15 kpc from the center (Fig. 49). These fragments appear to be part of an outer gas disk that rotates with 225 km s^{-1} in the same direction as the stellar body, i.e., at about a right angle to the central gas disk. This shared rotation between the outer gas and the bulk of the old stars makes it now appear unlikely that at least this outer gas stems from a small companion galaxy having fallen into a pre-existing elliptical. Rather, there seems to have been a major disk galaxy which collided and merged with another gas rich disk galaxy. The two merging galaxies were probably strongly inclined to each other and, since their remnant is an elliptical, were of comparable mass. Supporting this view is the observed kinematics of 433 planetary nebulae in NGC 5128 (Hui et al. 1995). The axis of rotation of these planetaries appears misaligned by $39° \pm 10°$ from the minor axis of the isophotes, a classical signature of strongly triaxial galaxies (Binney 1985). In short, then, NGC 5128 shows many of the trappings of a major-merger remnant (see also the chapters by Dr. Barnes).

5.3 Oddly Rotating Stellar Cores in Ellipticals

The discovery of oddly rotating cores in many elliptical galaxies has provided us with yet another kinematic signature of past major mergers.

The first such core was discovered in NGC 5813, where Efstathiou et al. (1982) noted that "the [stellar] rotation curve is quite unusual. The peak rotation velocity of 89 ± 7 km s^{-1} is reached at ∼3″ from the center, the rotational velocity then falls rapidly and remains roughly constant at 8 ± 2 km s^{-1} between 10″ and 80″". Though Kormendy (1984) argued that a small, fast spinning elliptical might have fallen into a large, slowly rotating elliptical in NGC 5813, the merger origin of this kinematic signature was not generally recognized until oddly rotating cores were discovered in many more giant ellipticals later during the decade (Franx & Illingworth 1988; Jedrzejewski & Schechter 1988; Bender 1988b, 1990). In the following, I first discuss the properties of these cores and then the various hypotheses put forward to explain the cores' origin.

Figure 50 displays the major-axis rotation curves of two E1 galaxies measured from stellar absorption lines. Note that in both galaxies the central regions rotate differently from the outer regions. Whereas in NGC 3608 the stellar core rotates in the opposite sense from the rest of the galaxy, in NGC 4494 the core rotates in the same sense, but locally faster than the surrounding stars (Jedrzejewski & Schechter 1988). Similar, oddly rotating stellar cores have been found in about 20%–30% of all giant ellipticals (e.g., Franx et al. 1989b). Some of these cores counterrotate, others corotate, and yet others spin at right or skewed angles to their host galaxy. Because of their odd rotation, these cores are often referred to as "kinematically decoupled cores" or "kinematic subsystems".

These cores vary in size and mass, but have radii typically of order 1 kpc and masses of $\sim 10^9$–$10^{10}\ \mathcal{M}_\odot$. Each of the three counterrotating cores studied in detail and under good seeing conditions (in IC 1459, NGC 5322, and NGC 4365) appears to be *disky* (e.g., Bender 1990; Surma & Bender 1995). Also, the core

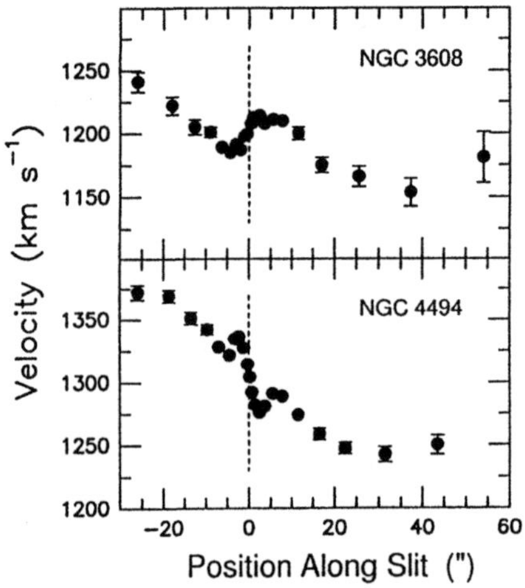

Fig. 50. Mean stellar velocities along the major axes of two E1 galaxies. Note the counterrotating core in NGC 3608 and the corotating, but kinematically distinct core in NGC 4494. Diagrams after Jedrzejewski & Schechter (1988).

boundary as defined by the kinematics often coincides with a marked break in the Mg-abundance gradient (Bender & Surma 1992). Finally, when studied at high resolution with the *Hubble Space Telescope* many cores of ellipticals show intricate dust disks and, occasionally, even spiral-like structure (Kormendy et al. 1994; Lauer et al. 1995).

How, then, did these cores form? Various formation hypotheses have been advanced, of which the four most important are as follows: (1) A pre-existing elliptical cannibalized a faster-spinning and denser small elliptical (Kormendy 1984; Balcells & Quinn 1990); (2) a pre-existing elliptical accreted a gas-rich companion galaxy, whose gas formed a disky core (Franx & Illingworth 1988); (3) oddly rotating cores are natural byproducts of major mergers between two spirals and form from the combined gas masses (Schweizer 1990; Hernquist & Barnes 1991); and (4) such cores are not physical, but apparent structures due to projection effects in triaxial galaxies (Statler 1991).

The disky nature of the best-studied oddly rotating cores argues against Hypotheses 1 and 4 and favors gaseous dissipation instead. Gas settling into a disk and subsequently forming stars seems the only feasible way to place an oddly rotating *disk of stars* deep in a potential well. (Stellar disks falling in would be shredded and violently relaxed.) The masses of these disky cores ($\sim 10^9$–$10^{10}\,\mathcal{M}_\odot$) then suggest that at least one of the two merging galaxies had to be a nearly full-grown spiral in order to contribute $\gtrsim 10^{10}\,\mathcal{M}_\odot$ of gas (star formation being rarely more than 10% efficient). My own favorite hypothesis has been that *two* disk galaxies may normally be involved, of which at least one is gas rich (Schweizer 1990). I was led to this hypothesis, which of course mirrors the Toomre (1977b) hypothesis of elliptical formation, when I noticed the striking resemblance between the major- and minor-axis velocity profiles of IC 1459 (Franx & Illingworth 1988) and of the merger remnant NGC 7252.

Figure 51 displays the velocity profiles of these two galaxies one above the other for easy comparison. Note that the line-of-sight velocities measured in

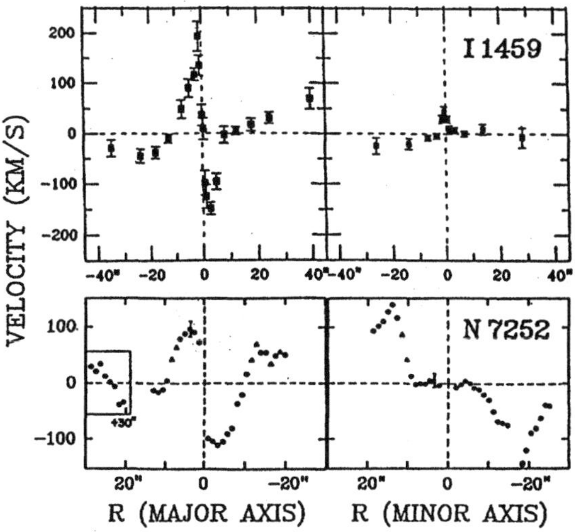

Fig. 51. Comparison between kinematics of stars in IC 1459 (E3) and ionized gas in NGC 7252. Left panels show major-axis rotation, right panels minor-axis motions. Note the oddly rotating cores in both galaxies. IC 1459 data from Franx & Illingworth (1988), diagram from Schweizer (1990).

IC 1459 refer to stars, while those in NGC 7252 refer to the ionized gas. In both galaxies, the major-axis profiles show that the central disks rotate in the opposite sense from the surrounding matter. Along the minor axes, the disks show no rotation as expected, but chaotic gas motions are observed in NGC 7252 beyond the disk boundary. As described in Chap. 3, NGC 7252 appears to be in the process of building itself a kinematically distinct *stellar* core. All the right ingredients are there: $\sim 8\times 10^9\,\mathcal{M}_\odot$ of molecular gas, spiral structure probably excited by gas falling back from the tidal tails at the rate of 4–$8\,\mathcal{M}_\odot\,\mathrm{yr}^{-1}$, and star formation at a rate of at least 1–$2\,\mathcal{M}_\odot\,\mathrm{yr}^{-1}$ (see Fig. 72 of the central "minispiral" in Chap. 7). Indeed, numerical simulations of mergers involving mutually inclined disk galaxies with gas seem to produce counterrotating central gas disks in about 1/4 of all cases (Hernquist & Barnes 1991; Barnes & Hernquist 1996; and Chap. 8 by Dr. Barnes in this volume).

To test this core-building hypothesis, we need further observations and a better theoretical understanding of how exactly the H I returning from the tidal tails finds its way into the core. Why does this H I disappear at the base of the tails (Hibbard et al. 1994), and how does it get converted into molecular form in a central gas disk (Wang et al. 1992)? Observational signatures to look for are shock fronts and, perhaps, small regions of intense X-ray emission.

The disky cores lead, I believe, to yet another interesting conclusion. As stated above, such disky *stellar* cores are likely to be disrupted during major mergers. Therefore, the existence of such cores in present-day ellipticals suggests that no major dissipationless mergers occurred after the dissipational mergers that formed these ellipticals. Since disky cores seen nearly face-on are difficult to detect, the 20%–30% detection rate for oddly rotating cores implies that $\gtrsim$50% of all ellipticals feature such cores. Therefore, one is nearly forced to conclude that probably the *majority* of elliptical galaxies formed in *one* major disk–disk merger, as already envisaged by TT.

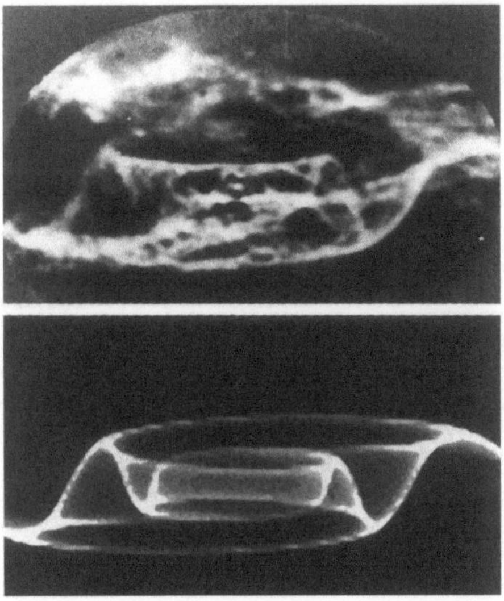

Fig. 52. (*top*) Dust lanes in the S0 galaxy NGC 4753 and (*bottom*) model of a 15° inclined gas disk strongly twisted by differential precession. The model reproduces the observed dust lanes in detail. Photomontage courtesy of Steiman-Cameron et al. (1992).

5.4 Counterrotating Gas in Disk Galaxies

Even in disk galaxies with their normally well-ordered rotation observers are increasingly finding counterrotating parts. Such telltale kinematic signatures appear to decrease in frequency from S0 galaxies to Sb galaxies, and none is known yet in any Sc galaxy. Interestingly, this trend along the Hubble sequence coincides with the decreasing relative importance of bulges, leading to the suspicion that perhaps even bulges owe their formation to past accretions or mergers. This section discusses counterrotating gas in disk galaxies, while §5.5 describes evidence for counterrotating stars.

The fraction of S0 galaxies with gas rotating in the opposite sense from the stellar disk is remarkably high. In a spectroscopic study of 15 S0 galaxies with emission lines, Bertola et al. (1992) find three whose ionized gas counterrotates. There can be little doubt that such gas traces some past accretion or merger. However, since accretions presumably happen at random angles, there must be a similar fraction of galaxies where the accreted gas *co*rotates with the disk stars. Hence, in ~40% of S0 galaxies with ionized gas this gas is of likely external origin. Even higher estimates of the fraction of S0 galaxies affected by past mergers follow from a literature survey of 33 S0 galaxies measured for rotation. Of these 33 S0 galaxies, 14 show corotating disks of ionized gas, 3 show skewed rotation of the gas, and fully 9 show counterrotation. Hence, at least ~ 21 $(= 9 + 3 + 9)$ of these galaxies, or 2/3 of the sample, experienced accretions or mergers.

If early-type disk galaxies experienced such events so commonly, one should occasionally find examples that show direct evidence of ongoing gaseous settling. There are several such cases among the transition objects between S0 galaxies and S0 galaxies with polar rings (see Category C objects in Whitmore et al. 1990). For example, the highly warped disk galaxy NGC 660 is clearly in a transient configuration, and its gaseous settling has led to enhanced star formation

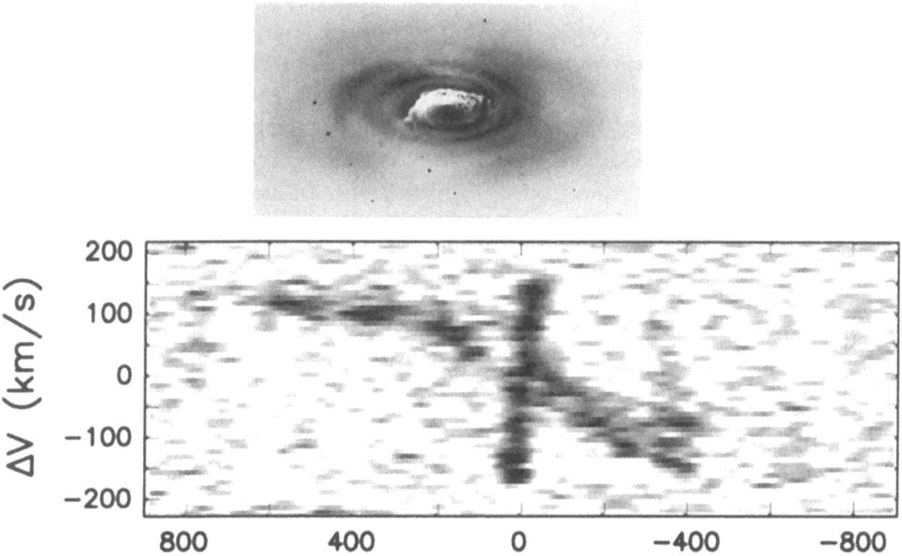

Fig. 53. Photograph of NGC 4826 by Ritchie and H I velocity profile along major axis. Note signature of two nested, counterrotating parts of H I disk. Photograph courtesy of Sandage & Bedke (1994), diagram from Braun et al. (1994).

and intense infrared emission (Benvenuti et al. 1976; van Driel et al. 1995). Perhaps the most spectacular example of a settling gas-and-dust disk is observed in NGC 4753 (Steiman-Cameron et al. 1992). Figure 52 compares a highly processed CCD image of this galaxy with the model of a 15° inclined gas disk twisted by 3.8π of differential precession over a factor of 7 in radius. This model reproduces the observed system of dust lanes in great detail. It implies that a merger occurred a few rotation periods ago and that the accreted gas will settle over most of the disk within a few more rotation periods.

An especially interesting galaxy with a counterrotating gas disk is NGC 4826, sometimes referred to as the "Evil Eye" galaxy (see Fig. 53). This Sab spiral is interesting for two reasons: It represents one of the latest Hubble types known to display a strong kinematic signature of a past merger, and its extended gas disk counterrotates only in its outer parts. This interesting fact was not immediately appreciated. A study of the H I kinematics with the VLA revealed that, while most of the H I disk rotates in a well-behaved manner, a central region of radius $\sim 50'' \approx 2\,\mathrm{kpc}$ rotates in the opposite direction (Braun et al. 1992; all numbers here converted to $H_0 = 50$). In a transition region extending from about $50''$ to $200''$ the velocity field displays severe distortions. Overall, the H I disk extends out to $11' \approx 24\,\mathrm{kpc}$ and contains $1.1 \times 10^9\,\mathcal{M}_\odot$ of gas (H_2 included). The kinematically distinct central region of 2 kpc radius contains about half of the total gas mass, most of it in molecular form. Whereas the surface density of gas is a high $\sim 55\,\mathcal{M}_\odot\,\mathrm{pc}^{-2}$ in this central disk, it is about 100 times lower further out (Braun et al. 1994). Figure 53 shows the H I rotation profile of NGC 4826 along the major axis and, twice enlarged, an optical photograph. The kinematic signature of the two counterrotating parts of the H I disk is striking.

The surprise is the direction in which the *stellar* disk rotates (Rubin 1994; Braun et al. 1994). This disk rotates in the same direction as the inner H I and opposite to the outer H I. Hence, it must be the outer H I that was accreted. Yet, in the outer stellar disk, fewer than 5% of the stars (if any at all) rotate opposite to the bulk of the others (Rix et al. 1995). From this finding, Rix et al. conclude that the intruder was a dwarf galaxy with $\lesssim$10% the mass of the main galaxy. The details of this past merger, however, remain unclear. In the transition region where the gas velocities reverse, ionized gas is falling in at ~100 km s^{-1} (Rubin 1994). Hence, the intruder may have deposited considerably more gas than is now detected in the outer disk. Also, its stars may have been scattered over most of the halo of NGC 4826, leading to a large velocity dispersion and nondetection. It would hardly seem surprising to learn that the intruder had 10%–20% the mass of the present NGC 4826. In light of claims made about the ease with which galactic disks get disrupted by even minor mergers (Tóth & Ostriker 1992), perhaps the most interesting fact about NGC 4826 is that its stellar disk survived the past merger apparently intact.

5.5 Counterrotating Stars in Disk Galaxies!

Occasionally, Nature produces even more extraordinary disk galaxies, such as NGC 4550 (E7/S0). At any given radius in this galaxy, approximately one half of the disk stars orbit around the center in one direction, while the other half orbit in the opposite direction (Rubin, Graham, & Kenney 1992). This discovery by my colleague Vera Rubin came as a complete surprise. While analyzing spectra of Virgo Cluster galaxies to study their central kinematics, she noticed that NGC 4550 features not only a disk of ionized gas counterrotating to the stars, but also a stellar disk with doubled absorption lines that indicate stars counterrotating to each other! This finding seemed so strange that Rubin spent more than a year checking and rechecking the data and her interpretation before deciding to publish the result.

Figure 54 shows a region of the spectrum of NGC 4550 centered on the Mg triplet of absorption lines. The *crossed lines* form a striking kinematic signature, as sketched to the left (Rubin 1993). From the measured line strengths, rotation velocities, and velocity dispersions there can be little doubt that the two populations of counterstreaming stars are roughly equal and form two cospatial, though counterrotating disks (Rubin et al. 1992). As it turns out, another group of astronomers had also obtained spectra of this galaxy and had processed them routinely, but had failed to find the extraordinary kinematic signature. After learning of Rubin's discovery, Rix et al. (1992) analyzed their spectra with software modified to recognize multiple-component absorption lines and quickly confirmed the existence of two counterrotating stellar populations in the disk of NGC 4550. Figure 55 shows their line-of-sight velocity distributions measured at five distances from the center along the major axis. The double Gaussians show fits to the observed double profiles, while the broader single Gaussians show fits

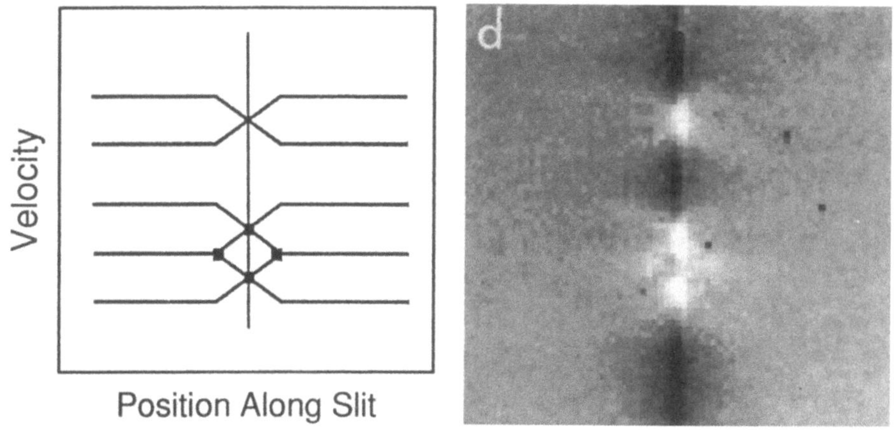

Fig. 54. Crossed stellar absorption lines in spectrum taken along the disk of NGC 4550 (E7/S0). Wavelength increases upward. The Mg triplet of two counterrotating stellar disk populations is sketched to the left. Illustration after Rubin (1993).

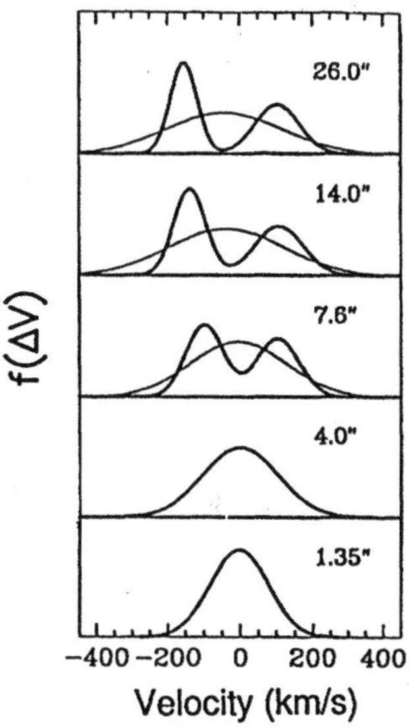

Fig. 55. Line-of-sight velocity distributions in disk of NGC 4550 measured at five distances from the center. Note doubling of the velocity distribution with increasing distance (*toward top*) and failure of single-Gaussian fits. Diagram from Rix et al. (1992).

with the old software. As is obvious from this figure, if one has software that expects and fits only one Gaussian to the data, one misses the most interesting phenomenon. This is exactly what happened at first. There is a lesson to be learned from this experience: no matter how fancy your measuring software is, ***do inspect your data visually!***

As observers have begun analyzing spectra of disk galaxies with increasingly sophisticated software, more examples of counterrotating stars are being found. For example, in the Sb galaxy NGC 7217 an analysis of the line-of-sight velocity distributions reveals that about 70% of the disk stars rotate one way and 30%

the other way (Kuijken 1993; Merrifield & Kuijken 1994). And in the Sb galaxy NGC 7331, the whole central bulge appears to counterrotate relative to the disk (Prada et al. 1996). This bulge has long been known to be boxy, and one proposed mechanism for generating boxy bulges is the resonant scattering of disk stars into the halo by a bar (Combes et al. 1990; Pfenniger & Friedli 1991). Though this mechanism is supported by N-body simulations (e.g., Martinet 1995), it seems to me that at least in the case of NGC 7331 this mechanism cannot have created the boxy bulge because there are no counterrotating disk stars. Instead, the counterrotating bulge strongly points toward some merger having formed either the bulge or the disk in a second event.

The most vexing problem is to explain how any merger could lead to two cospatial, but counterrotating disks of stars. At present, there are no good explanations. Perhaps gas slowly fed into a preexisting stellar disk might do the trick, but Nature hardly seems that purposeful or gentle. More likely some merger event may have led a galaxy through a polar-ring stage, and the gas may have subsequently settled into a counterrotating disk that formed new stars. What remains perplexing, however, is that in both NGC 4550 and NGC 7217 the merging galaxies seem to have contributed comparable masses of disk material. Given the claimed fragility of stellar disks (Tóth & Ostriker 1992), how exactly did this process work? One way out would be to envisage the formation of two counterrotating *gaseous* disks, but could they have been cospatial and have coexisted long enough to form all the observed stars? Anyone able to answer such questions and solve the puzzle of NGC 4550 is likely to earn, if not a job, then at least *fame!*

5.6 Summary

The main results of this chapter are as follows:

- Two distinct spin systems in one single galaxy are the kinematic signature of a "Second Event", i.e., of a past accretion or merger.
- S0 galaxies with polar rings probably formed through mergers. Galaxies with relatively small polar rings may have experienced accretion over their poles, while in galaxies with extended polar rings it may be the S0 bodies that are the secondary structures ("failed bulge" hypothesis).
- Accretions and mergers at a skewed angle to the principal plane of a galaxy lead to gas settling through differential precession and dissipation (e.g., NGC 4753).
- In elliptical galaxies, oddly rotating *stellar* cores suggest major past disk–disk mergers followed by dissipation and star formation (cf. NGC 7252). Extended, often warped *gas* disks are mostly of external origin and may be indicative of major mergers as well (e.g., inner disk and outer H I fragments of NGC 5128).
- In S0 to Sb galaxies, counterrotating disks of gas and stars offer us glimpses of mergers and accretions that must have reshaped these galaxies, probably moving them toward earlier types in Hubble's morphological sequence.

6 Formation of E and S0 Galaxies

During the past decade, it has become increasingly clear that early-type galaxies are not as uniformly old as was previously believed. Their structural and kinematic complexities point toward mergers having either formed or at least reshaped them (see Chaps. 4 and 5). Table 3 below recapitulates the main properties of E and S0 galaxies that point toward such past mergers. For example, the $r^{1/4}$ light distribution of an elliptical is indicative of violent relaxation and is, therefore, consistent ("C") with a major merger event in the galaxy's past history, while a pair of tails provides more direct evidence ("Y") for a past merger involving two disk galaxies.

This evidence of various sorts for past mergers leads quite naturally to three important questions: (1) What fraction of E and S0 galaxies did form through major mergers? (2) Does this fraction depend on the environment? And (3), in any given E or S0 galaxy, when did the last major merger occur? The present Chapter addresses these three questions. Yet, given the rudimentary state of our knowledge, the answers presented below are to be regarded as preliminary.

In order to try to answer the above questions, we need mainly two kinds of information. First, we need statistics about the various types of merger signatures in E + S0 galaxies. And second, we also need to develop a more detailed understanding of how merger remnants evolve with time. Specifically, if many early-type galaxies formed through delayed mergers as envisaged by Toomre (1977a; and TT), they should contain evidence for ancient starbursts of different ages. The search for such evidence is developing rapidly, and results obtained to date by different research groups appear often to contradict each other. For relatively recent collections of views held by the experts, see the proceedings of the conferences on the *Structure, Dynamics, and Chemical Evolution of Elliptical*

Table 3. Clues to the Formation of E and S0 Galaxies

Property	Indicative of	Evidence for Merger?[a]	
		E	S0
$R^{1/4}$ light distribution	Violent relaxation	C	(C)
Twist, minor-axis rotation	Triaxiality & pressure support	C	—
Boxiness, X-structure	Preferential orbit filling	Y	C
Ripples	External disk material	Y	Y?
Tail(s)	Merged disk galaxy(ies)	Y	—
Counterrot./warped gas disk	Gas of external origin	Y	Y
Oddly rotating stellar core	Dissipation & star formation	Y	—
Counterrotating stars	"Second event"	C	Y
Polar ring	"Second event"	Y	Y

[a] C = Consistent with; Y = Yes, evidence for past merger.

Table 4. Fine Structure in 74 Field E and S0 Galaxies

Type of Fine Structure	E + E/S0 ($N_{\rm gal} = 36$)	S0 + S0/Sa ($N_{\rm gal} = 38$)
Ripples	56%	32%
"Jets" (= plumes, tails, etc.)	22%	21%
Boxy isophotes	31%	32%
X-Structure	8%	11%
Ripples/Jets/Boxy/X-structure	69%	53%

Galaxies (Danziger et al. 1993), on *Fresh Views of Elliptical Galaxies* (Buzzoni et al. 1995), and on *The Nature of Elliptical Galaxies* (Arnaboldi et al. 1997). In this Chapter, I attempt to give at least a brief introduction to the vast field of population synthesis and to various ongoing efforts at age dating early-type galaxies, but I admittedly follow a rather personal path to some of the answers.

6.1 Fine-Structure Statistics and Index

Reliable statistics on signatures attributed to past mergers in E + S0 galaxies are still hard to come by. Studies of such signatures are often made on galaxy samples that are not specifically designed for statistical analysis. Some valuable, but fragmentary statistics on gas disks and oddly rotating stellar cores in ellipticals has already been discussed in Chap. 5. Perhaps the largest sample of early-type galaxies systematically searched for morphological signatures of past mergers is the sample of 74 E + S0 galaxies imaged by Patrick Seitzer and myself. Even this sample is incomplete, but at least its galaxies were picked at random from a well-defined parent sample of 145 nearby ($cz < 4000\,{\rm km\,s^{-1}}$) E and S0 galaxies north of $\delta = -15°$ and brighter than $B_{\rm T} = 13.5$. The one exception to the random selection process is that whenever there was a choice, we observed galaxies in the field or in small groups preferentially over those in the Virgo cluster, whence our sample consists nearly purely of non-cluster galaxies (Schweizer & Seitzer 1992, hereafter SS92). I shall call these objects "field" galaxies for short.

Table 4 presents statistics for four types of morphological fine structure found in these field galaxies and thought to be indicative of past mergers (SS92; Schweizer 1993). Note that *ripples* are detected in more than half of the field ellipticals and in one third of the field S0s. If we ask what fractions of the sample galaxies feature at least one of the four types of fine structure indicative of mergers, we find that about 2/3 of the Es and 1/2 of the S0s display such signatures (bottom line of Table 4). Therefore, the *majority* of field E + S0 galaxies seem to have experienced at least one merger during the past Hubble time. As discussed in §§4.3 and 4.5, this merger is likely to have been a *major* merger and may in many cases well be the event that formed these E and S0 galaxies.

A question of great interest is whether the fine structure correlates with other properties indicative of past mergers and starbursts such as, e.g., unusually blue *UBV* colors or enhanced Hβ absorption lines. To address this and related questions, Seitzer and I defined a fine-structure index Σ based on the four types of fine structure listed in Table 4, as follows:

$$\Sigma = S + \log(1 + n) + J + B + X\,, \tag{3}$$

where S is our visual estimate of the strength of the most prominent ripples ($S = 0-3$), n is the number of detected ripples ($n = 0-17$), J is the number of "jets" ($J = 0-4$), B is a visual estimate of the maximum boxiness of isophotes ($B = 0-3$), and X indicates the absence or presence of X-structure ($X = 0$ or 1). The use of visual estimates of fine structure may surprise, but is – given the great difficulties involved in measuring faint fine structure – a useful shortcut for exploratory studies.[18] Note that in the above definition of Σ, the adopted combination of visual estimates is not unique and may not even be optimal, though the strength parameters S and B were chosen so as to give roughly equal weight to ripples, boxiness, and "jets". With this definition, featureless ellipticals have an index $\Sigma = 0$, the field ellipticals of our sample have a median index $\Sigma_{\rm med} = 1.7$, and the most fine-structure rich elliptical, NGC 3610, has $\Sigma = 7.6$ (based on $S = 2$, $n = 3$, $J = 1$, $B = 3$, and $X = 1$). For comparison, the two merger remnants NGC 3921 and NGC 7252, which are $\lesssim$1 Gyr old and feature more fine structure than any elliptical, have indices of $\Sigma = 8.8$ and 10.1, respectively.

Now that we have some statistics of merger-induced fine structure in E + S0 galaxies and an index to roughly quantify this fine structure, we are ready to tackle the next question.

6.2 Evidence for Ancient Starbursts in E and S0 Galaxies

Is there evidence from the stellar populations in E + S0 galaxies that different galaxies formed at different epochs?

The reason for asking this question is, of course, the hypothesis of elliptical formation through major disk mergers. Merger remnants like NGC 3921 and NGC 7252 show post-starburst spectra with strongly enhanced Balmer absorption lines (see Fig. 27 and chapters by Dr. Kennicutt). Since even present-day mergers nearly always involve gas and trigger major starbursts, the same must be true of past mergers that involved less mature, more gas-rich galaxies. Thus, faded forms of the spectral signatures seen so prominently in recent merger remnants should be detectable at least among ellipticals that formed during, say, the past 5–7 Gyr.

[18] As current morphological classification systems for galaxies illustrate, visual type estimates can remain useful for long periods of time and are often hard to replace with more quantitative measures. For example, the term "Sc galaxy" evokes many interesting notions in our minds, while a measured bulge-to-disk ratio of $B/D = 0.02$ does less so.

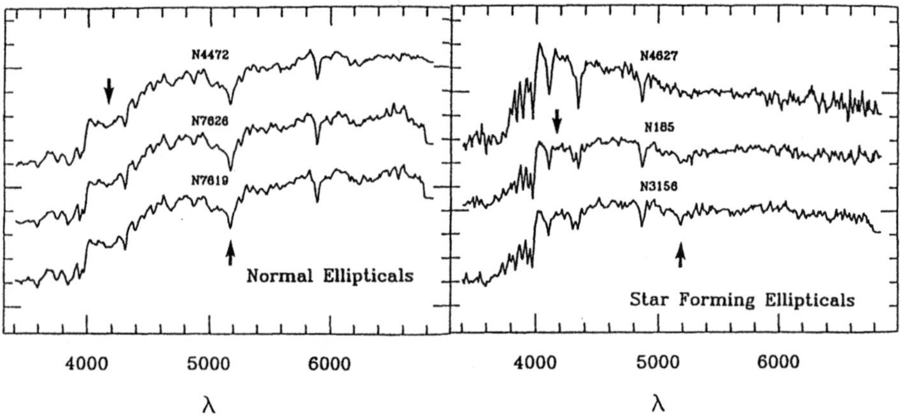

Fig. 56. Optical spectra of (a) three "normal" ellipticals and (b) three ellipticals with signs of relatively recent star formation (enhanced Balmer absorption, weak CN bands at λ4160, and weak Mg lines at λ5175). Diagrams after Gregg (1989a).

In fact, some elliptical galaxies with enhanced Balmer absorption lines have been known for many years. O'Connell (1976) pointed out that although the bulk of the stars in the nuclei of these ellipticals seemed to have formed 8–11 Gyr ago, star formation appeared to have continued in them "at a much reduced level until about 4 Gyr ago". In M32, the E2 companion of the Andromeda galaxy, even *intense* star formation seems to have occurred as little as 5 Gyr ago (O'Connell 1980). Further examples of such ellipticals were found by various investigators (Pickles 1985; Véron & Véron-Cetty 1985; Rose 1985; Bica & Alloin 1986, 1987). Figure 56 compares the spectra of three "normal" ellipticals with those of three ellipticals that seem to have formed stars until a few Gyr ago (Gregg 1989a). Notice the enhanced Balmer lines and the weak CN band and weak Mg absorption lines of the latter galaxies. Since A- to F-type stars show strong Balmer lines, but weaker CN bands and Mg lines than later-type stars, an admixture of A- and F-type stars to a population of old K-type stars can reproduce the observed galaxy spectra quite well. It is this well-known fact that spectral syntheses of early-type galaxies have attempted to exploit for the past quarter-century in order to estimate the relative numbers of old, intermediate-age, and young stars.

Correlations Between Fine Structure and Line Strengths

If indeed ellipticals formed through mergers, the fact that mergers both create fine structure and trigger starbursts suggests that the fine structure of ellipticals ought to correlate with spectral signatures of ancient starbursts.

A spectroscopic study of 137 "shell" galaxies from the Malin & Carter (1983) catalog produced some first evidence for such a correlation (Carter et al. 1988). Among these galaxies, about 15% – 20% show enhanced Balmer absorption lines, and some quite dramatically so. In extreme cases, more than half of the total

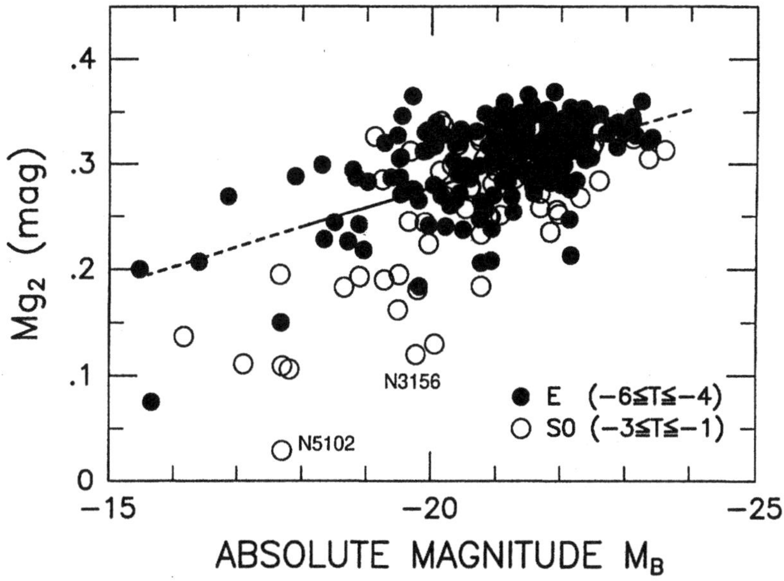

Fig. 57. Line-strength indices Mg_2 of 258 E + S0 galaxies plotted versus absolute magnitudes M_B. The *straight line* represents a least-squares fit to 241 galaxies with $-18 > M_B > -23$. The line-strength scatter is 3–4× larger than the mean observational errors, which are about the size of the data points.

light at $\lambda = 4000\,\text{Å}$ is estimated to be contributed by A-type stars. However, the results from this study are difficult to interpret statistically because a significant fraction of these shell galaxies are outright peculiar rather than being of type E or S0, and no comparison sample of normal E and S0 galaxies was observed. Nevertheless, this study is the first to have suggested that fine structure and population youth might indeed correlate.

This result is of particular interest because line strengths of normal E and S0 galaxies have long been known to show scatter significantly in excess of the observational errors (Faber 1977). Figure 57 illustrates this point. It shows the line-strength indices Mg_2 of 258 E + S0 galaxies measured at Lick Observatory (Faber et al. 1989) and here plotted versus the blue absolute magnitudes M_B. The sizes of the data points for the E and S0 galaxies correspond to the mean observational errors for the two kinds of galaxies ($\pm 1\sigma$). The figure shows that the vertical scatter in line strengths is about 3–4× larger than these mean observational errors. Therefore, most of the scatter must be real and of cosmic origin. The question then is: What causes this scatter? Traditionally, astronomers have interpreted the scatter as being due to abundance differences between galaxies. Yet, our above reasoning and the shell-galaxy results suggest that differences in the *mean age* of the stellar populations might contribute to this scatter as well. This suspicion is reinforced by outlier galaxies such as NGC 5102 and NGC 3156 (marked in Fig. 57), which both are well known to possess post-starburst spectra (Gallagher et al. 1975; Gregg 1989a,b).

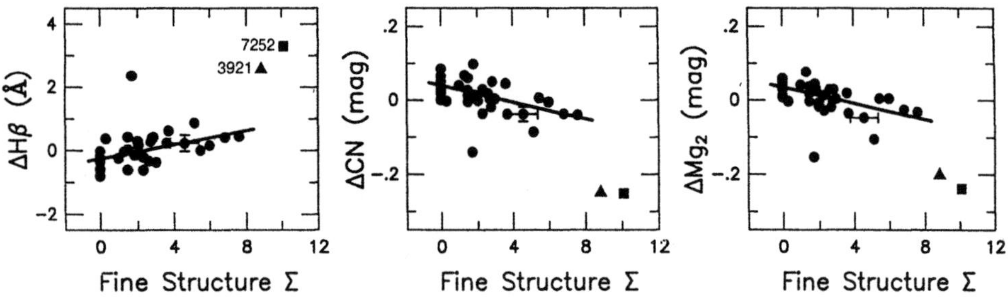

Fig. 58. Correlations between line-strength residuals and Σ for 36 field ellipticals and for the merger remnants NGC 3921 (*triangles*) and NGC 7252 (*squares*). The *straight lines* mark least-squares fits to the ellipticals. Note that $\Delta\mathrm{H}\beta$ correlates with Σ, whereas ΔCN and $\Delta\mathrm{Mg}_2$ anticorrelate. Diagrams after Schweizer et al. (1990).

A study of the vertical scatter of all field ellipticals with measured fine structure shows that, indeed, population age seems to be a major contributor to the vertical scatter (Schweizer et al. 1990). Figure 58 displays the line-strength residuals $\Delta\mathrm{H}\beta$, ΔCN, and $\Delta\mathrm{Mg}_2$ for 36 field ellipticals and NGC 3921 and NGC 7252 plotted versus each galaxy's fine-structure index Σ. These residuals were computed by subtracting from the line strength for each galaxy the mean line strength of all E + S0 galaxies at the same absolute magnitude, the mean itself having been determined by a linear least-squares fit like the one plotted in Fig. 57. As Fig. 58 shows, the line-strength residuals clearly correlate with fine structure: While the Hβ line strength increases with increasing Σ, the CN and Mg_2 indices decrease. These correlations are significant at the $4\sigma - 5\sigma$ level. The strengthening of Hβ with increasing fine structure is exactly what Carter et al. (1988) had observed in shell galaxies, but is now seen in a pure sample of field ellipticals picked at random. Again, since enhanced Balmer lines are a sign of younger stellar populations, the correlation suggests that the most recently merged systems (as indicated by a high Σ) have the youngest stellar populations, while those that merged long ago (low or zero Σ) have the oldest ones. The weakening of the CN and Mg_2 indices as Hβ strengthens can be explained naturally and quantitatively as being due to the dilution of these absorption features with continuum light from the younger stars. Detailed calculations show that <50% of the observed vertical scatter in the line-strength vs. absolute-magnitude relations is due to metallicity variations, while >50% of this scatter is caused by variations in the mean age of the stellar populations.

Correlations Between Fine Structure and *UBV* Colors

If the above conclusion that age differences are the main driver for the scatter of line strengths is correct, then the *UBV* colors of E + S0 galaxies should display a similar behavior since they measure mostly the continuum light and are sensitive to young stars.

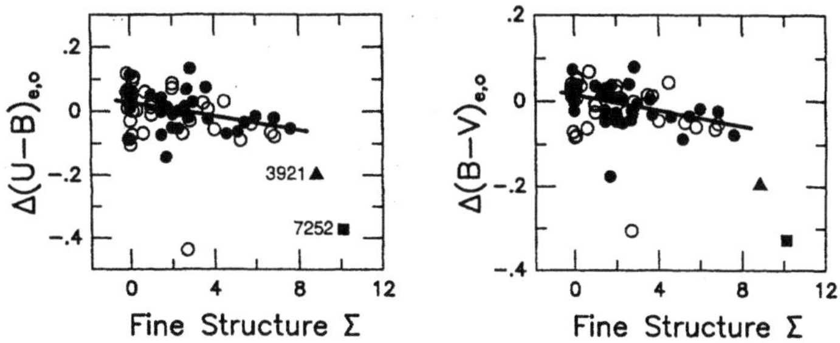

Fig. 59. Correlations between *UBV* color residuals and fine-structure index Σ for 69 field E and S0 galaxies and for NGC 3921 and NGC 7252. *Solid lines* mark mean relations for the E + S0s. Note that the color residuals, and by implication the colors themselves, get systematically bluer as Σ increases. Diagrams after SS92 and Schweizer (1996).

Interestingly, the color–magnitude relations for E + S0 galaxies have long been known to also show more vertical scatter than can be explained through observational errors alone (Sandage & Visvanathan 1978). An analysis of this scatter similar to the one for line strengths reveals that the color residuals again correlate with fine structure (SS92). Figure 59 shows the color residuals $\Delta(U-B)_{e,0}$ and $\Delta(B-V)_{e,0}$ of 69 field E + S0 galaxies and of NGC 3921 and NGC 7252 plotted versus the fine-structure index Σ. These residuals pertain to integrated *UBV* color indices within an *effective* aperture (i.e., one containing 50% of the total blue light of a galaxy). These indices have been corrected for foreground reddening. As the figure illustrates, E and S0 galaxies with much fine structure are systematically bluer than those without. Though the S0 galaxies show slightly more residual scatter than the ellipticals, the correlations with fine structure for both types of galaxies are indistinguishable.

It is easy to see that the bluer colors of high-Σ galaxies are caused by admixtures of younger stars rather than by metallicity variations. We know that fine structure reflects past mergers. To the extent that mergers induce starbursts, they tend to increase the mean metallicity of stellar populations. Therefore, if metallicity were the main driver of the color scatter, merger remnants with their higher metallicities should have redder colors. Yet, they appear systematically bluer. Hence, as for the line strengths, it is mainly mean-age variations that cause the color scatter observed among E + S0 galaxies of similar luminosity.

The degree to which the color residuals track the line-strength residuals is impressive. Figure 60 plots $\Delta(B-V)_{e,0}$ versus ΔMg_2 for 35 E and 21 S0 galaxies from the SS92 sample with both kinds of residuals available, and for NGC 3921 and NGC 7252 for comparison. The correlation for the E + S0 galaxies (*solid line*) is strong (correlation coefficient $r = 0.85$, for Es alone $r = 0.88$) and yields three interesting results: (1) The same agent that causes the scatter of line strengths must also cause the scatter of *UBV* colors, as expected if this agent is the mean population age. The fact that the young merger remnants NGC 3921 and NGC 7252 lie so close to the extrapolated regression line leaves little doubt that population age is indeed this agent. (2) The observed *UBV*-

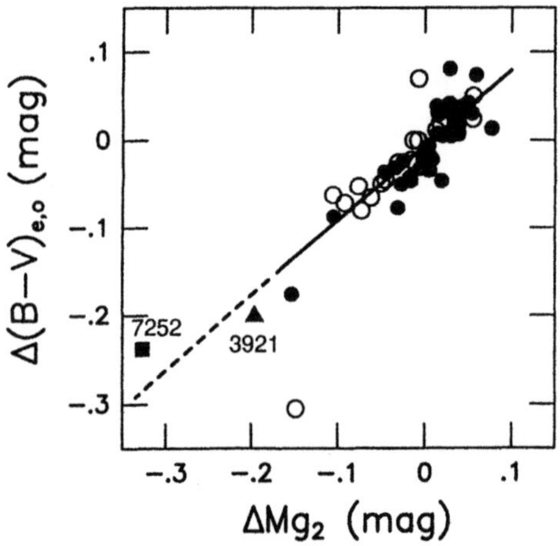

Fig. 60. Color residuals $\Delta(B\text{–}V)_{e,0}$ for 56 E (*dots*) and S0 (*open circles*) galaxies plus NGC 3921 and NGC 7252, plotted versus line-strength residuals ΔMg_2. Note the strong correlation between the two kinds of residuals. Diagram after SS92.

color variations cannot reflect internal reddening due to dust since they correlate with the line-strength indices, which are independent of reddening.[19] And (3), since $(B-V)_{e,0}$ refers to 50% of the total galaxian light while the Mg_2 index is typically measured through apertures containing only 10% – 15% of the total light, the color and line-strength scatter reflects mean-age variations in the *global* stellar populations, rather than just in the nucleus. This result agrees with the observed absence of significant age gradients within elliptical galaxies (Davies et al. 1993) and within NGC 7252 (Chap. 3).

Note that diagrams similar to Fig. 60 can easily be plotted for hundreds of E + S0 galaxies since the computation of the various residuals does not involve any knowledge of the fine structure. Plotting and analyzing such diagrams should yield interesting kinds of new information.

Dependence of Colors and Line Strengths on the Environment

The studies of line-strength and color scatter described above deal mostly with *field* E and S0 galaxies. At first sight, the conclusion that this scatter reflects differences in the mean age of the stellar populations may appear to be in conflict with a large number of mostly older studies claiming that all E + S0 are uniformly old. However, many of these studies were mostly of ellipticals in the Virgo and Coma clusters (e.g., Bower et al. 1992). Could it be that early-type galaxies in the field tend to be younger than those in clusters?

Recent studies are indeed beginning to uncover systematic line-strength and color differences between early-type galaxies in the field and in clusters, and even between galaxies in the outskirts and in the cores of clusters. For example, a multivariate statistical analysis of a large sample of E galaxies finds that field

[19] This result pretty much invalidates the claim, made in a growing number of recent papers (e.g., Wise & Silva 1996), that color gradients in early-type galaxies may reflect mainly the effects of extended dust distributions.

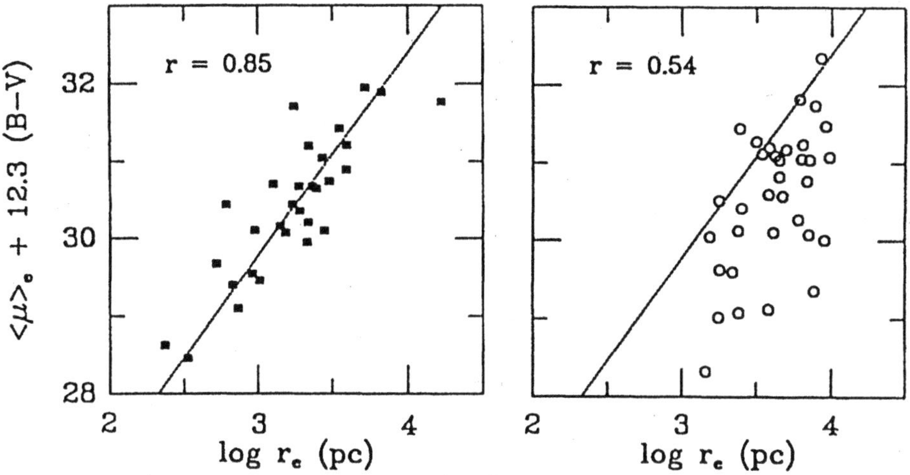

Fig. 61. Linear combination of mean surface brightness $\langle\mu\rangle_e$ and color $B-V$ plotted versus effective radius r_e for ellipticals in clusters (*squares, left*) and in the field (*open circles, right*). The line in both panels shows the mean relation for cluster Es. Note the systematic differences between field and cluster ellipticals. Diagrams from de Carvalho & Djorgovski (1992).

ellipticals tend to have bluer colors, weaker Mg lines, and higher surface brightnesses than cluster ellipticals of the same luminosity (de Carvalho & Djorgovski 1992). Figure 61 compares a linear combination of the mean surface brightness $\langle\mu\rangle_e$ and color index $B-V$ for ellipticals in clusters (*left* panel) and in the field (*right* panel). This linear combination is plotted against the logarithm of the effective radius r_e, and a fit to the cluster ellipticals is shown in both panels. Notice how the field ellipticals show more scatter among themselves and fall below the cluster sequence. Their bluer colors, higher surface brightnesses, and weaker Mg lines (not shown) all suggest that these field ellipticals contain younger stellar populations than cluster ellipticals do.

Even within the Coma cluster, differences can be discerned between E + S0 galaxies in the outskirts and near the center. Galaxies located in the outskirts feature systematically weaker Mg lines and show more scatter in line strengths, again suggesting that they may contain stellar populations that are younger and vary more in age (Guzmán et al. 1992). In fact, early-type galaxies with enhanced Hβ absorption and, thus, unmistakable signs of aging starbursts seem to congregate in an extended area about 40′ southwest of the center (Caldwell et al. 1993), indicating that perhaps a subgroup of galaxies having experienced more recent mergers is presently falling into the main cluster.

Overall, the evidence is strong that mean-population-age differences do exist among E + S0 galaxies and that they depend on the environment. In the field, these mean-age differences are directly linked to morphological signatures of mergers. In clusters, the situation is less clear since these differences could be linked either to mergers (directly or indirectly) or conceivably also to multiple interactions having not yet led to mergers (Moore et al. 1996).

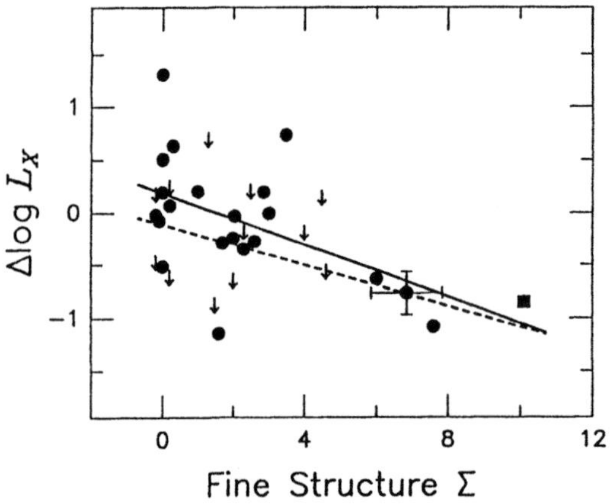

Fig. 62. Correlation between residuals $\Delta \log L_X$ and fine-structure index Σ for 34 early-type galaxies (*dots, arrows*) and NGC 7252 (*square*). *Solid line* marks regression for 22 X-ray detected galaxies, *dashed line* regression for 34 galaxies (incl. 12 non-detections). Note that the L_X-residuals anticorrelate with Σ.

Correlations Between Isophotal Shape, Fine Structure, Radio Luminosity, and X-ray Luminosity

Various other correlations also suggestive of effects of past mergers in E + S0 galaxies have been found.

For example, the radio and X-ray luminosities of elliptical galaxies seem to correlate with the central isophotal shapes (Bender et al. 1987, 1989). Whereas disky ellipticals ($C_4 > 0$, see Chap. 4) have uniformly low radio luminosities L_R, boxy ellipticals ($C_4 < 0$) scatter in radio luminosity and can reach very high L_R. Similar, but weaker relations seem to also exist for the X-ray luminosities L_X. These correlations suggest that the space-filling orbits associated with boxiness lead to more radio and X-ray emission, while the more nearly two-dimensional orbits associated with diskiness lead to little such emission. However, it is not clear to me why this apparent dichotomy should imply that boxy ellipticals are merger remnants, while disky ellipticals are not, as has been widely interpreted (e.g., Bender 1990). Since Bender's isophotal analyses have been restricted to the central regions of ellipticals, the shape parameter C_4/a (where a is the semimajor axis) used by him is a measure of the inner structure of an elliptical, rather than of the global structure or of effects directly related to mergers. If there were a 1 : 1 correlation between boxiness and mergers, then the L_X–C_4/a correlation would suggest that merger remnants are much more X-ray luminous than non-remnants.

Yet, in an unpublished study with Fabbiano and Seitzer we found the exact opposite. Figure 62 displays L_X-residuals for 34 E + S0 galaxies and for NGC 7252 plotted versus the fine-structure index Σ. These residuals were determined from L_X–M_B relations for early-type galaxies in a manner completely analogous to the line-strength and color residuals. Galaxies detected in X-rays are plotted as dots, while those with only upper limits to L_X are plotted as downward-pointing arrows. Obviously, the residuals $\Delta \log L_X$ of the 34 E + S0 galaxies *anti*correlate with Σ. The anticorrelation is significant at the $2\sigma - 3\sigma$ level, depending on the weights one assigns to the upper limits. This anticorrelation suggests that E + S0 galaxies with rich fine structure may have systematically *lower*, rather than higher, X-ray luminosities. Since these galaxies appear to have experienced mergers relatively recently, they must have been either depleted of hot gas through starburst-induced winds or borne with less hot gas because they involved less gas-rich galaxies than ancient mergers did.

This result is at odds with the results by Bender et al. (1989). Personally, I believe that the fine-structure index Σ is a better indicator of past merger activity than the isophotal shape parameter C_4/a. Obviously, with new *ROSAT* data becoming available, the various correlations with L_X and the conclusions drawn from them can soon be tested in greater detail.

6.3 Age Dating E and S0 Galaxies

If indeed field E + S0 galaxies are the products of delayed mergers, one ought to be able to date their formation epochs. Such age dating will eventually provide the acid test for the hypothesis of elliptical formation through mergers. Since the early 1990s, there has been a tremendous revival of interest in this fundamental issue.

The problems to be overcome are not only of a technical nature, but also psychological and rooted in the history of the subject. In introducing his concept of two stellar populations, Baade (1944) proposed that the stellar population of early-type galaxies is "closely related to, if not identical with, that of the globular clusters". Therefore, as subsequent research showed all globular clusters in the Milky Way to be very old, a dogma arose that all E + S0 galaxies are old as well. In a landmark paper, Searle, Sargent, & Bagnuolo (1973) demonstrated that the *UBV* colors of galaxies can be explained if (1) all galaxies are uniformly 15 Gyr old, (2) their star formation rates decline exponentially with time, and (3) in Sc galaxies this decline occurs with time constants of $\tau \approx 10$ Gyr, while in early-type galaxies it occurs with shorter time constants. This paper deeply influenced the field and seemed, unintentionally, to support the belief that all ellipticals are old. This is why, when O'Connell (1976, 1980) and others found ellipticals with signs of prolonged and even relatively recent star formation, serious controversy arose and many of the issues remain hotly contested.

To be sure, there are both fundamental difficulties and serious technical problems when one tries to age date E and S0 galaxies from their integrated spectra.

The fundamental difficulties arise from the many free parameters buried in the past star-formation histories of these galaxies and from the complexities of stellar evolution. To interpret the present-day spectrum of a composite stellar population, one needs to know the past star-formation rate SFR(t), initial mass function IMF(t), and chemical abundance $Z(t)$ as functions of time. The notion of a mean chemical abundance $Z(t)$ itself is an oversimplification, since different elements may form at different rates. In addition, advanced stages of stellar evolution that are still poorly understood play a pivotal role, determining the enrichment of the interstellar medium through stellar winds and supernova ejecta as well as spectral parameters such as the continuum shape and line strengths.

There are also practical difficulties and technical problems. For example, the spectral lines of galaxies appear smeared and blended by the velocity dispersion of the stars. Also, many older data sets lack reliable information about the shape of the spectral continuum. But perhaps the greatest challenge has arisen from the presence of apparent age–metallicity and age–burst-strength degeneracies, as the following subsection discusses (see also O'Connell 1986, 1994).

Age Dating Stellar Populations from Line Strengths

The integrated spectra of composite stellar populations in galaxies are sensitive to both age and metallicity. Unfortunately, at the relatively low dispersions typically used in extragalactic spectroscopy, the effects of decreasing age are difficult to disentangle from those of increasing metallicity. As Worthey (1992, 1994) has reemphasized, the optical spectrum of a 5 Gyr old population of a given metallicity looks remarkably similar to that of a 15 Gyr old population of half that metallicity. This is especially true if the exact shape of the continuum is not known with high accuracy and one judges a spectrum mostly by its metal lines. It is this apparent *age–metallicity degeneracy* that has made the determination of ages of old populations so difficult and has fueled controversy over the ages of stars in early-type galaxies for more than a decade. Yet, the situation is not quite as bad as might appear because both the Balmer lines and the continuum behave distinctly differently from the metal lines and can, therefore, be used to isolate age effects (Burstein et al. 1984; Bica & Alloin 1986; Schweizer et al. 1990; Worthey 1994).

Figure 63 shows a modern diagram of Hβ line strengths plotted versus a combined Mg + Fe metallicity index for 23 ellipticals (González 1993). Superposed on the data points is an age–metallicity grid for single-burst populations computed by Worthey (1992). Notice that the galaxies (*data points*) scatter relatively little in metallicity [Fe/H], but a great deal in single-burst age. Eighteen of the 23 ellipticals have metallicities within the narrow range $-0.05 \le$ [Fe/H] $\le +0.30$, as measured through an aperture of radius $r_e/2$, while their single-burst ages range from about 3 Gyr to over 15 Gyr. These ages represent upper limits to the epochs of the last significant star formation, since old stellar populations are presumably mixed in with the younger populations. Therefore, even though these single-burst ages should not be interpreted too literally, Fig. 63 supports the result from the line-strength–fine-structure correlations that large mean-age variations do occur in early-type galaxies and contribute significantly to the vertical scatter in the line-strength–luminosity relations (§6.2).

Observations of higher spectral resolution hold promise for overcoming the problem posed by the apparent age–metallicity degeneracy. Work in this area is progressing rapidly. For example, Jones & Worthey (1995) find that a high-resolution Hγ-line index ($\Delta\lambda$ = 3.74 Å) is an excellent age diagnostic, while the broader index "Fe4668" ($\Delta\lambda$ = 85 Å) measures mostly the abundance of the metals Fe, Ti, Cr, Mg, and Ni. Applied to M32, these two indices now clearly date the last major starburst in this nearby elliptical to $\lesssim$7 Gyr ago, thus confirming the claims by O'Connell (1980) and Rose (1994), the latter

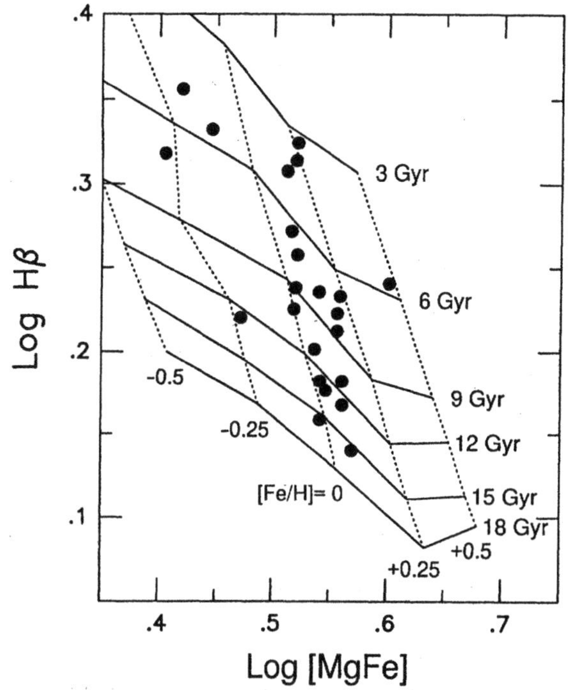

Fig. 63. Hβ line-strength indices for 23 E galaxies plotted versus metallicity index [MgFe] (*data points*). A grid of isochrones (*solid lines*) and isofers[20] (*dashed*) for single-burst model populations is superposed. Diagram after González (1993).

also based on narrow-line indices. The price one pays for breaking the apparent age–metallicity degeneracy with higher-resolution spectroscopy is that one needs larger telescopes, longer exposure times, and accurate corrections for the effects of velocity dispersion and Balmer emission-line contamination.

The second serious problem encountered when trying to age-date stellar populations in E + S0 galaxies is the apparent *age–burst-strength degeneracy.* When dealing with mixtures of old and young stellar populations, it is difficult to spectrally distinguish between the effects of relatively recent weak starbursts and older stronger starbursts (Bica et al. 1990; Schmidt et al. 1991). Figure 64 illustrates the problem. The upper spectrum stems from a starburst of 300 Myr duration viewed immediately after burst completion and contributing 40% of the light at 4000 Å, with the remaining 60% contributed by an old stellar population. The lower spectrum stems from a starburst of 500 Myr duration that ended 1.5 Gyr ago and contributes 80% of the light at 4000 Å. The ratio of the two spectra, plotted at the bottom of the figure, shows that only the K line of CaII differs significantly between the two composite spectra. Thus, this line is a valuable diagnostic for helping break the age–burst-strength degeneracy (Leonardi & Rose 1996).

In short, age dating the formation of E + S0 galaxies from their composite spectra is still in its infancy. The challenge is to break the two apparent degeneracies involving age, metallicity, and burst strength. Progress can be made

[20] *Isofer:* a line of constant iron abundance.

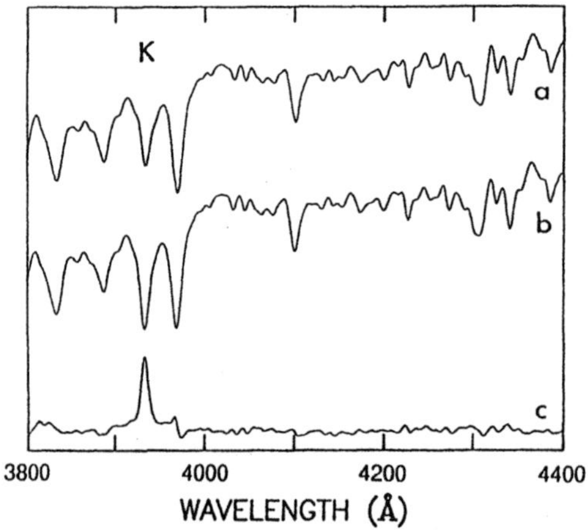

Fig. 64. Spectra from two simulated mixed-age stellar populations: (a) Zero-age starburst population mixed with old population in ratio 40:60 of light output at 4000 Å; (b) 1.5 Gyr old starburst mixed with old population in ratio 80:20 at same wavelength; and (c) ratio of the above two spectra. Diagram from Leonardi & Rose (1996).

by using higher spectral resolution, measuring the continuum or *UBVRIJHKL* color indices with high accuracy, and computing families of two-burst models to interpret the observations. Also, many of the uncertainties due to stellar evolution can be reduced significantly by using observations of star clusters, rather than of individual stars, to build spectral libraries (e.g., Bica & Alloin 1986, 1987; Bica et al. 1994). With new cluster-evolution models of different metallicities now available (Bruzual & Charlot 1993, 1996; Charlot et al. 1996), major progress seems finally within reach, and I strongly recommend this research area to anyone looking for an interesting thesis subject.

Heuristic Merger Ages from *UBV* Colors

Even though reliable stellar-population ages of E + S0 galaxies determined from spectra seem still several years in the future, one can estimate rough formation ages from *UBV* colors. The basis for this method is the finding that the vertical scatter in the color–magnitude relations correlates with fine structure and is determined mainly by age variations (§6.2). Since the evolution of population colors with time is well understood, one can translate the color scatter into rough ages if one makes some assumptions about the past star formation history of E + S0 galaxies (SS92).

Figure 65 presents the basic idea of this method, to be refined in Figs. 66 and 67. Assume that all galaxies began forming 15 Gyr ago, but formed stars at different, exponentially declining rates as envisaged by Searle et al. (1973). In a $(U-B,B-V)$ diagram, these model galaxies will evolve between the two limiting paths shown in Fig. 65a. The upper limiting path is that followed by a galaxy with constant star formation rate (SFR), while the lower path is that followed by a galaxy with a single, instantaneous starburst at the beginning (Charlot & Bruzual 1991). At times 1, 5, 10, and 15 Gyr, all galaxies with different exponential SFRs lie along one of the plotted isochrones. For comparison, Fig. 65b shows the reduced color indices $(U-B)_{e,0}^{-21}$ and $(B-V)_{e,0}^{-21}$ of field E + S0 galaxies

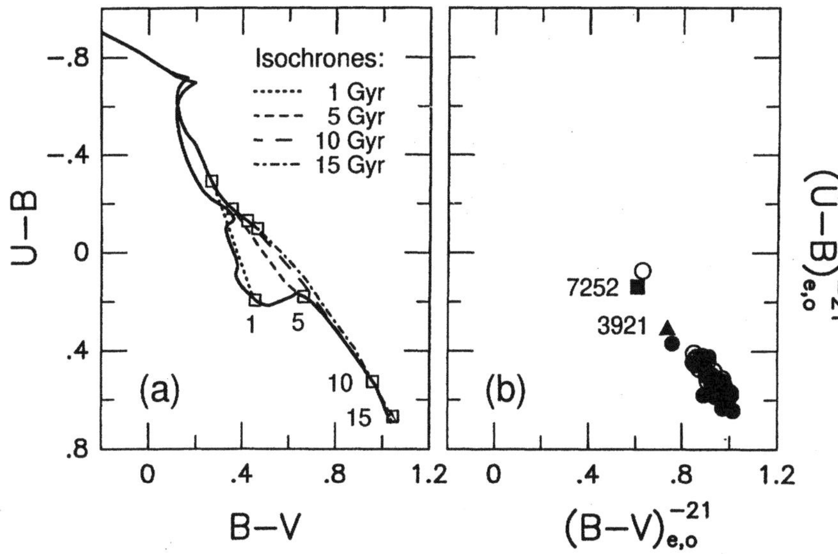

Fig. 65. (a) Evolutionary paths of model galaxies in two-color diagram compared with (b) reduced colors of E + S0 galaxies (*dots + circles*) and two merger remnants (*marked*). Upper solid curve shows path of an aging galaxy with constant star formation rate, lower curve that of an aging single-burst galaxy; dotted and dashed lines represent isochrones as listed. Diagram after SS92.

and the merger remnants NGC 3921 and NGC 7252.[21] Obviously, these galaxies form a sequence in the two-color diagram. In the past it was thought that this sequence reflects small differences in the decline of the SFR among the galaxies. However, the location of galaxies like NGC 3921 and NGC 7252 in this diagram reminds us that remnants of mergers having experienced second bursts of star formation lie along the same sequence (Larson & Tinsley 1978). Hence, if early-type galaxies formed through mergers we need to consider *two-burst models* of star formation to interpret the observed color variations along the sequence.

Figure 66 shows the star formation rate as a function of time for two such models. In both examples, the SFR first declines exponentially with a time constant $\tau_1 = 6\,\mathrm{Gyr}$ characteristic of Sb galaxies (Searle et al. 1973) and then experiences a second burst – presumed to be triggered by a merger – beginning at the epoch $t_\mathrm{m} = 10\,\mathrm{Gyr}$. This second burst features an exponentially declining SFR with a different time constant, τ_2, and converts a fraction ϵ of the remaining gas into stars. As the figure illustrates, these simple two-burst models feature four

[21] The superscript "–21" indicates that the color index of any given galaxy has been reduced to that of a galaxy of absolute magnitude $M_B = -21$ and same vertical color scatter. This trick of reducing the colors to a standard absolute magnitude is useful in that it eliminates systematic color variations that depend on luminosity and, hence, mainly on the mean metallicity (for details, see SS92).

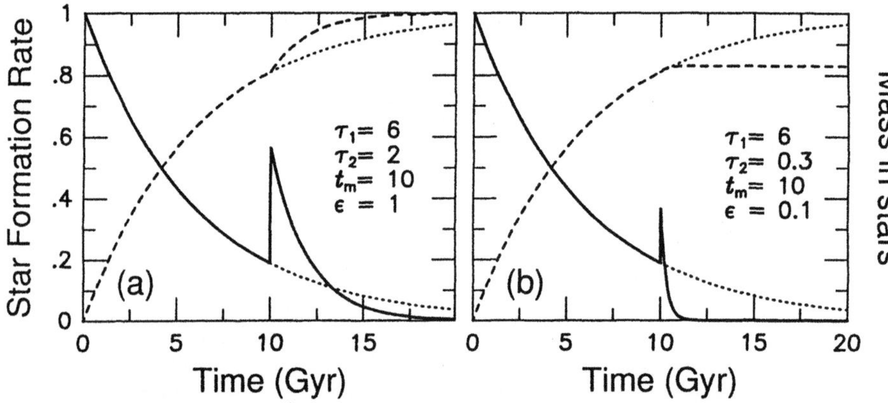

Fig. 66. Two-burst models of star formation; plotted as functions of time are the relative SFR (*thick lines*) and the relative mass in stars (*dashed lines*). The exponential declines before $t_m = 10$ Gyr are characteristic of Sb galaxies, while the second bursts beginning at t_m are typical of bursts occurring during mergers. Diagram after SS92.

parameters (τ_1, τ_2, t_m, ϵ) and represent, in essence, the presumed star formation histories of two disk galaxies that merge into a single early-type galaxy at time t_m. Note that these models are self-consistent in accounting for the gradual conversion of gas to stars, but do not include chemical evolution.

The calculation of *UBV* colors for such models is straightforward. Figure 67 shows a sample of evolutionary paths in the ($U-B$,$B-V$) diagram. Panel (a) illustrates the effects of varying the merger epoch t_m, while Panel (b) shows the effects of varying the Hubble type of the pre-merger disk galaxies via the parameter τ_1. All paths end at 15 Gyr, where the solid square represents the location of galaxies at the current epoch. The *UBV* colors were computed from the colors of evolving instantaneous starbursts (Charlot & Bruzual 1991) by convolving the fluxes with the SFR. In each panel, the left-most path appears at its proper position in the two-color diagram, while paths to the right have been shifted successively by +0.2 mag in *B–V* for increased clarity. A detailed survey of the parameter space of these two-burst models shows that the largest variations in the *UBV* colors of present-day merger remnants are due to variations in the merger epoch (t_m) and Hubble types of the merging galaxies (via τ_1), as illustrated in the figure. Within limits set by observations, the parameters τ_2 and ϵ affect the colors at 15 Gyr much less.

When interpreted with these simple two-burst models, the reduced *UBV* colors of E + S0 galaxies yield a ***heuristic merger age*** for each galaxy. The term "heuristic" indicates that this age depends on the morphological types of the disk galaxies that merged. Since these types are not known at present, we can only compute ages in an exploratory fashion by assuming the Hubble types of the merger components. Figure 68 displays two distributions of heuristic merger ages computed on the assumption that the merging disk galaxies were Sb–Sb pairs (*left* panel) or Sc–Sc pairs (*right* panel). The ages are plotted in the form

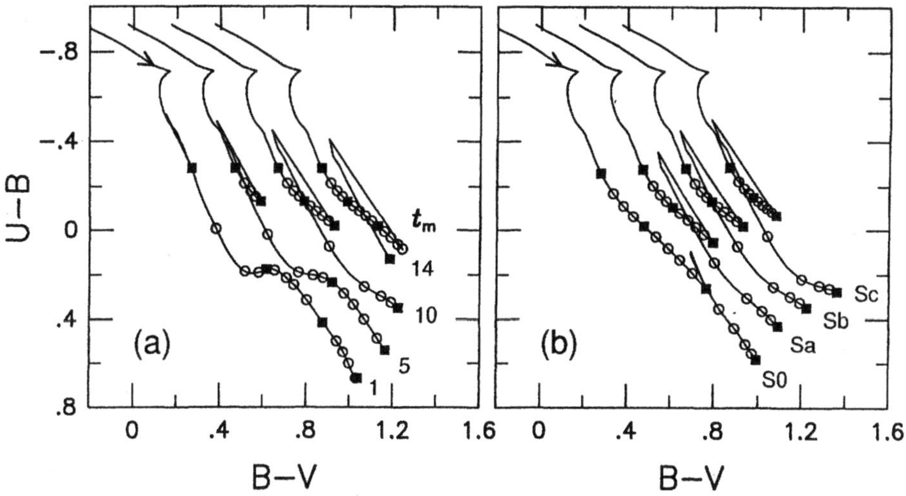

Fig. 67. Evolutionary paths of model mergers in two-color diagram. *Solid squares* along paths mark times 1, 5, 10, and 15 Gyr. (a) Paths for mergers occurring at $t_m = 1$, 5, 10, and 14 Gyr; (b) paths for merging disks of Hubble types S0, Sa, Sb, and Sc (corresponding to $\tau_1 = 2$, 4, 6, and 10 Gyr); for details, see text. Diagrams after SS92.

"Age = 15 Gyr − t_m" and were computed for $\epsilon = 0.033$. The two panels show histograms of these merger ages for 65 field E + S0 galaxies. Note that the age distribution for presumed Sb–Sb mergers is probably the more realistic of the two, since the mean Hubble type of present-day galaxies is close to Sb. This distribution and Figs. 66 and 67 yield several important results.

First, if the field E + S0 galaxies indeed formed through disk mergers, their *UBV* colors suggest that this process must have been prolonged, lasting 5–10 Gyr or 1/3 to 2/3 of the age of the universe. Second, the hypothesis of elliptical formation through major disk mergers is entirely consistent with the observed, relatively uniform *UBV* colors of early-type galaxies. And third, the colors and line strengths of present-day E + S0 galaxies are influenced not only by the time t_m when the last major merger occurred, but also by the Hubble types of the merged components. These types determine the amount of gas available for the last starburst at the time of merging (cf. NGC 3921 and NGC 7252 [§3.3]).

Besides these general results, the heuristic merger ages based on *UBV* colors also point toward individual ellipticals likely to be either dynamically young ($\lesssim$7 Gyr) remnants of major mergers (e.g., NGC 1700, 3610, 4125, 4915, and 5322) or at least galaxies rejuvenated through accretions accompanied by significant starbursts (e.g., NGC 596, 3640, and 5018). Clearly, these ellipticals should be prime targets for further studies aimed at dating their stellar contents.

As efforts at age dating early-type galaxies progress, it will be important to clearly define the various ages that are being measured. For example, the spectroscopic method of age dating with single-burst models described above (esp. Fig. 63) yields some sort of luminosity-weighted mean population ages.

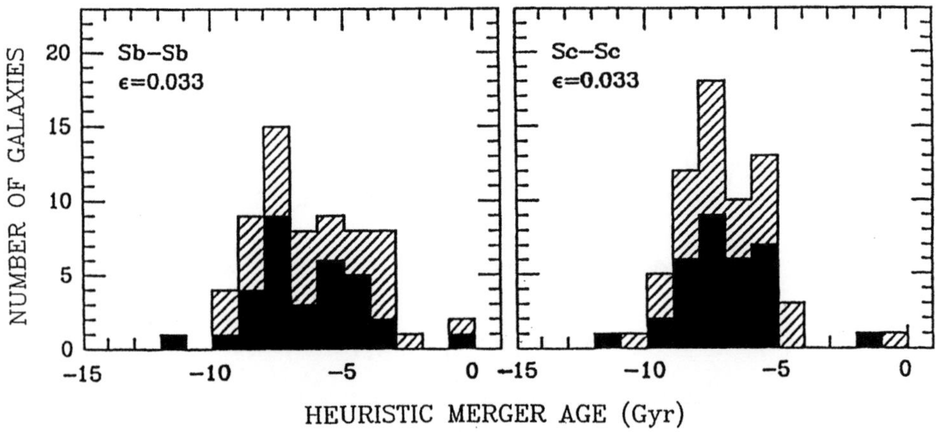

Fig. 68. Distribution of heuristic merger ages for 65 field E + S0 galaxies. Ages shown were computed for presumed (*left*) Sb–Sb mergers and (*right*) Sc–Sc mergers. Age "−15" marks big bang, age "0" the present time. Ellipticals are shown black, S0 galaxies hatched. Note prolonged period of E + S0 formation suggested by these ages. Diagram after SS92.

On the other hand, the heuristic merger ages derived from *UBV* colors and two-burst models represent estimates of the age of the second starburst (and presumed merger). Clearly, these two ages need not agree in general. A third age of interest would be the true mean age of stars in any given galaxy. As Fig. 66 suggests, in any relatively young merger remnant this true mean age of stars will tend to be significantly older than the merger age because the majority of stars formed in the disk galaxies *before* the merger. It is for this reason that I have called galaxies like NGC 3921, NGC 7252, and ellipticals with blue colors and enhanced Balmer absorption lines "dynamically young". Their formation age, and hence dynamical state, is clearly younger than the mean age of their stars.

Interestingly, semi-analytical models of galaxy formation in a cosmological setting with hierarchical mergers confirm the prolonged formation periods of early-type galaxies and reproduce the observed small scatter in the color–magnitude and line-strength–luminosity relations (Kauffmann 1996). These models also yield valuable information concerning the fractions of the visual luminosity of present-day galaxies contributed by stars of different ages. In luminous *cluster* ellipticals, most of the visual light stems from stars that formed at relatively high redshifts ($z \gtrsim 2$), often well before the final merger. In similarly luminous *field* ellipticals, most of the visual light stems from stellar populations that are on average ~4 Gyr younger than those in cluster ellipticals (mean ages of 5–8 Gyr). Finally, in merger-less spirals the visual light is dominated by stars formed during the past few Gyr, as is well known.

The studies by SS92 and Kauffmann (1996) make it very clear why age-dating elliptical galaxies from spectra has been so difficult and controversial. Because the bulk of the stars formed before the final mergers, these ellipticals *look* more uniform and old than they really are!

6.4 Delayed Formation of Bulges

There is growing evidence that even some bulges of disk galaxies may form with a significant delay and often through mergers.

The most direct evidence for a recent formation stems from bulges that are blue and feature post-starburst spectra. The bulge of the S0 galaxy NGC 5102 is a famous example. Its color indices are $(U-B)_0 = 0.18$ and $(B-V)_0 = 0.67$ within a radius of $13'' \approx 300\,\text{pc}$ (Pritchet 1979), and its spectrum shows strongly enhanced Balmer absorption lines out to about the same radius (Gallagher et al. 1975). The inferred age of the starburst is about 400 Myr. Therefore, NGC 5102 must have either just formed or at least enhanced its bulge. Yet, this galaxy also possesses an exponential disk extending over ~4 scale lengths, showing that the disk was not destroyed by whatever process formed the bulge. The present bulge-to-disk ratio is roughly 1/3 in blue light (Yoshizawa & Wakamatsu 1975; van Driel 1987) and will diminish to $B/D \approx 1/10$ over the next few Gyr as the bulge starburst fades. Therefore, the recent bulge-building episode in this galaxy was clearly significant.

Other early-type disk galaxies feature similarly blue bulges and show often direct evidence of recent mergers. Some are S0 galaxies with polar rings (see §5.1), while others are normal S0 to Sb galaxies that feature ripples (see §4.4). As pointed out before, in many of these likely merger remnants, the disks seem to have survived the merger events relatively intact. This evidence and the even more powerful evidence provided by counterrotating gas and stellar disks call into question any claims, based on semi-analytical arguments, that stellar disks are easily disrupted by even minor mergers (Tóth & Ostriker 1992).

Presumably, bulge formation occurs episodically and often concurrent with disk growth. This must certainly be true for bulge building induced by bar instabilities in disks, as proposed by Combes et al. (1990; see also Pfenniger & Friedli 1991; Martinet 1995). But it must also be true for certain mergers, as the case of Arp 220 illustrates. This remnant features an impressive $\sim 6 \times 10^{10}\,\mathcal{M}_\odot$ of molecular gas (for $H_0 = 50$), of which only about 2/3 are concentrated near the center (Scoville et al. 1991). The remaining 1/3, or about $2 \times 10^{10}\,\mathcal{M}_\odot$, of H_2 is distributed in an extended, 10 kpc diameter disk that coincides with the optical dust lane apparently bisecting the main body. At present, we cannot predict the efficiency with which this gas will turn into stars. Yet, given its present large extent and cool temperature, this disk of H_2 seems likely to survive the central starburst and form stars at a more leisurely rate. If so, Arp 220 may one day appear as a Sombrero-like Sa galaxy whose bulge formed more than 10 Gyr after the big bang and whose disk formed even later.

In light of such complex formation histories, age dating bulges will presumably be even more difficult than the age dating of ellipticals. Yet, it forms a challenging and noble task for astronomers of the next century.

6.5 Summary

The main results of this chapter are as follows:

- There is much evidence that many, and probably the *majority* of, field E + S0 galaxies formed through major mergers. Such evidence comes both from statistics of their fine structure and from indicators of past starbursts, such as enhanced Hβ absorption and bluish *UBV* colors.
- Correlations between line-strength residuals, color residuals, and the fine-structure index Σ suggest that the main cause of the vertical scatter in the line-strength–luminosity and color–magnitude relations is the varying *mean stellar-population age*, while metallicity variations are a secondary influence.
- The mean ages of stellar populations seem to be younger in field E + S0 galaxies than in cluster members, and younger in the outskirts of the Coma cluster than in its core. These mean ages also seem to vary more among galaxies in the field and cluster outskirts than among galaxies at the Coma cluster center.
- Age dating the stellar populations of early-type galaxies from their composite spectra is difficult, but becoming feasible. The main challenges are posed by the apparent age–metallicity and age–burst-strength degeneracies. These degeneracies can be broken through observations at higher spectral resolution, measurements of the continuum or of color indices, and better modeling of multiple starbursts likely to have occurred. It seems now established that in M32 the last major starburst occurred $\lesssim$7 Gyr ago.
- Two-burst models of star formation applied to the *UBV* colors of field E + S0 galaxies suggest that merger-induced second (or last) bursts occurred during 1/3 – 2/3 of the age of the universe. These models also explain why age-dating early-type galaxies from spectra is so difficult and often controversial. Because the bulk of their stars formed before the final mergers, these E and S0 galaxies appear more uniformly old than they really are.

7 Formation of Globular Clusters

Globular clusters are like crown jewels adorning the halos of many galaxies, including that of the Milky Way. Each of these clusters contains typically 10^5–10^6 stars packed into a volume of order (10 pc–30 pc)^3. Astronomers have long faced the challenge of explaining how such tightly packed clusters of stars may have formed. One of their early successes in the 1950s was the demonstration, via stellar-evolution theory, that stars in globular clusters are coeval and that Milky Way globulars are of an age comparable to that of the universe itself. Before long, astronomers extrapolated that globular clusters in all galaxies are very old and searched for an explanation in terms of early formation mechanisms. Yet, evidence has grown – first slowly and then since about 1990 at a rapidly accelerating rate – that globular clusters tend to form wherever vigorous star formation occurs and, especially, in starbursts triggered by galaxy interactions and mergers. It is this new evidence that the present chapter concentrates on. We shall end up with a simple, though clearly tentative recipe for making globular clusters. First, however, let us briefly review the status of globular-cluster formation theories around 1990. This brief review will help us understand some of the problems with previous formation scenarios and appreciate the phenomenal progress made during the past few years.

7.1 Status Around 1990

Three Types of Globular-Cluster Formation Scenarios

Until about a decade ago, the paucity of observational evidence concerning the formation of globular clusters led to a profusion of proposed formation scenarios. Any such scenario can be assigned to one of three broad categories, depending on whether the clusters formed before, during, or after the formation of their host galaxies (Fall & Rees 1988).

Primary formation scenarios place the birth of globular clusters before protogalactic collapse occurred. Perhaps the most influential such scenario was that proposed by Peebles & Dicke (1968), who noted that the smallest gravitationally unstable clouds produced right after recombination had Jeans masses of about 10^5–$10^6\,\mathcal{M}_\odot$ and may have been the progenitors of present-day globular clusters. This would, of course, have explained why the Milky Way globulars seemed all so old. Yet, as our knowledge of globular-cluster systems in other galaxies has increased, this scenario has been found to be seriously flawed in at least two respects: (1) Properties of globular-cluster systems (e.g., the mean metallicity) vary systematically with the properties of the host galaxies, so how could the clusters have formed long before the hosts? And (2), globular-cluster systems are more centrally concentrated than the dark matter, which seems to be in conflict with the clusters having formed very early.

Secondary formation scenarios assume that globular clusters formed simultaneously with their host galaxies. Various such scenarios describe the growth of cool dense clouds in pressure equilibrium with hot gas during the collapse

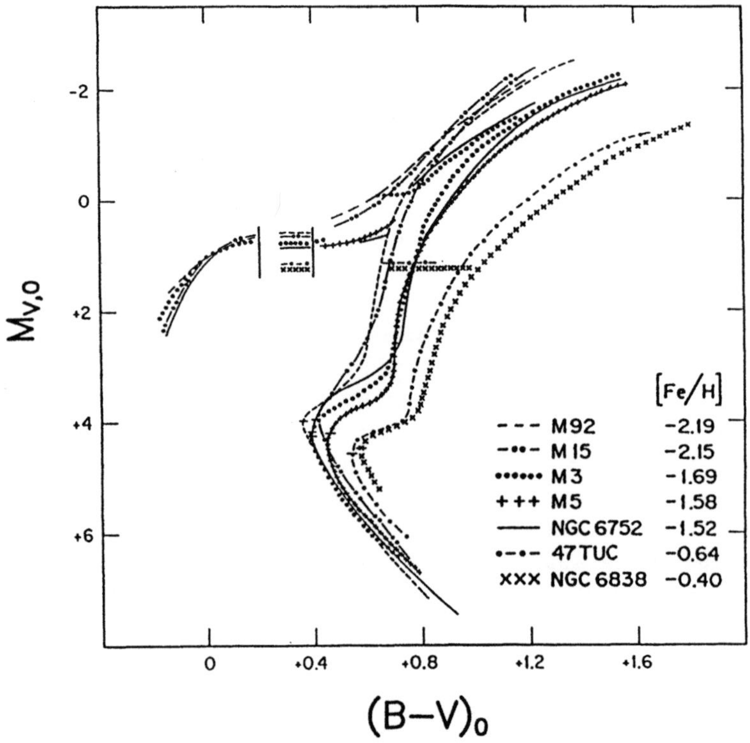

Fig. 69. Composite color–magnitude diagram for seven globular clusters of different metallicities in Milky Way. Diagram from Sandage (1982).

of a protogalaxy (e.g., Fall & Rees 1985, 1988; Larson 1987, 1988). Two of the difficulties with these scenarios are: (1) In order to avoid rapid cooling and fragmentation the protocluster gas *must* have low metallicity, yet there exist globulars with near-solar metallicity. And (2), it is difficult to understand why the growing protocluster clouds do not disperse after the first few supernovae explode in them.

Tertiary formation scenarios assume that globular clusters form either after their host galaxies or separately from them. These scenarios represent a mixed bag, and the boundary between them and secondary-formation scenarios is not always obvious. We shall consider the following four proposed scenarios: (1) Globular cluster formation during major mergers of galaxies (Schweizer 1987; Ashman & Zepf 1992); (2) formation in protogalactic fragments (Searle & Zinn 1978); (3) formation as nuclei of former dwarf galaxies (Zinnecker, Keable, & Dunlop 1988; Freeman 1990); and (4) formation in cooling flows (Fabian et al. 1984). Surprising as it may be, during the past few years observational evidence has rapidly grown in favor of the first three of these scenarios, while there is now pretty strong evidence against the fourth.

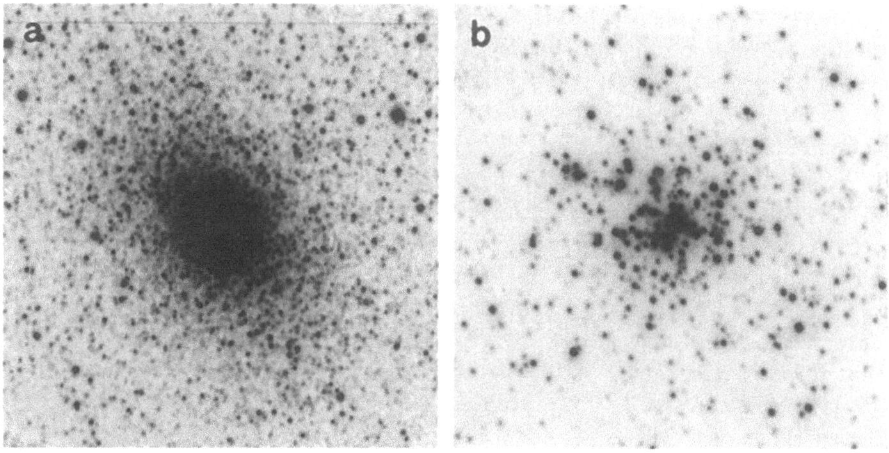

Fig. 70. (a) Young globular cluster NGC 1978 in LMC (ground-based image) and (b) nascent globular cluster R136 (with *HST*). Images courtesy of Douglas L. Welch and Hunter et al. (1995).

Globular Clusters in the Local Group

As stated already, the age-dating of globular clusters belonging to our Milky Way was one of the great astrophysical successes of the 1950s and 1960s and a triumph for stellar-evolution theory (Sandage & Schwarzschild 1952; Sandage 1953). Figure 69 shows color-magnitude diagrams of seven well-observed globular clusters stylized and superposed into a single diagram. From the individual diagrams Sandage (1982) determined cluster ages of 17 ± 2 Gyr. Perhaps under the influence of Peebles & Dicke's (1968) primary formation scenario, this interesting result soon solidified into a near-dogma that all globular clusters in the Milky Way and anywhere in the universe were of this same old age. Yet, in his Ph. D. thesis, Bruce Carney (1978, 1980) claimed significant age differences between different Milky Way globulars, thus incurring the wrath of the Gods. But he was right, and today there is good evidence for an age spread of ~25% among our globulars, with the youngest clusters about 11 Gyr and the oldest 14–16 Gyr old (e.g., Hesser 1995; Chaboyer et al. 1996a).

Besides these modest age differences among Milky Way globulars, there has long been other evidence that globular clusters are not all uniformly old. Specifically, in our Local Group the Large and Small Magellanic Clouds, M31, and M33 all seem to contain intermediate-age globulars, and some contain even young globulars. Best known are the LMC's "populous young clusters" (e.g., Searle et al. 1980), which some astronomers refuse to call globulars because of their youth.[22] Figure 70 shows images of a populous young cluster and of a nascent such cluster in the LMC. Except for their young ages, the populous clusters have

[22] For an informative, color-coded map of the distribution of young and old globulars in the LMC, see van den Bergh (1992).

properties nearly identical to those of Milky Way globulars, i.e., similar numbers of stars, similar radii, and similar masses (e.g., Mateo 1992). Hence, in accordance with the saying "If it looks like a duck, walks like a duck, and talks like a duck, then it *is* a duck", there can be no question by now that these populous young clusters with ages of about 10 Myr to 2 Gyr indeed are *globulars.*

In one lucky case, we are able to observe an extremely young globular cluster shortly after its formation. At the center of the famous 30 Doradus complex of OB stars and ionized gas lies the compact cluster R136, shown in Fig. 70b as observed with *HST*. The core of this cluster, named R136a, is so compact that from the ground it was first mistaken for a single, supermassive star! Yet, new CCD observations and a detailed review of all published data showed convincingly that R136a is the center of a dense star cluster just born from the surrounding gas, a cluster that by its properties is a likely globular (Kennicutt & Chu 1988). An excellent description of this object and its surroundings, based on the best ground-based observations, is given by Walborn (1991). From recent observations with the repaired *Hubble Space Telescope* Hunter et al. (1995) determine a cluster age of 3–4 Myr and find a normal, Salpeter-like initial mass function that is *not* top-heavy. They conclude that the "integrated properties of R136 are consistent with its being comparable to a rather small globular cluster when such clusters were the same age as R136". Two of the brightest stars in this cluster observed with *HST* and the Goddard High-Resolution Spectrograph have spectral types O3f/WN and WN4-w, respectively, and show line profiles indicative of huge mass-loss rates of $\sim 2\times10^{-5}\,\mathcal{M}_\odot\,\mathrm{yr}^{-1}$ (Heap et al. 1994). Therefore, extremely young globulars like R136 may experience rapid dynamical evolution, and some of them may disintegrate in the process (for a brief, but excellent review, see Trimble & Leonard 1996).

The Specific Globular-Cluster Frequency

The question of young versus old globular clusters became an issue when Harris & van den Bergh (1981) introduced a quantity called the specific globular-cluster frequency S_N and used it to argue that ellipticals could not have formed from mergers of spirals. This frequency S_N is defined as the number of globular clusters in a galaxy divided by the galaxy's luminosity in units of $M_V = -15$. Figure 71 shows S_N plotted as a function of Hubble type for about 45 galaxies (Harris 1991). Clearly, the specific globular-cluster frequency increases systematically from $S_N \approx 1$ for late-type spirals to $S_N \approx 4$ for ellipticals and reaches values as high as $S_N \approx 15-20$ in some cD galaxies in rich clusters. It is this systematic increase that Harris & van den Bergh used to argue against a merger formation of ellipticals. Simply put, they argued that adding the contents of two spirals does not change S_N, whence remnants of spiral–spiral mergers cannot be ellipticals.[23] Imagine that two similar spirals with 150 globular clusters each

[23] At first, I did not take seriously the Harris & van den Bergh (1981) argument, because the determination of S_N seemed so difficult and uncertain. For example, after van den Bergh (1979) claimed that NGC 5128 had very few, if any, globular clusters Graham

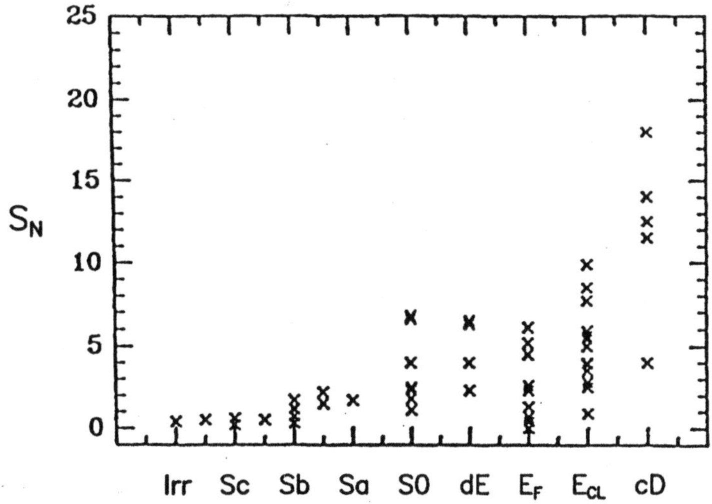

Fig. 71. Specific globular-cluster frequency S_N for about 45 galaxies plotted as a function of their Hubble type (E_F: field E; E_{CL}: cluster E). Diagram from Harris (1991).

merge. Then, the above argument goes, the resulting merger remnant will have 300 globulars and twice each spiral's luminosity, whence its specific frequency S_N will be the same as that of each spiral before. This reasoning wrongly assumes that globular clusters cannot form long after the big bang.

Instead, as we have just seen, there is much evidence for continued globular-cluster formation in Local Group galaxies. In addition, the presence of extremely luminous candidate clusters in the merger remnants NGC 1316 (Schweizer 1980) and NGC 7252 (Schweizer 1982) has long suggested that new star clusters may form as part of the enhanced star formation during mergers. Therefore, I proposed that with their high molecular-gas contents and starbursts major mergers of spiral galaxies may offer near-ideal conditions for cluster formation and may produce many new globulars (Schweizer 1987). If so, the observed increase of S_N from spirals to ellipticals would seem to simply be a natural consequence of major mergers.

& Phillips (1980) announced the discovery of a first globular in it, then Hesser et al. (1984) reported observations of 20 spectroscopically confirmed globulars, and the present-day estimate is that NGC 5128 possesses ~1600 globulars (G. Harris et al. 1984)! Therefore, individual determinations of S_N must be taken with a grain of salt. By now, however, CCD observations have established the systematic increase of S_N toward early Hubble types reasonably well.

Zepf & Ashman (1993) point out that part of this systematic increase reflects the decrease of the mean mass-to-light ratio of galaxies from late to early Hubble types. Therefore, they propose replacing S_N with the number T_N of globular clusters per unit *mass*. This number increases by a factor of only about 2–2.5 between late-type spirals and field ellipticals.

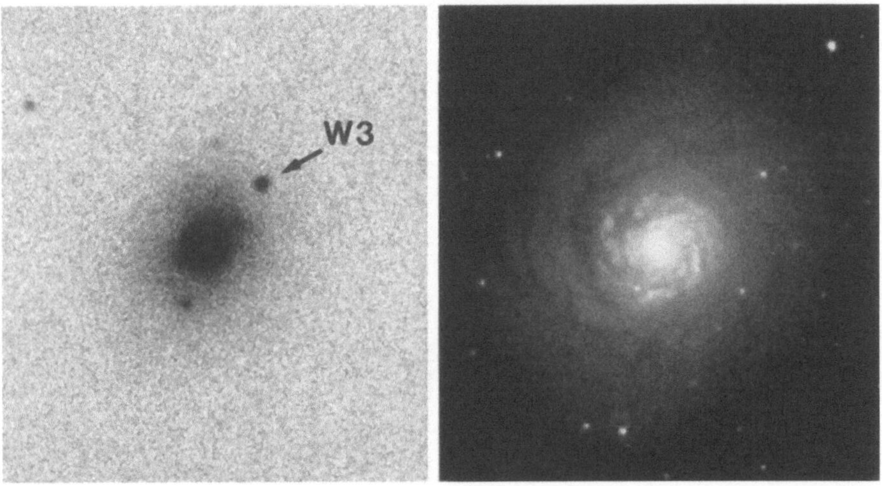

Fig. 72. Candidate globular clusters in NGC 7252 as photographed from the ground (*left*, Schweizer 1982) and imaged with *HST* (*right*, Whitmore et al. 1993). Brightest cluster W3 appears in upper-right corner of *HST* image. Arrow points exactly East.

7.2 Young Globular Clusters in Merger Remnants

Beginning in 1991, systems of young globular clusters apparently formed during galactic mergers have been discovered in rapid succession and have helped transform our view of globular-cluster formation. We first discuss young globulars found in four merger *remnants*, where the evidence for recent globular-cluster formation is perhaps most clear-cut.

The first such remnant is NGC 3597, a peculiar S0 galaxy with tell-tale ripples and at least one tidal tail. On ground-based images of this galaxy, Lutz (1991) discovered 14 blue point-like objects that cluster around the center and have absolute visual magnitudes $M_V = -11.5$ to -14.5. Because of their high luminosity, these objects cannot be individual stars, but must be star clusters instead. Though he could not spatially resolve these candidate clusters, Lutz noted that they measure less than about $1\farcs0$ or 100 pc in diameter and are unlikely to be major H II regions or OB associations. Inferring rough ages from their measured colors, he predicted that over the next 10 Gyr the clusters would fade to about a median $M_V = -10$ to -11, thus resembling very luminous old globulars like ω Cen.

The second remnant is NGC 1275, the peculiar galaxy at the center of the Perseus cluster. On images taken with the Planetary Camera of *HST*, Holtzman et al. (1992) discovered 64 blue pointlike objects within 5 kpc from the nucleus. The high luminosities ($M_V = -11$ to -15.4), blue colors, and small sizes ($\lesssim$15 pc) of these objects again point to their being young globular clusters. This discovery was completely unexpected and was reported by news media world-wide. The only problem lies in the interpretation. Although NGC 1275 fea-

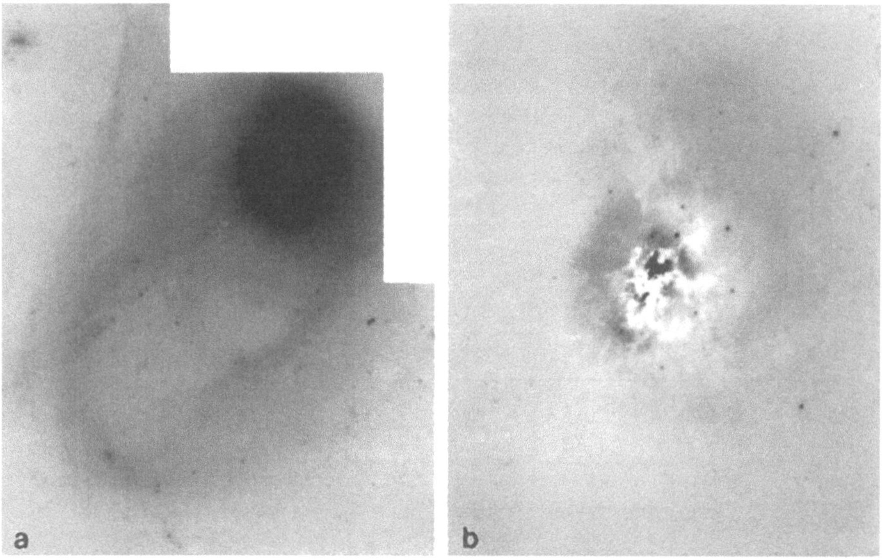

Fig. 73. Candidate globular clusters and stellar associations in NGC 3921 imaged with *HST*. (a) Field covered by WFPC2 and (b) by Planetary Camera, masked to better show clusters. Images from Schweizer et al. (1996).

tures ripples suggestive of a past merger, it also lies at the center of a claimed cooling flow in the Perseus cluster. Holtzman et al. favor the hypothesis that the young candidate globulars formed during the merger, while others claim that these clusters formed in the cooling flow (e.g., Richer et al. 1993). A decision between the two competing hypotheses will have to come from new photometric observations with the repaired *Hubble Space Telescope*. If the clusters all have similar ages, their formation is likely due to a merger-induced starburst; if, on the other hand, their ages spread over 1 Gyr or more, then continued formation in the cooling flow may be a possibility. Note, however, that ten other cD galaxies at the centers of cooling-flow clusters show no young clusters (see Sect. 7.4), whence NGC 1275 would have to be a unique case of cluster formation in a cooling flow. This seems rather unlikely.

The third merger remnant with candidate young globulars is our old friend NGC 7252. It is in this galaxy that I had noticed a suspicious grouping of 6–7 blue knots suggesting very luminous young star clusters (Schweizer 1982). Figure 72 shows these knots on a short-exposure photograph taken in 1976 with the 4-m telescope at Cerro Tololo. The frame outlines the area imaged with *HST* and reproduced to the right (Whitmore et al. 1993). On the *HST* image, notice the beautiful "minispiral", which coincides with the central disk of molecular and ionized gas, and the many surrounding point-like sources. Within this and three adjacent fields imaged with the Planetary Camera, Whitmore et al. identified 53 young star clusters, of which 16 lie within the minispiral (the "inner

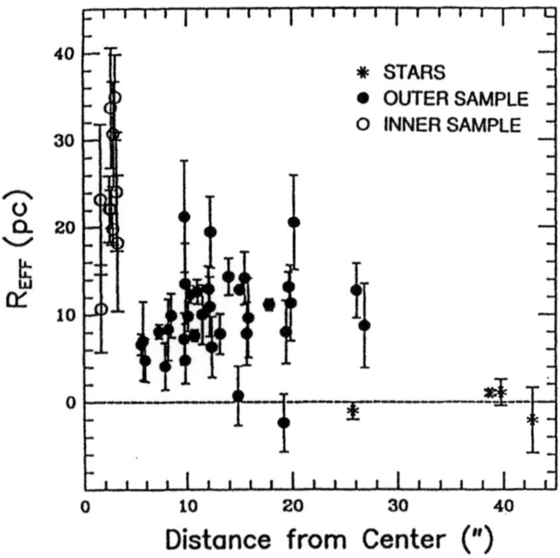

Fig. 74. Effective radii for candidate star clusters in NGC 7252 and for four foreground stars, measured with *HST*. Diagram from Whitmore et al. (1993).

sample") and 37 lie beyond its limits (the "outer sample"). The objects within the minispiral are clearly more fuzzy than those in the outer sample and may be OB associations, while, as we shall see, the outer objects are candidate young globulars.

Finally, the fourth merger remnant is NGC 3921, in which our *HST* team has just recently identified 102 candidate globulars and 49 separate candidate stellar associations (Schweizer et al. 1996). Figure 73 shows the whole area imaged by the Wide-Field and Planetary Camera 2 of the repaired *HST* and, enlarged to the right, the image formed by the high-resolution Planetary Camera alone. This latter image covers a central area of about 7 kpc radius in NGC 3921 (for $H_0 = 75$) and shows 58 candidate globulars and 3 candidate stellar associations brighter than $V = 26$ ($M_V = -8.5$).

What has been learned from all these *HST* observations, and how do we deduce that most of the blue point-like sources in these merger remnants are young globular clusters?

Besides the high luminosities and blue colors indicative of young ages, it is the small radii of these objects, measured with *HST*, that provide one of the strongest clues concerning their nature. Figure 74 shows half-light or effective radii $R_{\rm eff}$ for all candidate clusters in NGC 7252 and for four stars measured in the same field of view. Most candidate clusters are marginally resolved. Notice that the clusters of the inner sample are too large to be globular clusters, the median half-light radius of Milky Way globulars being only 3 pc. The clusters of the outer sample, on the other hand, appear about of the right size, especially when we consider that the radii plotted in the figure were measured with the un-repaired *HST* and are based on a Hubble constant of $H_0 = 50\,\mathrm{km\,s^{-1}\,Mpc^{-1}}$. Using $H_0 = 75$ would make the radii 33% smaller, and new measurements with the repaired *HST* indeed yield median cluster radii of 3–5 pc in several remnants (Schweizer et al. 1996; Miller et al. 1997).

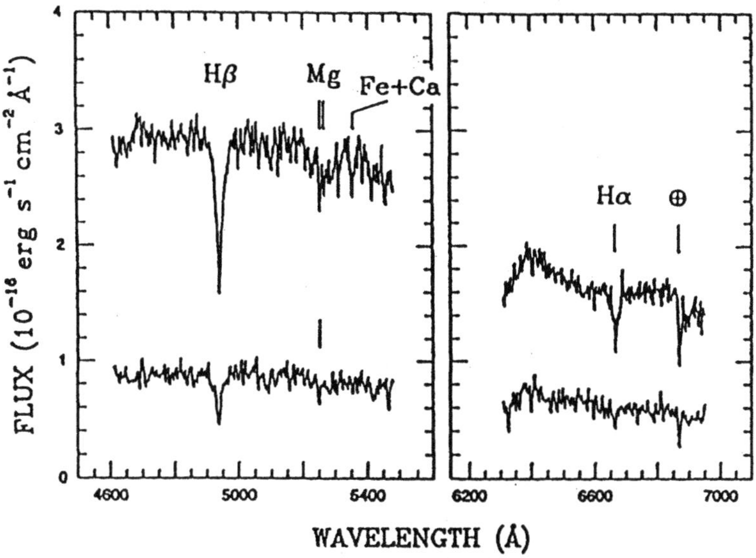

Fig. 75. Spectra of candidate globular clusters W3 (*top trace*) and W30 (*bottom*) in NGC 7252, obtained with Palomar 5-m telescope. Note broad Balmer absorption lines indicative of young, A-type stars. Diagram from Schweizer & Seitzer (1993).

Small half-light radii are necessary, but not sufficient, to establish the globular cluster nature of these objects. We must also demonstrate that these candidate clusters are gravitationally bound, which means that they must be at least several internal crossing times old. Typical crossing times in Milky Way globulars are about 3–5 Myr. Hence, if the candidate clusters are older than ~20 Myr, then they are gravitationally bound and likely globulars. The strongest evidence for ages in excess of 20 Myr comes from spectra of the two bright candidate clusters W3 and W30 in NGC 7252 (Schweizer & Seitzer 1993). As Fig. 75 shows, these clusters feature no emission lines whatsoever, implying that they lack stars of spectral types O3 – B1 and are, therefore, older than 20 Myr. Indeed, the broad Balmer absorption lines Hα and Hβ show that most of the cluster light stems from A-type stars indicative of cluster ages ~30–500 Myr. Therefore, at least these two luminous clusters are gravitationally bound and very likely young globulars. Color indices $V-I$ measured with *HST* yield similar ages for most other blue pointlike objects in both NGC 7252 and NGC 3921, suggesting that the vast majority of these objects are young globular clusters formed *during* the recent mergers there.

The absolute magnitudes of clusters W3 and W30 in NGC 7252 are extraordinary: $M_V = -17.1$ and -15.5 for $H_0 = 50$, or 1 mag less for $H_0 = 75$. If cluster W3 were at the distance of ω Cen in our own Galaxy, it would appear as bright as Venus at maximum brightness and would cast soft shadows here on Earth! Cluster evolution models by Bruzual & Charlot (1993) predict fadings of 4–7 mag in V over the next 15 Gyr. Hence, both W3 and W30 will be very

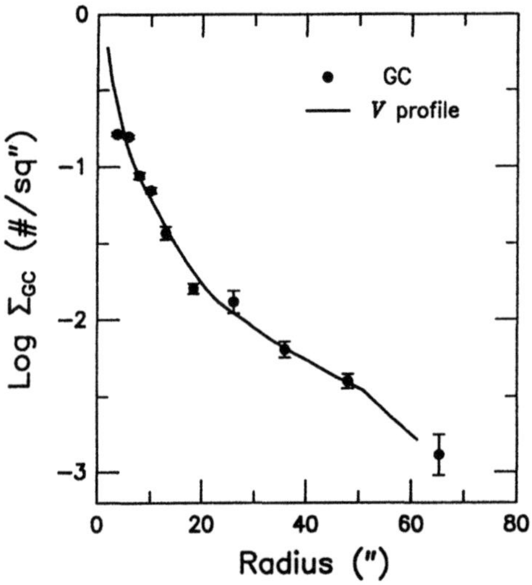

Fig. 76. Projected radial distribution of candidate globular clusters in NGC 3921, compared with V surface-brightness profile. Diagram after Schweizer et al. (1996).

luminous globular clusters even when they are as old as present-day Milky Way globulars. A spectrum of the cluster H1 in NGC 1275 leads to similar conclusions and suggests that that young cluster is of roughly solar metallicity (Zepf et al. 1995b).

In both NGC 7252 and NGC 3921, the candidate globular clusters are distributed similarly to the background light of the galaxy (Whitmore et al. 1993; Schweizer et al. 1996). Figure 76 compares the projected radial distributions of candidate globulars (data points) and V light (solid line) in NGC 3921. The close agreement suggests that the clusters or their progenitors experienced the same violent relaxation as did the stars during the merger of the two disks galaxies. Interestingly, systems of old globular clusters in elliptical galaxies closely follow the background light as well.

How has the merger in NGC 3921 affected the specific globular-cluster frequency S_N? From the 102 observed candidate globulars, we estimate that at least 110 new globulars more luminous than $M_V = -8.5$ formed. (The number of fainter globulars is unknown.) From the H I distribution, the component galaxies that merged appear to have been a gas-rich Sc and a gas-poor S0 or Sa (Hibbard & van Gorkom 1996). Their luminosities multiplied by the specific frequencies appropriate for these Hubble types suggest that together the two galaxies contributed about 270 ± 60 old globulars. Therefore, the number of globulars seems to have increased by $\gtrsim$40% during the merger, from an estimated 270 ± 60 globulars before to $\gtrsim$380 globulars after the merger. The inferred specific frequency is $S_N \gtrsim 0.7$ at present and will, barring further mergers or significant cluster destruction, increase to $S_N \gtrsim 1.4$, 1.8, and 2.9 over the next 2, 5, and 15 Gyr as the galaxy fades. Hence, within a few Gyr this merger remnant will begin to appear as a normal elliptical with a globular-cluster population similar to that of E's in low-density environments (Schweizer et al. 1996).

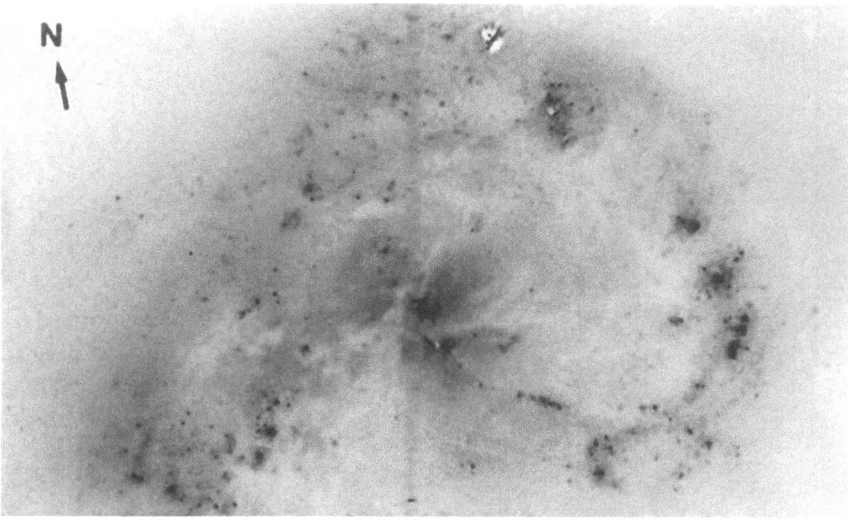

Fig. 77. Part of disk of NGC 4038 imaged with *HST*. Note many compact star clusters, some thought to be candidate young globulars. From the ground, unresolved groups of clusters appear as luminous "knots". Figure from Whitmore & Schweizer (1995).

7.3 Young Globular Clusters in Merging and Starburst Galaxies

Although the presence of many young globular clusters in merger remnants weakens arguments against the merger formation of elliptical galaxies based on S_N and is revolutionizing some of our views on globular-cluster formation, we need to look at *ongoing* mergers and starburst galaxies to learn more about the physics of cluster formation.

One ongoing merger that Whitmore and I have observed with *HST* is NGC 4038/39 ("The Antennae"). The disks of the two colliding galaxies are full of truly giant H II regions, which Rubin et al. (1970) studied spectroscopically and of which Malin (1992) shows a beautiful color photograph. Obviously, both galaxies are experiencing a major burst of star formation. Figure 77 shows part of a V image of the disks obtained with the Wide-Field Camera of *HST* before the repair. Notice that the "knots" that were seen coincident with the H II regions from the ground break up into many smaller objects. Since these objects have a mean $\langle M_V \rangle = -11$ and the brightest reach $M_V = -15$, they are much brighter than individual stars and must be star clusters. We have found a total of 700 candidate young clusters, of which perhaps as many as half may be young globulars (Whitmore & Schweizer 1995).

Typically, a knot seen in $1'' - 1\rlap{.}''5$ seeing from the ground as a single unit breaks up into about one dozen individual star clusters, many of which have properties similar to the candidate young globulars in NGC 3921 and NGC 7252. These *HST* images of "The Antennae" and similar images of other starburst galaxies suggest that globular clusters may tend to form in groups from large molecular-gas complexes.

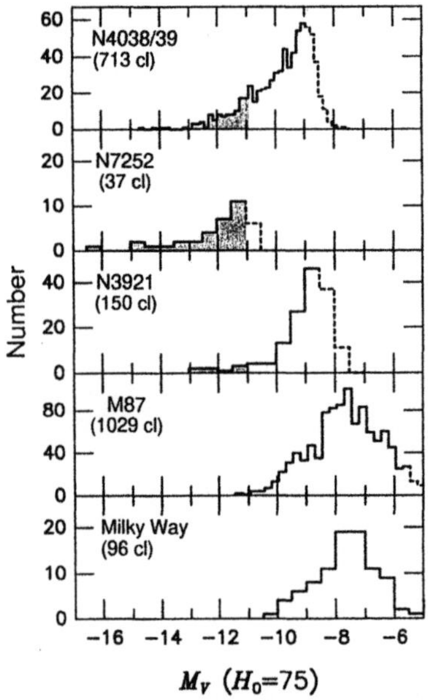

Fig. 78. Luminosity functions of globular-cluster systems in three merger galaxies, in M87, and in Milky Way. Note excess numbers of globulars brighter than $M_V = -11$ in merger systems. Diagram from Schweizer et al. (1996).

An interesting study of about a dozen starburst galaxies has been conducted by Meurer et al. (1995) with *HST* and the Faint Object Camera in the ultraviolet ($\lambda \approx 2200\,\text{Å}$), where *HST* reaches its best resolution. Hundreds of star clusters have been identified in these galaxies, yielding several new insights concerning star and cluster formation. First, the general surface brightness in the UV seems to peak at about the same value in all observed starbursts, suggesting that in intensely star-forming regions there may be a self-limiting process at work. Second, star clusters seem to form mainly in regions with a surface brightness within 1.5 mag from that peak value. Third, the effective radii of these clusters are very similar to those of Milky Way globulars, suggesting that many clusters may be protoglobulars. And fourth, the fraction of the total UV light emitted by clusters is about 20%, indicating that cluster formation is indeed an important part of any starburst.

How do the luminosity functions and color distributions of the young globular clusters in merging and merged galaxies compare with those of old globulars in M87 and the Milky Way?

Figure 78 displays the luminosity functions of the globular-cluster systems in NGC 4038/4039 (ongoing merger), NGC 7252 ($1.0^{+1.0}_{-0.5}$ Gyr old remnant), NGC 3921 (0.7 ± 0.3 Gyr old remnant), M87 (E with old metal-rich globulars), and the Milky Way (Sb with old metal-poor globulars). Note that the dashed parts of the top three luminosity histograms indicate incompleteness of the luminosity functions due to faint magnitudes, whence the real turnovers of the luminosity functions have not yet been observed in these three cluster systems. This figure emphasizes that young globular clusters manifest themselves through their high luminosities. In old cluster systems like our own and that of M87, clusters of $M_V = -11$ are exceedingly rare. The Milky Way's most luminous cluster is ω

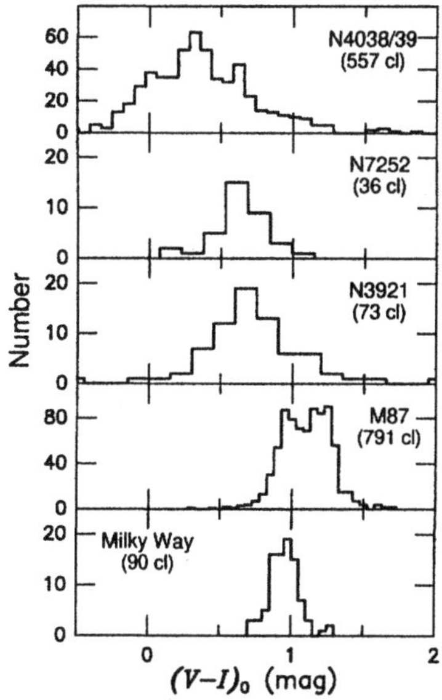

Fig. 79. Color distributions of globular-cluster systems in three merger galaxies, in M87, and in Milky Way. Diagram from Schweizer et al. (1996).

Cen with $M_V = -10.3$, and among the thousands of clusters in M87 there are only 0.3% more luminous than $M_V = -11$. Yet, in NGC 4038/39 about 25% of all clusters exceed that limit. In NGC 7252, where we do not know the percentage, the most luminous candidate globulars are 250 times more luminous than any in M87! Even among the only 150 candidate globulars plotted in NGC 3921, eight exceed that limit. These high luminosities, then, are a sure sign of youth, since clusters fade with age.

A second item of interest is that over the observable range in luminosity the luminosity functions of young clusters are well described by power laws of the form $\Phi(L)\mathrm{d}L \propto L^{-\alpha}\mathrm{d}L$, where $1.7 \lesssim \alpha \lesssim 2.1$ (Whitmore & Schweizer 1995; Schweizer et al. 1996). It is not known at present how these luminosity functions of young globular-cluster systems evolve with time to become the lognormal luminosity functions (i.e., Gaussians in M_V) characteristic of old globular-cluster systems. Presumably, some of the fainter clusters dissolve due to internal processes (e.g., loss of stars, stellar winds) or external processes (e.g., tidal shocks).

And third, notice a strange difference between the clusters of NGC 7252 and NGC 3921. Although the former galaxy is likely to be the older of the two remnants, its most luminous globular clusters are about 10× more luminous than the brightest clusters in the latter. One reason for this difference may be the different precursor galaxies involved in these two mergers. Whereas it was two gas-rich Sc galaxies that collided and merged in NGC 7252, only one of the two colliding galaxies in NGC 3921 was gas rich, while the other seems to have been gas poor (Hibbard & van Gorkom 1996). Therefore, we may be seeing some first signs that the gas contents of merging galaxies imprint the newly formed globular-cluster systems.

Figure 79 intercompares the color distributions of candidate globular clusters in the same five galaxies. Plotted are the numbers of globulars as function of color index $V{-}I$. In old globular-cluster systems, like those of M87 and the Milky Way, we know that the color is related to cluster metallicity. However, we also know that as the metallicity decreases toward zero the cluster colors cannot get much bluer than those of Milky Way globulars. Yet, the figure shows that the clusters of NGC 4038/39, 7252, and 3921 are very significantly bluer than those of M87 and the Milky Way. Therefore, we conclude again that on average these clusters must be much younger than those of M87 or the Milky Way. It is from these color distributions that one can derive age distributions, if one knows the mean metallicity.

7.4 Globular Clusters in Ellipticals and cD Galaxies

The next logical question is: Can one find *intermediate-age* globular clusters in elliptical galaxies? Several *HST* programs are currently underway to address this question, but have yielded no firm answers yet. In NGC 3610, the E with the most fine structure in the sample observed by Schweizer & Seitzer (1992), there appear to be a few excessively luminous globular clusters, while in NGC 1700, another E with fine structure indicative of dynamical youth, there appear to be a few excessively blue globulars (Whitmore et al. 1997). The main difficulty in interpreting these observations in terms of cluster ages is the lack of spectroscopic information on the metallicities of the candidate globulars. Because of the well-known age–metallicity degeneracy, accurate ages can only be determined once metallicities have been measured. Very large telescopes like the Keck and VLT reflectors will be needed to obtain spectra of these faint clusters. Meanwhile, however, there is older evidence gained from ground-based observations that sheds some light on the question of globular-cluster formation in early-type galaxies.

Figure 80 displays the mean metallicities of globular-cluster systems in 14 galaxies plotted versus the absolute magnitudes of these galaxies. Most of the observational data were obtained by Brodie & Huchra (1991), but the figure is as plotted by Harris (1991), who added a solid line representing the mean metallicity of the host galaxies. There is a clear correlation between the globular-cluster and galaxy metallicities, indicating that the globular clusters must have formed *no earlier* than their host galaxies. Hence, as pointed out in §7.1, primary cluster-formation scenarios are unsustainable. However, I believe that this figure presents a major puzzle: Why should it be that globular clusters are systematically more metal-poor than their host galaxies?

Based on the merger-formation hypothesis for ellipticals and the observations of young globulars in merger remnants, one might predict that at least the most recently formed globulars would be among the most metal-rich objects in these galaxies and should, therefore, be at least as metal rich as the average stellar background. This is not what Fig. 80 seems to show. There are several possible solutions to this apparent contradiction. First, the mean galaxy metallicities refer to the centers of the galaxies, whence the offset between the clusters and their host galaxies could simply be a measure of metallicity gradients between the galaxy centers and material at the mean distances of the clusters from the

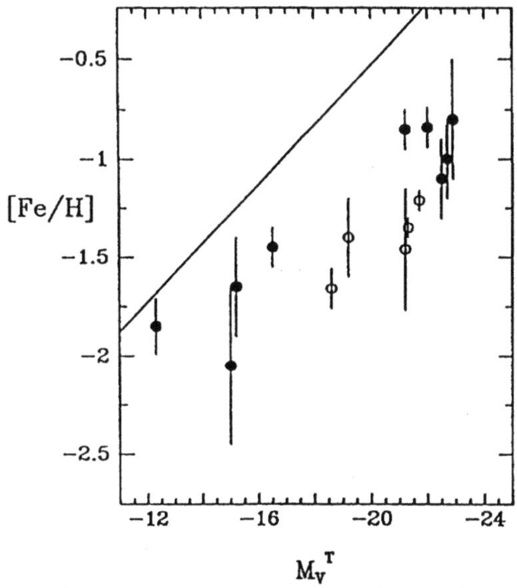

Fig. 80. Mean metallicity of globular-cluster systems plotted as a function of absolute visual magnitude of host galaxy, as explained in text. Data from Brodie & Huchra (1991), diagram from Harris (1991).

center. Second, cluster systems may consist of a mix of intermediate-age and old clusters (see below), with the old clusters indeed much more metal-poor than the mean galaxy background. And third, part of the offset in Fig. 80 may reflect *younger ages* of the globular clusters than assumed in the abundance determinations (Harris 1991). In light of the recent discoveries about globular-cluster formation in merging galaxies, this intriguing third possibility deserves serious further consideration.

Recent discoveries of bimodal color distributions in the globular-cluster systems of elliptical galaxies strongly hint at episodic cluster formation in the past (e.g., review by Zepf 1997). Figure 81 shows the bimodal distributions of the color index $(V-I)_0$ for globulars in M87 (Whitmore et al. 1995) and NGC 3923 (Zepf et al. 1995a). Many elliptical galaxies seem to have cluster systems with such bimodal, and occasionally even multimodal, color distributions. The detection of bimodality is difficult from the ground because of the relatively large photometric errors (e.g., NGC 3923 in Fig. 81), but is becoming increasingly easy with new observations from space (e.g., M87, observed with *HST*). The scale at the top of Fig. 81 shows the cluster metallicity inferred from $(V-I)_0$ on the simplifying assumption that all clusters are equally old. It seems likely that the redder, more metal-rich globulars in M87 and NGC 3923 formed later than the bluer, more metal-poor clusters. Just how much later will again have to be determined via spectroscopic measurements of the cluster metallicities. Of course, the suspicion is that the redder, presumably younger clusters formed during a merger. If this hypothesis is correct, globular clusters – which contain single-burst stellar populations – may turn out to be *useful objects to help age-date the last major merger in their host ellipticals.*

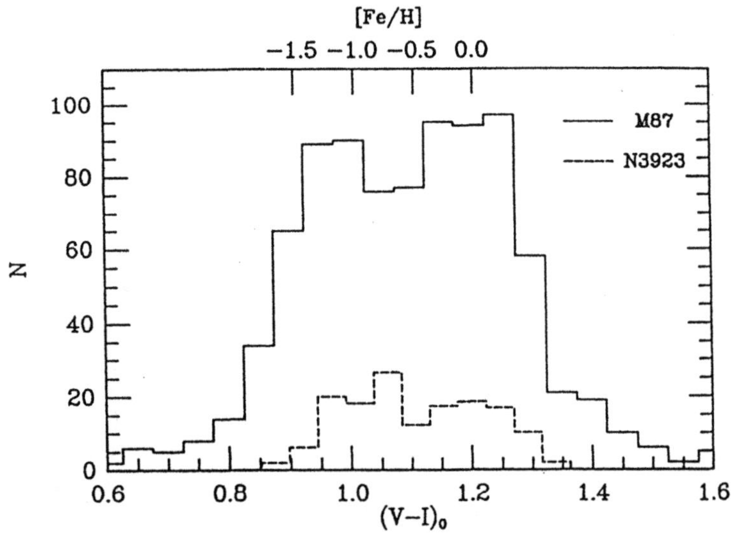

Fig. 81. Color distributions of globular-cluster systems of M87 (Whitmore et al. 1995) and NGC 3923 (Zepf et al. 1995a). Metallicity scale is at the top. Note that distributions are bimodal, suggesting episodic cluster formation. Diagram from Zepf (1997).

The one place where globular clusters seem *not* to form at present is cD galaxies in "cooling-flow" clusters. After the discovery of young globulars in NGC 1275 (Holtzman et al. 1992), searches for similar young clusters were conducted in about a dozen "cooling-flow" cDs (Harris et al. 1995; Gallagher & Holtzman 1994). These searches have yielded no young globulars except for those in NGC 1275 and in one other cD with luminous matter and dust indicative of a recent accretion. In addition, none of the expected correlations between numbers of globulars, inferred mass-flow rates, and X-Ray luminosities have been found. Therefore, the evidence is now strong that globular clusters do not form in cooling flows (Harris et al. 1995). Instead, the high specific globular-cluster frequencies observed in cD galaxies ($S_N \approx$ 6–20) suggest that these galaxies formed early and – if from mergers at all – then from mergers of still very gaseous components (Whitmore 1996).

7.5 Globular Clusters via Satellite Accretion

Given all the enthusiasm for globular-cluster formation during mergers, it is important to remember that some other formation scenarios have also been proposed and have found support from recent observations. Therefore, there may be different ways for globular clusters to form and end up in their host galaxies. Foremost among these alternative formation scenarios is the one proposed by Searle & Zinn (1978), whereby globular clusters may form in "gas fragments" that later merge to form bulges and halos like those of our Milky Way. Zinnecker et al. (1988) and Freeman (1990) have proposed a variation on this theme, whereby globular clusters may be the nuclei of former dwarf galaxies accreted by larger galaxies.

The recent discovery of a dwarf galaxy in Sagittarius that contains four well-known globular clusters and appears to be in the process of being accreted by the Milky Way (Ibata, Gilmore, & Irwin 1994) provides evidence that both of these proposed mechanisms may be at work. The globular cluster M54 lies at the center of this dwarf, has the same radial velocity, and is the second most luminous globular in the Milky Way. It may indeed be the nucleus of the Sagittarius dwarf that is being accreted. The other three globulars associated with this dwarf, Arp 2 and Terzan 7 and 8 (Da Costa & Armandroff 1995; Sarajedini & Layden 1995) have been known for a while to be younger than most halo globulars and may be just the sort of globulars "pre-cooked" in fragments (today we would say: dwarf companions) that Searle & Zinn envisaged being accreted by our Galaxy. (For more details about the Sagittarius dwarf, see §8.3.) The lesson to be learned from this recent discovery is that whenever we study an individual globular cluster, its age may have little to do with the age of the bulk of the stars in the host galaxy and may also differ from the time when it joined the host galaxy through accretion!

7.6 A Recipe for Making Globular Clusters

Can we condense what we have learned from observations of globular clusters in merger remnants, starbursts, and cD galaxies into a recipe for globular-cluster formation?

We mentioned in §7.3 that the luminosity functions of young cluster systems seem to be well represented by power laws. Figure 82 displays the luminosity function of 700 clusters observed in NGC 4038/39 (Whitmore & Schweizer 1995). The curved line represents a power law in luminosity fit to the counts by least squares. As the figure shows, this power law,

$$\Phi(L)\,\mathrm{d}L \propto L^{-1.78\pm0.05}\,\mathrm{d}L\,, \tag{4}$$

fits the data quite well down to the completeness limit around $V = 22.5$ ($M_V \approx -10$). Similar power laws with exponents of about -1.6 to -2.1 hold for young clusters in the Milky Way, the LMC (Elson & Fall 1985), NGC 3921 (Schweizer et al. 1996), and various starburst galaxies (Meurer et al. 1995). These power laws and their exponents must contain an important clue concerning the cluster formation process.

As various astronomers and, most recently, Harris & Pudritz (1994) have emphasized, it appears striking that all these luminosity functions of young-cluster systems are very similar to the *mass* function of *giant molecular clouds*. In the Milky Way, this mass function is well represented by a power law of the form

$$\Phi(\mathcal{M})\,\mathrm{d}\mathcal{M} \propto \mathcal{M}^{-1.63\pm0.12}\,\mathrm{d}\mathcal{M}\,. \tag{5}$$

Not only are the exponents of the luminosity and mass functions identical to within the combined errors, but the most massive giant molecular clouds ($\mathcal{M} \approx 8 \times 10^6\,\mathcal{M}_\odot$) are only about 2–3 times more massive than the most

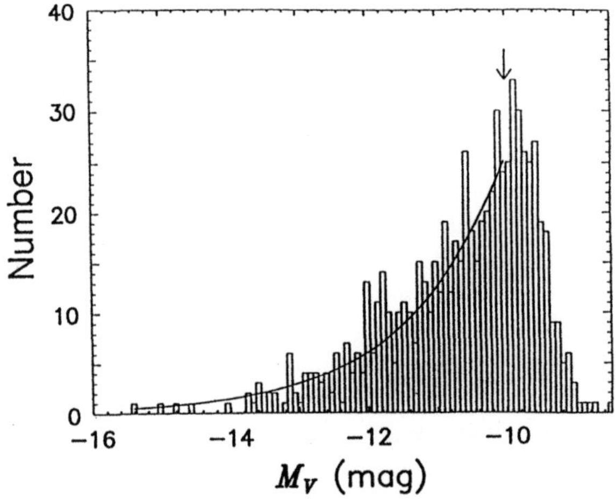

Fig. 82. Luminosity function of about 700 young star clusters in NGC 4038/39 observed with *HST*. Curved line represents a power law in luminosity; arrow marks completeness threshold. Diagram from Whitmore & Schweizer (1995).

massive globular clusters ($\mathcal{M} \approx 3 \times 10^6\, \mathcal{M}_\odot$). This is probably not a mere coincidence. Since giant molecular clouds are a standard component of the gas in spiral galaxies and since star formation always involves molecular gas, it seems only natural that in starbursts triggered by mergers these giant molecular clouds turn into star clusters and, the most massive among them, into globular clusters. What may surprise is that the star-formation efficiency seems to be of the order of 30% or higher. This has led various authors to postulate the existence of *supergiant* molecular clouds. Yet, as Jog & Solomon (1992) first pointed out, the sudden over-pressure generated in the surrounding gas by a starburst may be quite sufficient to induce both the collapse of normal giant molecular clouds and very efficient star formation in them. This mechanism, then, is probably responsible for the globular cluster formation observed in mergers. Note that the physics of gas involved in present-day mergers is not that different from the gas physics long envisaged to have held during the presumed collapse of protogalaxies.

We are now ready to present our recipe, tentative as it may be:

RECIPE FOR MAKING GLOBULAR CLUSTERS

- Take a mixture of *gas* and *giant molecular clouds* (e.g., in two colliding spirals)
- Use a *pressure cooker* (e.g., potential well of merging galaxies)
- *Heat rapidly* to raise the pressure (e.g., starburst)

The giant molecular clouds will collapse from overpressure, will form stars efficiently, ...

et voilà vos amas globulaires!

7.7 Summary

In the present chapter, we have reviewed the growing evidence that globular clusters are a natural by-product of intense star formation. Some of the main results are as follows:

- About a half dozen merger galaxies are now known to contain significant populations of young ($\lesssim$1 Gyr) globular clusters.
- In three age-dated disk–disk mergers (NGC 4038/39, 7252, & 3921), these globular clusters seem to have formed *during* the merger.
- The likely progenitors of globular clusters are giant molecular clouds in merging spirals, triggered into collapse and efficient star formation by overpressure of the surrounding, starburst-heated gas (Jog & Solomon 1992).
- Merger-induced globular-cluster formation provides a natural explanation for the increased specific frequency S_N observed in ellipticals.
- Bi- and multi-modal color distributions of globular-cluster systems support the notion of episodic cluster formation in elliptical galaxies. This suggests the use of globular clusters for age-dating the last major merger episodes in these galaxies.
- Some globular clusters are "pre-cooked" in companion galaxies and later accreted with these companions by larger host galaxies.

8 Interactions and Mergers in the Local Group

In the present chapter, we take a stroll through our Local Group and discuss the various signs of interactions and mergers that astronomers have found in its member galaxies over the past 40 years. Personally, I confess having often thought that – perhaps with the exception of the Milky Way – Local Group galaxies did not contribute much to our understanding of galaxy evolution. Rather, astronomers seem to have seen reflected in these nearby galaxies their emerging understanding of galaxy evolution, gained from more distant galaxies showing more extreme forms of interactions. Another way to express this is that until recently there seemed to be precious little *solid* evidence for interactions and mergers in the Local Group. This situation, however, has changed dramatically with the discovery of the dwarf galaxy Sagittarius I (Ibata, Gilmore, & Irwin 1994). This dwarf is falling into the Milky Way and can certainly today be counted as one of the best pieces of evidence for galaxy growth through accretion.

Lynden-Bell (1994) presents a nice review of the genesis of the Local Group. Look at his figures showing the locations of Local Group galaxies projected onto the plane of the Milky Way. There are two obvious subgroups, one centered on our own Milky Way and the other on M31. Within the former, the proximity of the three most massive members – the LMC, SMC, and the Milky Way itself – is striking. Similarly, M32 and NGC 205 are close satellites of M31, and even relatively massive M33 lies remarkably close to M31. However, there are also outliers like NGC 6822 and IC 10 that may or may not be counted as members of the Milky-Way and M31 subgroups, respectively. Obviously, the orbits and interactions of Local Group members must be pretty complicated!

8.1 The Milky Way: ELS (1962) versus Searle–Zinn (1978)

Two distinctly different scenarios for the formation of the Milky Way have been proposed by Eggen, Lynden-Bell, & Sandage (1962, = ELS) and Searle & Zinn (1978). Both have deeply influenced astronomers' thinking, and it is important to understand their origins, merits, and problems.

By the late 1950s, astronomers knew that most elements heavier than hydrogen and helium form in stars and that the mean metallicity of stars has been increasing with time. From a study of 221 stars with known space velocities, ELS found two interesting correlations, shown in Fig. 83. First, the velocity components W of stars perpendicular to the Milky Way plane appear to correlate with ultraviolet excess $\delta(U{-}B)$, which itself is a measure of low blanketing and, therefore, low metallicity. Whereas stars with $\delta(U{-}B) = 0$ (solar abundance) all seem to be moving in the disk of the Milky Way, stars with large UV excesses (low metallicities) show a wide range of vertical motions, up to $>250\,\mathrm{km\ s^{-1}}$, indicating that on their galactic orbits they rise up to $Z_{\mathrm{max}} \approx 10\,\mathrm{kpc}$ into the halo. And second, the orbital eccentricities e of stars appear to correlate strongly with UV excess, stars of low metallicity following very eccentric, plunging orbits. Thus,

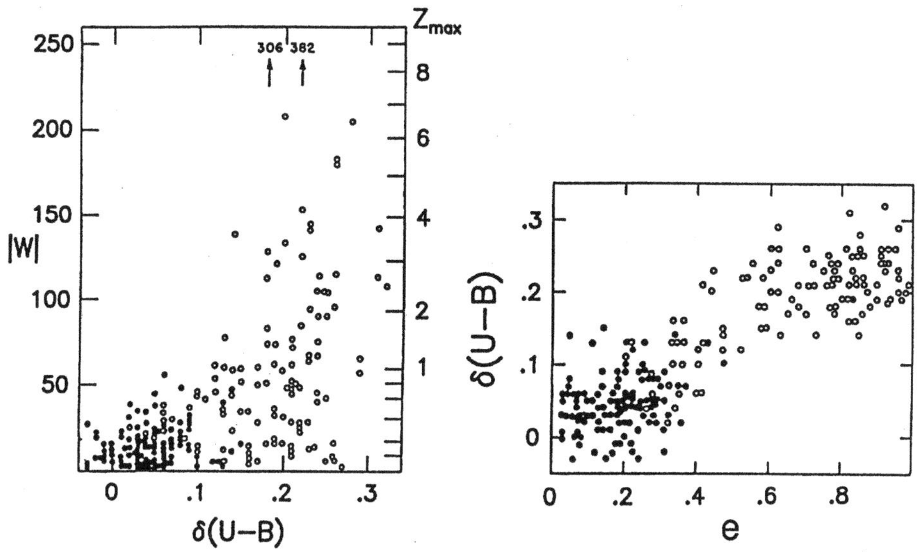

Fig. 83. Correlations between ultraviolet excess $\delta(U-B)$ and kinematics for a sample of 221 halo stars. Diagrams from Eggen, Lynden-Bell, & Sandage (1962).

ELS firmly established the existence of a close link between stellar abundance and kinematics, confirming previous work (see Roman 1995).

They interpreted these correlations in terms of the dynamics of a rapidly collapsing galaxy, suggesting that the oldest stars formed from gas falling toward the galactic center and collapsing from the halo onto the plane. Their conclusion that the collapse occurred in only a few times 10^8 years was very influential, but has not quite stood the test of time.

More than a decade later, Searle and Zinn were curious to see if globular clusters show similar correlations as individual halo stars do. Determining cluster metallicities from spectral scans of 177 red giant stars in 19 halo globulars and supplementing their sample with cluster data measured by others, Searle & Zinn (1978) found *no* abundance gradient among globular clusters from galactocentric distances of 8 kpc on outward, as shown here in Fig. 84. They also noticed that various cluster parameters, such as those indicative of horizontal-branch morphology, vary significantly more among the outer globulars than among the inner ones, suggesting that halo globulars may have a broader range of ages. From this evidence they concluded that the collapse of our Galaxy must have been slower and more chaotic than envisaged by ELS and may have involved delayed crashes of major gas fragments. Thus, *the period of halo formation probably lasted a few Gyr.* At present, this scenario seems amply supported by evidence that (1) the range of globular-cluster ages is about 11–15 Gyr (or 25% of the age of the universe) and (2) a dwarf companion of the Milky Way, Sgr I, is being accreted and brings with it four globular clusters.

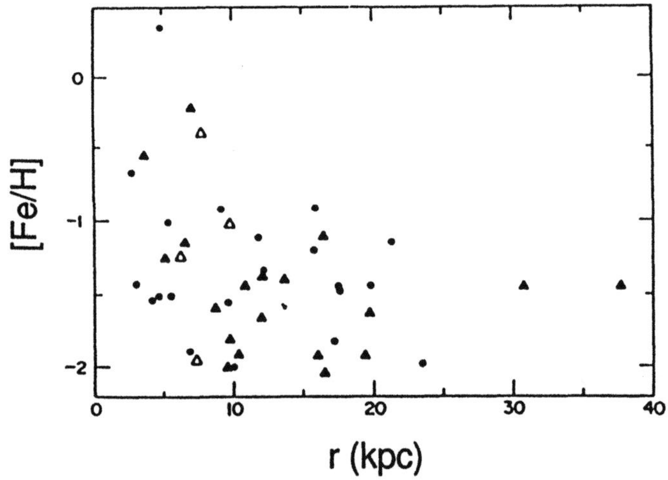

Fig. 84. Abundances of globular clusters in Milky Way plotted versus Galactocentric distance r. Note absence of abundance gradient at $r > 8\,\mathrm{kpc}$. Diagram from Searle & Zinn (1978).

8.2 The Milky Way's Warp

The thin layer of neutral hydrogen that pervades the Milky Way disk is remarkably flat inside the solar circle. Beyond, however, it warps down toward the LMC and up on the opposite side. The discoverers of this two-sided warp, Burke (1957) and Kerr (1957), both realized that the gravitational force of the Magellanic Clouds at their present location is insufficient to explain the warp. An old map often reproduced in textbooks shows the H I layer warping up to 800 pc above and below the plane (Oort, Kerr, & Westerhout 1958). Modern maps show the warp to be even more extreme and asymmetric: At $R \gtrsim 18\,\mathrm{kpc}$ the H I layer reaches heights of >3 kpc above the plane in the North, but of only ~1 kpc below the plane in the South, where it then bends back toward the plane again (Burton 1988).

The Galactic warp has also been detected at other wavelengths. It is seen in the molecular gas via CO sources (Wouterloot et al. 1990), in the dust layer via *IRAS* fluxes (Sodroski et al. 1987), and even in the *stellar* disk via excess star counts in fields that include the warp (Carney & Seitzer 1993).

There have been many distinguished efforts at formulating a theory for this warp and for similar warps observed in other galaxies. Yet, as Binney (1992) concludes in his excellent review, the origin of these warps remains "a fascinating puzzle".

Early ideas that a *past* interaction, during which the Magellanic Clouds would have approached the Milky Way to 20–25 kpc, could have produced the H I warp (Toomre 1972; Fujimoto & Sofue 1976; Lin & Lynden-Bell 1977) now appear untenable since the necessary models force the Clouds to orbit too fast and too close to survive for 15 Gyr (e.g., Lin, Jones, & Klemola 1995; see §8.6).

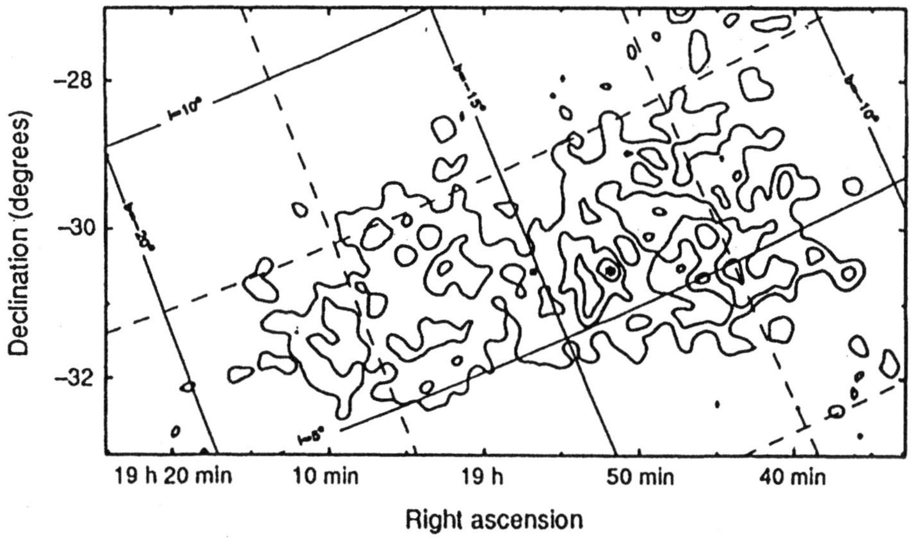

Fig. 85. Isopleths of dwarf galaxy Sgr I. Diagram from Ibata, Gilmore, & Irwin (1994).

For over a decade, some favorite hypotheses for explaining warps have revolved around misalignments between disks and their surrounding dark halos (e.g., Lynden-Bell 1965; Toomre 1983; Dekel & Shlosman 1983). Yet, even these hypotheses face problems due to the relatively rapid damping of bending modes. In an important new paper, Nelson & Tremaine (1995) study the interaction between a warp and halo and conclude that dynamical friction either dampens or excites warps on a relatively short time scale. Thus, it appears that warps like the one in the Milky Way must be excited either recently or continuously.

8.3 Accretion (?) of the Sagittarius Dwarf

The discovery of the dwarf galaxy Sgr I is a prime example of astronomical serendipity. While reducing spectra of Milky-Way bulge stars obtained with the Anglo-Australian Telescope and its multi-fiber spectrograph, Ibata et al. (1994, 1995) noticed a striking excess of stars with radial velocities close to $+140\,\mathrm{km\ s^{-1}}$. After checking that no reduction errors had been made, they realized that this excess implies the presence of some stellar system either behind or in front of the bulge, at only $\sim 15°$ from the Galactic center and in the midst of rich star clouds in Sagittarius. Using automated plate scanning to derive color–magnitude diagrams over several $6° \times 6°$ fields covered by the Southern Sky Survey, they managed to derive isopleths of this hidden system by plotting the excess of horizontal-branch star images, as shown in Fig. 85. These results, first published in a 2^{+}-page Letter to *Nature*, represent a remarkable tour de force.

The dwarf Sgr I appears centered at $\alpha = 18^{\mathrm{h}}50^{\mathrm{m}}$, $\delta = -30°$ (l=5°, b=−15°) and lies about 17 kpc beyond the Galactic center as seen from Earth. With an

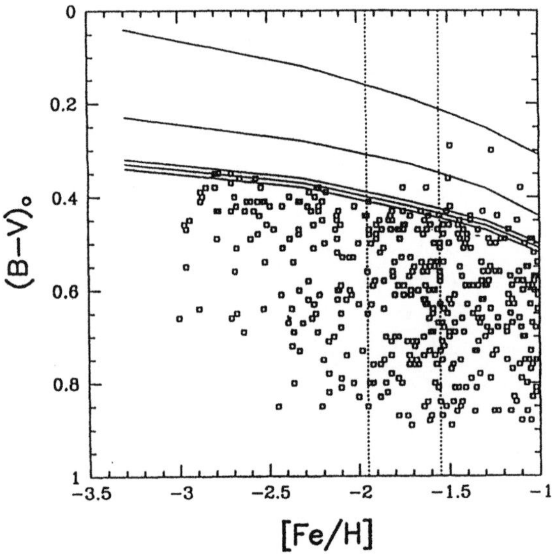

Fig. 86. Distribution of reddening-corrected index $(B-V)_0$ and metallicity for 477 halo stars. Superposed isochrones are for (*top to bottom*) 5, 10, 15, 16, and 17 Gyr. Diagram from Unavane, Wyse, & Gilmore (1996).

approximate size of 1.8×4.4 kpc ($4° \times 10°$) and absolute magnitude $M_V \approx -13$, Sgr I is the largest dwarf spheroidal galaxy near the Milky Way and the third or fourth most luminous companion after the LMC, the SMC, and perhaps also the Fornax dwarf (Ibata et al. 1995; Mateo et al. 1995). Its elongated shape and orientation nearly perpendicular to the plane of the Milky Way suggest tidal stretching. It moves with $V_r = +172 \pm 2\,\mathrm{km\,s^{-1}}$ relative to the Galactic center either toward or away from the Galactic plane. Model simulations suggest that Sgr I has suffered serious tidal disruption, presumably mostly from its last passage (Velázquez & White 1995). However, its current passage may still not be the last (Johnston et al. 1995).

The dominant stellar population of Sgr I is ~10 Gyr old, and there may be an additional weak population of younger age (Mateo et al. 1995). Stellar metallicities range between [Fe/H] ≈ −1.2 and −0.5, which seems remarkably high for a galaxy of $M_V \approx -13$. Perhaps the most interesting population component are four *globular clusters* that all appear to belong to Sgr I: M54, Arp 2, Ter 7, and Ter 8 (Da Costa & Armandroff 1995; Sarajedini & Layden 1995). Their ages and metallicities have long attracted attention and range between (19 Gyr, −2.0) and (9 Gyr, −0.4) (Chaboyer et al. 1996b; Sarajedini & Layden 1995). It seems nearly unavoidable that with the accretion of Sgr I on the present or next passage, these four globulars will become part of our Milky Way cluster system. Hence the great interest in them and in the fact that the most luminous cluster, M54, lies smack at the center of Sgr I and may be its nucleus (§7.5).

How many accretions similar to that of Sgr I took place *after* the initial assembly of the Milky Way, say during the past 10 Gyr? To address this interesting question, Unavane, Wyse, & Gilmore (1996) plotted a fascinating diagram, here reproduced as Fig. 86, for 477 halo stars observed by Carney et al. (1994). These high-proper-motion stars have known space velocities and metallicities.

Unavane et al. selected stars with [Fe/H] < -1 and plotted their [Fe/H] versus the reddening-corrected $(B-V)_0$. Also plotted in Fig. 86 are isochrones that mark the bluest color at the main-sequence turn-off as a function of metallicity for ages of 5, 10, and 15–17 Gyr. Any star *blue*ward of such an isochrone is *younger* than the isochrone age. As the figure shows, most stars lie redward of the 15 Gyr isochrone, indicating that the average halo star is very old. However, there is a sprinkling of stars between the 10 Gyr and 15 Gyr isochrones, tracing the declining accretion rate of that period. Finally, three stars between the 5 Gyr and 10 Gyr isochrones suggest that by then accretion had slowed to a trickle. Unavane et al. interpret these numbers as indicating that $\lesssim$6 Sagittarius- or Fornax-like dwarfs (or $\lesssim$60 Carina-like dwarfs) have been accreted during the past 10 Gyr. These upper limits depend sensitively on where exactly the bulk of an accreted dwarf gets deposited. Since the Milky Way halo has $M_V \approx -17.5$ or about half the luminosity of the LMC, we conclude that accretion events of the kind now apparently happening with Sgr I have been relatively rare in the recent past.

8.4 Milky Way Halo: Retrograde Orbits and Ghostly Streams

There have long been hints of a chaotic assembly of the Milky Way halo in the form of retrograde orbits of individual halo stars (Oort 1965). Moreover, "moving groups" of stars traveling from and into the halo have been found both near the Sun (e.g., Eggen 1958–1989; Soderblom & Mayor 1993) and at heights of 4–5 kpc above the plane (Majewski et al. 1994), tracing dispersed clusters of coeval stars.

Even whole groups of globular clusters display strange kinematics. Rodgers & Paltoglou (1984) discovered that, as judged from radial velocities, globular clusters with $-1.7 <$ [Fe/H] < -1.4 have a retrograde mean motion, a result confirmed by van den Bergh (1993). Along a similar line, Zinn (1993) separated Milky Way globulars into two groups according to horizontal-branch morphology and, apparently, also age. These two age groups have different spatial distributions, line-of-sight velocity dispersions, and mean orbital motions. Figure 87 shows radial velocities plotted as a function of the angle ψ between the line of sight and the direction of galactic rotation at the cluster. For 46 old globulars, the mean rotation velocity is $\langle V_{\rm rot}\rangle = +70 \pm 22\,{\rm km\,s^{-1}}$, while for 19 younger globulars $\langle V_{\rm rot}\rangle = -64 \pm 70\,{\rm km\,s^{-1}}$. Ahead of the discovery of Sgr I, Zinn concluded that the younger clusters may have formed in companion galaxies that were later accreted.

The existence of several dSph galaxies in the plane of the Magellanic Stream (Lynden-Bell 1976; Kunkel & Demers 1976; Kunkel 1979) and of a separate Fornax–Leo–Sculptor Stream (Lynden-Bell 1982) seems to present corroborating evidence in support of this picture of disrupted companions. The latter stream was originally defined by the four dwarf spheroidals Fornax, Leo I and II, and Sculptor, which lie along a big circle around the Galactic center. It is now thought to consist of 6 dSph's plus, perhaps, a few youngish globular clusters (Majewski 1994; Fusi Pecci et al. 1995).

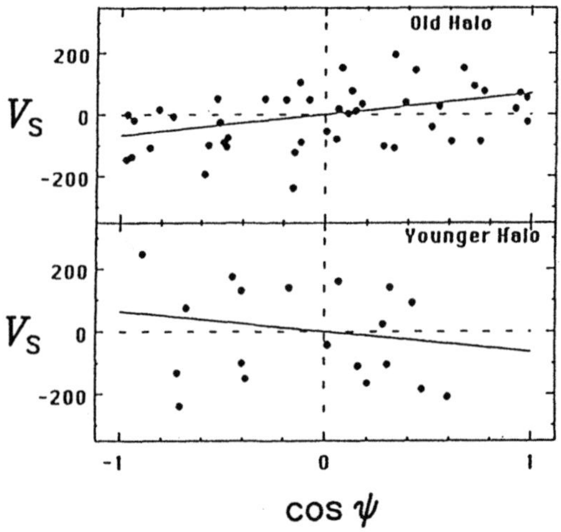

Fig. 87. Radial velocities V_s of 46 old and 19 younger globular clusters plotted versus $\cos\psi$. Velocities are as measured at Sun's position by fictitious observer that would be at rest with respect to Galactic center. Note retrograde mean motion of younger clusters. Diagram from Zinn (1993).

Such tentative evidence for streams of presumed debris and the discovery of the dwarf Sgr I have led to an increasingly intense search for "ghostly streams" (Lynden-Bell & Lynden-Bell 1995). The name of the game is to look for apparent streams of globular clusters and dwarf spheroidals in a common plane centered on the Galactic center (not the Sun). One does this by finding multiple intersections of great circles at right angles to individual objects ("polar paths"). These intersections define the pole of a common plane. As Fig. 88 suggests for objects within two distance ranges from the Galactic center, there are many candidate multiple intersections. The Lynden-Bells then use a radial-energy continuity condition to further narrow down the choices. They make predictions about the proper motions of 22 objects suspected to be associated with various streams and point out that a proper motion measured for even only one object will help fix both the orientation of its candidate plane and the sense of rotation in that plane. With Hipparcos and *HST* observations promising proper motions of unprecedented accuracy, astronomers may be poised to unravel the merging history of our Milky Way.

8.5 High-Latitude A V Stars, Thick Disk, and Youngish Bulge

As the importance of mergers in galaxy evolution has become better understood, there have been many claims of observational evidence for past mergers in the Milky Way beyond those discussed above. Three of the more important pieces of claimed evidence are here discussed briefly and critically. For details and different perspectives, see the reviews by Freeman (1995), Gilmore (1995), and Wyse (1995) in the Proceedings of IAU Symposium No. 164.

Perhaps the best additional evidence for past accretions in the Milky Way may be the high-latitude A V stars (Rodgers 1971, Rodgers et al. 1981). These are metal-rich A-type main-sequence stars with ages of typically 200–600 Myr and located at heights of 1–4 kpc above the plane. Since these stars are too

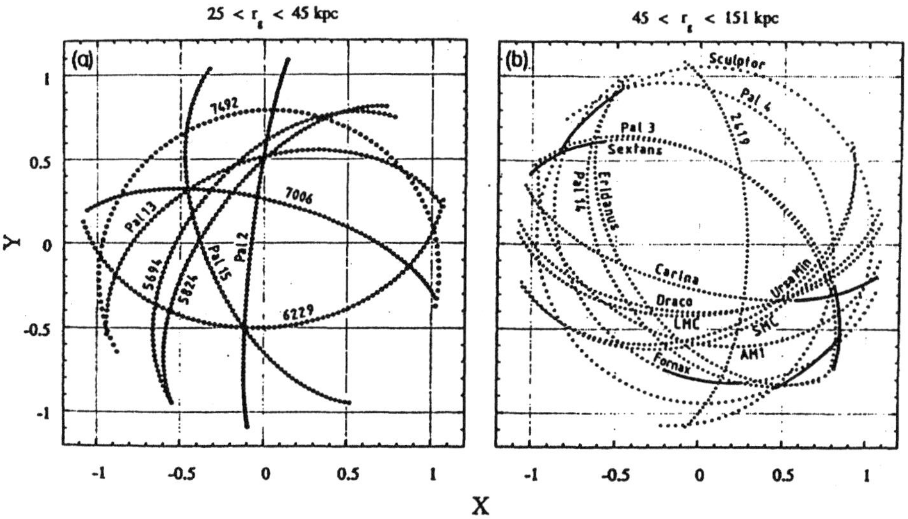

Fig. 88. Polar paths of globular clusters and dwarf spheroidals at different distances from Galactic center: (a) $25 < r_g < 45$ kpc and (b) $45 < r_g < 151$ kpc. Diagrams from Lynden-Bell & Lynden-Bell (1995).

young to have been accelerated stochastically that far out of the plane, they are thought to represent evidence for a galaxy that was accreted about 0.6–0.7 Gyr ago (Lance 1988).

The thick disk of our Milky Way is also often said to have been stirred up by an ancient merger (Gilmore et al. 1989). This disk is a concept evolved from the concept of the Intermediate Population II, worked out at the Vatican Conference (O'Connell 1958). It has an exponential scale height of ~1.0 kpc, a metallicity $-1.0 \lesssim$ [Fe/H] $\lesssim -0.5$, and a mean rotation velocity of 190 km s^{-1}, and is estimated to contain 5% – 10% of the luminous mass of the Milky Way. The main evidence cited for a merger origin is that the thick disk seems to show no vertical abundance gradient between 0.5 kpc and 3 kpc above the plane (Gilmore et al. 1995). Personally, I remain skeptical of this rather weak evidence and believe that different interpretations might have been chosen if we did not know of mergers occurring in other galaxies. For example, during the 1960s the Intermediate Population II was being regarded as good evidence in favor of a monolithic collapse.

The discovery that the Milky Way bulge contains stars of ages ~5–12 Gyr (e.g., Holtzman et al. 1993; Carney et al. 1995 and refs. therein) has again led some astronomers to claim a past merger. However, many other interpretations, including prolonged accretion of gas, seem possible. Perhaps the most serious challenge to any merger interpretation is posed by an alternative bulge-building mechanism: Vertical disk instabilities can be excited by bars and can kick disk material into distributions strongly resembling box- and peanut-shaped bulges (Combes et al. 1990; Pfenniger & Friedli 1991). This mechanism is well documented and supported by N-body simulations, as reviewed by Martinet (1995).

8.6 The Magellanic Clouds and Stream

After the Milky Way warp, the Magellanic Stream is the second major signature of a possible past interaction in the Local Group that has been known for over two decades. Despite its well-defined and generally accepted properties, its interpretation has had a long and tortuous history.

It all began with a strange discovery by Nannielou Dieter (1965), who found a cloud of neutral hydrogen near the South Galactic Pole (SGP) with negative velocities of up to $-90\,\mathrm{km\,s^{-1}}$ in it.[24] Mapping the area around the SGP with higher sensitivity, Wannier & Wrixon (1972) discovered that Dieter's cloud was only the brightest part of a 60° long H I filament that extended from the SGP (near their southern horizon) past the equator to the North and featured radial velocities decreasing from $-60\,\mathrm{km\,s^{-1}}$ near the SGP to $-400\,\mathrm{km\,s^{-1}}$ at the northern end. In Australia, Mathewson, Cleary, & Murray (1974) used the Parkes 60-ft telescope to extend the mapping southward and discovered with surprise that this H I filament reaches all the way to the Magellanic Clouds, where it seems to connect to the region *between* the LMC and SMC. They named the now 120° long filament the "Magellanic Stream" and suggested that it may be a tidal tail or bridge torn from the Clouds during a recent interaction with the Milky Way. A beautiful H I map of this Stream based on their and many later observations is shown in Mathewson (1985).

We now know that the Magellanic Stream lies in a plane approximately centered on the Milky Way and nearly perpendicular to it. Therefore, if the Stream consists of orbiting material, it moves in a near-polar orbit. The H I velocities relative to the Galactic Standard of Rest range from $-200\,\mathrm{km\,s^{-1}}$ near the Stream's northern end to $+80\,\mathrm{km\,s^{-1}}$ near the Clouds, where they connect smoothly onto the Clouds' velocities. Finally, the Stream contains about $10^9\,\mathcal{M}_\odot$ of H I, but either no or very few stars.

Even before the Stream was discovered, Toomre (1972) had attempted to explain the Milky Way's warp through a past tidal interaction with the Magellanic Clouds. To do so, he had had to (1) shrink the size of the Milky Way, moving the Sun from 10 kpc to 8 kpc radius, (2) lower V_{rot} to 185–220 $\mathrm{km\,s^{-1}}$, and (3) assume that the LMC had reached a perigalacticon of only $\sim$20 kpc. Following the discovery of the Magellanic Stream, several new models were proposed along the same lines, all putting the Clouds presently near their apogalacticon (e.g., Lin & Lynden-Bell 1977; Fujimoto & Sofue 1977). However, these early models had serious problems. They failed to reproduce the high negative velocities of the Stream and, by placing the Clouds near apogalacticon, led to orbital periods so short ($\lesssim$1 Gyr) that it was hard to understand why the Clouds would not have been tidally disrupted already.

Later models placed the Clouds near *peri*galacticon and added a massive dark halo to the Milky Way to yield higher negative velocities (e.g., Murai & Fujimoto 1980; Lin & Lynden-Bell 1982.) They brought significant improvement,

[24] I remember some of us Berkeley graduate students thinking that the sky was falling in!

but could no longer produce the warp, the Clouds now never coming close enough to exert the required strong torque. All these models placed the Clouds at the Stream's leading edge, thus making the Stream a *tidal tail.*

Recent models also place the Magellanic Clouds near perigalacticon, but use the LMC's proper motion of $\mu_\alpha = 0\farcs120 \pm 0\farcs028$ century^{-1}, $\mu_\delta = 0\farcs026 \pm 0\farcs027$ century^{-1} (Jones et al. 1994) as an additional constraint, and explain the Stream as a tidal tail generated during the *last* encounter about 1.0–1.5 Gyr ago (e.g., Lin et al. 1995; Gardiner & Noguchi 1996). In these models, the Clouds reach apogalactica at $r > 100\,\mathrm{kpc}$, and it becomes more plausible that some dwarf spheroidals may be tidal debris.

The paper by Gardiner & Noguchi (1996) is especially instructive because it makes detailed comparisons between the model simulations and observations, thus returning to the high standards set by Toomre & Toomre (1972) and Lin & Lynden-Bell (1977). Figure 89 shows the observed H I distribution of the Magellanic Stream and the corresponding model view, while Fig. 90 compares the measured H I velocities (*diamonds*) with the model velocities. Notice that the model reproduces the Stream's apparent H I distribution reasonably well, but reproduces the "leading arm" of H I on the Clouds' opposite side only with considerable distortion. The model velocities, too, agree in general with the observed velocities. Specifically, the high negative Stream velocities now pose no problem any more. However, the model velocities fit the "leading-arm" H I velocities only poorly, again leaving room for further improvement.

Though less detailed in comparing their model with the observations, Lin et al. (1995) show interesting figures of the past and future orbits of the LMC and SMC. According to their model, the Cloud orbits have slowly shrunk in the past, with the LMC and SMC forming a binary system. This binary is near perigalacticon at present and will break up in the near future. Because of its relatively large mass, the LMC will then continue descending and will merge with the Milky Way in about 7–8 Gyr, while the less massive SMC will orbit much longer. There is reason to believe that the merger of the LMC will be considerably more spectacular than the present accretion of Sgr I, adding $\sim$1/4 mag to the luminosity of the Milky Way in the long run (Tremaine 1976) and probably considerably more for several 100 Myr. Unfortunately, none of us will be around to check on this prediction!

There is also a class of ram-pressure models that try to explain the Magellanic Stream as consisting of gas stripped during the most recent encounter of the Clouds with the Milky Way (e.g., Mathewson et al. 1977; Mathewson 1985; Moore & Davis 1994). These models all tend to suffer from the same problem: It is nearly impossible to impart to the stripped gas the highly negative velocities that are observed. Several such models stem from a period when the ubiquity of massive dark halos was not yet fully appreciated. More recent such models have been justified on grounds that the Stream may be too narrow for a tidal feature and contains no stars (e.g., Moore & Davis 1994). Observationally, Hα emission may have been detected at the leading edge of the H I distribution and has been presented as evidence of ram-pressure stripping there (Weiner & Williams 1996).

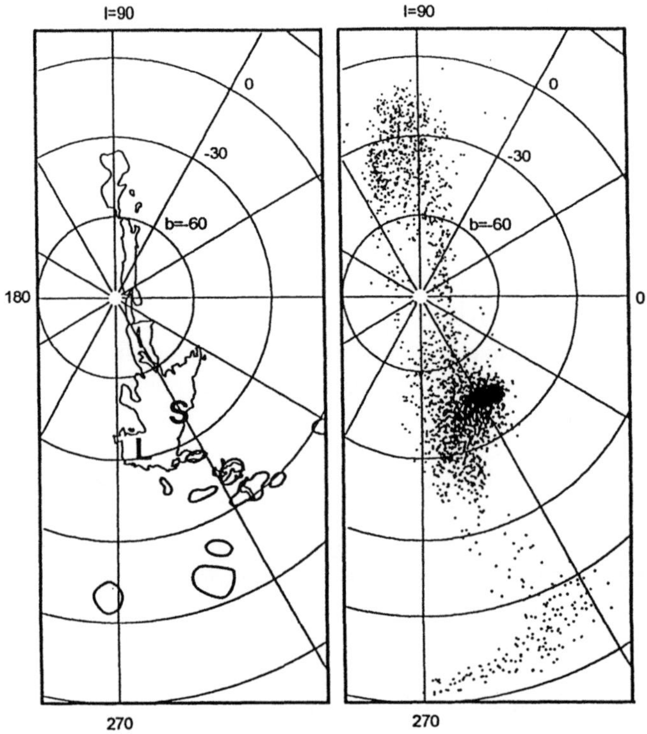

Fig. 89. The Magellanic Stream as (*left*) observed in H I on the sky and (*right*) reproduced by an N-body simulation. Figure from Gardiner & Noguchi (1996), with letters L and S added to mark locations of LMC and SMC, respectively.

One cannot and should not dismiss such alternative models out of hand. Yet, at present they all look a bit contrived to me. They lack anything approaching the tidal-interaction models' impressive ability to reproduce the main kinematic and morphological signatures of the Magellanic Stream.

8.7 Interactions in the M31 Subgroup

This subgroup consists of M31 itself and, roughly in order of increasing projected distance: the well-known companions M32 and NGC 205, the dwarf spheroidals And I – And III, the pair of dwarf ellipticals NGC 147 and NGC 185, M33 at a projected distance of $15° \approx 200\,\mathrm{kpc}$, plus perhaps some more distant galaxies like IC 10 (Lynden-Bell 1994, esp. Fig. 7). Signs of interactions similar to those observed in and around the Milky Way abound, but have been studied less and are generally much less well understood.

Any large-field photograph of M31 shows not only the satellites M32 and NGC 205, but also the warped disk (e.g., Sandage 1961) and, when properly masked, the strikingly misaligned bulge. The misalignment against the major axis of the outer disk is of order 10° (see Hodge 1992). What created this bulge

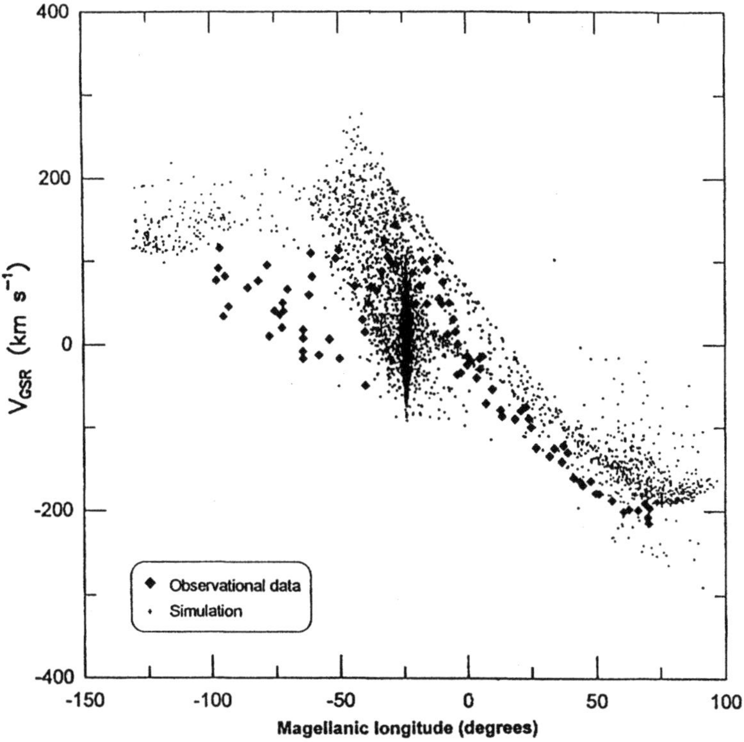

Fig. 90. Measured radial velocities of H I in Magellanic Stream (*diamonds*) compared with model velocities from N-body simulation by Gardiner & Noguchi (1996).

misalignment and the warp? Recent interactions with M32 or NGC 205, or a past passage of M33? We do not know at present, though several papers have been written on this subject.

Past interactions with the nearest satellites are likely. Note that M32 features one of the youngest stellar populations known in any bona fide elliptical (Chap. 6), which could imply that it gobbled up a significant amount of gas from M31 a few Gyr ago. Better yet, NGC 205 shows direct, unmistakable evidence of a tidal interaction: As Zwicky (1959) noticed already, and as here reproduced in Fig. 91, its outer isophotes are distorted into an integral-sign shape that is typical of strong tides. Photographs of the same phenomenon are shown in Kormendy (1982). It is generally assumed that NGC 205's tidal interaction was with M31, but it could conceivably also have been with M32. Proper motions of M32 and NGC 205 relative to M31 should prove invaluable in trying to model these satellites' orbits and past interactions.

Perhaps not coincidentally, M33 is warped even more strongly than M31. On deep blue-light photographs one notices a steep drop in surface brightness to the east–southeast. On large-scale plates resolving this galaxy into individual stars, one can further see that it is the unresolved stellar disk that appears like

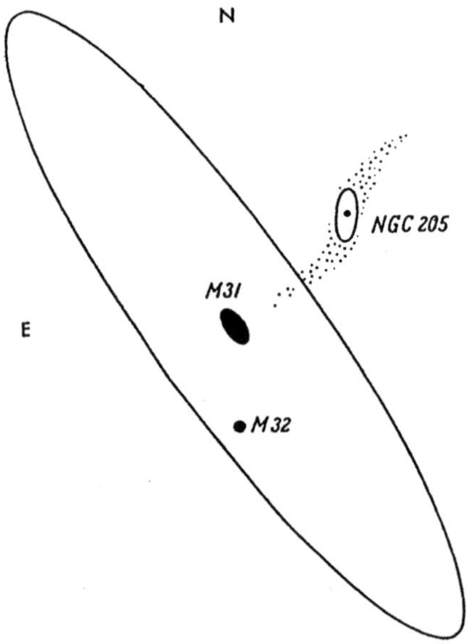

Fig. 91. Schematic drawing of M31 and satellites M32 and NGC 205. Latter appears tidally distorted. Figure from Zwicky (1959).

cut off with a knife in a direction parallel to *P.A.* ≈ 24°, while bright stars appear sprinkled across this boundary (see, e.g., Plate 1 in Sandage & Bedke 1988). Where the surface-brightness drop occurs, the H I layer warps sharply away from us, whence the apparent boundary may be caused by dust extinction in this layer. Figure 92 shows the H I distribution across the face of M33, which was mapped with the Arecibo telescope (Corbelli et al. 1989). Notice the strong, integral-shaped distortion. As the figure shows, the H I distribution is still fitted quite well by the old tilted-ring model developed by Rogstad et al. (1976) to explain their own H I observations. In this model, the tilt of successive H I rings is severe, reaching 40° in the outermost ring compared to the central part. The cause of the M33 warp is usually presumed to be a past interaction with M31, yet details are not known.

In addition to this multiple, though tentative evidence of interactions among members of the M31 subgroup, evidence has also been found for intermediate-age stars in the bulges of both M31 and M33 (Rich & Mould 1991; Mighel & Rich 1995). However, the recently discovered double nucleus of M31 is probably *not* evidence for a past merger, contrary to what was thought right after its discovery, as we now discuss.

8.8 The Double Nucleus of M31: Merger or not?

The discovery of a double nucleus in M31 with the *Hubble Space Telescope* even before its repair created much excitement and seemed at first to point toward a recent merger (Lauer et al. 1993). There is a bright optical peak P1 that is separated from a fainter optical peak P2 by a projected $0''.49$ (1.8 pc). Surprisingly, it is the latter, fainter P2 that lies at the optical and kinematic center of

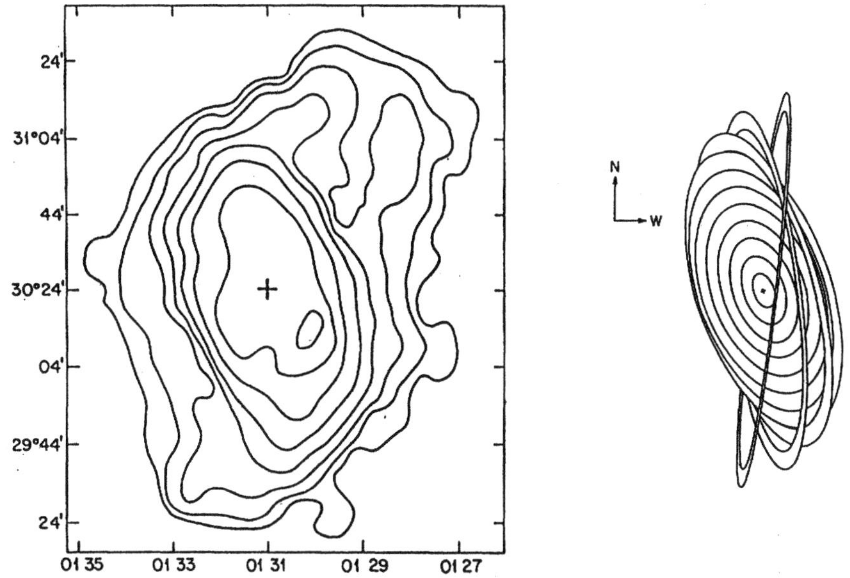

Fig. 92. (*Left*) Observations and (*right*) tilted-ring model of H I distribution in M33. Diagrams from Corbelli et al. (1989) and Rogstad et al. (1976).

the bulge (to within ~0.″05). Since the color index $V-I$ is nearly constant both around and between the two nuclei, the possibility that dust extinction creates the appearance of two nuclei seems virtually excluded. Also, it appears that P2 is dynamically hotter than P1 and harbors a UV- and radio-bright non-thermal source (King et al. 1995), again pointing to P2 as being the true center of M31 and perhaps harboring a massive black hole. Lauer et al. conclude that "P1 is more likely to be a cannibalized galactic nucleus".

The problem with this interpretation is that under most circumstances the orbit of P1 should decay very fast. Assuming that P2 is the true nucleus, the orbital period at a radius of 1″ (3.6 pc) is ~80,000 yr, and the frictional decay time is $60/\mathcal{M}_6$ Myr, where $\mathcal{M}_6$ is the mass of P1 in units of $10^6\,\mathcal{M}_\odot$. Some arguments have been made that if there were no relative motion between P1 and the surrounding stars, there would be no friction, but – given the high central $\sigma_v \approx 220\,\mathrm{km\ s^{-1}}$ – this seems quite implausible.

A remarkably simple model for this apparent double nucleus has been put forward by Tremaine (1995). He proposes that a *thick eccentric disk* of stars surrounds the true center of M31, which indeed may harbor a Black Hole (BH). In this model, the bright off-center source P1 is the apocenter region of this eccentric disk. For the model to work, the central object (BH?) at P2 must be smaller than 1 pc so that the potential felt by the stellar disk is nearly Keplerian. The disk's eccentricity may then be excited by dynamical friction from the bulge. Remarkably, a simple realization of this model with three nested elliptical ringlets (a = 0.″5–1.″5, e = 0.44–0.04) reproduces both the observed brightness distribution and all available spectroscopic observations in detail. Tremaine points out that various possible mechanisms may keep the stellar orbits aligned within the eccentric disk. Perhaps the most likely mechanism has to do with "discrete nonlinear Eigenmodes of a self-gravitating disk orbiting in the external field of a black hole and bulge".

The upshot of this simple model is that the merger hypothesis put forward by Lauer et al. (1993) does not seem necessary to explain the double nucleus of M31. Instead, this nucleus seems to involve a small disk of stars around a black hole, forced into being eccentric and thick by a tug between the strong gravitational force from the BH and the weaker gravitational interaction with the surrounding bulge.

8.9 Motions Within and Around the Local Group

In a classic paper, Kahn & Woltjer (1959) pointed out that – if there really was a big bang – then M31 and our Milky Way were probably falling toward each other for the first time. This well-known timing argument has been elaborated on many times since (e.g., Lynden-Bell & Lin 1977). It links together the mass of the Local Group, the distance between us and M31, and the net approach velocity of these two galaxies, currently pegged at $-123 \pm 20\,\mathrm{km\ s^{-1}}$. Due to vast increases in computing power, it is becoming increasingly feasible to consider the motions of all Local Group galaxies in a cosmological setting. A pioneer of such calculations has been Jim Peebles (e.g., 1990).

In a recent paper, Peebles (1994) pursues in greater detail than ever before a study of possible galaxy motions within and around the Local Group. It turns out that the Local Group galaxies cannot be considered in isolation because they are influenced by nearby groups of galaxies such as, e.g., the M81 and Centaurus groups. Peebles uses numerical methods and the action variational principle to try placing galaxies at their present positions in space and give them their present radial velocities relative to the Milky Way. The initial conditions are set by requiring near homogeneity of the universe at high redshifts. It is not the details of Peebles's method that interest us here, but its promise for furthering our understanding of past evolution. Figure 93 shows two representative orbit solutions for a low-density, flat model universe. The orbits of the various galaxies are shown projected onto a plane related to celestial coordinates as follows: The X-axis points toward (α=0, δ=0), the Y-axis toward a point 90° away on the celestial equator, and the Z-axis toward North. Each orbit begins at high redshift near the label for the galaxy, and the present position of the galaxy is at the box. Notice how the Milky Way (MW) and M31 are falling toward each other in a coordinate system that, of course, is growing with the universe.

From such calculations we can, at present, only learn some general features about how the Local Group may have evolved. Once proper motions will become available for Local Group galaxies (perhaps in 10–20 years from now), they will add valuable constraints, and we may then hope to reconstruct the past history of the Local Group in more detail. Hopefully, at that time the many signatures of past interactions – like the warps of the Milky Way, M31, and M33, the tidal

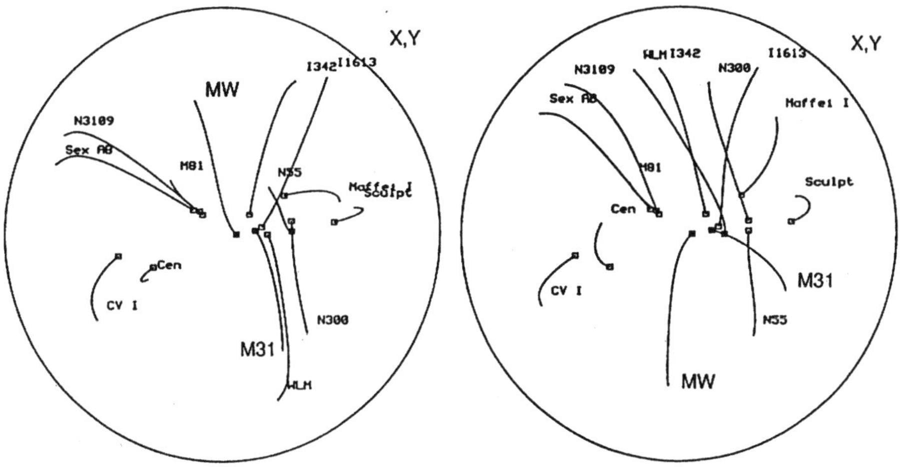

Fig. 93. Two representative orbit solutions for galaxies in Local Group and environs; see text for details. Diagrams from Peebles (1994).

distortion of NGC 205, and the Magellanic Stream – will help narrow down the choices between various possible past histories. Whether we shall gain a truly detailed understanding of what must, in essence, have been a stochastic process is uncertain. Yet, unraveling the mysteries of the grandiose Celestial Spiel around us will remain one of the most noble tasks that future generations of astronomers can tackle.

8.10 Summary

The main results of this chapter are as follows:

- In the Milky Way, there is strong evidence for an ongoing accretion of the Sgr I dwarf. Weaker evidence for past accretions comes in the form of retrograde globular clusters and stars, the Fornax–Leo–Sculptor Stream, and high-latitude A V stars. Whether the thick disk, warp, and intermediate-age bulge present compelling evidence in favor of past mergers is unclear at present.
- The Magellanic Stream appears to be a tidal tail torn from the Magellanic Clouds during one of their most recent encounters with the Milky Way. Models predict that the LMC will merge with our Milky Way within the next Hubble time.
- There are many signs of tidal interactions in members of the M31 subgroup (warps, a misaligned bulge, integral-sign distortions), but the details of these interactions remain unclear. The double nucleus of M31 is probably *not* due to a merger, but instead to an eccentric stellar disk surrounding a black hole.
- Proper motions of globular clusters and Local Group galaxies hold great promise for improving our understanding of past interactions and mergers in both the Milky Way and the Local Group.

9 Interactions and Mergers at High Redshift

Until a few years ago, progress in our understanding of galaxy evolution from observations of objects at high z seemed slow and tedious. Granted, the enormous efforts spent on observing Quasi-Stellar Objects were yielding interesting results, and counts of galaxies to ever fainter magnitudes were yielding puzzles, like the famous one concerning the apparent excess of faint blue objects. Yet, ten years ago it seemed at least to me that our understanding of galaxy evolution was progressing much more rapidly through observations of relatively nearby objects. This situation has changed dramatically of late, and work on high-redshift objects is now rapidly begetting results.

A number of technical advances have contributed to this change of pace: (1) The increased light-gathering power of new giant telescopes like Keck; (2) the increased sensitivity and spectral coverage of instruments at many wavelengths, including mm-wave radio dishes; and (3), the increased resolution achieved with some ground-based telescopes and, most dramatically, with the repaired *Hubble Space Telescope*. The latter telescope has revolutionized our ability to study distant galaxies and clusters. Its new capabilities and, especially, its December 1995 observations of the "Hubble Deep Field" have led astronomers to rush to the new frontier.

This rush forms the first subject of the present lecture. Making use of recent results obtained with *HST*, we then discuss interactions and mergers in QSOs, galaxy clusters, and distant field galaxies. Finally, we present current ideas about the merger rate as a function of z and close with a brief assessment of past successes and future challenges.

9.1 Rush to the Frontier

During 1995 December 18–30, Williams et al. (1996) observed the Hubble Deep Field (HDF) for 10 days and in four passbands with the WFPC2 camera of *HST*. The resulting color images, one of which is shown in Plate 3, have received much publicity and reveal fainter galaxies than ever seen before. What is unique and a true first in the history of optical astronomy is that these limit images – obtained with the most powerful telescope for the purpose – were put on the World-Wide Web and released immediately to the general public. Hence, anyone interested in the HDF can access these images from anywhere in the world and do research with them. For astronomers this opportunity is not unlike that perceived by East-Coast settlers of the 19th century when the news of gold being found in California reached them. These settlers joined the gold rush in droves. Like California then, the Hubble Deep Field and its high-z nuggets beckon now.

However, like anyone thinking of joining a gold rush, we must do our best to inform ourselves, both to protect ourselves from false rumors and to avoid some of the risks that inevitably lure along the paths to frontier territory. One way to prepare for the trip back in time is to learn what was known before the rush to high z began. As an excellent general reference, I recommend the "Saas-Fee" lectures by Sandage, Kron, & Longair (1995).

Table 5. Redshift, cosmic time, and gas content

z	$(t_0-t)/t_0$	t (Gyr)	$\mathcal{M}_{\text{gas}}/\mathcal{M}_{\text{stars+gas}}$ Type Sb	Type Sc
0.0	0.00	15.0	0.08	0.22
0.2	0.24	11.4	0.15	0.32
0.5	0.46	8.2	0.25	0.44
1.0	0.65	5.3	0.41	0.59
2.0	0.81	2.9	0.62	0.75
5.0	0.93	1.0	0.84	0.90

To sharpen our expectations, let us first take a look at the relations between redshift z, cosmic time t, and the fractional gas content of galaxies. For a flat universe ($q_0 = 1/2$), cosmic time relates to redshift as

$$\frac{t}{t_0} = \frac{1}{(1+z)^{3/2}}, \tag{6}$$

where $t_0 \approx 15\,\text{Gyr}$ for $H_0 \approx 50$. The fractional gas content, $\mathcal{M}_{\text{gas}}/\mathcal{M}_{\text{stars+gas}}$, can be computed once the star formation rate in galaxies is known as a function of time. We follow Searle, Sargent, & Bagnuolo (1973) in assuming that this rate declined exponentially,

$$\frac{\mathrm{d}\mathcal{M}_{\text{stars}}}{\mathrm{d}t} = C\mathrm{e}^{-t/\tau}, \tag{7}$$

with a time constant of $\tau = 6\,\text{Gyr}$ for Sb and $10\,\text{Gyr}$ for Sc spirals. Table 5 gives, for a number of representative redshifts, the fractional lookback time $(t_0 - t)/t_0$, cosmic time t, and fraction of luminous matter in Sb and Sc spirals that remains in the form of gas. Notice that, at $z = 0.54$ in such a flat universe, we are looking back to objects half as old as nearby objects. Also, at that redshift an average Sb galaxy contains 4× more gas than at present, and an average Sc galaxy nearly twice. Therefore, we should not be surprised to find distant galaxies in a significantly more gaseous state than nearby ones, as indeed has been found. Next, notice that by redshift $z = 2$ we are looking back 81% in time toward the big bang: Objects have an age of barely 3 Gyr and Sb and Sc galaxies still consist mostly of gas. Finally, if we can find objects at $z = 5$, about the distance of the most remote QSOs known, then in this universe we are looking at objects only 1.0 Gyr old.

Among the many risks in high-z territory are the following three:

(1) It is easy to underestimate the effects of cumulative absorption with increasing z, and it has taken us a long time to realize this. Due to the many gas clouds producing the "Lyman forest" in QSO spectra and to occasional absorbing gas disks of greater optical depth, known as "Lyman-limit" systems, the ultraviolet light from distant objects gets increasingly absorbed as we observe to higher and higher z. This effect, known as the *Lyman valley* (Møller & Jakobsen

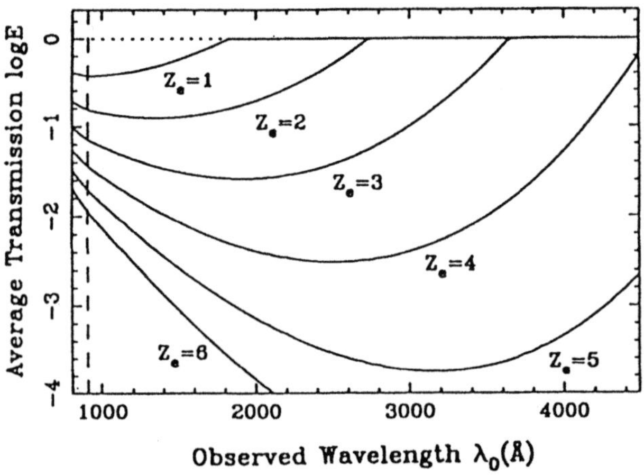

Fig. 94. Average transmission of ultraviolet universe out to large emission redshifts z_e as a function of observed wavelength, showing the "Lyman valley". Diagram from Longair (1995), after Møller & Jakobsen (1990).

1990; Longair 1995), is illustrated in Fig. 94. If we observe near $\lambda = 4000\,$Å, it prevents us from seeing *any* object at (emission) redshift $z_e \gtrsim 5$, and even at $z_e \approx 4$ we can expect to see only few objects.

(2) With increasing z, we may also see more "mirages". As more objects fall near our line of sight, the chances for gravitational lensing increase and we may, at least for a while, be fooled like travelers in the desert. For example, the galaxy FSC 10214+4724 at $z = 2.28$ generated much excitement after being discovered as an intense source of CO emission. Initially, its total luminosity appeared to be of order $10^{14}\,L_\odot$. Yet, after dozens of papers were published on its extraordinary properties, Close et al. (1995) announced their discovery of a gravitational lens that creates this mirage, a galaxy at $z = 0.42$ with a velocity dispersion of $210\,\mathrm{km\,s^{-1}}$. When corrected for the lensing, the luminosity of FSC 10214+4724 decreases by a factor of 27 to a still hefty, but no longer extraordinary, value of $L \approx 3.7 \times 10^{12}\,L_\odot$. This, then, is clearly a pitfall to be avoided on trips to high z.

(3) Finally, some experiences in high-z land are considered outright "failures". A prominent one is the failure, so far, to find primeval ellipticals with huge starbursts. Such collapsing spheroidal objects, first predicted by Meier (1976), have proven exceedingly hard to find (Pritchet 1994). Is this due to dust, to other difficulties, or perhaps – as at least some of us would suspect – to ellipticals forming not monolithically as envisaged by Meier and Pritchet, but through time-delayed mergers as proposed by Toomre (1977a)?

9.2 Interactions and Mergers in QSOs

The hypothesis that interactions and mergers may fuel galactic centers was put forth by Toomre & Toomre (1972) in a short section entitled "Stoking the Furnace", and was given additional theoretical underpinning by Gunn (1979) in his famous lecture about "Feeding the Monster". Stockton's (1990) excellent review of this subject is well worth reading. As he could state even then, there is little doubt that many QSOs occur in gravitationally interacting or merging galaxies. For a few such systems, there is some hard evidence. However, if one asks whether *most* or *all* quasars occur in interacting galaxies, the evidence is weaker and mostly of a rough statistical nature.

Following Stockton's (1978, 1982) discovery that QSOs seem to have an excess of companion galaxies, much work has been done to show that – indeed – QSOs are often interacting (~80%), have close companions, and even feature tidal tails in ~1/3 of all cases (e.g., Hutchings 1987; Hutchings et al. 1988, 1992). Until recently, such studies were generally limited to relatively nearby QSOs ($z \lesssim 0.6$). Although the statistics often look impressive, they rarely represent strong evidence. Stockton points out that a serious weakness is the frequent lack of proper comparison samples. Redshifts have been measured for very few companions. And when they have been, as e.g. in the case of a $z = 2.76$ QSO observed by Steidel et al. (1991), they have often been found to differ by a relatively large amount from the redshift of the QSO (~1000 km s^{-1} in that case). Thus, some of these "companions" may be chance superpositions or perhaps fellow cluster members, and may not be interacting with the QSO.

As mentioned already, hard evidence for QSOs occurring in interacting or merging galaxies does exist in a few lucky cases, mostly found by Stockton and his collaborators.

A first impressive case is the QSO 3CR 249.1 at $z = 0.31$, which features two long tails of fluorescing gas (Stockton & MacKenty 1983). Figure 95 displays the evidence: When imaged through a narrow-band interference filter centered on the redshifted [O III] λ5007 line, the two tails show nicely over a projected length of ~55 kpc (tip to tip, for $H_0 = 75$ and $q_0 = 1/2$), while in the line-free continuum they are invisible. At the time, Stockton & MacKenty reasoned that any gas in tidal tails – exposed to the fierce luminosity of the quasar – would nearly have to be ionized. Today, after the demonstration that most of the H I gas in mergers resides in the tidal tails (Hibbard & van Gorkom 1996), one is more inclined than ever to accept their reasoning. Also, close inspection of the continuum image shows that the host galaxy of this QSO consists of two blobs rather than one, suggesting an ongoing merger of two disks.

A second case of hard evidence in favor of an ongoing, though late-stage merger exists in 3C48, the first optically identified QSO at $z = 0.37$. Much work has been done on the host galaxy, including the demonstration that its spectrum features strong Balmer absorption lines and that it contains an extended region of ionized gas (Boroson & Oke 1982, 1984). Stockton & Ridgway (1991) added two important details: (1) In this case the long filament that looks like a tidal tail emits mainly continuum light, whence it is probably a *stellar* tail. And (2), careful subtraction of the quasar light reveals a continuum-emitting blob at 1″ distance from the QSO. This blob is likely the center of a galaxy merging with the QSO host. The projected linear separation between it and the host is a mere 4 kpc. Given the obvious isophotal distortions of this system, I suspect that this is the way NGC 3921 (see Chap. 3) may have looked just a few 100 Myr ago.

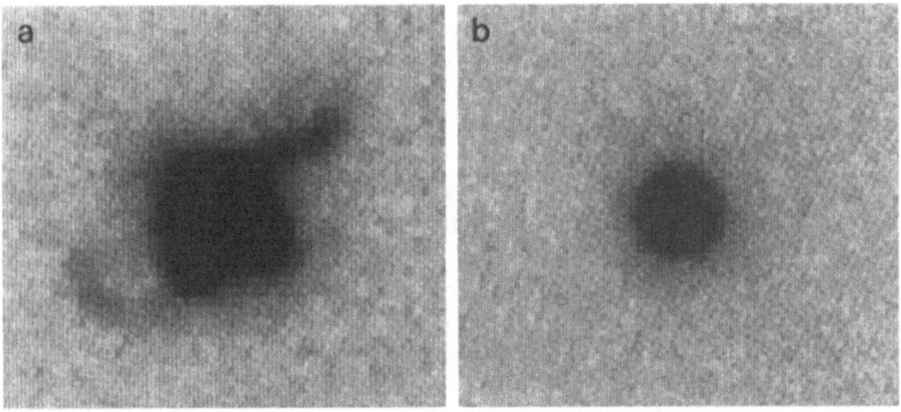

Fig. 95. Gaseous tails of QSO 3CR 249.1 ($z = 0.31$). Images taken through (a) a 30 Å bandwidth filter centered on the redshifted [O III] λ5007 line and (b) a 300 Å filter covering the nearby line-free continuum. Images from Stockton & MacKenty (1983).

A third good case for a quasar in an ongoing merger or recent remnant is that of the X-ray-variable object OX 169. There, a nearly straight "jet" has long been known to protrude from the host galaxy, more to one side (~10″) than to the other. A spectrum of this "jet" shows unambiguously that it consists of intermediate-age stars and, therefore, is a tidal tail probably seen from nearly within its plane (Stockton & Farnham 1991). The second remarkable feature of this QSO is its variable and double Balmer emission lines. Over a period of six years, the broad components of these double lines have varied by different amounts, thus excluding any interpretation of the double profile in terms of self-absorption. Although some other interpretations cannot be ruled out, Stockton & Farnham make a strong case that we are seeing *two* broad emission-line regions in this one object. If so, we must be observing this system just shortly before the two nuclear regions merge. In short, it is this kind of detailed evidence available for only a few QSOs that has allowed us to "hang" all the weaker statistical evidence on some good, strong nails.

Against this background, it came as a big surprise when in early 1995 John Bahcall and collaborators announced that a majority of QSOs observed with *HST* did not show any associated host galaxies (see also Bahcall et al. 1994). They called these QSOs "naked" and got much press coverage for it. Yet, this is one example where the rush to the frontier may have been too hasty.

Through ground-based infrared observations of many of the same QSOs, McLeod & Rieke (1994b) had already shown that most are not naked at all. In fact, out of 26 high-luminosity QSOs forming a complete sample and imaged at 1.6 μm, at least 23 show the associated host galaxy unmistakably. By re-analyzing Bahcall's *HST* observations, McLeod & Rieke (1995) were able to demonstrate that even on the *HST* images the evidence for host galaxies is stronger than found by Bahcall et al. The main problem with *HST* is that the telescope has a very slow focal ratio, making it ill suited for observing features of low surface brightness (Hutchings 1995). Also, the first claims of naked QSOs were based on images obtained with *HST* before its repair. One interesting result from the ground-based IR work is the confirmation that QSOs of *high* luminosity occur nearly always in E-type galaxies, while those of *low* luminosity tend to occur in disk galaxies (McLeod & Rieke 1994a, b).

Fortunately for peace among astronomers, Bahcall and co-workers have increasingly been finding host galaxies of various kinds associated with their QSO sample. One good example is PKS 2349−014, which is yet another QSO showing signatures of a late-stage merger and resembling the remnant NGC 3921 (Bahcall et al. 1995). At present it is not clear whether a small companion is merging or whether, as I believe, the "wisps" of this QSO are tidal tails indicating a major merger. In the latter case, the "companion" may be the nucleus of the second galaxy. Again, more work is needed to decide such issues, but the obvious misalignments of the various components of this object make a merger of some sort seem nearly certain. Interestingly, Disney et al. (1995) find that out of 20 QSOs scheduled for imaging with the repaired *HST*, the first four already imaged are all luminous ellipticals with close companions ($z = 0.24$, 0.26, 0.29, and 0.50). Thus, it is with considerable relief that Joshua Roth, editor of the News Notes in *Sky & Telescope*, could proclaim "Quasars: Not Naked After All" (Roth 1996).

9.3 The Cluster Environment

Two properties of galaxy clusters have long nourished our hopes for a better understanding of galaxy evolution, but have proven surprisingly difficult to interpret: the morphology–density relation and the Butcher-Oemler effect.

The fact that the mean morphological type of galaxies correlates strongly with distance from the cluster center has been known for half a century. Near cluster centers, galaxies are almost exclusively of type E or S0, while in the outskirts spirals prevail (e.g., Oemler 1974). Dressler (1980) formalized this correlation into the morphology–density relation, showing that the fractions of ellipticals and spirals depend primarily on the local density rather than on cluster richness or concentration. Ever since, the question has been whether the morphologies of cluster galaxies reflect mainly initial conditions or evolutionary effects.

Butcher & Oemler's (1978, 1984) discovery that clusters at $z = 0.2 - 0.4$ contain a higher fraction of blue galaxies than clusters at $z \approx 0$ was first received with skepticism, but has stood the test of time unscathed. The introduction of CCD detectors in the early 1980s yielded higher photometric accuracy and efficiency, and helped establish the reality of the Butcher-Oemler (B-O) effect beyond any reasonable doubt. In addition, Dressler & Gunn (1982, 1983) demonstrated through spectroscopy that many of the blue galaxies are indeed members of the B-O clusters, some showing signs of star formation and others signs of past starbursts ("E+A spectra", see Fig. 27 and chapters by Dr. Kennicutt).

Dressler & Gunn bet on ram-pressure stripping being responsible for the increased fraction of blue galaxies, but Couch & Sharples (1987) showed convincingly that *major starbursts*, rather than just snuffing out of star formation,

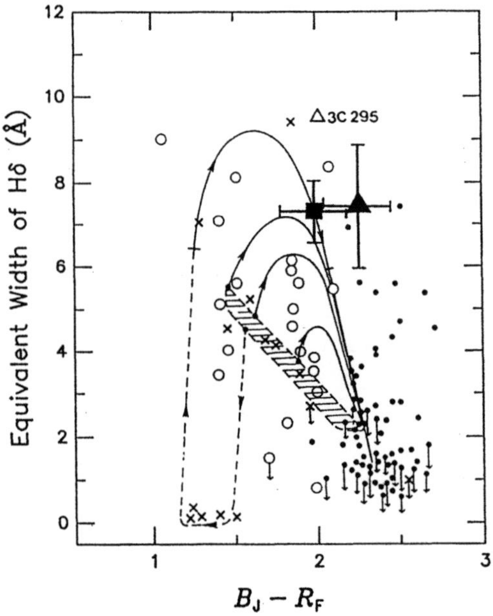

Fig. 96. Equivalent width of Hδ absorption line plotted vs. color index $B - R$ for cluster galaxies at $z = 0.31 - 0.46$ (*small symbols*) and merger remnants NGC 3921 (*triangle with error bars*) and NGC 7252 (*square*); for details, see text. Diagram by Couch & Sharples (1987), adapted by Schweizer (1996).

had to be at work to explain the most extreme E+A spectra. Their Hδ vs. $B-R$ diagram for cluster galaxies, shown here as Fig. 96, illustrates the point. In this figure, cluster spirals lie along the cross-hatched strip. If star formation is simply snuffed out, such spirals will evolve along the solid tracks leaving the strip in the direction of increasing Hδ equivalent width. Maximum widths reached in this manner are 6–7 Å. If, on the other hand, a major starburst of the kind triggered by gravitational interactions and mergers occurs, then such spirals will evolve along paths similar to the dashed-and-then-solid track, reaching Hδ equivalent widths of 9–10 Å and redder colors than otherwise possible. Couch & Sharples conclude that reddish cluster galaxies with strong Hδ must be post-starburst galaxies. For comparison, I have plotted the Hδ equivalent widths of the merger remnants NGC 3921 and NGC 7252 versus $B-R$, which strongly supports the conclusion by Couch & Sharples and also suggests that at least some high-z galaxies with E+A spectra may be remnants of major, locally recent mergers (Schweizer 1996).

There has been a long-standing prejudice against galaxy mergers occurring in rich clusters because of the high velocity dispersions observed in such clusters. Yet, evidence for interactions and mergers in clusters started trickling in during the 1980s and has recently increased dramatically due to new images from *HST*. In an effort to understand the nature of the blue galaxies in B-O clusters, Lavery and collaborators (1988, 1990, 1992, 1994) imaged established cluster members both blue and red under superior seeing conditions from Mauna Kea. One of their first results, shown in Fig. 97, was that there is a clear excess of nearby companions around blue cluster galaxies when compared with red galaxies, at a level of significance of 98%. Also, as images of improved quality were obtained with the high-resolution camera on the 2.2-m telescope, it became clear that

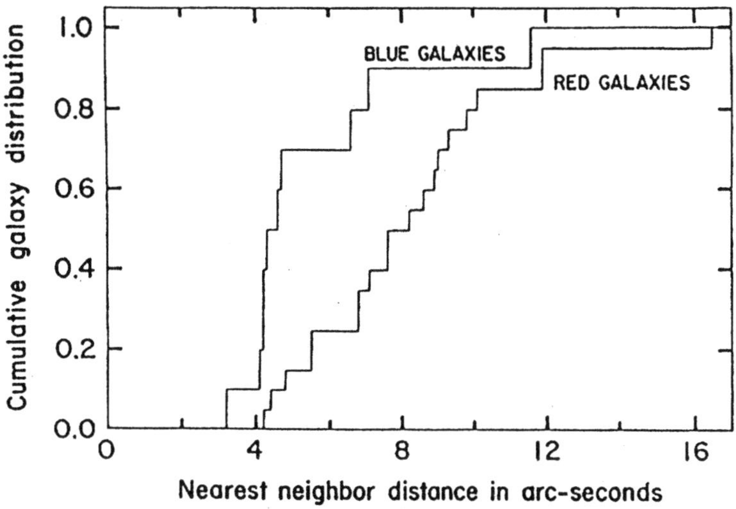

Fig. 97. Cumulative blue- and red-galaxy distributions plotted vs. distance to nearest neighbor. Blue and red galaxies are members of three clusters at $z \approx 0.2$. Note excess of nearby companions around blue galaxies. Diagram from Lavery & Henry (1988).

many of the blue, multiple cluster members show clear signs of tidal interactions or mergers.

A multitude of new images of distant clusters obtained with *HST* has begun to lend support to Lavery & Henry's (1988) contention that interactions and mergers may contribute significantly to the B-O effect. *HST* images of galaxies in Cluster AC 114 at $z = 0.31$ leave little doubt that strong tidal interactions are occurring among them (Couch et al. 1994). Therefore, something must be wrong with the notion that high velocity dispersions in clusters prevent interactions from being effective. In fact, Roos & Aarseth (1982) showed long ago through numerical simulations that "the upper end of the [galaxy] mass spectrum is most strongly affected by the merging process occurring predominantly during the expansion and subsequent collapse of the cluster".

Similarly, pre-repair *HST* observations of Cluster 0393+4713 at $z = 0.41$ suggest that most cluster galaxies are of normal Hubble types, but that there is also a significant number of confirmed cluster members which appear to be undergoing tidal interactions or mergers (Dressler et al. 1994a). Improved, post-repair *HST* images of the same cluster yield further evidence for increased interactions, leading Dressler et al. (1994b) to venture the guess that interactions and mergers in the recent past may be responsible for the disappearance of late-type spirals by the present epoch. These authors suggest that another evolutionary mechanism may also be provided by the destabilization of disks in cluster galaxies as their halos are tidally distorted or removed through gravitational encounters. This idea, discussed already by Oemler (1992), has recently been taken up and popularized under the concept of "galaxy harassment" by Moore et al. (1996). This concept does not appear to be fundamentally new. Yet, a better understanding of the relative damages inflicted upon cluster galaxies by many cumulative, high-velocity encounters and by a few slow-but-strong interactions and mergers is clearly desirable.

Even in nearby clusters there has, of course, long been some evidence for past mergers having contributed to the formation of the E + S0 populations. First, some of the best-known E + S0 galaxies with oddly rotating cores are located in the Virgo cluster (e.g., NGC 4365, M86=NGC 4406, and NGC 4550). Second, several well-known E + S0 galaxies with ripples are also located in Virgo (e.g., M89=NGC 4552 and M85=NGC 4382). Notice that these galaxies tend to be highly luminous, as indicated by their having Messier numbers. Third, we know of no structural differences between E + S0 galaxies in clusters and in the field, where evidence for their formation through mergers is strong. Why, then, should the clusters' galaxies have formed in a different manner? And fourth, many pairs of interacting spirals exist even in some nearby clusters (e.g., Hercules) and are bound to eventually merge.

In summary, the evidence is mounting that interactions and mergers may play a *major* role in galaxy evolution in clusters. In fact, Barger et al. (1996) argue that in three much studied galaxy clusters at $z \approx 0.31$, now imaged with *HST*, there is evidence for about 30% of all galaxies having undergone major starbursts during the past ~2 Gyr, many being spheroidal and likely merger remnants (i.e., those with very strong Hδ absorption).

9.4 Mergers in Distant Field Galaxies

Be forewarned: In our rush to the frontier, we are now entering a smoke-filled tavern in gold-rush territory. In this tavern, it is hard to see clearly, the crowd is noisy, and all kinds of people are screaming their results at us in hopes of getting our attention.

Two main problems plague current studies of field galaxies at high redshifts ($z \gtrsim 0.2$).

First, there is the problem of unknown redshifts. Whereas redshifts for QSOs and cluster galaxies are generally known (though one may have to worry about some contamination from fore- and background objects), the situation is much worse for galaxies in the general field. We know that these galaxies must have a mix of redshifts, yet very few of these redshifts have been determined so far. The reason is, of course, that measuring redshifts of objects at $V \gtrsim 22$mag is time consuming even with the largest telescopes. In the absence of known redshifts, we are then facing the perennial dilemma one has when dealing with categories of objects at unknown distances. Are most of these objects at similar distances or do they cover a large distance range? And in either case, are these objects predominantly nearby and of low luminosity, or predominantly distant and of high luminosity?

Second, there is the problem of the shifting rest-frame wavelength as we observe objects at increasingly high z. The difficulty here is that we are observing in fixed passbands dictated by the telescope and camera (e.g., mostly in V and I with *HST*), whence with increasing z we are observing objects at shorter and shorter rest-frame wavelengths. At ultraviolet wavelengths many galaxies look very different from in the optical. Does a distant faint galaxy look ragged because of a recent interaction or because star-forming regions in it are patchy and dominate its ultraviolet radiation?

Given such uncertainties, many different opinions have developed concerning the importance of evolutionary effects in field galaxies at high z. Two extreme views are as follows. From the well-known apparent excess of faint blue galaxies discovered by Kron (1980), many astronomers conclude that *there was much merging between $z = 0.3$ and 0*. However, Kron (1995) himself thinks that this explanation is unlikely to be correct. Instead, he suspects problems with the poorly known UV properties of average field galaxies and with the models used for interpreting the observations. Supporting his skepticism is growing evidence that the mean redshift of the faint blue galaxies may be higher than previously thought, perhaps as high as $z \approx 0.5$–1 (Glazebrook et al. 1995a,b). A completely opposite view is that *very little merging may have taken place between $z \approx 1.5$ and 0*. Such lack of dynamical evolution appears to be suggested by studies of galaxies found via their absorption lines in QSO spectra. Steidel et al. (1995) have used a color- and redshift-independent technique to study the properties of such "QSO-absorber" galaxies. They find little number-density evolution for normal field galaxies from $z \approx 1.5$ to 0 (Steidel & Dickinson 1995). A problem with this method may be that the QSO absorbers may tend to be high-luminosity galaxies and need not necessarily be related in any way to the faint blue galaxies.

Some generally accepted results are emerging despite the above uncertainties. Many studies now agree that the average star formation rate (SFR) in galaxies increases with increasing z, and at $z \gtrsim 1$ reaches levels well beyond those known in the local universe. For example, measured strengths of the [O II]λ3727 emission-line doublet indicate that $\text{SFR}(z=1.1) \approx 3\times \text{SFR}(z=0.4)$ (Cowie et al. 1995a). A second apparently robust result found by several studies is that there seem to exist many relatively normal-looking galaxies at $z \approx 1$. Combining spectroscopic observations with the Keck telescope by Forbes et al. (1996) and by himself, Koo (1996) finds that for a morphologically representative sample of some 30 galaxies at $I \approx 22$–24 mag the redshifts lie predominantly in the range $z \approx 0.6$–1.1. The broad-band colors of these galaxies are surprisingly normal when compared to the corresponding rest-frame colors of nearby galaxies. A third important result from Koo's and similar studies is that unless there is much reddening, which – of course – could be the case in more gaseous objects, at least some of these galaxies must have formed most of their stars at $z > 2$. It is tempting to identify the latter galaxies as objects corresponding to the oldest E + S0 galaxies in the local universe, but observed at 1/3 the age (i.e., ~5 Gyr).

Important new insights concerning the structure and evolution of high-z field galaxies have been gained from observations of the *Hubble Deep Field* (HDF) made with *HST*. The HDF is a $2\overset{'}{.}5 \times 2\overset{'}{.}5$ patch of sky at $\alpha = 12^h36^m49\overset{s}{.}4$, $\delta = +62°12'58''$ (J2000) that was observed in 1995 December during 150 orbits and through four filters corresponding roughly to *UBVI*. With a total exposure of 34 hours in I, these observations reach 3 mag deeper than the deepest ground-

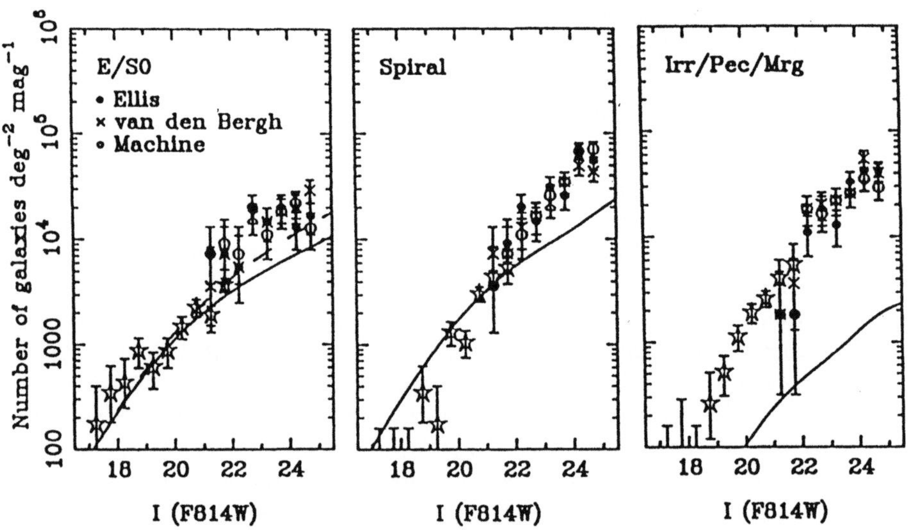

Fig. 98. Number–magnitude relations for three morphological categories of galaxies from the Medium-Deep Survey (*pentagrams*) and HDF; symbols identify source of classification of HDF galaxies as labeled. *Solid curves* are model predictions for no dynamical evolution and $\Omega = 1$, while *dashed curve* in E/S0 panel shows effect of assuming $\Omega = 0.1$. Notice the dramatic excess of Irr/Pec/Merger galaxies over model predictions based on local number counts. Diagrams from Abraham et al. (1996).

based observations ever achieved before and 1 mag deeper than the previously deepest *HST* observations. Plate 3 shows a color image covering about 1/4 of the whole field, while Plate 4 displays enlargements of three subareas ("details") of the HDF. In the top detail, notice that there are both very blue and quite red galaxies. There are also one or two linear "chains", which Cowie et al. (1995b) tentatively identified as a new morphological class of objects and suggested to be galaxies in formation at $z = 0.5 - 3$. However, relatively few of these "chains" are visible over the whole HDF, and even these few may well be disk galaxies seen edge-on in their rest-frame UV light. In the middle and bottom details, notice the nice pairs of interacting galaxies and a small group of four compact objects. These and other enlarged sections of the HDF leave little doubt that *there are many tidal interactions going on at redshifts of* $z \approx 1 - 1.5$. The question, of course, is how the number of interactions per comoving volume compares with the number in the local universe.

A fascinating first study of the morphology of galaxies in the HDF has been published by Abraham et al. (1996). Classifying all objects brighter than $I = 25$ mag both visually and by automated methods, these authors find good agreement among the various classifications and derive two main results, illustrated here in Fig. 98. First, plots of the numbers of objects of different morphological classes as a function of magnitude I show that E + S0 galaxies and spirals in both the Medium-Deep-Survey and the HDF form continuous sequences roughly in agreement with the predictions of low-density cosmological models ($\Omega \approx 0.1$), shown by a dashed line. Note that at the faintest magnitudes the E + S0 counts appear to flatten out or even drop slightly, while the spiral counts continue to increase steeply. Abraham et al. speculate that this difference may be indicative of many spirals at high z having not yet merged to form E + S0's. The second result is perhaps more extraordinary: There appear to be 1–2 orders of magnitude more Irr/Pec/Merger objects in the magnitude interval $I = 22 - 25$ than expected from the local universe! In fact, near the limit at $I = 25$ Abraham et al. find that fully 40% of all objects fall into the Irr/Pec/Merger category, suggesting that "the conventional Hubble classification system no longer provides an adequate description of the morphological characteristics of a high fraction of field galaxies". If truly ~40% of all galaxies near the limit are of the type Irr/Pec/Merger, then I think even skeptics will have to agree that the average field galaxy near $z \approx 1$ is different from the typical local field galaxy. Of course, this interpretation rests on the view that most of these galaxies are not nearer-by dwarfs. And this view itself rests on first spectroscopic results from Keck that do seem to indicate higher average redshifts. Clearly, the next few years are going to be exciting, with much progress to be expected in our understanding of faint field galaxies and of the interactions and mergers among them.

9.5 The Merger "Rate"

These recent observations of high percentages of interacting and merging systems among the faintest galaxies beg the question: What is the merger rate as a function of z?

Notice that I have put the word "Rate" in the above title in quotation marks. This is to indicate that often, when people talk of the merger rate, they in fact discuss some other, though related quantity. Given the vast range in galaxy masses and luminosities, one obviously should specify just what kind of mergers the rate is supposed to refer to. Many attempts have been made to determine *some kind* of merger rate, but I regard only perhaps two such attempts as yielding reliable results.

Concerning the *local* merger rate, I believe that Toomre's (1977a) estimate of the rate of mergers between major, near-equal disk galaxies is still the best, partially because it is so specific as to which objects are being considered. Toomre points out a remarkable coincidence between the number of merger remnants expected from ongoing disk–disk mergers and the about 800 E + S0 galaxies that can be found among the 4000^{+} galaxies in the New General Catalogue (NGC). Within the distance range covered by the NGC, there are ~10 two-tailed, ongoing disk–disk mergers known. The median "age" of these systems (i.e., time since the strong interaction began) is ~0.5 Gyr (Toomre & Toomre 1972), whence – if the universe is ~12.5 Gyr old and the merger rate had remained constant – about 250 remnants should have formed from such mergers up to the present. However, as Fig. 99 illustrates, the universe's higher density in the past suggests a higher past merger rate, declining as a function of time t like $t^{-5/3}$ if the binding energies of distinct galaxy pairs had a flat distribution. With such a rate history, there

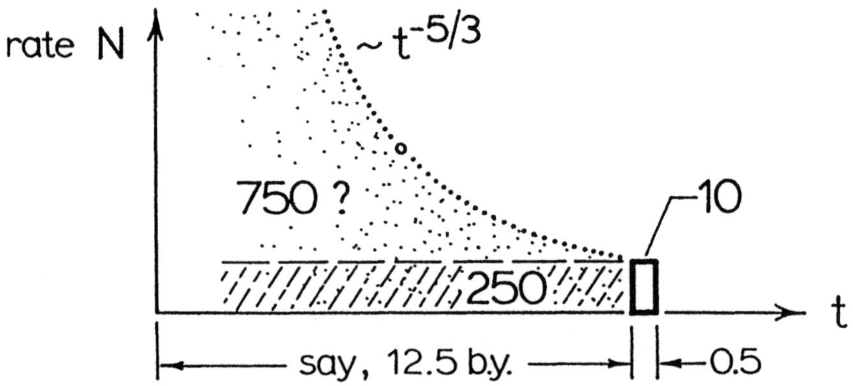

Fig. 99. Toomre's (1977a) estimate of numbers of past merger remnants.

should have been about 3× more disk–disk mergers than a linear extrapolation would indicate, leading to an estimated 750 merger remnants among present-day NGC galaxies. This number is, indeed, remarkably close to the 800 or so E + S0 galaxies in the Catalogue. In short, according to Toomre's estimate ~20% of the nearby luminous galaxies have experienced at least one major, reshaping merger during their past history.

Improving on this estimate is difficult, though several attempts have been made. For example, by inspecting Palomar Sky Survey prints Fried (1988) has found that few field galaxies seem to show morphological perturbations, but fully 30% of galaxies in groups do and 16% of galaxies in clusters. From this statistics he concludes that every galaxy not in the field has undergone significant tidal interaction during the past Hubble time. Similarly, Keel & Wu (1995) have tried to improve on Toomre and find a merger rate of 0.33 per spiral and Hubble time, though based on a rather haphazard sample of candidate merger remnants. The main difficulty is that any sample other than equal-disk mergers is not nearly as clear-cut in its signatures. For example, many people would call M51 an ongoing merger of mass ratio about 1/3. Yet, when will the two components merge: in 2, 5, or 10 Gyr? We simply do not know the answer even for this well studied system. This, of course, makes it difficult to include such objects meaningfully in any derivation of a merger rate.

However, luckily an *upper limit* to the merger rate integrated over the past can be derived as follows. From disk-to-bulge ratios observed in nearby galaxies, Schechter & Dressler (1987) find that roughly equal amounts of mass are located in bulges (incl. ellipticals) and in disks. Therefore, in a simplified scenario in which all gas collapsed into disks and all spheroids are completely due to mergers, the maximum amount of galactic structure affected by mergers would be of the order of 50% by mass. This upper limit underlines that at least mergers of nearly equal-mass galaxies cannot be too frequent, or else they would have destroyed all disks.

Finally, even without necessarily knowing the merger rate in any true, all-encompassing sense, many authors have attempted to study its dependence on redshift. If the rate R_{mrg} is expressed in the form of a power law, $R_{\mathrm{mrg}} \propto (1+z)^m$, then the exponent has been found to be $m = 2.5$ from Toomre's (1977a) argument (in a $\Omega = 1$ universe) and variously $m \approx 1.5-3$ (with error bars typically of ± 2) from statistics of binary galaxies as a function of z. These kinds of estimates were first made by Zepf & Koo (1989), but have been repeated several times since (e.g., Burkey et al. 1994; Yee & Ellingson 1995). However, given the extremely uncertain time scale for "mergers" like M51, I have little faith in these kinds of estimates. Note that even though the various estimates seem to agree within their combined error bars, the differences between rates with different exponents increase rapidly with z. Already at a modest $z = 1$, mergers with a $m = 1.5$ dependence would be 2.8× more frequent than locally, while mergers with $m = 3$ would be 8× more frequent. In fact, with the original estimate of $m = 3 \pm 2.5$ by Zepf & Koo (corresponding to binary galaxies increasing like $[1+z]^{4\pm 2.5}$, see Burkey et al. 1994), the estimated range of the merger rate at $z = 1$ lies between 1.4× and 45× the local rate!

9.6 One Victory, Three Challenges

Instead of summarizing this last Chapter, it seems more appropriate to briefly assess the past successes and future challenges of a galaxy-formation scenario in which major mergers play a significant role. I would like to claim one victory for the merger hypothesis and point out three challenges.

The victory, it seems to me, is that the delayed formation of a significant fraction of all E + S0 galaxies through major mergers is now supported through considerable amounts of observational as well as theoretical evidence. That is, *prolonged* galaxy building and *delayed* transformations of galaxies form now an integral part of our thinking about galaxy formation and evolution.

Perhaps our most serious challenge is to answer the following questions: Is Hubble's morphological sequence really mostly *a sequence of decreasing merger damage*, as the presently available evidence suggests? If so, how unique or how varied are the evolutionary paths by which, e.g., Sb galaxies of similar luminosity have reached their present-day position in the Hubble sequence?

A second major challenge we have to address is: If merging is prevalent, especially in an early, gaseous form, why are there so many bulge-less late-type galaxies even in not-so-low density regions (e.g., M33 in the Local Group)? How do relatively massive, nearly pure disks like M101 assemble *without suffering major damage ?*

As our third and more long-term challenge, I suggest we adopt the following questions: Are there mechanisms other than merging that are crucial in stimulating star formation early in the life of most major galaxies? What *relative roles* do major mergers, minor accretions, disk instabilities, and, e.g., jet-induced star formation (the "alignment effect", see McCarthy 1993) play?

References

Abraham, R.G., et al. 1996, MNRAS 279, L47
Alladin, S.M. 1965, ApJ 141, 768
Amram, P., Marcelin, M., Boulesteix, J., & le Coarer, E. 1992, A&A 266, 106
Appleton, P.N., & Struck-Marcell, C. 1987, ApJ 312, 566
Appleton, P.N., & Struck-Marcell, C. 1996, Fundamentals Cosmic Phys. 16, 111
Arnaboldi, M., Da Costa, G.S., & Saha, P. (eds.) 1997, *The Nature of Elliptical Galaxies* (A.S.P., San Francisco)
Arnaboldi, M., Oosterloo, T., Combes, F., Freeman, K.C., & Koribalski, B. 1997, AJ 113, 585
Arp, H. 1962, ApJ 136, 1148
Arp, H. 1966, *Atlas of Peculiar Galaxies* (California Institute of Technology, Pasadena); also: ApJS 14, 1
Arp, H. 1969, A&A 3, 418
Ashman, K.M., & Zepf, S.E. 1992, ApJ 384, 50
Athanassoula, E., & Bosma, A. 1985, ARA&A 23, 147
Baade, W. 1944, ApJ 100, 137
Bahcall, J.N., Kirhakos, S., & Schneider, D.P. 1994, ApJ 435, L11
Bahcall, J.N., Kirhakos, S., & Schneider, D.P. 1995, ApJ 447, L1
Balcells, M., & Quinn, P.J. 1990, ApJ 361, 381
Balick, B., & Heckman, T. 1981, A&A 96, 271
Barger, A.J., et al. 1996, MNRAS 279, 1
Barnes, J.E. 1988, ApJ 331, 699
Barnes, J.E. 1992, ApJ 393, 484
Barnes, J. 1994, in *The Formation and Evolution of Galaxies*, ed. C. Muñoz-Tuñón & G. Sánchez (Cambridge Univ. Press, Cambridge), p. 399
Barnes, J.E., & Hernquist, L.E. 1991, ApJ 370, L65
Barnes, J.E., & Hernquist, L. 1992, ARA&A 30, 705
Barnes, J.E., & Hernquist, L. 1996, ApJ 471, 115
Bender, R. 1988a, A&A 193, L7
Bender, R. 1988b, A&A 202, L5
Bender, R. 1990, in *Dynamics and Interactions of Galaxies*, ed. R. Wielen (Springer, Berlin), p. 232
Bender, R., Döbereiner, S., & Möllenhoff, C. 1987, A&A 177, L53
Bender, R., & Surma, P. 1992, A&A 258, 250
Bender, R., Surma, P., Döbereiner, S., Möllenhoff, C., & Madejsky, R. 1989, A&A 217, 35
Benvenuti, P., Capaccioli, M., & D'Odorico, S. 1976, A&A 53, 141
Bernlöhr, K. 1993, A&A 268, 25
Bertola, F. 1981a, Sky & Telesc. 61, 380
Bertola, F. 1981b, in *The Structure and evolution of Normal Galaxies*, ed. S.M. Fall & D. Lynden-Bell (Cambridge Univ. Press, Cambridge), p. 13
Bertola, F., Bettoni, D., Buson, L.M., & Zeilinger, W.W. 1990, in *Dynamics and Interactions of Galaxies*, ed. R. Wielen (Springer, Heidelberg), p. 249
Bertola, F., Bettoni, D., Rusconi, L., & Sedmak, G. 1984, AJ 89, 356
Bertola, F., Buson, L.M., & Zeilinger, W.W. 1992, ApJ 401, L79
Bertola, F., & Capaccioli, M. 1975, ApJ 200, 439
Bertola, F., Galletta, G., & Zeilinger, W.W. 1985, ApJ 292, L51
Bica, E., & Alloin, D. 1986, A&A 162, 21

Bica, E., & Alloin, D. 1987, A&AS 70, 281

Bica, E., Alloin, D., & Schmidt, A. 1990, MNRAS 242, 241

Bica, E., Alloin, D., & Schmitt, H.R. 1994, A&A 283, 805

Binney, J. 1976, MNRAS 177, 19

Binney, J. 1978a, MNRAS 183, 779

Binney, J. 1978b, Comments in Ap. 8, 27

Binney, J. 1985, MNRAS 212, 767

Binney, J. 1992, ARA&A 30, 51

Binney, J., Kormendy, J., & White, S.D.M. 1982, *Morphology and Dynamics of Galaxies*, ed. L. Martinet & M. Mayor (Geneva Observatory, Sauverny)

Binney, J., & Petrou, M. 1985, MNRAS 214, 449

Binney, J., & Tremaine, S. 1987, *Galactic Dynamics* (Princeton Univ. Press, Princeton), esp. Chap. 7

Borne, K.D., Balcells, M., Hoessel, J.G., & McMaster, M. 1994, ApJ 435, 79

Borne, K.D., et al. 1996, in *Science with the Hubble Space Telescope – II*, ed. P. Benvenuti, F.D. Macchetto, & E.J. Schreier (Space Telescope Science Institute, Baltimore), p. 239

Borne, K.D., & Richstone, D.O. 1991, ApJ 369, 111

Boroson, T.A., & Oke, J.B. 1982, Nat 296, 397

Boroson, T.A., & Oke, J.B. 1984, ApJ 281, 535

Borra, E.F., & Arp, H.C. 1974, private communication

Bosma, A., Smith, R.M., & Wellington, K.J. 1985, MNRAS 212, 301

Bower, R.G., Lucey, J.R., & Ellis, R.S. 1992, MNRAS 254, 601

Braun, R., Walterbos, R.A.M., & Kennicutt, R.C. 1992, Nat 360, 442

Braun, R., Walterbos, R.A.M., Kennicutt, R.C., & Tacconi, L.J. 1994, ApJ 420, 558

Brodie, J.P., & Huchra, J.P. 1991, ApJ 379, 157

Bruzual, A.G., & Charlot, S. 1993, ApJ 405, 538

Bruzual, A.G., & Charlot, S. 1996, electronic distribution via ftp from *urania.astroscu.unam.mx*

Burke, B.F. 1957, AJ 62, 90

Burkey, J.M., Keel, W.C., Windhorst, R.A., & Franklin, B.E. 1994, ApJ 429, L13

Burstein, D., Faber, S.M., Gaskell, C.M., & Krumm, N. 1984, ApJ 287, 586

Burton, W.B. 1988, in *Galactic and Extragalactic Radio Astronomy*, 2nd edition, ed. G.L. Verschuur & K.I. Kellermann (Springer, Berlin), p. 295

Butcher, H., & Oemler, A. 1978, ApJ 219, 18

Butcher, H., & Oemler, A. 1984, ApJ 285, 426

Buzzoni, A., Renzini, A., & Serrano, A. (eds.) 1995, *Fresh Views of Elliptical Galaxies* (A.S.P., San Francisco)

Caldwell, N., Kirshner, R.P., & Richstone, D.O. 1986, ApJ 305, 136

Caldwell, N., Rose, J.A., Sharples, R.M., Ellis, R.S., & Bower, R.G. 1993, AJ 106, 473

Cannon, R.D., Lloyd, C., & Penston, M.V. 1970, Observatory 90, 153

Caon, N., Capaccioli, M., & D'Onofrio, M. 1993, MNRAS 265, 1013

Capaccioli, M., Caon, N., & Rampazzo, R. 1990, MNRAS 242, 24p

Capaccioli, M., & Longo, G. 1994, A&ARev 5, 293

Carney, B.W. 1978, Ph.D. thesis, Harvard University

Carney, B.W. 1980, ApJS 42, 481

Carney, B.W., Fulbright, J.P., Terndrup, D.M., Suntzeff, N.B., & Walker, A.R. 1995, AJ 110, 1674

Carney, B.W., Latham, D.W., Laird, J.B., & Aguilar, L.A. 1994, AJ 107, 2240

Carney, B.W., & Seitzer, P. 1993, AJ 105, 2127

Carter, D. 1978, MNRAS 182, 797
Carter, D. 1987, ApJ 312, 514
Carter, D., Prieur, J.-L., Wilkinson, A., Sparks, W.B., & Malin, D.F. 1988, MNRAS 235, 813
Chaboyer, B., Demarque, P., Kernan, P.J., & Drauss, L.M. 1996a, Sci 271, 957
Chaboyer, B., Demarque, P., & Sarajedini, A. 1996b, ApJ 459, 558
Charlot, S., & Bruzual, A.G. 1991, ApJ 367, 126
Charlot, S., Worthey, G., & Bressan, A. 1996, ApJ 457, 625
Christodoulou, D.M., Katz, N., Rix, H.-W., & Habe, A. 1992, ApJ 395, 113
Close, L.M., Hall, P.B., Liu, C.T., & Hege, E.K. 1995, ApJ 452, L9
Clutton-Brock, M. 1972, Ap&SS 17, 292
Colina, L., & Borne, K.D. 1995, ApJ 454, L101
Combes, F. 1994, in *The Formation and Evolution of Galaxies*, ed. C. Muñoz-Tuñón & G. Sánchez (Cambridge Univ. Press, Cambridge), p. 317
Combes, F., Debbasch, F., Friedli, D., & Pfenniger, D. 1990, A&A 233, 82
Combes, F., Rampazzo, R., Bonfanti, P.P., Prugniel, P., & Sulentic, J.W. 1995, A&A 297, 37
Contopoulos, G. 1956, Zeitschrift f. Ap. 39, 126
Corbelli, E., Schneider, S.E., & Salpeter, E.E. 1989, AJ 97, 390
Couch, W.J., & Sharples, R.M. 1987, MNRAS 229, 423
Couch, W.J., Ellis, R.S., Sharples, R.M., & Smail, I. 1994, ApJ 430, 121
Cowie, L.L., Hu, E.M., & Songaila, A. 1995a, Nat 377, 603
Cowie, L.L., Hu, E.M., & Songaila, A. 1995b, AJ 110, 1576
Da Costa, G.S., & Armandroff, T.E. 1995, AJ 109, 2533
Danziger, I.J., Zeilinger, W.W., & Kjär, K. (eds.) 1993, *Structure, Dynamics and Chemical Evolution of Elliptical Galaxies* (ESO, Garching)
Davies, R.L., Sadler, E.M., & Peletier, R.F. 1993, MNRAS 262, 650
de Carvalho, R.R., & Djorgovski, S. 1992, ApJ 389, L49
Dekel, A., & Shlosman, I. 1983, in *Internal Kinematics and Dynamics of Galaxies*, ed. E. Athanassoula (Reidel, Dordrecht), p. 187
de Mello, D.F., et al. 1995, A&A 297, 331
de Vaucouleurs, G. 1953, MNRAS 113, 134
de Vaucouleurs, G. 1974, in *The Formation and Dynamics of Galaxies*, ed. J.R. Shakeshaft (Reidel, Dordrecht), p. 1 and footnote on p. 6
de Vaucouleurs, G., & de Vaucouleurs, A. 1975, ApJ 197, L1
de Vaucouleurs, G., de Vaucouleurs, A., & Corwin, H.G. 1976, *Second Reference Catalogue of Bright Galaxies* (Univ. Texas Press, Austin) (RC2)
de Zeeuw, P.T. 1994, in *The Formation and Evolution of Galaxies*, ed. C. Muñoz-Tuñón & G. Sánchez (Cambridge Univ. Press, Cambridge), p. 231
Dieter, N.H. 1965, AJ 70, 552
Disney, M.J., et al. 1995, Nat 376, 150
Doyon, R., et al. 1994, ApJ 437, L23
Dressler, A. 1980, ApJ 236, 351
Dressler, A., & Gunn, J.E. 1982, ApJ 263, 533
Dressler, A., & Gunn, J.E. 1983, ApJ 270, 7
Dressler, A., Oemler, A., Butcher, H.R., & Gunn, J.E. 1994a, ApJ 430, 107
Dressler, A., Oemler, A., Sparks, W.B., & Lucas, R.A. 1994b, ApJ 435, L23
Dubinski, J., Mihos, J.C., & Hernquist, L. 1996, ApJ 462, 576
Duc, P.-A. 1995, Ph.D. thesis, Université Paris VI
Duc, P.-A., & Mirabel, I.F. 1994, A&A 289, 83

Duncan, J.C. 1923, ApJ 57, 137
Dupraz, C., Casoli, F., Combes, F., & Kazès, I. 1990, A&A 228, L5
Dupraz, C., & Combes, F. 1986, A&A 166, 53
Dupraz, C., & Combes, F. 1987, A&A 185, L1
Efstathiou, G., Ellis, R.S., & Carter, D. 1982, MNRAS 201, 975
Eggen, O.J. 1958, MNRAS 118, 65
Eggen, O.J. 1989, Fund. Cosmic Phys., 13, 1
Eggen, O.J., Lynden-Bell, D., & Sandage, A.R. 1962, ApJ 136, 748 (ELS)
Elmegreen, B.G., Elmegreen, D.M., & Seiden, P.E. 1989, ApJ 343, 602
Elson, R.A., & Fall, S.M. 1985, PASP 97, 692
Eneev, T.M., Kozlov, N.N., & Sunyaev, R.A. 1973, A&A 22, 41
Evans, D.S. 1952, MNRAS 112, 606
Faber, S.M. 1977, in *The Evolution of Galaxies and Stellar Populations*, ed. B.M. Tinsley & R.B. Larson (Yale Univ. Observatory, New Haven), p. 157
Faber, S.M. (ed.) 1987, *Nearly Normal Galaxies* (Springer Verlag, New York)
Faber, S.M., et al. 1989, ApJS 69, 763
Faber, S.M., & Jackson, R. 1976, ApJ 204, 668
Fabian, A.C., Nulsen, P.E.J., & Canizares, C.R. 1984, Nat 310, 733
Fall, S.M., & Rees, M.J. 1985, ApJ 298, 18
Fall, S.M., & Rees, M.J. 1988, in *The Harlow-Shapley Symposium on Globular Cluster Systems in Galaxies*, ed.J.E. Grindley & A.G. D. Philip (Kluwer, Dordrecht), p. 323
Forbes, D.A., Phillips, A.C., Koo, D.C., & Illingworth, G.D. 1996, ApJ 462, 89
Fort, B.P., Prieur, J.-L., Carter, D., Meatheringham, S.J., & Vigroux, L. 1986, ApJ 306, 110
Fosbury, R.A.E., & Hawarden, T.G. 1977, MNRAS 178, 473
Fosbury, R.A.E., & Wall, J.V. 1979, MNRAS 189, 79
Franx, M. 1988, Ph.D. thesis, University of Leiden
Franx, M., & Illingworth, G.D. 1988, ApJ 327, L55
Franx, M., Illingworth, G., & de Zeeuw, T. 1991, ApJ 383, 112
Franx, M., Illingworth, G., & Heckman, T. 1989a, AJ 98, 538
Franx, M., Illingworth, G., & Heckman, T. 1989b, ApJ 344, 613
Franx, M., van Gorkom, J.H., & de Zeeuw, T. 1994, ApJ 436, 642
Freeman, K.C. 1990, in *Dynamics and Interactions of Galaxies*, ed. R. Wielen (Springer, Heidelberg), p. 36
Freeman, K.C. 1995, in *Stellar Populations*, ed. P.C. van der Kruit & G. Gilmore (Kluwer, Dordrecht), p. 119
Fried, J.W. 1988, A&A 189, 42
Fried, J.W., & Schulz, H. 1983, A&A 118, 166
Fritze-von Alvensleben, U., & Gerhard, O.E. 1994, A&A 285, 775
Fujimoto, M., & Sofue, Y. 1976, A&A 47, 263
Fujimoto, M., & Sofue, Y. 1977, A&A 61, 199
Fusi Pecci, F., Bellazzini, M., Cacciari, C., & Ferraro, F.R. 1995, AJ 110, 1664
Gallagher, J.S., Faber, S.M., & Balick, B. 1975, ApJ 202, 7
Gallagher, J.S., & Holtzman, J.A. 1994, BAAS 26, 1489
Gardiner, L.T., & Noguchi, M. 1996, MNRAS 278, 191
Gerhard, O.E. 1981, MNRAS 197, 179
Gerhard, O.E. 1983a, MNRAS 202, 1159
Gerhard, O.E. 1983b, MNRAS 203, 19p
Gilmore, G. 1995, in *Stellar Populations*, ed. P.C. van der Kruit & G. Gilmore (Kluwer, Dordrecht), p. 99

Gilmore, G., Wyse, R.F.G., & Jones, J.B. 1995, AJ 109, 1095
Gilmore, G., Wyse, R.F.G., & Kuijken, K. 1989, ARA&A 27, 555 (esp. Sect. 4)
Glazebrook, K., Ellis, R., Santiago, B., & Griffiths, R. 1995a, MNRAS 275, L19
Glazebrook, K., et al. 1995b, MNRAS 273, 157
González, J.J. 1993, Ph.D. thesis, University of California, Santa Cruz
Goudfrooij, P. 1994, Ph.D. thesis, University of Amsterdam
Goudfrooij, P., et al. 1994, A&AS 104, 179
Graham, J.A. 1979, ApJ 232, 60
Graham, J.A., & Phillips, M.M. 1980, ApJ 239, L97
Graham, J.R., et al. 1990, ApJ 354, L5
Gregg, M.D. 1989a, ApJS 69, 217
Gregg, M.D. 1989b, ApJ 337, 45
Gunn, J.E. 1979, in *Active Galactic Nuclei*, ed. C. Hazard & S. Mitton (Cambridge Univ. Press, Cambridge), p. 213
Gunn, J.E. 1987, in *Nearly Normal Galaxies*, ed. S.M. Faber (Springer Verlag, New York), p. 455
Guzmán, R., Lucey, J.R., Carter, D., & Terlevich, R.J. 1992, MNRAS 257, 187
Harris, W.E. 1991, ARA&A 29, 543
Harris, G.L.H., Hesser, J.E., Harris, H.C., & Curry, P.J. 1984, ApJ 287, 175
Harris, W.E., Pritchet, C.J., & McClure, R.D. 1995, ApJ 441, 120
Harris, W.E., & Pudritz, R.E. 1994, ApJ 429, 177
Harris, W.E., & van den Bergh, S. 1981, AJ 86, 1627
Heap, S.R., et al. 1994, ApJ 435, L39
Hernquist, L. 1992, ApJ 400, 460
Hernquist, L., & Barnes, J.E. 1991, Nat 354, 210
Hernquist, L., & Quinn, P.J. 1988, ApJ 331, 682
Hernquist, L., & Quinn, P.J. 1989, ApJ 342, 1
Hernquist, L., & Spergel, D.N. 1992, ApJ 399, L117
Hernquist, L., & Weil, M.L. 1993, MNRAS 261, 804
Hesser, J.E. 1995, in *Stellar Populations*, ed. P.C. van der Kruit & G. Gilmore (Kluwer, Dordrecht), p. 51
Hesser, J.E., Harris, H.C., van den Bergh, S., & Harris, G.L.H. 1984, ApJ 276, 491
Heyl, J.S., Hernquist, L., & Spergel, D.N. 1994, ApJ 427, 165
Hibbard, J.E. 1995, Ph.D. Thesis, Columbia University
Hibbard, J.E., Guhathakurta, P., van Gorkom, J.H., & Schweizer, F. 1994, AJ 107, 67
Hibbard, J.E., & Mihos, J.C. 1995, AJ 110, 140
Hibbard, J.E., & van Gorkom, J.H. 1996, AJ 111, 655
Higdon, J.L. 1995, ApJ 455, 524
Higdon, J.L. 1996, ApJ 467, 241
Higdon, J.L., Smith, B.J., Lord, S.D., & Rand, R.J. 1995, ApJ 438, L79
Hodge, P. 1992, *The Andromeda Galaxy*, (Kluwer, Dordrecht), Chap. 4 and Fig. 12.15
Holmberg, E. 1941, ApJ 94, 385
Holtzman, J.A., et al. 1992, AJ 103, 691
Holtzman, J.A., et al. 1993, AJ 106, 1826
Horellou, C., Casoli, F., Combes, F., & Dupraz, C. 1995, A&A 298, 743
Hubble, E. 1926, ApJ 64, 321
Hubble, E. 1936, *The Realm of the Nebulae* (Yale Univ. Press, New Haven), Chap. 2
Hui, X., Ford, H.C., Freeman, K.C., & Dopita, M.A. 1995, ApJ 449, 592
Hunsberger, S.D., Charlton, J.C., & Zaritsky, D. 1996, ApJ 462, 50
Hunter, D.A., et al. 1995, ApJ 448, 179

Hutchings, J.B. 1987, ApJ 320, 122
Hutchings, J.B. 1995, Nat 376, 118
Hutchings, J.B., Johnson, I., & Pyke, R. 1988, ApJS 66, 361
Hutchings, J.B., & Neff, S.G. 1992, AJ 104, 1
Ibata, R.A., Gilmore, G., & Irwin, M.J. 1994, Nat 370, 194
Ibata, R.A., Gilmore, G., & Irwin, M.J. 1995, MNRAS 277, 781
Illingworth, G. 1977, ApJ 218, L43
Jarvis, B. 1987, AJ 94, 30
Jedrzejewski, R.I. 1987a, MNRAS 226, 747
Jedrzejewski, R.I. 1987b, in *Structure and Dynamics of Elliptical Galaxies*, ed. T. de Zeeuw (Reidel, Dordrecht), p. 37
Jedrzejewski, R., & Schechter, P.L. 1988, ApJ 330, L87
Jog, C.J., & Solomon, P.M. 1992, ApJ 387, 152
Johnston, K.V., Spergel, D.N., & Hernquist, L. 1995, ApJ 451, 598
Jones, B.F., Klemola, A.R., & Lin, D.N.C. 1994, AJ 107, 1333
Jones, L.A., & Worthey, G. 1995, ApJ 446, L31
Jørgensen, I. 1993, Ph.D. thesis, Copenhagen University
Joseph, R.D., & Wright, G.S. 1985, MNRAS 214, 87
Kahn, F.D., & Woltjer, L. 1959, ApJ 130, 705
Kauffmann, G. 1996, MNRAS 281, 487
Kaufman, M., et al. 1989, ApJ 345, 674
Keel, W.C., & Wu, W. 1995, AJ 110, 129
Kennicutt, R.C. 1990, in *Paired and Interacting Galaxies*, ed. J.W. Sulentic & W.C. Keel, NASA Conf. Publ. 3098, p. 269
Kennicutt, R.C., & Chu, Y.-H. 1988, AJ 95, 720
Kent, S.M. 1983, ApJ 266, 562
Kerr, F.J. 1957, AJ 62, 93
Kibblewhite, E.J. 1969, Ph.D. thesis, Cambridge University
King, I.R. 1976, *The Universe Unfolding* (Freeman, San Francisco), p. 82
King, I.R. 1978, ApJ 222, 1
King, I.R., Stanford, A., & Crane, P. 1995, AJ 109, 164
Knapp, G.R., Turner, E.L., & Cunniffe, P.E. 1985, AJ 90, 454
Koo, D.C. 1996, in *New Light on Galaxy Evolution*, ed. R. Bender & R.L. Davies (Kluwer, Dordrecht), p. 217
Kormendy, J. 1977, ApJ 218, 333
Kormendy, J. 1982, in *Morphology and Dynamics of Galaxies*, ed. L. Martinet & M. Mayor (Geneva Observatory, Sauverny), p. 113
Kormendy, J. 1984, ApJ 287, 577
Kormendy, J., & Djorgovski, S. 1989, ARA&A 27, 235
Kormendy, J., et al. 1994, in *Dwarf Galaxies*, ed. G. Meylan & P. Prugniel (ESO, Garching), p. 147
Kormendy, J., & Norman, C.A. 1979, ApJ 233, 539
Kron, R.G. 1980, ApJS 43, 305
Kron, R.G. 1995, in *The Deep Universe*, ed. B. Binggeli & R. Buser (Springer, Berlin), p. 233
Kuijken, K. 1993, PASP 105, 1016
Kunkel, W.E. 1979, ApJ 228, 718
Kunkel, W.E., & Demers, S. 1976, Greenwich Obs. Bull., 182, 241
Lacey, C., & Cole, S. 1993, MNRAS 262, 627
Lake, G., & Dressler, A. 1986, ApJ 310, 605

Lance, C.M. 1988, ApJ 334, 927
Larkin, J.E., Armus, L., Knop, R.A., Matthews, K., & Soifer, B.T. 1995, ApJ 452, 599
Larson, R.B. 1987, in *Nearly Normal Galaxies*, ed. S.M. Faber (Springer, New York), p. 26
Larson, R.B. 1988, in *The Harlow-Shapley Symposium on Globular Cluster Systems in Galaxies*, ed. J.E. Grindley & A.G.D. Philip (Kluwer, Dordrecht), p. 311
Larson, R.B. 1990, PASP 102, 709
Larson, R.B., & Tinsley, B.M. 1978, ApJ 219, 46
Lauer, T.R. 1983, Ph.D. thesis, University of California, Santa Cruz
Lauer, T.R. 1985a, ApJS 57, 473
Lauer, T.R. 1985b, MNRAS 216, 429
Lauer, T.R., et al. 1993, AJ 106, 1436
Lauer, T.R., et al. 1995, AJ 110, 2622
Laustsen, S., & West, R.M. 1980, Journal Ap. & Astron. 1, 177
Lavery, R.J. 1990, in *Dynamics and Interactions of Galaxies*, ed. R. Wielen (Springer, Heidelberg), p. 30
Lavery, R.J., & Henry, J.P. 1988, ApJ 330, 596
Lavery, R.J., & Henry, J.P. 1994, ApJ 426, 524
Lavery, R.J., Pierce, M.J., & McClure, R.D. 1992, AJ 104, 2067
Leonardi, A.J., & Rose, J.A. 1996, AJ 111, 182
Lester, D.F., & Gaffney, N.I. 1994, ApJ 431, L13
Liller, M.H. 1961, ApJ 132, 306
Lin, D.N.C., Jones, B.F., & Klemola, A.R. 1995, ApJ 439, 652
Lin, D.N.C., & Lynden-Bell, D. 1977, MNRAS 181, 59
Lin, D.N.C., & Lynden-Bell, D. 1982, MNRAS 198, 707
Lindblad, B. 1956, Stockholm Obs. Ann., Vol. 19, No. 2
Longair, M.S. 1995, in *The Deep Universe*, ed. B. Binggeli & R. Buser (Springer, Berlin), p. 317
Lowe, S.A., Roberts, W.W., Yang, J., Bertin, G., & Lin, C.C. 1994, ApJ 427, 184
Lutz, D. 1991, A&A 245, 31
Lynden-Bell, D. 1965, MNRAS 129, 299
Lynden-Bell, D. 1967, MNRAS 136, 101
Lynden-Bell, D. 1976, MNRAS 174, 695
Lynden-Bell, D. 1982, Observatory, 102, 201
Lynden-Bell, D. 1994, in *The Formation and Evolution of Galaxies*, ed. C. Muñoz-Tuñón & F. Sánchez (Cambridge Univ. Press, Cambridge), p. 85
Lynden-Bell, D., & Lin, D.N.C. 1977, MNRAS 181, 37
Lynden-Bell, D., & Lynden-Bell, R.M. 1995, MNRAS 275, 429
Lynds, R., & Toomre, A. 1976, ApJ 209, 382
Mahoney, J.H., van der Hulst, J.M., & Burke, B.F. 1987, MIT Radio Astr. Contrib. No. 3; see also: Mahoney, J.H., Burke, B.F., & van der Hulst, J.H. 1987, in *Dark Matter in the Universe*, ed. J. Kormendy & G.R. Knapp (Reidel, Dordrecht), p. 94
Majewski, S.R. 1994, ApJ 431, L17
Majewksi, S.R., Munn, J.A., & Hawley, S.L. 1994, ApJ 427, L37
Malin, D.F. 1977, *AAS Photo Bull.*, No. 16, p. 14
Malin, D.F. 1979, Nat 277, 279
Malin, D. 1992, QJRAS 33, 321
Malin, D.F., & Carter, D. 1980, Nat 285, 643
Malin, D.F., & Carter, D. 1983, ApJ 274, 534
Malin, D.F., Quinn, P.J., & Graham, J.A. 1983, ApJ 272, L5

Marcelin, M., Lecoarer, E., Boulesteix, J., Georgelin, Y., & Monnet, G. 1987, A&A 179, 101
Marston, A.P., & Appleton, P.N. 1995, AJ 109, 1002
Martinet, L. 1995, Fundamentals Cosmic Phys., 15, 341
Mateo, M. 1992, in *The Stellar Populations of Galaxies*, ed. B. Barbuy & A. Renzini (Kluwer, Dordrecht), p. 147
Mathewson, D. 1985, Sci. Am. 252, 106
Mathewson, D.H., Cleary, M.W., & Murray, J.D. 1974, ApJ 190, 291
Mathewson, D.H., Schwarz, M.P., & Murray, J.D. 1977, ApJ 217, L5
McCarthy, P. 1993, ARA&A 31, 639 (esp. Sect. 5)
McGaugh, S.S., & Bothun, G.D. 1990, AJ 100, 1073
McLeod, K.K., & Rieke, G.H. 1994a, ApJ 420, 58
McLeod, K.K., & Rieke, G.H. 1994b, ApJ 431, 137
McLeod, K.K., & Rieke, G.H. 1995, ApJ 454, L77
Mees, C.E.K., & James, F.H. 1966, *The Theory of the Photographic Process* (3rd ed.; Macmillan, New York), p. 495
Meier, D.L. 1976, ApJ 207, 343
Merrifield, M.R., & Kuijken, K. 1994, ApJ 432, 575
Merritt, D., & de Zeeuw, T. 1983, ApJ 267, L19
Meurer, G.R., et al. 1995, AJ 110, 2665
Michard, R. 1985, A&AS 59, 205
Mighel, K.J., & Rich, R.M. 1995, AJ 110, 1649
Mihos, J.C., Bothun, G.D., & Richstone, D.O. 1993, ApJ 418, 82
Mihos, J.C., & Hernquist, L. 1994, ApJ 431, L9
Mihos, J.C., & Hernquist, L. 1996, ApJ 464, 641
Mihos, J.C., Walker, I.R., Hernquist, L., Mendes de Oliveira, C., & Bolte, M. 1995, ApJ 447, L87
Miller, B.W., Whitmore, B.C., Schweizer, F., & Fall, S.M. 1997, AJ, submitted
Mirabel, I.F., Dottori, H., & Lutz, D. 1992, A&A 256, L19
Mirabel, I.F., Lutz, D., & Maza, J. 1991, A&A 243, 367
Møller, P., & Jakobsen, P. 1990, A&A 228, 299
Moore, B., & Davis, M. 1994, MNRAS 270, 209
Moore, B., Katz, N., Lake, G., Dressler, A., & Oemler, A. 1996, Nat 379, 613
Muñoz-Tuñón, C., & Sánchez, G. (eds.) 1994, *The Formation and Evolution of Galaxies* (Cambridge Univ. Press, Cambridge)
Murai, T., & Fujimoto, M. 1980, PASJ 32, 581
Nelson, R.W., & Tremaine, S. 1995, MNRAS 275, 897
Norris, R.P. 1988, MNRAS 230, 345
O'Connell, D.J.K. (ed.) 1958, *Stellar Populations* (Vatican Observatory, Vatican City & North Holland Publ., Amsterdam)
O'Connell, R.W. 1976, ApJ 206, 370
O'Connell, R.W. 1980, ApJ 236, 430
O'Connell, R.W. 1986, in *Stellar Populations*, ed. C.A. Norman, A. Renzini, & M. Tosi (Cambridge Univ. Press, Cambridge), p. 167
O'Connell, R.W. 1994, in *The Nuclei of Normal Galaxies: Lessons from the Galactic Center*, ed. R. Genzel & A.I. Harris (Kluwer, Dordrecht), p. 255
Oemler, A. 1974, ApJ 194, 1
Oemler, A. 1992, in *Clusters and Superclusters of Galaxies*, ed. A.C. Fabian (Kluwer, Dordrecht), p. 29

Oort, J.H. 1965, in *Galactic Structure*, ed. A. Blaauw & M. Schmidt (Univ. Chicago Press, Chicago), p. 455
Oort, J.H., Kerr, F.T., & Westerhout, G. 1958, MNRAS 118, 379
Ostriker, J.P. 1980, Comments in Ap. 8, 177
Peebles, P.J.E. 1990, ApJ 362, 1
Peebles, P.J.E. 1994, ApJ 429, 43
Peebles, P.J.E., & Dicke, R.H. 1968, ApJ 154, 891
Peletier, R.F. 1989, Ph.D. thesis, University of Groningen
Peletier, R.F., Davies, R.L., Illingworth, G.D., Davis, L.E., & Cawson, M. 1990, AJ 100, 1091
Pence, W.D. 1986, ApJ 310, 597
Pfenniger, D., & Friedli, D. 1991, A&A 252, 75
Pfleiderer, J. 1963, Zeitschrift f. Ap. 58, 12
Pfleiderer, J., & Siedentopf, H. 1961, Zeitschrift f. Ap. 51, 201
Pickles, A.J. 1985, ApJ 296, 340
Prada, F., Gutiérrez, C.M., Peletier, R.F., & McKeith, C.D. 1996, ApJ 463, L9
Prieur, J.-L. 1988, ApJ 326, 596
Prieur, J.-L. 1990, in *Dynamics and Interactions of Galaxies*, ed. R. Wielen (Springer, Berlin), p. 72
Pritchet, C. 1979, ApJ 231, 354
Pritchet, C.J. 1994, PASP 106, 1052
Quinn, P.J. 1982, Ph.D. thesis, Australian National University, Canberra
Quinn, P.J. 1984, ApJ 279, 596
Reshetnikov, V.P., & Combes, F. 1994, A&A 291, 57
Rich, R.M., & Mould, J.R. 1991, AJ 101, 1286
Richer, H.B., Crabtree, D.R., Fabian, A.C., & Lin, D.N.C. 1993, AJ 105, 877
Richter, O.-G., Sackett, P.D., & Sparke, L.S. 1994, AJ 107, 99
Rix, H.-W., Franx, M., Fisher, D., & Illingworth, G. 1992, ApJ 400, L5
Rix, H.-W., Kennicutt, R.C., Braun, R., & Walterbos, R.A.M. 1995, ApJ 438, 155
Rix, H.-W., & White, S.D.M. 1992, MNRAS 254, 389
Rodgers, A.W. 1971, ApJ 165, 581
Rodgers, A.W., Harding, P., & Sadler, E. 1981, ApJ 244, 912
Rodgers, A.W., & Paltoglou, G. 1984, ApJ 283, L5
Rogstad, D.H., Wright, M.C.H., & Lockhart, I.A. 1976, ApJ 204, 703
Roman, N.G. 1995, in *Stellar Populations*, ed. P.C. van der Kruit & G. Gilmore (Kluwer, Dordrecht), p. 31
Roos, N., & Aarseth, S.J. 1982, A&A 114, 41
Rose, J.A. 1985, AJ 90, 1927
Rose, J.A. 1994, AJ 107, 206
Roth, J. 1996, Sky & Telesc. 91, 12
Rots, A.H. 1975, A&A 45, 43
Rubin, V.C. 1993, *Mercury*, 22, 109
Rubin, V.C. 1994, AJ 107, 173
Rubin, V.C., Ford, W.K., & D'Odorico, S. 1970, ApJ 160, 801
Rubin, V.C., Graham, J.A., & Kenney, J.D.P. 1992, ApJ 394, L9
Sandage, A.R. 1953, AJ 58, 61
Sandage, A. 1961, *The Hubble Atlas of Galaxies* (Carnegie Institution of Washington, Washington, D.C.)
Sandage, A. 1982, ApJ 252, 553

Sandage, A., & Bedke, J. 1988, *Atlas of Galaxies Useful for Measuring the Cosmological Distance Scale* (NASA, Washington, DC)

Sandage, A., & Bedke, J. 1994, *The Carnegie Atlas of Galaxies* (Carnegie Institution of Washington, Washington, D.C.)

Sandage, A.R., Kron, R.G., & Longair, M.S. 1995, *The Deep Universe*, ed. B. Binggeli & R. Buser (Springer, Berlin)

Sandage, A.R., & Schwarzschild, M. 1952, ApJ 116, 463

Sandage, A., & Visvanathan, N. 1978, ApJ 225, 742

Sanders, D.B. 1990, in *Dynamics and Interactions of Galaxies*, ed. R. Wielen (Springer Verlag, Berlin), p. 459

Sanders, D.B., Scoville, N.Z., Sargent, A.I., & Soifer, B.T. 1988a, ApJ 324, L55

Sanders, D.B., et al. 1988b, ApJ 325, 74

Sanders, D.B., Scoville, N.Z., & Soifer, B.T. 1991, ApJ 370, 158

Sarajedini, A., & Layden, A.C. 1995, AJ 109, 1086

Schaller, G., Schaerer, D., Meynet, G., & Maeder, A. 1992, A&AS 96, 269

Schechter, P.L., & Dressler, A. 1987, AJ 94, 563

Schechter, P.L., & Gunn, J.E. 1978, AJ 83, 1360

Schechter, P.L., Sancisi, R., van Woerden, H., & Lynds, C.R. 1984a, MNRAS 208, 111

Schechter, P.L., Ulrich, M.-H., & Boksenberg, A. 1984b, ApJ 277, 526

Schiminovich, D., van Gorkom, J.H., van der Hulst, J.M., & Kasow, S. 1994, ApJ 423, L101

Schiminovich, D., van Gorkom, J.H., van der Hulst, J.M., & Malin, D.F. 1995, ApJ 444, L77

Schmidt, A.A., Copetti, M.V.F., Alloin, D., & Jablonka, P. 1991, MNRAS 249, 766

Schombert, J.M. 1987, ApJS 64, 643

Schombert, J.M., & Wallin, J.F. 1987, AJ 94, 300

Schombert, J.M., Wallin, J.F., & Struck-Marcell, C. 1990, AJ 99, 497

Schweizer, F. 1976a, ApJS 31, 313

Schweizer, F. 1976b, BAAS 8, 496

Schweizer, F. 1978, in *Structure and Properties of Nearby Galaxies*, ed. E.M. Berkhuijsen & R. Wielebinski (Reidel, Dordrecht), p. 279

Schweizer, F. 1980, ApJ 237, 303

Schweizer, F. 1982, ApJ 252, 455

Schweizer, F. 1983, in *Internal Kinematics and Dynamics of Galaxies*, ed. E. Athanassoula (Reidel, Dordrecht), p. 319

Schweizer, F. 1986, Sci 231, 227

Schweizer, F. 1987, in *Nearly Normal Galaxies*, ed. S.M. Faber (Springer, New York), p. 18

Schweizer, F. 1990, in *Dynamics and Interactions of Galaxies*, ed. R. Wielen (Springer, Berlin), p. 60

Schweizer, F. 1993, in *Structure, Dynamics and Chemical Evolution of Elliptical Galaxies*, ed. I.J. Danziger, W.W. Zeilinger, & K. Kjär (ESO, Garching), p. 651

Schweizer, F. 1996, AJ 111, 109

Schweizer, F., et al. 1990, ApJ 364, L33

Schweizer, F., & Ford, W.K. 1985, in *New Aspects of Galaxy Photometry*, ed. J.-L. Nieto (Springer, Berlin), p. 145

Schweizer, F., Miller, B.W., Whitmore, B.C., & Fall, S.M. 1996, AJ 112, 1839

Schweizer, F., & Seitzer, P. 1988, ApJ 328, 88

Schweizer, F., & Seitzer, P. 1992, AJ 104, 1039 (SS92)

Schweizer, F., & Seitzer, P. 1993, ApJ 417, L29

Schweizer, F., van Gorkom, J.H., & Seitzer, P. 1989, ApJ 338, 770
Schweizer, F., Whitmore, B.W., & Rubin, V.C. 1983, AJ 88, 909
Scorza, C., & Bender, R. 1990, A&A 235, 49
Scorza, C., & Bender, R. 1995, A&A 293, 20
Scoville, N.Z., Sargent, A.I., Sanders, D.B., & Soifer, B.T. 1991, ApJ 366, L5
Searle, L., Sargent, W.L.W., & Bagnuolo, W.G. 1973, ApJ 179, 427
Searle, L., Wilkinson, A., & Bagnuolo, W.G. 1980, ApJ 239, 803
Searle, L., & Zinn, R. 1978, ApJ 225, 357
Seitzer, P., & Schweizer, F. 1990, in *Dynamics and Interactions of Galaxies*, ed. R. Wielen (Springer, Berlin), p. 270
Sérsic, J.L., & Agüero, E.L. 1972, Ap&SS 19, 387
Shane, W.W. 1980, A&A 82, 314
Shlosman, I. (ed.) 1994, *Mass-Transfer Induced Activity in Galaxies* (Cambridge Univ. Press, Cambridge)
Simonson, G.F. 1982, Ph.D. thesis, Yale University
Soderblom, D.R., & Mayor, M. 1993, AJ 105, 226
Sodroski, T.J., Dwek, E., Hauser, M.G., & Kerr, F.J. 1987, ApJ 322, 101
Stanford, S.A., & Bushouse, H.A. 1991, ApJ 371, 92
Statler, T.S. 1991, ApJ 382, L11
Statler, T.S., Smecker-Hane, T., & Cecil, G. 1996, AJ 111, 1512
Stauffer, J.R. 1982a, ApJ 262, 66
Stauffer, J.R. 1982b, ApJS 50, 517
Steidel, C.C. & Dickinson, M. 1995, in *Wide Field Spectroscopy and the Distant Universe*, ed. S.J. Maddox & A. Aragon-Salamanca (World Scientific, London), p. 349
Steidel, C.C., Pettini, M., & Hamilton, D. 1995, AJ 110, 2519
Steidel, C.C., Sargent, W.L.W., & Dickinson, M. 1991, AJ 101, 1187
Steiman-Cameron, T.Y., Kormendy, J., & Durisen, R.H. 1992, AJ 104, 1339
Stockton, A. 1974, ApJ 187, 219
Stockton, A. 1978, ApJ 223, 747
Stockton, A. 1982, ApJ 257, 33
Stockton, A. 1990, in *Dynamics and Interactions of Galaxies*, ed. R. Wielen (Springer, Heidelberg), p. 440
Stockton, A., & Farnham, T. 1991, ApJ 371, 525
Stockton, A., & MacKenty, J.W. 1983, Nat 305, 678
Stockton, A., & Ridgway, S.E. 1991, AJ 102, 488
Struck, C., Appleton, P.N., Borne, K.D., & Lucas, R.A. 1996, AJ 112, 1868
Surma, P., & Bender, R. 1995, A&A 298, 405
Theys, J. 1973, Ph.D. thesis, Columbia University
Theys, J.C., & Spiegel, E.A. 1976, ApJ 208, 650
Theys, J.C., & Spiegel, E.A. 1977, ApJ 212, 616
Theys, J.C., Spiegel, E.A., & Toomre, J. 1972, PASP 84, 851
Tinsley, B.M., & Larson, R.B. (eds.) 1977, *The Evolution of Galaxies and Stellar Populations* (Yale Univ. Observatory, New Haven)
Toomre, A. 1972, QJRAS 13, 266
Toomre, A. 1977a, in *The Evolution of Galaxies and Stellar Populations*, ed. B.M. Tinsley & R.B. Larson (Yale Univ. Observatory, New Haven), p. 401
Toomre, A. 1977b, in *The Evolution of Galaxies and Stellar Populations*, ed. B.M. Tinsley & R.B. Larson (Yale Univ. Observatory, New Haven), p. 418
Toomre, A. 1977c, ARA&A 15, 437

Toomre, A. 1978, in *The Large Scale Structure of the Universe*, ed. M.S. Longair & J. Einasto (Reidel, Dordrecht), p. 109
Toomre, A. 1981a, in *The Structure and Evolution of Normal Galaxies*, ed. S.M. Fall & D. Lynden-Bell (Cambridge Univ. Press, Cambridge), p. 111
Toomre, A. 1981b, private commun., see Schweizer, F. 1983, in *Internal Kinematics and Dynamics of Galaxies*, ed. E. Athanassoula (Reidel, Dordrecht), p. 319 and Fig. 3
Toomre, A. 1983, in *Internal Kinematics and Dynamics of Galaxies*, ed. E. Athanassoula (Reidel, Dordrecht), p. 177
Toomre, A. 1990, in *Dynamics and Interactions of Galaxies*, ed. R. Wielen (Springer, Berlin), p. 292
Toomre, A., & Toomre, J. 1972, ApJ 178, 623 (TT)
Toomre, A., & Toomre, J. 1973, Sci. Am. 229, 38
Tóth, G., & Ostriker, J.P. 1992, ApJ 389, 5
Tremaine, S.D. 1976, ApJ 203, 72
Tremaine, S. 1981, in *The Structure and Evolution of Normal Galaxies*, ed. S.M. Fall & D. Lynden-Bell (Cambridge Univ. Press, Cambridge), p. 67
Tremaine, S. 1995, AJ 110, 628
Trimble, V., & Leonard, P.J.T. 1996, PASP 108, 8
Tubbs, A.D. 1980, ApJ 241, 969
Ulrich, M.-H. 1975, PASP 87, 965
Unavane, M., Wyse, R.F.G., & Gilmore, G. 1996, MNRAS 278, 727
van Albada, T.S., Kotanyi, C.G., & Schwarzschild, M. 1982, MNRAS 198, 303
van den Bergh, S. 1979, PASP 91, 639
van den Bergh, S. 1982, PASP 94, 459
van den Bergh, S. 1992, Sky & Telesc. 83, 508
van den Bergh, S. 1993, AJ 105, 971
van der Hulst, J.M. 1979a, A&A 71, 131
van der Hulst, J.M. 1979b, A&A 75, 97
van der Kruit, P.C. 1976, A&A 49, 161
van Driel, W. 1987, Ph.D. thesis, University of Groningen
van Driel, W., et al. 1995, AJ 109, 942
van Gorkom, J.H., Schechter, P.L., & Kristian, J. 1987, ApJ 314, 457
Velázquez, H., & White, S.D.M. 1995, MNRAS 275, L23
Véron, P., & Véron-Cetty, M.-P. 1985, A&A 145, 433
Visser, H.C.D. 1980, A&A 88, 149
Vorontsov-Velyaminov, B.A. 1959, *Atlas and Catalogue of Interacting Galaxies*, Part I (Sternberg Institute, Moscow State Univ., Moscow)
Vorontsov-Velyaminov, B.A., & Krasnogorskaja, A.A. 1962, *Morphological Catalogue of Galaxies*, Part I (Moscow State Univ., Moscow)
Wagner, S.J. 1990, in *Dynamics and Interactions of Galaxies*, ed. R. Wielen (Springer, Berlin), p. 244
Wakamatsu, K.-I., & Nishida, M.T. 1987, ApJ 315, L23
Walborn, N.R. 1991, in *Massive Stars in Starbursts*, ed. C. Leitherer et al. (Cambridge Univ. Press, Cambridge), p. 145
Walker, M.F., & Chincarini, G. 1967, ApJ 147, 416
Wang, Z., Schweizer, F., & Scoville, N.Z. 1992, ApJ 396, 510
Wannier, P., & Wrixon, G.T. 1972, ApJ 173, L119
Weiner, B.J., & Williams, T.B. 1996, AJ 111, 1156
White, S.D.M. 1979, MNRAS 189, 831

Whitmore, B.C. 1996, in *Clusters, Lensing, and the Future of the Universe*, ed. V. Trimble & A. Reisenegger (A.S.P., San Francisco), p. 85
Whitmore, B.C., & Bell, M. 1988, ApJ 324, 741
Whitmore, B.C., et al. 1990, AJ 100, 1489
Whitmore, B.C., McElroy, D.B., & Schweizer, F. 1987, ApJ 314, 439
Whitmore, B.C., Miller, B.W., Schweizer, F., & Fall, S.M. 1997, AJ, submitted
Whitmore, B.C., & Schweizer, F. 1995, AJ 109, 960
Whitmore, B.C., Schweizer, F., Leitherer, C., Borne, K., & Robert, C. 1993, AJ 106, 1354
Whitmore, B.C., Sparks, W.B., Lucas, R.A., Macchetto, F.D., & Biretta, J.A. 1995, ApJ 454, L73
Wielen, R. (ed.) 1990, *Dynamics and Interactions of Galaxies* (Springer Verlag, Berlin)
Wilkinson, A., Sharples, R.M., Fosbury, R.A.E., & Wallace, P.T. 1986, MNRAS 218, 297
Williams, R.E., et al. 1996, AJ 112, 1335
Williams, T.B., & Schwarzschild, M. 1977, BAAS 9, 319
Williams, T.B., & Schwarzschild, M. 1979, ApJ 227, 56
Wise, M.W., & Silva, D.R. 1996, ApJ 461, 155
Worthey, G. 1992, Ph.D. thesis, University of California, Santa Cruz
Worthey, G. 1994, ApJS 95, 107
Wouterloot, J.G.A., Brand, J., Burton, W.B., & Kwee, K.K. 1990, A&A 230, 21
Wright, A.E. 1972, MNRAS 157, 309
Wright, G.S., James, P.A., Joseph, R.D., & McLean, I.S. 1990, Nat 344, 417
Wyse, R.F.G. 1995, in *Stellar Populations*, ed. P.C. van der Kruit & G. Gilmore (Kluwer, Dordrecht), p. 133
Yabushita, S. 1971, MNRAS 153, 97
Yee, H.K.C., & Ellingson, E. 1995, ApJ 445, 37
Yoshizawa, M., & Wakamatsu, K. 1975, A&A 44, 363
Young, J.S., Kenney, J., Lord, S.D., & Schloerb, F.P. 1984, ApJ 287, L65
Young, J.S., Xie, S., Kenney, J.D., & Rice, W.L. 1989, ApJS 70, 699
Yun, M.S., Ho, P.T.P., & Lo, K.Y. 1994, Nat 372, 530
Zepf, S.E. 1997, Astroph. Lett. Commun., in press
Zepf, S.E., & Ashman, K.M. 1993, MNRAS 264, 611
Zepf, S.E., Ashman, K.M., & Geisler, D. 1995a, ApJ 443, 570
Zepf, S.E., Carter, D., Sharples, R.M., & Ashman, K.M. 1995b, ApJ 445, L19
Zepf, S.E., & Koo, D.C. 1989, ApJ 337, 34
Zinn, R. 1993, in *The Globular Cluster – Galaxy Connection*, ed. G.H. Smith & J.P. Brodie (A.S.P., San Francisco), p. 38
Zinnecker, H., Keable, C.J., & Dunlop, J.S. 1988, in *The Harlow Shapley Symposium on Globular Cluster Systems in Galaxies*, ed. J.E. Grindlay & A.G.D. Philip (Kluwer, Dordrecht), p. 603
Zwicky, F. 1941, in *Theodore von Karman Anniversary Volume, Contrib. to Applied Mechanics and Related Subjects* (California Inst. Technology, Pasadena), p. 137
Zwicky, F. 1953, Phys. Today 6, 7
Zwicky, F. 1956, Ergebnisse d. exakten Naturw. 29, 344
Zwicky, F. 1958, L'Astronomie, 72, 285
Zwicky, F. 1959, Handbuch d. Physik 53, 373
Zwicky, F., & Herzog, E. 1966, *Catalogue of Galaxies and of Clusters of Galaxies* (California Inst. Technology, Pasadena), Vol. III, p. 338
Zwicky, F., & Humason, M.L. 1961, ApJ 133, 794

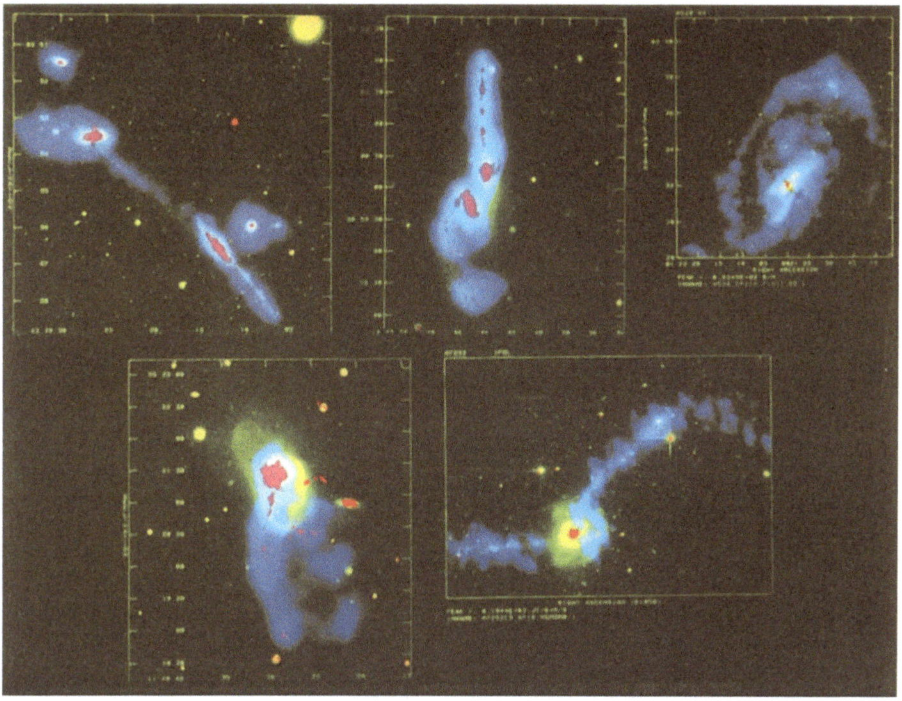

Plate 1. False-color composite images of five galaxies forming a merger sequence, with HI distributions shown in blue, HII distributions in pink–red, and optical images (passband R) in yellow–green–white. From top left to bottom right: Arp 295, NGC 4676, NGC 520, NGC 3921, and NGC 7252. Montage courtesy of J. E. Hibbard & J. H. van Gorkom (1996).

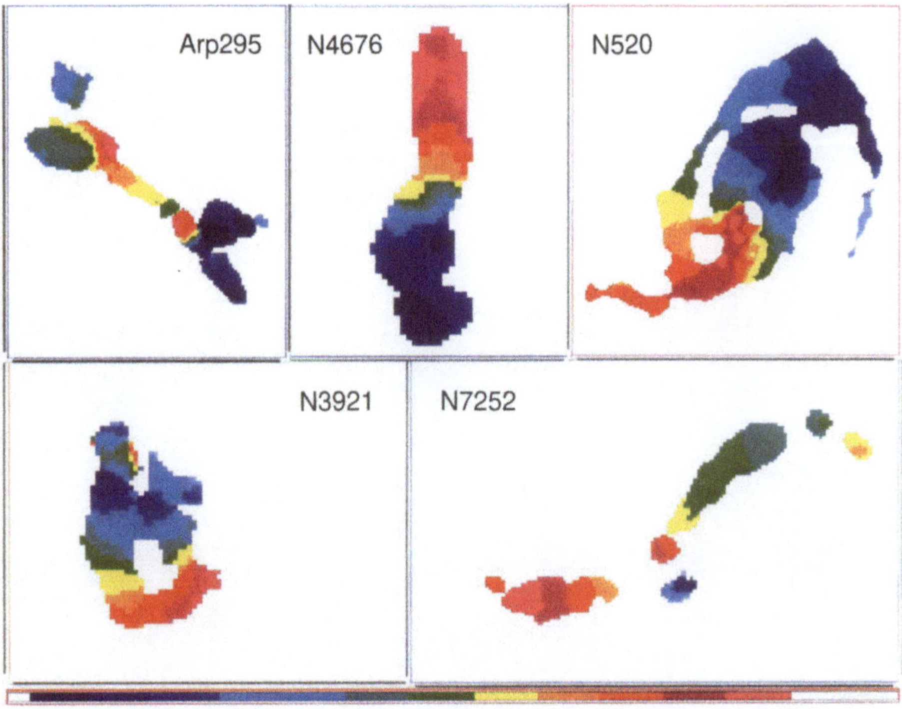

Plate 2. Color-coded velocity maps of five galaxies forming a merger sequence, derived from H I observations with the VLA. From top left to bottom right, the galaxies (and displayed heliocentric radial velocity ranges) are: Arp 295 (6500 – 7150 km s^{-1}), NGC 4676 (6300 – 6900 km s^{-1}), NGC 520 (2050 – 2450 km s^{-1}), NGC 3921 (5780 – 5970 km s^{-1}), and NGC 7252 (4600 – 4880 km s^{-1}); cf. with Plate 1. Blue colors indicate blue-shifted radial velocities relative to systemic velocity, while red colors indicate red-shifted velocities. Montage courtesy of J. E. Hibbard & J. H. van Gorkom (1996).

Plate 3. Color rendition of a $1'.2 \times 1'.2$ portion of the Hubble Deep Field (HDF), made from *HST* observations in passbands B (33 hr), V (30 hr), and I (34 hr). Photo courtesy of R. E. Williams, the HDF team, and NASA.

Plate 4. Three enlarged sections ("details") of the Hubble Deep Field image. Photo courtesy of R. E. Williams, the HDF team, and NASA.

Dynamics of Galaxy Interactions

Joshua E. Barnes

Institute for Astronomy, University of Hawai'i,
2680 Woodlawn Drive, Honolulu, HI 96822, USA

Preface

Interacting galaxies are well-understood in terms of the effects of gravity on stars and dark matter. The details of star formation resist any such capsule summary. Nonetheless, the two are linked by the bursts of star formation observed in interacting galaxies. As "house theoretician" for a winter school on this subject, my job was to shed light on this linkage. I've tried to summarize the growing body of results on dynamics of galactic encounters, and to emphasize where encounters may create conditions for rapid star formation. This presentation is idiosyncratic; I make no claim to have completely surveyed the field! For a brave attempt at such a survey, see Barnes & Hernquist (1992a).

These lectures were originally interwoven with François Schweizer's on Interacting Galaxies and Rob Kennicutt's on Star Formation. They have been shamelessly revised to stand as a self-contained unit, closely related to the contributions of my co-authors but no longer commingled with them. I've also taken the opportunity to expand some subjects, to include some new results, and to eliminate some material which now seems out of place.

Here then is the revised outline. § 1 provides a historical and theoretical background of galaxy interactions. Techniques for simulating the stellar and gas dynamics of interacting galaxies are discussed in § 2 and § 3, respectively; readers more interested in results than in methods may be content to skim these chapters. § 4 gets down to business, discussing tidal encounters between galaxies and the response of disks to tides. § 5 describes attempts to model some well-known systems: The Antennae, The Mice, and The Whirlpool. Orbit decay and the dynamics of merging are covered in § 6, and the stellar-dynamical structure of merger remnants in § 7. Finally, § 8 examines the dynamics of gas inflows in interacting disk system, and the distribution and kinematics of gas in merger remnants. A last lecture largely concerned with dynamical evolution of compact groups has been dropped; for an up-to-date discussion, see Barnes (1997).

The lectures included several computer-generated videos of galactic interactions and orbit structures. These animations and other images are available from `http://www.ifa.hawaii.edu/~barnes/saas-fee.html`.

1 Theory of Interacting Galaxies

Classical physics needs two things to predict the future: dynamical laws and initial conditions (Laplace 1820). Both of these play a critical role in understanding the nature of interacting galaxies. The classical laws of gravity and motion are a natural starting point for an astrophysical discussion of collisions between galaxies. But without the right initial conditions it's impossible to account for observations of galactic interactions.

1.1 The Role of Gravity

Today it may seem obvious that gravity, and specifically *gravity acting on stars*, creates the bridges, tails, and other striking features of interacting galaxies. But this was not always obvious. Astronomers of yesterday had nothing like the enormous volume of high-quality imaging and spectroscopic data in the wide range of wavebands now available. Nor did they have detailed theories of stellar evolution and nucleosynthesis, or accurately specified cosmological models. They had isolated observations, some of remarkably high quality, and bold, energetic theorizing. But their theorizing was hampered by one very specific blind-spot: sheer disbelief that gravity could produce structures as extended and narrow as those in interacting galaxies.

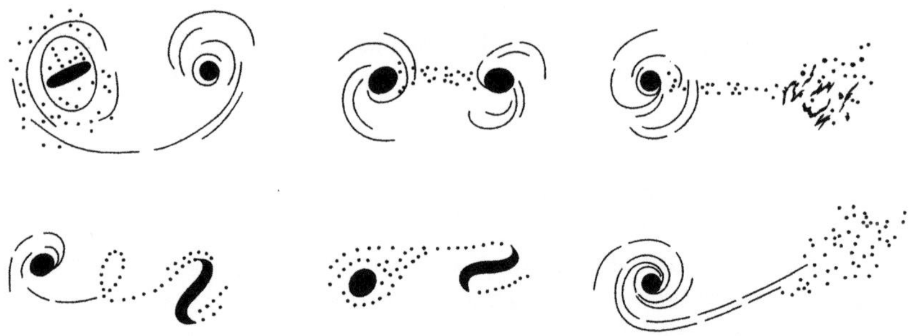

Fig. 1. Filamentary features in interacting galaxies, as sketched by Zwicky (1959).

At least one highly imaginative astronomer got it nearly right. Zwicky (1959) emphasized that "it is clouds, filaments, and jets of *stars* which are ejected massively from galaxies in collision". He also recognized the importance of "large-scale tidal effects". But Zwicky regarded the elongated filaments of interacting galaxies (Fig. 1) as evidence for non-gravitational effects, and speculated that "electromagnetic actions... may materially contribute to the internal viscosity of stellar systems".

Discussions recorded at IAU Symposium 15, "Problems of Extragalactic Research", offer an interesting historical perspective on this issue. Vorontosov-Vel'yaminov (1961) called for new physics to explain bridges and tails; Gold advocated magnetic forces but noted that they would not restrain matter "once... in stellar form"; Harwit proposed that the stellar component would then break up into lumps resembling chain galaxies; Hoyle asked about the relationship between spiral arms and filaments; Zwicky affirmed to G. Burbidge that bridges are dynamically young but contain old stars; finally, the Lindblads came closest to the mark in describing "streams of matter" extracted from galaxies by tidal forces. In sum, there was general agreement that bridges and tails posed a puzzle, but *no* agreement on the solution – though several key elements were already on the table.

Ultimately, the hypothesis that *gravity governs galaxy interactions* stands or falls on its success at accounting for observations of colliding galaxies. But very simple estimates show that no known force is likely to deflect stars from their free-fall trajectories; dynamically young bridges and tails containing old stars can only, it seems, be created by gravity.

1.2 Holmberg's Work on Tidal Interactions

Two decades before IAU Symposium 15, Holmberg (1941) had already used two key ingredients – stars and gravity – in a numerical study of galaxy interactions. But Holmberg did not realize that his model could account for the extended features of interacting systems; he had the right laws, but the wrong initial conditions.

"Clustering Tendencies". Holmberg was trying to explain the origins of groups and clusters of galaxies. He had in mind the mutual *tidal capture* of galaxies during initially hyperbolic passages – that is, passages in which the galaxies approach each other with sufficient energy to escape to infinity, but lose energy during the encounter and consequently become bound to one another. Starting with a uniform "sea" of galaxies with peculiar velocities of a few hundred kilometers per second, Holmberg imagined that successive captures could build up groups and clusters.

This picture has a fundamental flaw: the estimated encounter rates are quite low. Suppose bright galaxies are randomly distributed, with number densities of $n \simeq 10^{-2}h^3\,\mathrm{Mpc}^{-3}$ and peculiar velocities of $v \simeq 300\,\mathrm{km\,s}^{-1} \simeq 0.3\,\mathrm{Mpc\,Gyr}^{-1}$; let the capture cross-section be $\sigma = \pi(30h^{-1}\,\mathrm{kpc})^2$. Then the time required for a typical galaxy to be captured is $(nv\sigma)^{-1} \simeq 10^5 h^{-1}\,\mathrm{Gyr} \simeq 10^4 H_0^{-1}$, so quite independent of the adopted value of $H_0 \equiv 100h\,\mathrm{km\,s}^{-1}\,\mathrm{Mpc}^{-1}$, only about one galaxy in ten thousand will undergo tidal capture. For this reason, Holmberg favored *static* world-models. He did note that in an expanding universe, groups and clusters could arise due to initial irregularities in the expansion, but regarded this as an unlikely possibility.

Similar estimates still occasionally surface in the literature and have been used to claim that tidal interactions in the field are too infrequent to be relevant for most galaxies. The flaw in these estimates is the assumption that galaxies are initially uncorrelated.

Holmberg's Simulations. To test the idea that close encounters could lead to tidal capture, Holmberg did numerical experiments. Each galaxy was represented by 37 mass points arranged in concentric circles and traveling around a common center. Holmberg recognized that the evaluation of gravitational forces between these mass points represented the most computationally intensive part of the problem. His solution was to use the inverse-square behavior of *light* to mimic that of gravity; the mass points were represented by light bulbs, and a photocell was used to measure the magnitude and direction of the light. With forces thus determined, the trajectories of the mass points were calculated by graphical integration.

Using this numerical technique, Holmberg simulated planar encounters of disk galaxies and estimated the efficiency of tidal capture for various approach velocities, minimum separations, and rotation directions. He correctly noted that the maximum tidal distortion occurs *after* the passage and called attention to the spiral arms developed as the disks passed one another. But some of Holmberg's results are puzzling. To begin with, he did not report any trouble with bar instabilities, which should have developed rapidly in the "pure disk" systems he studied. Yet more curious is his result that tidal capture was more effective when the disks counter-rotated with respect to their orbital motion. Alas, the published description of Holmberg's experiments does not provide enough detail to resolve these puzzles.

In retrospect, it remains surprising that Holmberg did not discover the tidal origins of bridges and tails. While his theoretical discussion emphasized hyperbolic passages, he also studied parabolic encounters, and rather naturally used these slower passages to illustrate tidal deformations. Holmberg cautioned that the small number of mass points used and the approximations made in calculating their trajectories precluded a detailed study of the tidal response. It is also possible that he simply did not follow the galaxies long enough to observe the development of extended tidal features.

1.3 "Galactic Bridges and Tails"

Following Holmberg, P.O. Lindblad (1960) had some success at reproducing interacting galaxies with mechanical models (B. Lindblad 1961). But these studies emphasized extraction of stars from "narrow resonance regions" due to the tidal influence of a companion galaxy in a periodic orbit. Pfleiderer & Siedentopf (1961) and Pfleiderer (1963) investigated the excitation of spiral patterns by tides between disk galaxies, concluding that chance encounters between unbound field galaxies are far too rare to produce a significant population of spiral galaxies. In passing, they did present perhaps the first plots of bridge and tail building in tidal encounters, but this work was not followed up until the end of the decade.

The early 1970s marked a watershed in the theory of interacting galaxies. On one side of this watershed, Pikel'ner (1968) argued for "regular magnetic fields", while Arp (1971) continued to call for new physics. On the other side, a new stream of thought was started by Pfleiderer & Siedentopf and developed by Yabushita (1971), Wright (1972), Clutton-Brock (1972a,b), Eneev, Kozlov, & Sunyaev (1973), and especially Toomre & Toomre (1972; hereafter TT72). This last paper, "Galactic Bridges and Tails" is famous for many reasons, including its superb illustrations of tidal interactions. But the Toomres' real contributions, in my view, were their clear statement of the necessary conditions, their lucid descriptions of bridge and tail making, their simple but convincing models of several well-known interacting systems, and their discussion of the "broader issues" which lie at the heart of the subject even today.

The computational aspects of this work were not novel; most studies followed Pfleiderer & Siedentopf in treating each galaxy as a disk of test-particles initially circling a central point-mass. This restricted 3-body technique has some serious limitations; for example, the central masses move along Keplerian trajectories, precluding orbit decay. But these limitations are not critical in the earlier stages of tidal encounters, and as the Toomres noted, a "deliberately simple-minded" technique maintained attention on essential physics instead of numerical details.

Fundamentals of Bridge and Tail Building. One key element of the new work was the treatment of bridges and tails as transient *kinematic* phenomena. This neatly sidestepped the problem of dynamically stabilizing these structures with magnetic fields, gas pressures, or forces yet unknown, a problem repeatedly discussed in the decade following IAU Symp. 15 (e.g. Burbidge, Burbidge, & Hoyle 1963, Zasov 1968). TT72 noted that the success of restricted 3-body simulations explicitly demonstrated the essentially kinematic nature of bridges and tails; even mundane *self-gravity is not needed* for their existence!

A second key element was to focus on *parabolic* encounters, in which the galaxies approach each other with speeds comparable to those acquired falling from rest at infinity. Such encounters were only used as limiting cases in earlier investigations; Holmberg and Pfleiderer both concentrated on hyperbolic passages, while the Lindblads apparently studied elliptical or circular orbits. In contrast, TT72 argued that "brief but violent tidal forces" on galaxies during close, nearly-parabolic encounters are effective at creating extended tidal *relics* of past interactions.

A third key was the Toomres' point that the narrow, ribbon-like forms of bridges and tails develop naturally when tides act on cold disks of stars. Tidal forces on hot, pressure-supported objects – for example, elliptical galaxies – produce broad features, since the random velocities of individual stars are comparable to the velocities imparted by tides. This was not appreciated by Tashpulatov (1969, 1970), who tried to account for the narrowness of tidal features by arguing that they are extracted from very localized regions of donor galaxies.

A fourth key was the realization that *direct* passages, in which the victim disk lies in the orbit plane and rotates the same direction as the galaxies pass one another, create a "near-resonance or matching of... orbital [angular] speeds" (TT72) between stars in the victim and its companion – though in parabolic passages this resonance is quite broad since tidal forcing is strong for only a fraction of the disk's rotation period. Direct passages thus produce far more

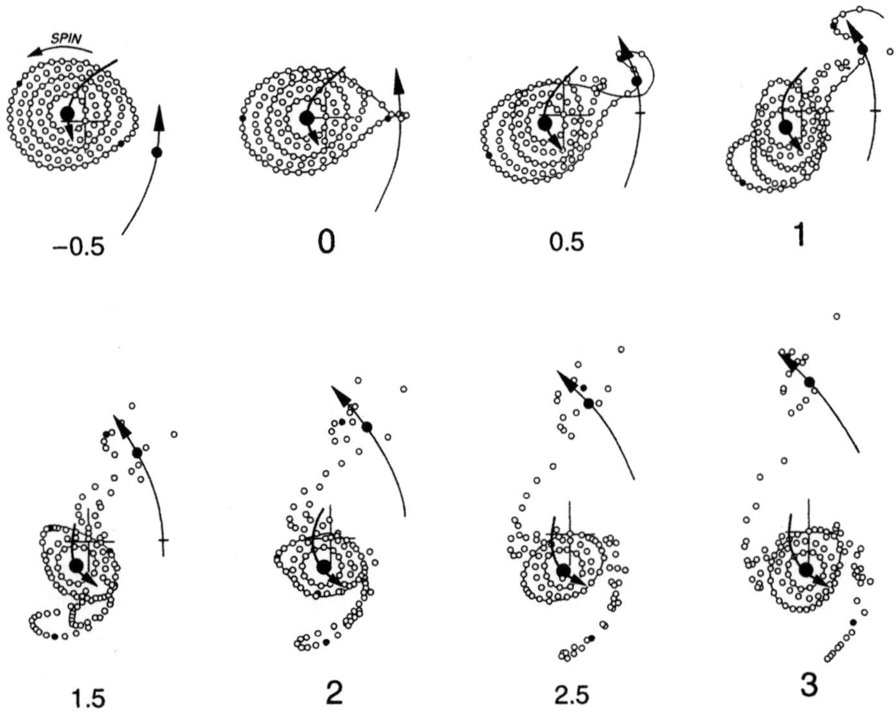

Fig. 2. Direct parabolic passage of a quarter-mass companion, and the ensuing tidal response. Adapted from TT72.

extended tidal features than their *retrograde* counterparts, in which the disk rotates the other way.

Once these key points are recognized, the actual production of tidal features is almost trivial. Bridges result when gravity extracts stars from the side of a disk galaxy nearer to its passing companion. Those created in equal-mass encounters develop rapidly but soon become quite tenuous as stars are captured by the companion or fall back to the galaxy from whence they came. If the mass of the companion is reduced to about one quarter that of the primary, the resulting bridges are longer-lived and can become impressively thin as they stretch between the separating galaxies (see Fig. 2).

Tails, on the other hand, more typically result from equal-mass encounters. As TT72 showed, tails are launched when stars on the far sides of disks fail to "turn the corner" with the rest of their parent galaxies. To be sure, the quarter-mass passages used to illustrate bridge-building also produced narrow "counter-arms" on the far sides of their disks; this is scarcely unexpected since tidal forces are *intrinsically* two-sided. But true tails which escape to infinity generally require rather violent interactions.

Some Related Issues. As the Toomres noted, the explanation of bridges and tails as kinematic relics of tidal interactions had further implications for both peculiar and normal galaxies.

First, close and nearly-parabolic encounters could only occur often enough to explain the observed number of interacting systems if there exists a large population of binary galaxies with highly eccentric long-period orbits. At any given time, then, only that fraction of this population which had recently had a perigalacticon would exhibit tidal features, while the rest would appear more or less normal. Indeed, TT72 proposed that many interacting pairs were on their *first* such passage, having loitered near apogalacticon for most of the past $\sim 10\,\mathrm{Gyr}$.

Second, the orbits of interacting galaxies rapidly decay after each close passage. TT72 envisioned that orbital energy and angular momentum would be transferred to tail-making material, and then carried to infinity. The actual story is complicated by interactions between dark halos, but the general implication – orbits decay until the actors tumble into a single heap of stars – could not have been much closer to the mark. And this points to "nothing less than the delayed formation of some elliptical galaxies – or at least of major stellar halos" (TT72).

Third, the relatively blue colors, high surface brightnesses, and gas-rich bodies of some interacting systems indicate that significant inflows of interstellar material occur during tidal interactions, a process TT72 dubbed "stoking the furnace" of star formation. This simple idea – which gathered much observational support, starting with the work of Larson & Tinsley (1978) – neatly anticipates much of the very subject of this volume!

Fourth, such "unusually fine spirals" as M 51, NGC 7753, and even M 81 may owe their bisymmetric, grand-design patterns to density waves induced by tidal interactions. This may sound like a return to Pfleiderer & Siedentopf's investigations of tidal spirals, but it is rather plausible in that all three of these famous galaxies possess close and evidently interacting companions.

And as the Toomres acknowledged, some of these issues call for "proper N-body calculations" in place of restricted three-body models. Clutton-Brock (1972a,b) used a potential expansion method (see § 2.2) to approximate the self-gravity of non-axisymmetric two-dimensional disks. He found that cold rotating disks are intrinsically unstable, and chose to stabilize the disk models by significantly increasing the degree of random motions. Seeking to recover the thin tidal features reported by TT72, he then added a second component, dubbed "gas" by virtue of its low initial velocity dispersion rather than any responsiveness to pressure gradients. Tidal forces on the gas then produced reasonably thin bridges and tails, but Clutton-Brock did not discuss the issues of orbit decay and tidal spiral-making, though his calculations could in principle address these matters.

1.4 Dark Matter

Another related issue, *not* yet appreciated in 1972, was the role of dark matter. While it is widely accepted that gravity governs galactic dynamics, it is also acknowledged that the *visible* matter is insufficient to explain the dynamics of real galaxies, and that large amounts of dark matter may be present (e.g. Trimble 1987). Dark matter may exist in any number of forms, ranging from the black holes of 10^6 to $10^9\,M_\odot$ which seem to be present in some galactic nuclei (e.g. Kormendy & Richstone 1995) to hypothetical particles with masses of 10^{-6} eV. Different kinds of dark matter have various implications for galaxy formation, and by probing the structures of galaxy halos, interaction studies may provide constraints on the nature of dark matter. Nonetheless, all these forms of dark matter respond to gravity in *exactly* the same way.

2 Numerical Stellar Dynamics

Particle techniques for simulating stellar-dynamical systems are the subject of an extensive literature (e.g. Sellwood 1987, Hockney & Eastwood 1988). This section focuses first on the basic concepts, and then on some numerical algorithms suitable for simulations of galactic interactions. The limitations of such methods – primarily due to finite particle number – are examined. Finally, some ways of generating initial conditions and some tests of isolated disk galaxy models are described.

2.1 Collisionless Stellar Systems

In view of the arguments given in § 1, it seems reasonable to adopt as a working model of a galaxy a self-gravitating system of stars and particles of dark matter. The Newtonian equations of motion for such a system of $\mathcal{N}$ mass points are

$$\frac{\mathrm{d}\mathbf{r}_i}{\mathrm{d}t} = \mathbf{v}_i\,, \qquad \frac{\mathrm{d}\mathbf{v}_i}{\mathrm{d}t} = \sum_{j\neq i}^{\mathcal{N}} G m_j \frac{\mathbf{r}_j - \mathbf{r}_i}{|\mathbf{r}_j - \mathbf{r}_i|^3}\,, \tag{1}$$

where $\mathbf{r}_i$, $\mathbf{v}_i$, and m_i are the position, velocity, and mass, respectively, of the ith point, and G is the gravitational constant.

These equations provide a general description of the motion of $\mathcal{N}$ particles – too general, in fact, to be of much use in galactic dynamics! A *continuous* description of the system is more useful; instead of specifying $\mathcal{N}$ masses, positions and velocities, we will work with the mass distribution in the 6-dimensional phase-space $(\mathbf{r}, \mathbf{v})$. Thus, let

$$f(\mathbf{r}, \mathbf{v}, t)\,\mathrm{d}\mathbf{r}\,\mathrm{d}\mathbf{v} \equiv \text{mass in } \mathrm{d}\mathbf{r}\,\mathrm{d}\mathbf{v} \text{ at point } (\mathbf{r}, \mathbf{v}) \text{ and time } t\,. \tag{2}$$

To distinguish between stars and dark matter, write $f = f_{\mathrm{s}} + f_{\mathrm{d}}$, where the two functions on the right describe these separate components, respectively.

To find a dynamical equation for the distribution function, we assume that the flow of matter through phase space is governed by a smooth 6-dimensional vector field

$$(\dot{\mathbf{r}}, \dot{\mathbf{v}}) = (\mathbf{v}, -\boldsymbol{\nabla}\Phi)\,, \tag{3}$$

where $\Phi(\mathbf{r}, t)$ is the gravitational potential (see Fig. 3). Mass conservation implies that the change of the mass in the phase-space volume $\mathbf{d}\mathbf{r}\mathbf{d}\mathbf{v}$ is just due to inflow minus outflow, so

$$\frac{\partial f}{\partial t} + \frac{\partial}{\partial \mathbf{r}} \cdot (f\dot{\mathbf{r}}) + \frac{\partial}{\partial \mathbf{v}} \cdot (f\dot{\mathbf{v}}) = 0\,. \tag{4}$$

Substituting (3) then yields the collisionless Boltzmann equation or CBE,

$$\frac{\partial f}{\partial t} + \mathbf{v} \cdot \frac{\partial f}{\partial \mathbf{r}} - \boldsymbol{\nabla}\Phi \cdot \frac{\partial f}{\partial \mathbf{v}} = 0\,. \tag{5}$$

If $f = f_{\mathrm{s}} + f_{\mathrm{d}}$, then each component obeys (5) separately.

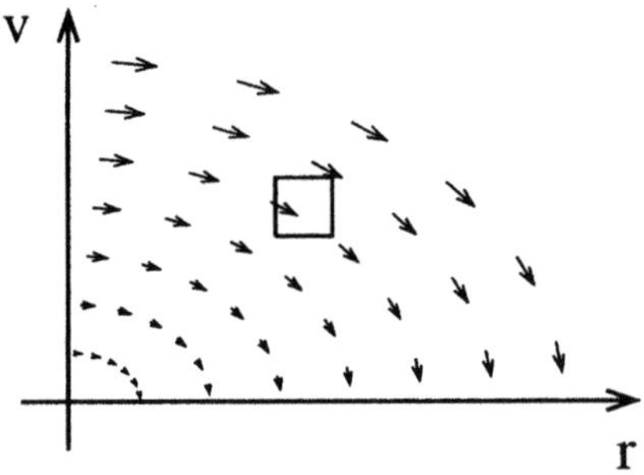

Fig. 3. Two-dimensional phase space with an elementary phase-space volume **drdv**. The arrows represent the flow field defined by (3).

This equation is complemented by Poisson's equation for the gravitational field,

$$\nabla^2 \Phi = 4\pi G \left(\int d\mathbf{v} f + \rho_{\rm ext} \right) , \tag{6}$$

where $\rho_{\rm ext}$ represents any mass not described by the distribution function f.

Several approximations have been made in going from (1) to (5) and (6). In particular, correlations between stars have been neglected, and the gravitational force due to a discrete collection of stars has been replaced with the gradient of a smooth potential. Both of these approximations are quite good in galactic contexts; near the Sun, for example, potential fluctuations due to nearby stars are $\sim 10^7$ times smaller than the net potential of the galaxy, and while multiple star systems are common, most remain safely bound and effectively isolated from other stars for the lifetime of the galaxy. True, the CBE is only a *model*, and one can imagine systems of 10^{12} stars which behave in ways that simply cannot be reproduced by solutions of (5) and (6). But these exceptions seem very contrived, and there is no good reason to think that the CBE is not highly applicable to real galaxies.

2.2 Simulating the Stars

It would take an impractically large grid to solve the time-dependent 3-D collisionless Boltzmann equation by finite-difference methods. N-body simulation is basically a Monte-Carlo method of solving this equation, with the number of bodies, N, governing the accuracy of the method (White 1982).

The basic idea behind Monte-Carlo methods is shown by the following procedure for approximating π. Draw a square of area A, and inscribe within it a circle of area $A_{\rm c}$; by simple geometry, $A_c = \pi A/4$. Now scatter n points independently and randomly within the square, and count the number $n_{\rm c}$ which fall within the circle. Since the expected number of points within any area is proportional to that area, the quantity $4n_{\rm c}/n$ approximates π, with a fractional uncertainty of order $n_{\rm c}^{-1/2}$.

As a method of calculating π, this procedure is very inefficient; for example, a trial with $n = 262144$ points yielded the estimate $\pi \simeq 3.138 \pm 0.007$. However, the error in a Monte-Carlo calculation depends not on the number of dimensions, but only on the number of points. Thus Monte-Carlo methods can outperform other numerical techniques in evaluating multidimensional integrals.

Representing $f(\mathbf{r}, \mathbf{v})$. To represent the mass distribution function $f(\mathbf{r}, \mathbf{v}, t_0)$ at some instant t_0 in a form suitable for Monte-Carlo calculations, one uses a set of N bodies, each possessing a mass m_i, position $\mathbf{r}_i$, and velocity $\mathbf{v}_i$, where $i = 1 \dots N$. In effect, the continuous distribution function is replaced with a set of δ-functions:

$$f(\mathbf{r}, \mathbf{v}) \to \sum_{i=1}^{N} m_i \, \delta(\mathbf{r} - \mathbf{r}_i) \, \delta(\mathbf{v} - \mathbf{v}_i), \; . \tag{7}$$

For this substitution to work, the expected mass of the bodies within any phase-space volume $\mathcal{V}$ must be equal to the integral of the distribution function over that volume; thus,

$$\int_{\mathcal{V}} \mathrm{d}\mathbf{r} \, \mathrm{d}\mathbf{v} \, f(\mathbf{r}, \mathbf{v}) = \left\langle \sum_{(\mathbf{r}_i, \mathbf{v}_i) \in \mathcal{V}} m_i \right\rangle , \tag{8}$$

where the angle brackets indicate an average over statistically equivalent realizations.

The simplest way to initialize bodies in accord with (8) is to pick phase-space coordinates by treating $f(\mathbf{r}, \mathbf{v})$ as a probability distribution; that is, select $(\mathbf{r}_i, \mathbf{v}_i)$ with probability proportional to $f(\mathbf{r}_i, \mathbf{v}_i)$, and assign all bodies the same mass $m_i = M/N$, where the total mass is $M = \int \mathrm{d}\mathbf{r}\mathrm{d}\mathbf{v} f$. Since bodies are selected independently, the actual number within any given volume $\mathcal{V}$ will have a Poissonian distribution about the mean. This scatter – the hallmark of a Monte-Carlo method – limits the accuracy of the calculation. More sophisticated ways of initializing bodies can reduce the scatter; for example, in a "quiet start" (Sellwood 1987), initial conditions are generated by dividing phase-space up into cells containing equal amounts of mass, and placing one body within each cell. Quiet start works well in one or two spatial dimensions, but it becomes harder to devise quiet initial conditions for 3-D calculations.

Such a pointillistic representation of the distribution function may be used to calculate integrals of $f(\mathbf{r}, \mathbf{v})$ over phase-space. Suppose that we wish to estimate the value of some observable q, defined as the integral

$$q \equiv \int \mathrm{d}\mathbf{r} \, \mathrm{d}\mathbf{v} \, f(\mathbf{r}, \mathbf{v}) \, Q(\mathbf{r}, \mathbf{v}) \, . \tag{9}$$

Using (8), this becomes

$$q \simeq \sum_{i}^{N} m_i \, Q(\mathbf{r}_i, \mathbf{v}_i) \, . \tag{10}$$

If bodies are selected independently, the fractional uncertainty in the estimated value of q is of order $N^{-1/2}$, just as in the the estimate of π above.

Dynamical Evolution. N-body representations are useful for other things besides Monte-Carlo integrations; in particular, they are easily projected into the future. This projection is accomplished by moving bodies along the flow field defined by (3). It is evident that this preserves the relation (8); thus starting with a valid realization of $f(\mathbf{r}, \mathbf{v}, t_0)$, the result is a realization of $f(\mathbf{r}, \mathbf{v}, t)$ at all $t > t_0$.

Now what to use for the potential, $\Phi(\mathbf{r}, t)$? In some cases the potential can be specified ahead of time; for example, within the context of the restricted 3-body calculations of TT72 and others, it is known a priori. But in a self-consistent calculation the potential is an unknown, to be estimated from the N-body representation of the mass distribution by using the bodies as the source term for (6); thus,

$$\nabla^2 \Phi = 4\pi G \sum m_i \, \delta(\mathbf{r} - \mathbf{r}_i) \, . \tag{11}$$

This yields the standard Newtonian equations of motion for N point masses (1). Formally there is nothing wrong with using these equations in a self-consistent N-body simulation. But in practice the singular potential wells associated with point masses create awkward numerical problems. These can be avoided by smoothing the N-body representation of the density field, for example via the substitution

$$\delta(\mathbf{r} - \mathbf{r}_i) \rightarrow \frac{3}{4\pi} \frac{\epsilon^2}{\left(|\mathbf{r} - \mathbf{r}_i|^2 + \epsilon^2\right)^{5/2}} \, , \tag{12}$$

where ϵ is a parameter with dimensions of length. The resulting equations of motion are

$$\frac{\mathrm{d}\mathbf{r}_i}{\mathrm{d}t} = \mathbf{v}_i \, , \qquad \frac{\mathrm{d}\mathbf{v}_i}{dt} = \sum_{j \neq i}^{N} G \, m_j \frac{\mathbf{r}_j - \mathbf{r}_i}{\left(|\mathbf{r}_j - \mathbf{r}_i|^2 + \epsilon^2\right)^{3/2}} \, . \tag{13}$$

The smoothing procedure incorporated into these equations is commonly called “Plummer softening”, since it effectively replaces each point mass with a little Plummer (1911) model; or equivalently, it replaces the potential of each point with the potential of a Plummer model. The latter interpretation leads to the phrase “softened potential”, but this is a confusing term because it's really the *density* which has been smoothed, and the potential then calculated exactly from this smoothed density. There seems to be no reason to also smooth bodies before calculating their *response* to the potential Φ via (3) as proposed by Dyer & Ip (1993); their claim that (13) is “inconsistent” is laid to rest if the smoothing step is kept logically distinct from the gravitational force calculation.

Besides taming singularities in the equations of motion, smoothing suppresses small-scale fluctuations due to the discrete nature of N-body representations. Discreteness fluctuations are the bane of collisionless N-body simulation, so some smoothing is a good thing. But smoothing *always* comes at a price in spatial resolution, and no useful amount of smoothing will completely eliminate the effects of discreteness (eg. Hernquist & Barnes 1990)! Moreover, it's unlikely that any alternate smoothing kernel will be radically better than (12), although smoothing kernels tailored to specific problems may offer modest advantages.

2.3 Force Calculation

The Monte-Carlo interpretation of N-body simulation places a premium on large values of N to reduce uncertainties. Thus N-body experimenters seek to run more bodies with the same fervor that observational astronomers seek to gather more photons. Starting with Holmberg's optical computations, much ingenuity has gone into the rapid evaluation of forces in N-body systems.

It should be emphasized that no single force calculation method is optimal for all applications. Hierarchical methods will be discussed here in the greatest detail since they are useful for galactic encounter simulations, which typically involve irregular mass distributions, require high spatial resolution, and demand many bodies. But no method will ever make N-body calculation "cheap"; faster computers and better algorithms simply shift the focus of attention to larger problems.

Direct Summation. Straightforward evaluation of the sum in (13) is robust, accurate, and completely general. As everyone knows, the cost of evaluating the force on all N bodies is $O(N^2)$. To some extent, direct summation methods can beat the high cost of force calculation by efficiently using individual time-steps (e.g. Ahmad & Cohen 1973, Aarseth 1985) and by implementing the force calculation in hardware (e.g. Sugimoto et al. 1990). Such improvements have kept direct summation remarkably competitive even as N has increased by several factors of 10.

Field Expansion. Field methods represent the potential and density as series expansions:

$$\Phi(\mathbf{r}) = \sum_k A_k \, \Phi_k(\mathbf{r}) \,, \qquad \rho(\mathbf{r}) = \sum_k A_k \, \rho_k(\mathbf{r}) \,, \tag{14}$$

where the A_k are expansion coefficients and the basis functions Φ_k and ρ_k are related by Poisson's equation,

$$\nabla^2 \Phi_k = 4\pi G \rho_k \,. \tag{15}$$

The basic procedure is to determine the expansion coefficients by fitting the density to the mass distribution, and then to obtain forces by differentiating the expansion of the potential. There are many ways to do this; for example, Cartesian grid methods use fast Fourier transforms (e.g. Sellwood 1987) while self-consistent field methods calculate overlap integrals (e.g. Hernquist & Ostriker 1992).

Speed is the main advantage of field methods; the time required to evaluate the force on all bodies is typically $O(N)$. If the basic geometry of the system is known ahead of time, a series method can be tailored to fit, and by deleting selected terms in the series expansion one can enforce various symmetries. Such "designer expansions" offer the smoothest potentials for a given N. But the finite set of basis functions used in the expansions imposes a limit on the spatial resolution of field methods. Most applications of field methods have focused on rather special problems where the geometry is relatively simple.

Hierarchical Evaluation. Hierarchical methods exploit the fact that higher-order multipoles of the gravitational field decay rapidly with respect to the dominant monopole term. Hence the long-range gravitational potential of a region can be approximated by $\Phi \propto 1/r$. In "tree" codes, this approximation is used to replace the sum over $N-1$ bodies in (13) with a sum over only $O(\log N)$ regions; such codes can be viewed as hierarchical variations on direct summation (Appel 1985, Barnes & Hut 1986). It is also possible to create hierarchical field methods; the Fast Multipole Method (Greengard & Rokhlin 1987) is an example. But compared to tree codes, such methods are less flexible and more complex to program; consequently they have not yet found widespread use in astrophysical N-body simulations.

Tree structures may be created by hierarchically grouping particles (Appel 1985, Jernigan 1985, Porter 1985, Press 1986) or by recursively subdividing space (Barnes & Hut 1986); the latter approach is a bit more widely used as it is easily implemented and reasonably fast. The usual procedure is to place a single cube, known as the "root" cell, around the entire system; this cell is recursively subdivided until each body is isolated in its own cell, as shown in Fig. 4. The gravitational force on a body can then be evaluated by starting with the root and recursively examining the cells it contains. In the original algorithm (Barnes & Hut 1986), the potential due to a given cell was approximated with a single $1/r$ term if $d > \ell/\theta$, where d is the distance between the body and the cell's center of mass, ℓ is the length of the cell, and the opening angle θ is a parameter typically less than unity; if this criterion is not satisfied then the subcells within the cell are examined instead, and this process is carried to as many levels as necessary. The parameter θ may be used to adjust the accuracy of the method, with smaller values yielding more accurate results at a greater computational expense.

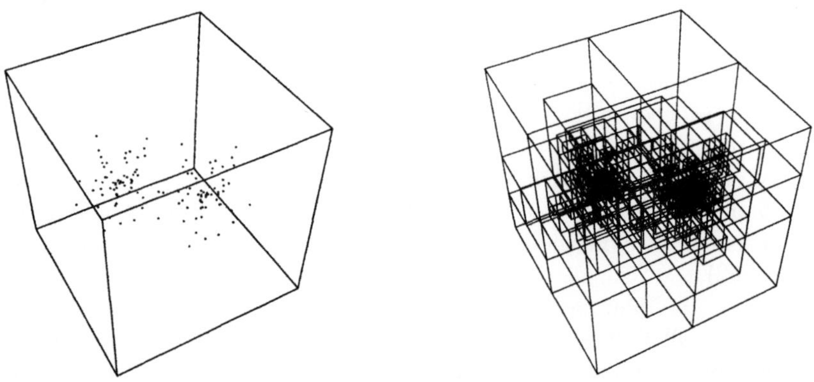

Fig. 4. On the left is a simple configuration of 128 bodies; on the right is the hierarchical box structure generated for this configuration.

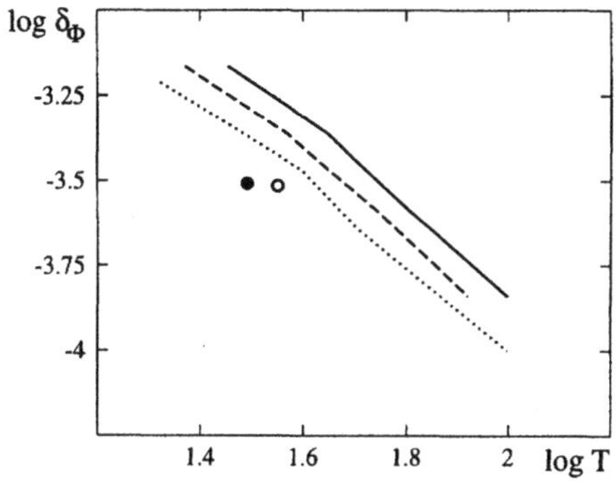

Fig. 5. Root-mean-square error in potential, δ_Φ, plotted against force computation time in minutes, T, for a Plummer model of $N = 262144$ bodies run on an SGI 4D/35 workstation. The solid and dashed lines show results using $d > \ell/\theta$, while the dotted line shows results using $d > \ell/\theta + \delta$. Circles are representative results from the new code; the filled circle shows results obtained using Newton's method to approximate square roots.

Tests show that average relative force errors of 10^{-3} can be obtained for $\theta \sim 0.5$ to 0.7 by including the quadrupole moment of each cell's gravitational field (Hernquist 1987; Barnes & Hut 1989). However, the simple criterion $d > \ell/\theta$ can fail catastrophically in rare circumstances where a cell's center of mass lies far from its geometric center (Salmon & Warren 1993). One cure for this problem is to replace the above criterion by $d > \ell/\theta + \delta$, where δ is the distance the cell's geometric center to its center of mass (Barnes 1994). This revised criterion improves the performance of the algorithm, and effectively fixes the problem described by Salmon & Warren for reasonable θ.

Fig. 5 serves two purposes. The first is to show how the accuracy of hierarchical methods, measured by the r.m.s. error in the computed potential, is related to the time required for force calculation. These results were obtained for a Plummer model with $N = 262144$; the codes were run on an SGI 4D/35 workstation, which is hardly a fast machine by present standards. The solid line shows how accuracy and time trade off as θ varys from 0.8 to 0.5 with the 1986 algorithm, while the dashed line shows results for a code which "threads" the tree with additional pointers so it can be scanned by an iterative procedure. The dotted line shows results from the revised criterion $d > \ell/\theta + \delta$, which tends to be more "careful" with cells containing irregular mass distributions. Finally, the two circles were obtained with a new code, adapted from an algorithm originally designed for a fine-grain parallel computer (Barnes 1986). In other codes, interactions for nearby bodies are independently determined, resulting in redundant computations. The geometrical criterion used in the new code allows this redundancy to be eliminated; other regularities permit most square-root evaluations to be replaced by two iterations of Newton's method.

The second message of Fig. 5 is that a decade's worth of work has produced only rather modest gains in the *speed* of hierarchical algorithms – evidently, the 1986 code was rather close to a local optimum! Nonetheless, much progress has been made at implementing hierarchical codes on different machine architectures, thus enabling N-body experiments to take advantage of new generations of vector and parallel computers (e.g. Barnes 1986, Hernquist 1987, Barnes 1990a, Hernquist 1990a, Makino 1990, Salmon 1990, Singh 1993, Olson & Dorband 1994, Dubinski 1996).

2.4 Time Integration

Many numerical integrators are available for sets of coupled ordinary differential equations, but not all respect the symmetric structure of systems like (3). One which does is the time-centered leap-frog. Let $\mathbf{r}_i^{[k]}$ be the position of the ith particle at the kth time-step, and let $\mathbf{v}_i^{[k+1/2]}$ be its velocity a half-step later. Then one leap-frog step is

$$\begin{aligned} \mathbf{r}_i^{[k+1]} &= \mathbf{r}_i^{[k]} + \Delta t\, \mathbf{v}_i^{[k+1/2]} , \\ \mathbf{v}_i^{[k+3/2]} &= \mathbf{v}_i^{[k+1/2]} + \Delta t\, \mathbf{a}_i^{[k+1]} , \end{aligned} \tag{16}$$

where Δt is the time-step and $\mathbf{a}_i^{[k+1]}$ is the acceleration of the ith particle at step $k+1$. Among other advantages, this method is simple to program and requires little memory. Moreover, *provided* that the time-step Δt is held constant, it is also globally accurate to second order and time-reversible.

A small problem with the leap-frog is the need to offset the position and velocity variables by half a time-step. One somewhat lazy solution is to split the velocity step:

$$\begin{aligned} \mathbf{v}_i^{[k+1/2]} &= \mathbf{v}_i^{[k]} + \frac{\Delta t}{2}\, \mathbf{a}_i^{[k]} , \\ \mathbf{r}_i^{[k+1]} &= \mathbf{r}_i^{[k]} + \Delta t\, \mathbf{v}_i^{[k+1/2]} , \\ \mathbf{v}_i^{[k+1]} &= \mathbf{v}_i^{[k+1/2]} + \frac{\Delta t}{2}\, \mathbf{a}_i^{[k+1]} ; \end{aligned} \tag{17}$$

this is equivalent to (16), since the relationship between $(\mathbf{r}_i^{[k]}, \mathbf{v}_i^{[k+1/2]})$ and $(\mathbf{r}_i^{[k+1]}, \mathbf{v}_i^{[k+3/2]})$ is the same for both. When used as an iterated mapping, (17) is equivalent to *starting* (16) with the linear approximation

$$\mathbf{v}_i^{[1/2]} = \mathbf{v}_i^{[0]} + \frac{\Delta t}{2}\, \mathbf{a}_i^{[0]} . \tag{18}$$

In effect, the resulting numerical solution "jump-starts" from a phase-space point offset in velocity by $O(\Delta t^2)$ from the specified $\mathbf{v}_i^{[0]}$ but is integrated correctly after this initial jump.

The main drawback to the leap-frog is that all bodies are advanced with the same time-step, which must be small enough to follow motions throughout the system. Considerable computation may be wasted if some bodies require much smaller time-steps than others. Relative to the leap-frog, the potential speed-up of an ideal individual time-step algorithm is $S = \max[\Delta t_i^{-1}]/\langle \Delta t_i^{-1} \rangle$, where Δt_i is the time-step for the $i^{\rm th}$ body and the maximum and average are taken over all bodies. In a spherical system composed of equal-mass bodies, each with a time-step proportional to the circular orbit period at its present radius, the speed-up can be calculated from the density profile. For a Plummer (1911) profile, $\rho \propto (r^2 + a^2)^{-5/2}$, the speed-up is only $S \simeq 2$ since most of the mass lies within the constant-density core, while the density profile $\rho \propto (r + a)^{-4}$ (Dehnen 1993) yields $S = 105/16 = 6.5625$. The speed-up is formally infinite for a mass model with a density profile diverging as $r \to 0$, but simulated density profiles are never singular after smoothing by (12). For mass profiles resembling the B/D/H galaxy model described in § 2.6, the speed-up is $S \simeq 7$.

Due to time-step scheduling constraints, a real code might not deliver more than half of the ideal speed-up, but given the relatively gradual improvement in force calculation shown by Fig. 5, it seems worth investigating individual time-step codes (e.g. Saha & Tremaine 1994). However, it's not trivial to preserve the *reversibility* of (17) with a variable time-step scheme, and reversibility implies many nice properties including conservation of phase-space density. Hut, Makino, & McMillan (1995) have proposed symmetrizing the time-step criterion with respect to the endpoints t and $t + \Delta t$, but this leads to an implicit relationship for Δt, which must be solved every time-step at a cost probably exceeding the modest speed-ups obtained above. Quinn et al. (in preparation) have described a reversible variant of the leap-frog integrator in which time-steps are "adjusted" by an operator which depends on the positions but not the velocities of bodies; if schemes of this kind prove practical, real although not enormous gains in the simulation of collisionless N-body systems may result.

2.5 Errors and Relaxation Effects

There are two kinds of uncertainties in N-body simulations of collisionless systems. First, the N-body equations (13) are integrated numerically with less-than-perfect accuracy. Second, even perfectly accurate integrations don't correspond to exact solutions of the collisionless Boltzmann (5) and Poisson (6) equations. Numerical errors are fairly easy to measure and control, but smoothing and discreteness effects have subtle implications for the interpretation of N-body simulations.

It's hard to rigorously test N-body codes because we don't have a wide range of exact solutions to compare with numerical integrations. To verify the correctness of the simulations, one should ideally show that all uncertainties can be made as small as desired, and that the results converge to a unique limit as the calculation is refined.

Numerical Errors. Errors in numerical solutions include approximations introduced by a hierarchical force calculation algorithm, truncation caused by using a finite time-step, and roundoff due to finite computer word-length. All can be

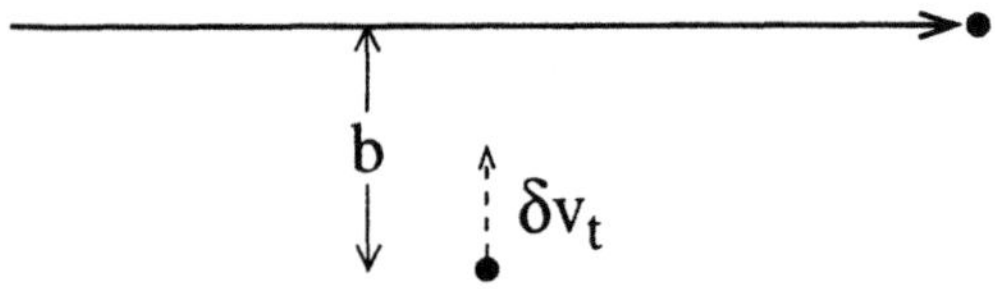

Fig. 6. Fast encounter between a field star and a test star with impact parameter b, seen in the test star's initial frame of reference. After the passage, the test star's velocity is δv_{t}.

treated as small perturbations introduced at every time-step; their cumulative effects can be gauged by monitoring the conservation of energy and momentum, or studied in more detail by running the same set of initial conditions with different time-steps and opening angles. Convergence testing shows that it is generally possible to constrain the uncertainty associated with numerical errors at a "reasonable" computational cost (e.g. Barnes & Hut 1989). But the interpretation of such tests is a tricky matter; a specific suite of convergence tests will be described in § 4.1.

Two-Body Relaxation. The discrete nature of the N-body representation (7) drives a slow evolution which is absent in continuous systems. The logical starting point for a discussion of this process is the theory of two-body relaxation worked out by Chandrasekhar (1942); here I summarize the treatment in Chapter 4 of Binney & Tremaine (1987; hereafter BT87).

A key element of Chandrasekhar's theory is that most two-body interactions are distant. For distant encounters, the impulse approximation, in which field stars move on unperturbed straight-line orbits, adequately describes the transfer of momentum to a test star. Let both stars have mass m, and let them approach each other along the linear relative orbit with velocity at infinity v and impact parameter b shown in Fig. 6. As a result of the encounter, the test star gains a transverse velocity

$$\delta v_{\mathrm{t}} = \frac{2Gm}{bv} \, . \tag{19}$$

The impulse approximation will be good as long as $\delta v_{\mathrm{t}} \ll v$; it breaks down for encounters in which $b \lesssim b_{\mathrm{min}} \equiv Gm/v^2 \simeq R/N$. In an N-body system, only $O(1)$ pair of bodies undergo such a close encounter per crossing time $t_{\mathrm{c}} \simeq (G\bar{\rho})^{-1/2}$, where $\bar{\rho}$ is the average mass density.

A second key element is that the effects of encounters accumulate incoherently, so the test star performs a random walk in velocity. The total deflection per crossing time t_{c} can thus be estimated by adding individual deflections in quadrature:

$$\Delta v_{\mathrm{t}}^2 = \sum \delta v_{\mathrm{t}}^2 \simeq 8N \left(\frac{GM}{Rv}\right)^2 \ln\left(\frac{R}{b_{\mathrm{min}}}\right) , \tag{20}$$

where the summation is over all encounters per crossing with impact parameters between b_{min} and the system radius R. The *relaxation time* is defined as the time over which the cumulative effects of these uncorrelated encounters effectively randomize the initial velocity of the test star; this time is

$$t_{\rm r} \equiv \left(\frac{v^2}{\Delta v_{\rm t}^2}\right) t_{\rm c} \simeq \frac{N}{8 \ln N} t_{\rm c} \,. \tag{21}$$

This theory, devised for star clusters, works just as well for N-body models if (19) is modified to take account of the effective size of field particles (White 1976). For Plummer smoothing (12) one may replace $b_{\rm min}$ by ϵ, in effect ignoring all encounters with impact parameters less than the smoothing length. The relaxation time then becomes

$$t_{\rm r} \simeq \frac{N}{8 \ln(R/\epsilon)} t_{\rm c} \,. \tag{22}$$

If smoothing is implemented in some other way then the evaluation of the relaxation time may be more complicated. In self-consistent field methods, for example, the smoothing depends on position and may impose certain symmetries; this further narrows the set of effective scattering encounters but does not alter the basic *nature* of the relaxation process.

Relaxation effects are driven by fluctuations in the gravitational potential of N-body systems. According to the theory just outlined, potentials measured at a fixed position in an equilibrium system should have a Gaussian distribution about the mean, with amplitudes equal to those obtained by calculating potentials in independent Monte-Carlo realizations of the mass distribution. To test this, I ran N-body simulations of a King (1966) model with a dimensionless central potential $W_0 = 5$ using $N = 4096$, 16384, and 65536 bodies, and measured the potential at 512 test positions distributed uniformly in radius. These potentials were compared with potentials evaluated at the same positions in a set of independently-generated realizations. The upshot is that the fluctuations in the self-consistent N-body models have the same amplitudes as those in the static realizations; in both cases, the amplitude scales like $N^{-1/2}$ as expected from Monte-Carlo statistics. Thus for King models – and presumably, for other highly stable equilibrium systems – Chandrasekhar's theory should accurately describe the relaxation process.

Collective Relaxation. The theory of two-body relaxation is based on the assumption that scattering bodies are *uncorrelated*. In some stellar systems, however, this assumption is known to fail. For example, Weinberg (1993) has discussed relaxation in homogeneous stellar systems with periodic boundary conditions. If the linear scale of the system is much smaller than the Jeans length then the relaxation rate is given by Chandrasekhar's formula, but if the system is only marginally stable to gravitational collapse then relaxation is much more rapid. In effect, fluctuations on scales comparable to the Jeans length are "amplified" by collective effects; the amplitudes of these fluctuations considerably exceed the amplitude of ordinary $\sqrt{N}$ fluctuations and they consequently dominate the evolution of nearly-unstable systems. N-body models of disk galaxies, to be discussed below, exhibit this kind of relaxation.

2.6 Initial Conditions

According to the prescription outlined in § 2.2, bodies can be independently initialized at time t_0 by drawing phase-space coordinates $(\mathbf{r}, \mathbf{v})$ with probability proportional to the distribution function $f(\mathbf{r}, \mathbf{v}, t_0)$. The job of generating $(\mathbf{r}, \mathbf{v})$ points with the specified $f(\mathbf{r}, \mathbf{v})$ is an elementary (if sometimes intricate) programming exercise; details are available in Knuth (1969, vol. II). More problematic is the task of specifying the initial distribution function for a system of interacting galaxies. The general practice since Holmberg (1941) has been to build equilibrium objects and throw them at each other. In general this seems fairly reasonable since many internal crossing times typically elapse before an encounter begins, though it probably fails to do justice to the outer regions of dark halos, which collapse on a time-scale comparable to the galaxies' orbital period.

The job of constructing equilibrium systems resembling real galaxies remains; here there is a considerable body of work, divided into several major categories. First, methods based on Jeans Theorem write $f(\mathbf{r}, \mathbf{v})$ as a function of the specific energy $E \equiv \Phi + \frac{1}{2}v^2$ and other integrals of motion (BT87, Ch. 4.4). Rather general methods of this kind are available for spherical systems with $f = f(E, J)$ (e.g. Dejonghe 1989) and axisymmetric systems with $f = f(E, J_z)$ (e.g. Quian et al. 1995). Second, orbit methods, following Schwarzschild (1979, 1993), represent the distribution function with a discrete collection of orbits. These methods work for triaxial systems and other cases in which no analytic approximation to a third integral is available. Third, methods based on taking velocity moments of the time-independent CBE do not completely specify the distribution function but do constrain parameters such as the dispersion and anisotropy (BT87, Ch. 4.2).

The techniques used in N-body experiments involve all of these categories, along with more ad hoc procedures. Moment methods are often used to construct the disk, and sometimes the bulge and halo as well (e.g. Negroponte & White 1983, Hernquist 1993a); such methods often work fairly well even though they don't yield true equilibria. In addition, N-body codes can be used to obtain approximate equilibria; for example, by adiabatically imposing the potential of a disk on a well-relaxed bulge and halo (e.g. Barnes & White 1984, Barnes 1988, 1992; hereafter B92). Apart from computational expense, the main drawback to this approach is that the outcome can't be easily anticipated; some trial-and-error may be needed to get the desired density profile. Recently introduced are pure Jeans Theorem models of multi-component galaxies, using $f = f_{\mathrm{b}}(E) + f_{\mathrm{h}}(E, J_z) + f_{\mathrm{d}}(E, J_z, E_z)$ where f_{b} describes the bulge, f_{h} the halo, and f_{d} the disk; here the energy of vertical motion E_z is used as an approximate third integral (Kuijken & Dubinski 1995). This general approach seems worth further exploration, although the specific choices for f_{b}, f_{h}, and f_{d} somewhat limit the range of models constructed. Finally, "made-to-measure" N-body systems may be constructed by orbit-based methods (Syer & Tremaine 1996); these may prove quite useful once "particle-splitting" schemes are implemented to reduce the range of assigned particle masses.

B/D/H Galaxy Models. Many of the new simulations presented here were constructed using a fairly simple procedure based on the moment method. An advantage of this method is that all mass distributions are specified in advance. Each model galaxy contains a bulge, described by a Hernquist (1990b) model with scale radius a_{b}, an exponential and isothermal disk, with scale length α^{-1} and constant scale height z_0, and a dark halo, described by a $\gamma = 0$ Dehnen (1993) model with scale radius a_{h}:

$$\rho_{\mathrm{b}} \propto \frac{1}{r(r+a_{\mathrm{b}})^3}, \quad \rho_{\mathrm{d}} \propto \mathrm{e}^{-\alpha R}\,\mathrm{sech}^2\left(\frac{z}{z_0}\right), \quad \rho_{\mathrm{h}} \propto \frac{1}{(r+a_{\mathrm{h}})^4}, \tag{23}$$

where $R \equiv \sqrt{x^2+y^2}$ is the cylindrical radius. All three components are assigned Gaussian velocity distributions, with parameters which are functions of position:

$$f_{\mathrm{b}} \propto \mathrm{e}^{-v^2/2\sigma_{\mathrm{b}}^2}, \quad f_{\mathrm{d}} \propto \mathrm{e}^{-v_{\mathrm{R}}^2/2\sigma_{\mathrm{R}}^2}\,\mathrm{e}^{-v_{\mathrm{z}}^2/2\sigma_{\mathrm{z}}^2}\,\mathrm{e}^{-(v_\phi-v_0)^2/2\sigma_\phi^2}, \quad f_{\mathrm{h}} \propto \mathrm{e}^{-v^2/2\sigma_{\mathrm{h}}^2}, \tag{24}$$

where σ_{b} and σ_{h} are the bulge and halo velocity dispersions, σ_{R}, σ_{z}, and σ_ϕ are disk velocity dispersions in the cylindrical, vertical, and azimuthal directions, respectively, and v_0 is the net rotation velocity of the disk.

The dispersion and velocity functions are computed by a modified version of the method outlined by Hernquist (1993a). First, since the disk is vertically self-gravitating, the equilibrium model for an isothermal slab (Spitzer 1942) gives $\sigma_{\mathrm{z}}^2(R) = \pi G \Sigma_{\mathrm{d}}(R) z_0$, where $\Sigma_{\mathrm{d}}(R)$ is the total surface density of the disk at radius R. By fiat, the radial dispersion is set to be twice the vertical dispersion, as it is locally in the Milky Way; thus $\sigma_{\mathrm{R}}^2 = 4\sigma_{\mathrm{z}}^2$. Taking the $v_{\mathrm{R}}v_\phi$ moment of the time-independent CBE relates the radial and azimuthal dispersions by $\sigma_\phi^2 = -[B/(A-B)]\sigma_{\mathrm{R}}^2$, where A and B are Oort's disk rotation parameters. Taking the v_{R} moment instead gives a relation between the net rotation velocity v_0 and the circular velocity v_{c} of the general form $(v_{\mathrm{c}} - v_0) \propto \sigma_{\mathrm{R}}^2$, analogous to the asymmetric drift seen in the Milky Way's disk. Finally, the dispersions σ_{b} and σ_{h} of the spherical bulge and halo are computed by neglecting the flattening of the disk potential and taking the v_{r} moment to derive spherical equations of "stellar hydrostatics". All of these moment equations appear in Ch. 4.2.1 of BT87.

Several further tricks are used in building N-body realizations of these models. The power-law tails of the bulge and halo are smoothly truncated at large r to avoid placing a small number of bodies at extremely large radii. The smoothing operation described by (12) is incorporated when calculating the gravitational potential of the bulge component; with some care, the simulated bulge can be made to follow a r^{-1} profile down to $r \sim 0.1\epsilon$. The bulge and halo velocity distributions are replaced by truncated Gaussians, with the same $\langle v^2 \rangle$ as the untruncated versions, to avoid generating unbound bodies,

For future reference, the parameters of these B/D/H models will be given in an arbitrary system of units with $G \equiv 1$. The bulge has mass $M_{\mathrm{b}} = 0.0625$ and scale radius $a_{\mathrm{b}} = 0.04168$; the disk has mass $M_{\mathrm{d}} = 0.1875$, scale length $\alpha^{-1} = \frac{1}{12}$, and scale height $z_0 = 0.007$; the halo has mass $M_{\mathrm{h}} = 1$ and scale radius $a_{\mathrm{h}} = 0.1$. Thus the dark halo is 80% of the total mass, the disk is 15%, and the bulge is 5%. A circular orbit at a radius $R = 3/\alpha$ has a period ~ 1 in these arbitrary simulation units. To roughly scale these models to the Milky Way, the simulation units of length, mass, and time can be equated to 40 kpc, $2.2 \cdot 10^{11}\,\mathrm{M}_\odot$, and 250 Myr, respectively.

Numerical Tests. In view of the litany of approximations just described, tests of B/D/H models are clearly required before they can be used for encounter simulations. The tensor virial theorem (BT87, Ch. 4.3) follows from the moment equations, so it should be obeyed exactly by their solution, and to $O(N^{-1/2})$ by N-body realizations generated from this solution; this provides a convenient check when debugging the routines used to generate initial conditions. But even complete compliance with the moment equations is not enough to assure that a system is actually close to equilibrium. This point can be checked by self-consistent N-body integrations.

Moreover, such tests are useful because it is generally difficult to assess the dynamical stability of disk galaxy models by any other means. The *local* response of a disk to an imposed perturbation depends on the velocity dispersion σ_R; a razor-thin stellar disk is locally stable if

$$Q \equiv \frac{\sigma_\mathrm{R}\kappa}{3.36 G\Sigma} > 1\,, \tag{25}$$

where κ is the epicyclic frequency and Σ is the surface density (Toomre 1964). The *global* response depends on the presence of a bulge or halo, the shape of the rotation curve, and the value of Q. Numerical work shows that pure disk systems with $Q \lesssim 2.5$ spontaneously turn into into tumbling bars after only a few rotations (Hohl 1971, Sellwood 1981, Efstathiou, Lake, & Negroponte 1982, Athanassoula & Sellwood 1986); the reality of this instability is supported by linear mode calculations (Zang 1976, Kalnajs 1978). Stability can be improved by embedding a disk in a massive dark halo (Ostriker & Peebles 1973), or by adding enough mass to the center to create an inner Lindblad resonance (Toomre 1981). The B/D/H models described above have massive dark halos, moderately cuspy bulges, and disks with $Q \gtrsim 1.5$; they should be stable, but testing is still appropriate.

The test presented here used a total of $N = 65536$ bodies: 8192 for the bulge, 24576 for the disk, and 32768 for the dark halo; the latter have four times the mass of the luminous bodies. Plummer smoothing with $\epsilon = 0.01$ length units was included in constructing the initial conditions and subsequently in calculating forces. Bodies were advanced by a leap-frog algorithm with a fixed time-step of $\Delta t = 1/128$ time units[1]. A run of 5 time units conserved energy to 0.2%.

Fig. 7 shows some results of this calculation. On the whole the appearance of the model changes little over the course of the run, suggesting that the initial conditions are close to equilibrium. In particular, this simulation does not exhibit

[1] The consistent use of powers-of-two for masses, particle numbers, and time-steps is deliberate; powers-of-two, along with their sums and products, can often be represented *exactly* as binary floating-point numbers.

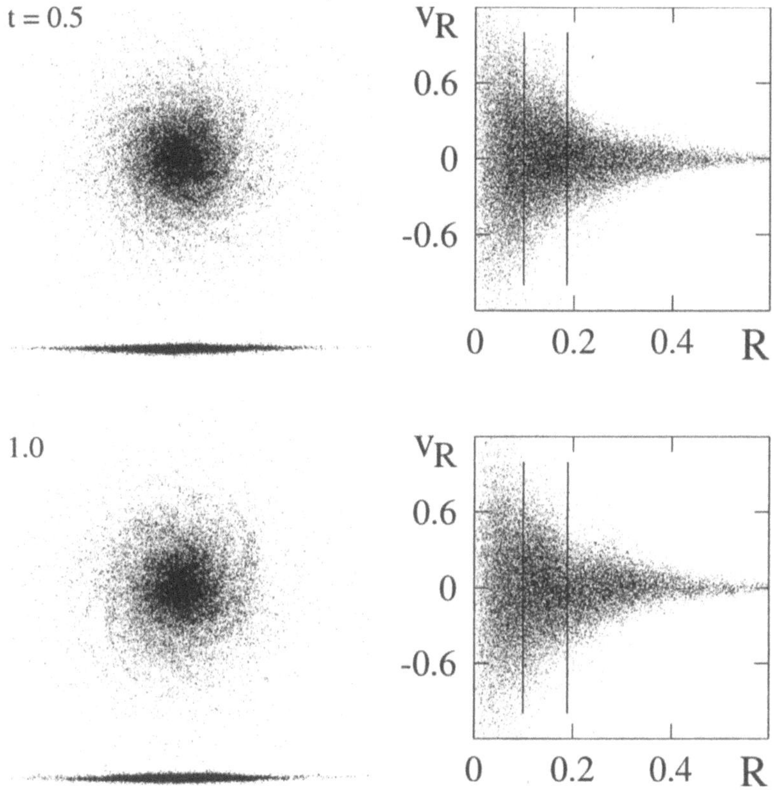

Fig. 7. Tests of a B/D/H galaxy model. The disk is viewed face on, edge on, and projected on the (R, v_R) plane at the times indicated; vertical lines divide the disk into equal thirds.

the outward-moving overdense rings mentioned by Kuijken & Dubinski (1995), though the handful of particles outside the envelope in the (R, v_R) plot at $t = 0.5$ show that the initial equilibrium was not perfect. But much of the evolution seen in this test would occur even if the initial conditions were as perfect as a Monte-Carlo N-body representation allows. This is because differentially-rotating disks contain a "swing-amplifier" which can significantly boost the amplitude of fluctuations (e.g. Toomre 1981). In disk simulations, discreteness provides a ready source of fluctuations of all shapes and sizes. Those fluctuations which initially have the form of leading spirals can be swing-amplified by an order of magnitude as they are sheared into trailing spirals. As a result, self-consistent disk simulations rapidly develop multi-armed trailing spiral patterns like those seen here.

This disk heats up due to collective relaxation as bodies scatter off the gravitational field of these short-lived spirals. Fig. 8 shows the resulting growth of

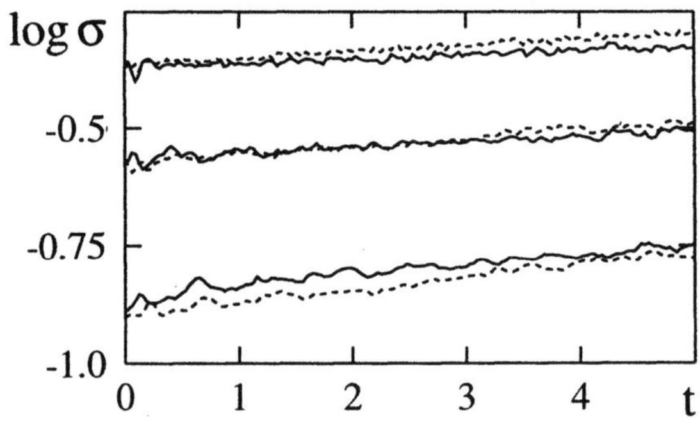

Fig. 8. Growth of velocity dispersions in a B/D/H galaxy simulation. From top to bottom, results are shown separately for the inner, middle, and outer thirds of the disk. Solid curves show $\langle v_R^2 \rangle^{1/2}$, while dotted curves show $2\langle v_z^2 \rangle^{1/2}$.

the disk's radial and vertical velocity dispersions $\langle v_R^2 \rangle^{1/2}$ and $\langle v_z^2 \rangle^{1/2}$; dispersions are plotted separately for the inner, middle, and outer thirds of the disk. The vertical dispersions have been multiplied by a factor of 2 in this plot so they may be more easily compared with the radial dispersions. Over ~ 5 rotation periods both radial and vertical dispersions grow by $\sim 30\%$. One may wonder why the *vertical* dispersion increases when a spiral pattern seems more likely to scatter stars in the disk plane. Part of the answer may be that the activity is not strictly confined to the initial plane of the disk; edge-on views at later times show that the disk develops transient corrugations.

3 Numerical Gas Dynamics

Diffuse interstellar material, in all its forms, probably amounts to $\sim$ 10% of the *stellar* mass of most galaxies, and an even smaller fraction of the total mass; one is tempted to view it as a minor perturbation on the underlying collisionless dynamics of stars and dark matter. But in regions of rapid star formation the fraction of interstellar material may be much greater, and even where diffuse matter does not dominate it still tends to play a large role in galactic interactions. Thus it is crucial to include the dynamics of this material in descriptions of interacting galaxies.

3.1 A Sketch of the Interstellar Medium

The interstellar medium (ISM) of our Galaxy has a number of distinct forms or *phases* with different atomic or molecular states, temperatures, spatial distributions, and dynamics (Spitzer 1978; McKee & Ostriker 1977). In our galactic neighborhood, roughly a third of the hydrogen has formed H_2 molecules within the coldest dense clouds ($T \sim 15\,\mathrm{K}$). Perhaps half of the hydrogen is in the form of single atoms which are either clumped into cold clouds ($T \sim 100\,\mathrm{K}$) or spread out in warm diffuse gas ($T \sim 8000\,\mathrm{K}$). This diffuse component is intermixed with a comparable amount of ionized hydrogen at similar temperatures. Finally, a small fraction of the ionized material has much higher temperatures ($T \sim 10^5$ to $10^7\,\mathrm{K}$).

With the important exception of the molecular clouds, the various phases of the ISM are in rough pressure equilibrium. This implies that the density of a phase is inversely proportional to its temperature, and indeed we find order-of-magnitude particle densities of $\sim 100\,\mathrm{cm}^{-3}$ in the cold atomic clouds, $\sim 1\,\mathrm{cm}^{-3}$ in the diffuse atomic and ionized gas, and $\sim 10^{-3}\,\mathrm{cm}^{-3}$ in the very hot plasma. Dense clouds, though they contain an appreciable fraction of the mass, occupy only a few percent of the volume, while the very hot gas, though containing only a few percent of the total mass, may occupy over half the volume.

The molecular clouds have densities which range upward from 100 to as much as $10^8\,\mathrm{cm}^{-3}$. These clouds are often found in giant cloud complexes with total masses of up to $10^7\,\mathrm{M}_\odot$. Unlike the cold atomic clouds, which would rapidly expand if not for the external pressure of the diffuse gas, molecular clouds appear to be confined by self-gravity.

In addition to various forms of atomic and molecular gas, the ISM includes magnetic fields and cosmic rays. Energy densities in these components are of order $1\,\mathrm{eV\,cm}^{-3}$, comparable to the thermal energy density of the gas; this strongly suggests that the various components of the ISM are in energy equipartition.

The ISM's makeup changes from place to place. Molecular and atomic clouds fall off rapidly with distance from the galactic plane, while the very hot gas forms an extended corona with a scale height of several kpc. Towards the center of the galaxy, the total fraction of molecular material dramatically increases. Even within the relatively tame environment of the Milky Way, the processes which transform one phase into another are incompletely understood.

Shocks. In galactic collisions the ISM will be subjected to various kinds of shocks. Shock speeds of $\sim 10^3\,\mathrm{km\,s^{-1}}$ occur during interpenetrating encounters of gas-rich spiral galaxies, as proposed by Spitzer & Baade (1951). Somewhat less extreme shocks form during tidal interactions; tidal bar-making creates flows which self-intersect with a fraction of the disk's circular velocity, while in tidal spirals such as M 51 and its "co-conspirator" M 81, streaming motions of $\sim 20\,\mathrm{km\,s^{-1}}$ are observed on the opposite sides of spiral shocks (e.g. Rots & Shane 1975, Rots et al. 1990).

A description of shock propagation in even a single phase of the ISM is quite complicated (e.g. Draine & McKee 1993). The gas can be treated as a magnetohydrodynamical (MHD) fluid, and jump conditions across the shock front may be derived from conservation laws. Most phases of the ISM are dense enough to cool rapidly after the shock passes; in such radiative shocks, isothermal jump conditions imply that the density of the post-shock gas increases by up to $\mathrm{O}(M^2)$, where M is the shock's Mach number. But astrophysical MHD shocks may also produce cosmic rays by first order Fermi acceleration as high-energy protons are repeatedly scattered by embedded magnetic field in compressive flows. The rough equality of the thermal and cosmic-ray energy densities in the ISM suggests that cosmic ray production is efficient at extracting energy from shocks; detailed estimates of cosmic-ray production by supernovae concur. Enhanced continuum radio emission in some interacting galaxies may be due to particles accelerated by this mechanism (e.g. van der Hulst & Hummel 1985, Condon et al. 1993).

Shocks in a multi-phase ISM are more complicated. The hot and warm phases, which together fill most of the space, create extended shock fronts when they collide – though only the fastest collisions can shock gas which is already at $T \sim 10^6\,\mathrm{K}$. After the shock passes the initially warm gas may fragment into clouds of cold atomic gas. Pre-existing cold clouds can burst right through the shock front with little loss of momentum; molecular clouds, in particular, move almost ballistically, with mean free paths of many kpc in the solar neighborhood.

3.2 Simulating the ISM

Any attempt to include interstellar material in galaxy simulations must forgo the rigor of purely stellar-dynamical models and adopt more phenomenological descriptions. A few galaxy collisions have been simulated using finite-difference gas codes (e.g. Mair et al. 1988) but the enormous range of length scales involved make such calculations difficult. In simulating systems containing both gas *and* collisionless material (stars and dark matter) it is natural to adopt particle-based representations for all components.

Sticky Particles. One approach is to modify an N-body code to include the general effects of gas dissipation in galactic systems; for example, in "sticky particle" calculations (e.g. Negroponte & White 1983, Combes & Gerin 1985, Noguchi 1988) the ISM is represented by finite-sized particles which collide with some loss of energy but travel ballistically between collisions. This approach captures something of the clumpy structure of the ISM in our galaxy, although it may be mistaken to identify gas particles too closely with interstellar clouds since the latter may form and dissolve on a time-scale comparable to a galactic crossing time.

Smoothed Particles. Another approach is to model the ISM as a compressible gas, obeying the equations

$$\begin{aligned}
&\frac{\partial \rho}{\partial t} + \nabla \cdot (\rho \mathbf{v}) = 0\,, \\
&\frac{\partial \mathbf{v}}{\partial t} + (\mathbf{v} \cdot \nabla)\mathbf{v} = -\nabla \Phi - \frac{1}{\rho}\nabla P\,, \\
&\frac{\partial u}{\partial t} + (\mathbf{v} \cdot \nabla)u = -\frac{P}{\rho}\nabla \cdot \mathbf{v} + \dot{u}
\end{aligned} \tag{26}$$

where ρ is the gas density, $\mathbf{v}$ is its velocity field, and u is its specific internal energy, and the pressure is given by the equation of state, $P = P(\rho, u)$. The quantity $\dot{u}$ is included to describe nonadiabatic effects such as cosmic-ray heating and radiative cooling. This model is hardly a comprehensive treatment of the ISM, though it may serve as a reasonable starting point for a description of the more diffuse phases of warm and hot gas.

By analogy with the method already used for the CBE, these equations can be solved by a Monte-Carlo-like technique known as "smoothed particle hydrodynamics" or SPH (Lucy 1977, Gingold & Monaghan 1977). SPH uses an interpolation procedure used to estimate continuous fluid variables from the particulate representation. For example, the estimated gas density is

$$\rho(\mathbf{r}) \simeq \sum_{i}^{N_{\rm sph}} m_i\, W(\mathbf{r} - \mathbf{r}_i, h)\,, \tag{27}$$

where $N_{\rm sph}$ is the number of particles used to represent the gas, and $W(\Delta\mathbf{r}, h)$ is a "smoothing kernel" with smoothing length h. This smoothing procedure is very similar to the smoothing operation (12), though a kernel with compact support is preferred for SPH calculations. The gas velocity $\mathbf{v}(\mathbf{r})$ and internal energy $u(\mathbf{r})$ may likewise estimated by expressions involving the velocities $\mathbf{v}_i$ and internal energies u_i of individual particles. In addition, gradients of these variables become expressions involving the derivative of the smoothing kernel.

A number of formulations of SPH have been developed (e.g. Gingold & Monaghan 1982, Benz 1990, Monaghan 1992, Nelson & Papaloizou 1993). The version used for the calculations presented in these lectures has been fully discussed elsewhere (Hernquist & Katz 1989, Barnes & Hernquist 1996; hereafter BH96), so only a schematic description will be given here. The equation of motion for gas particles is similar to (13), but contains two additional terms:

$$\frac{\mathrm{d}\mathbf{v}_i}{\mathrm{d}t} = (-\nabla \Phi)_i + \left(-\frac{1}{\rho}\nabla P\right)_i + \mathbf{a}_i^{\rm visc}\,. \tag{28}$$

Here the first term is the gravity on particle i, evaluated by an expression like the RHS of (13), the second term is an SPH estimate of the pressure, and the third term is the acceleration due to artificial viscosity. The internal energy of each particle is governed by

$$\frac{\mathrm{d}u_i}{\mathrm{d}t} = \left(-\frac{P}{\rho} \nabla \cdot \mathbf{v} \right)_i + \dot{u}_i + \dot{u}_i^{\mathrm{visc}} . \tag{29}$$

Here the first term represents $P\mathrm{d}V$ work on the gas, the second term includes nonadiabatic effects, and the third term represents energy deducted by artificial viscosity from convergent flows.

Artificial viscosity terms are needed because solutions of (26) develop shock discontinuities, whereas the SPH interpolation procedure aims to reproduce continuous functions. Artificial viscosity does the same job here as in older finite-difference codes, broadening shocks so that they can be resolved by the SPH interpolation. While it's not as precise as shock-capturing methods (e.g. Colella & Woodward 1984), numerical tests show that SPH can reproduce the jump conditions across shocks with considerable fidelity (e.g. Hernquist & Katz 1989, Steinmetz & Müller 1993). Most of the calculations described here used the artificial viscosity suggested by Monaghan & Gingold (1983), which strongly damps post-shock oscillations; a few experiments were run with a combination of bulk and von Neumann-Richtmyer viscosity to check that key results are not sensitive to the form of viscosity used.

Besides $P\mathrm{d}V$ work and effects of artificial viscosity, the gas may be cooled by radiative processes such as line and free-free emission and heated by cosmic rays and other radiation. These effects are included in the expression for $\dot{u}_i$ (BH96). At the typical gas densities found disk galaxies, cooling is rapid, and in simulations run with a standard cooling curve the gas cools and fragments into small lumps. Though this fragmentation process is physically plausible, it is computationally undesirable because it ruins the smooth representation of the gas and forces the integrator to take very short time-steps. Fragmentation can be prevented by suppressing radiative cooling below temperatures of 10^4 K (e.g. Hernquist & Katz 1989). Thus the gas in these simulations most closely approximates the warm phase of the ISM.

For the smoothing procedure to provide a reasonable approximation to the continuous gas density, the number of points contributing to the sum in (27) should not be too large or too small; thus SPH is often generalized to give each particle a different smoothing length h_i (e.g. Evrard 1988, Hernquist & Katz 1989, Monaghan 1992). This generalization is implemented so as to insure that forces between particles are properly antisymmetric (Hernquist & Katz 1989), but it neglects terms involving derivatives of h. Recent tests show that the neglect of these terms may introduce errors of $\sim 10\%$ in the internal energy or entropy (Hernquist 1993b). This is probably not a serious problem for simulations of galaxy collisions because cooling completely dominates the thermodynamic evolution of the gas. Nelson & Papaloizou (1993, 1994) have shown how to include the missing terms in a Hamiltonian formulation of SPH; their approach seems worth following in future implementations.

The Courant condition implies that the time-step Δt must be less than $\sim h/c_s$, where c_s is the sound speed. This constraint cannot easily be accommodated with a fixed time-step scheme like (17) because the maximum allowed Δt may vary by several orders of magnitude over the course of a simulation. Therefore, SPH codes use adaptive time-step schemes to integrate the trajectories of gas particles (e.g. Hernquist & Katz 1989, Monaghan 1992, Navarro & White 1993, Steinmetz 1996); the local time-step is set by a modified Courant condition which includes effects of viscosity (Hernquist & Katz 1989). The energy equation (29) must be solved implicitly if radiative cooling is included since cooling times are typically *much* shorter than the adopted Δt.

In practice, sticky particle, smoothed particle, and continuum methods often give similar results (e.g. Athanassoula 1994). This convergence comes about because of rather simple considerations. Pressure forces don't play a direct role on the large-scale dynamics of the ISM; both smoothed and sticky particles follow free-fall trajectories when not involved in shocks. When particles do get into shocks, they rapidly lose their kinetic energy while averaging their momentum with their neighbors. Thus both techniques basically reduce to momentum-conserving schemes under the conditions found in galactic simulations. Nonetheless, SPH provides somewhat more scope for physical reasoning about the simulations, since it converges to the ideal gas equations in the $N \to \infty$ limit.

3.3 Gas in B/D/H Models

Using the SPH methodology just described, Lars Hernquist and I have constructed self-consistent N-body/SPH models of gas-rich disk galaxies (Barnes & Hernquist 1991, 1992b, BH96, Hernquist & Barnes 1991). The model galaxies in these simulations differ from those described in § 2.6 in using King (1966) models with dimensionless central potentials $W_0 = 3$ for both the bulge and halo. N-body codes were used to relax the bulge and halo in the presence of the disk, following the procedure described in B92. These bulges have constant-density cores instead of power-law cusps, and cut off sharply at a fairly small radius. The halos are also contained within a finite radius instead of falling off as r^{-4} at large radii. Following Hernquist (1989), the gas is distributed with the same radial and vertical density profile as the disk stars. The bulge:disk:halo mass ratio is 1:3:16, and the gas amounts to 10% of the disk mass.

Fig. 9 presents separate views of the gas and stars in an isolated galaxy model. This simulation used 8192 particles to represent the gas, 4096 for the bulge, 16384 for the stellar disk, and another 16384 for the halo. Forces were calculated after smoothing using a spline kernel (Hernquist & Katz 1989) with smoothing length $\varepsilon = 0.015$. The force calculation accuracy and integration time-step were chosen such that total energy and angular momentum varied by approximately 0.02% and 0.1%, respectively, over the course of the simulation. Given other uncertainties – in particular, those due to effects of finite particle number – such inaccuracies are not likely to have significant consequences for these experiments. The spiral features in this model are more vivid than those in Fig. 7 because the smaller N used here increases the level of particle discreteness and because the

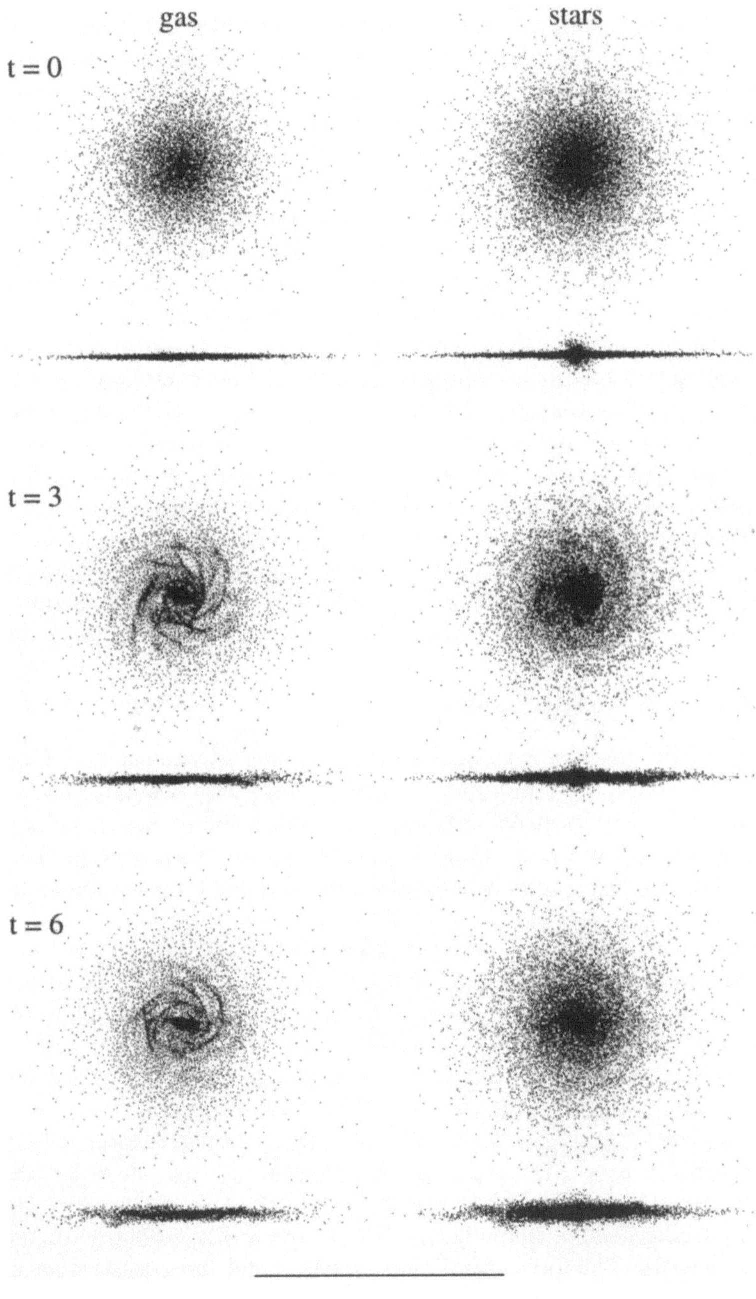

Fig. 9. Long-term evolution of a disk galaxy with gas (BH96). On the left is the gas, shown both face-on and edge-on, while on the right are similar views of the disk and bulge particles. Times since the start of the simulation are indicated, and the scale bar at the bottom is 1 unit long.

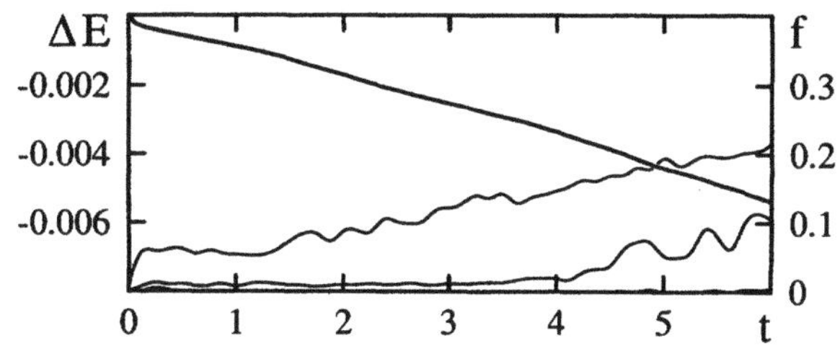

Fig. 10. Energy dissipation and density evolution for the isolated disk model shown in Fig. 9. The descending curve shows the change in binding energy due to dissipation; radiative losses per dynamical time amount to only one part in $\sim 10^3$ of the total binding energy. Ascending curves show gas fractions above densities of $\rho = 10^2$ and $10^{2.5}$ model units; these slowly increase as a result of dissipative evolution (BH96).

dynamically cold gas responds strongly to the gravitational perturbations of the swing-amplified spiral patterns.

Fig. 10 illustrates the slow evolution of this model. The descending line shows the net change in binding energy due to radiative losses, while the two rising curves represent gas fractions above densities of 10^2 and $10^{2.5}$, with one density unit corresponding to $0.0034\,\mathrm{M}_\odot\,\mathrm{pc}^{-3} \simeq 0.1\,\mathrm{cm}^{-3}$. There is an initial transient lasting about 0.1 time units, during which the gas collapses vertically to a scale height less than half that of the stars. After this the system dissipates very slowly, losing $\sim 0.1\%$ of its binding energy per time unit and gradually increasing its gas fraction at $\rho > 10^2$. At later times ($t \gtrsim 4.0$) the disk develops a weak bar, which results in a slightly higher dissipation rate and an increasing gas fraction at $\rho > 10^{2.5}$. Nonetheless, these changes are *much* smaller than the changes produced in collisions, implying that these initial disks are stable enough to be used in collision simulations.

4 Tidal Interactions

Bridges and tails can be understood as temporary kinematic features extracted from cold, rotating disks of stars by tidal forces during brief, roughly parabolic passages. A handful of elementary examples concocted by TT72, all involving in-plane parabolic passages, suffice to illustrate this explanation. But the amazing variety of peculiar galaxies described by Zwicky (1956, 1959), Vorontosov-Vel'yaminov (1958), and Arp (1966) has inspired numerical experimenters from the Toomres on to explore the parameter space of galactic encounters.

There are several reasons to emphasize self-consistent studies of tidal encounters. First is to confirm that self-gravity does not ruin the results of restricted three-body work, and second is to examine effects of tides on self-gravitating disks; as this chapter shows, self-gravity increases the diversity of tidal forms without contradicting TT72's basic message. A third reason is to build self-consistent models of real interacting galaxies, matching morphological and kinematic data to constrain the parameters and evolution of these systems; some examples are described in § 5. And a fourth reason is to study consequences of orbit evolution in self-consistent models, notably those including dark halos; these consequences are addressed in § 6 and 7.

4.1 Test-Particle Studies: Bridges and Tails

If the two galaxies are initially distinct yet isolated from their surroundings, they will – until they come close together – follow roughly Keplerian trajectories. Four parameters define their relative orbit: the shape, given by the eccentricity e, which is $e = 0$ for a circle, $e < 1$ for an ellipse, $e = 1$ for a parabola, and $e > 1$ for a hyperbola; the galaxy mass ratio $\mu \equiv M_2/M_1 \leq 1$; the pericentric separation $r_{\rm p}$ at closest approach along the initial Keplerian orbit; and the time $t_{\rm p}$ of the next (and first, if $e \geq 1$) such approach.

And if each galaxy starts out as an axisymmetric disk, two angles suffice to define its orientation with respect to the relative orbit, as shown in Fig. 11. Following TT72, I use the inclination angle i between the spin and orbit planes, ranging from $i = 0°$ for a direct, in-plane passage through $i = 90°$ for an over-the-pole passage to $i = 180°$ for a retrograde, in-plane passage. The other angle is the pericentric argument ω, measured in the orbit plane from the line of nodes to the separation vector at pericenter; the meaningful range of ω is $\pm 90°$.

As part of their study of tidal interactions, TT72 describe a survey of three-dimensional encounters more comprehensive than any self-consistent survey in the next quarter-century! The Toomres could do such a thorough job because the test-particle approach allows two important simplifications. First, the response of one disk does not depend on the other disk, so models can be run independently and superimposed after the fact, as TT72 noted with their Figs. 4 and 5. Second, the pericentric separation $r_{\rm p}$ can be effectively varied by either plotting or not plotting particles from successive rings in the initial disks; TT72 typically plotted particles initially within $0.6 r_{\rm p}$, but sometimes included particles out to $0.7 r_{\rm p}$ to show the effects of closer encounters.

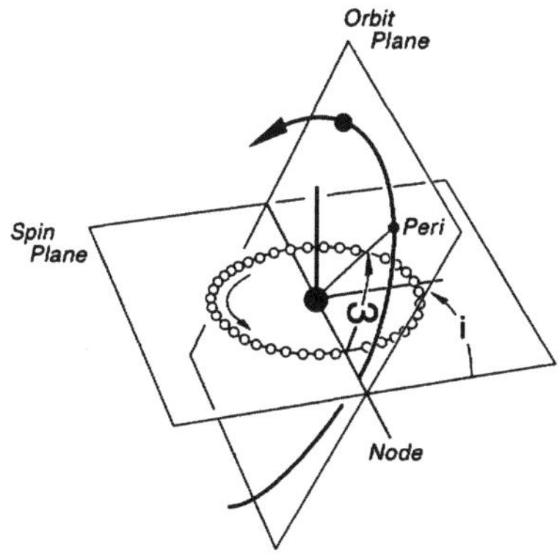

Fig. 11. Definition of the angles i and ω (TT72).

To survey bridge-making, TT72 narrowed the parameter space by focusing on parabolic passages ($e = 1$) of quarter-mass companions ($\mu = \frac{1}{4}$). They plotted the resulting configurations when the companion had traveled through exactly 90° since pericenter. This left only two parameters, the inclination i and pericentric argument ω, to describe each encounter. As i increases, the very open spiral pattern formed by the bridge and counter-arm (illustrated in Fig. 2, which has $i = 0°$) becomes less and less dramatic, and for $i > 90°$, a disk populated out to $0.6r_{\rm p}$ is only a bit distorted. The effects of varying ω proved more subtle; in highly inclined ($i \simeq 90 \pm 30°$) encounters, $\omega = 0°$ places the companion in the disk plane at pericenter, while $\omega = \pm 90°$ places it further from the disk at that instant, so it's "no great surprise" that tidal effects in high-inclination encounters are more striking for small $|\omega|$.

Viewed from directions other than face-on to the victim, these bridge-making simulations produced a wide range of shapes. Such tilted views show that only low-inclination ($i \lesssim 45°$) encounters yield true bridges, in which some particles are captured by the companion. For higher i the bridge does not actually connect, and the relative position of the bridge and companion are quite sensitive to viewing angle. And especially in high-inclination encounters, the bridge particles define thin, gently-curving ribbons which can appear very straight and narrow in projection.

To survey tail-making, TT72 focused on equal-mass ($\mu = 1$) passages while varying the eccentricity and viewing time. Proper tails escaping to infinity formed in parabolic passages with inclinations $i \lesssim 60°$ of disks extending to $0.6r_{\rm p}$, while higher inclinations produced tail-like counter-arms. Like bridges, tails are thin, ribbon-like structures, and perhaps even more than bridges, tails remain close to the spin planes of their parent disks. Tail morphology changes little with eccentricity for $0.6 \leq e \leq 1$, though slower passages increase tail length. Slow, highly-inclined passages, such as the $e = 0.6$, $i = 75°$ case shown in TT72's Fig. 17, produce striking tails but only modest bridges; such passages are promising building blocks for models of long-tailed systems like NGC 4038/9.

4.2 Self-Consistent Studies

A proper N-body simulation of a galactic encounter requires about a *million* times more floating-point operations than a TT72-style test-particle calculation. Even given the hyperinflation of computing power since the early 1970s, such experiments are expensive, a fact which influences the design of self-consistent encounter surveys. Many simplifications available with test-particle calculations are not valid here; the orientations of *both* disks, and their radii in terms of r_p, all must be specified beforehand. Consequently, self-consistent surveys are limited to just a few experiments, and each must count for as much as possible. But the rich dynamical behavior revealed by such studies makes their undertaking worthwhile (Gerhard 1981, Farouki & Shapiro 1981, Quinn 1982, Negroponte & White 1983, Barnes 1988, B92, Hernquist 1992, 1993c). A new self-consistent encounter survey is described below.

Survey Design. Since two galaxies are involved, two sets of angles must be given, and thus eight parameters are needed to describe the state of a pair of interacting disk galaxies: four for the relative orbit, and four for the initial orientations of the participants. The effect of varying the instant of pericenter t_p is obtained by examining a single simulation at various times, but seven parameters remain to be considered, and possibly varied from one simulation to the next. This survey focuses on initially *parabolic* encounters ($e = 1$). Because wide encounters are more expensive, these simulations employ close passages; in the units for B/D/H galaxies (see § 2.6), the cases presented here will have $r_p = 0.2$ length units ($2.4\alpha^{-1}$), although these will be compared with a small set of wider encounters from B92. But unlike that earlier study, this one will include encounters with a three-to-one mass ratio ($\mu = \frac{1}{3}$) as well as equal-mass collisions ($\mu = 1$).

Still to be chosen are the orientations of the two disks. The spin vector of each disk can lie anywhere on the surface of a sphere. Inscribe in this sphere a regular tetrahedron; its four vertices are as far apart as any four points on the sphere can get. These four points define four options for the spin vector for one disk, while the four possible spin vectors of the other disk are derived from the dual of the first tetrahedron, as shown in Fig. 12. This construction generates a set of 16 distinct pairs which span – very coarsely – the space of disk orientations.

This set is not minimal; the set of 12 distinct pairs obtained using the same tetrahedron for both disks is 25% smaller, but that set seemed too circumscribed to give much feeling for the range of possibilities. Nor is it unique, because the orientation of the first tetrahedron can be freely chosen. I aligned one vertex with the orbital axis, and placed another in the plane containing the orbital axis and the separation vector at pericenter, giving as options for the first disk the direct passage $i = 0°$ and three inclined passages with $i \simeq 71°$ and $\omega = -30°, 30°, 90°$. (Alar Toomre pointed out that it might have been better to chose $\omega = -60°, 0°, 60°$ since $\omega = 0°$ is most easily visualized, but neither choice is a better covering of the sphere.)

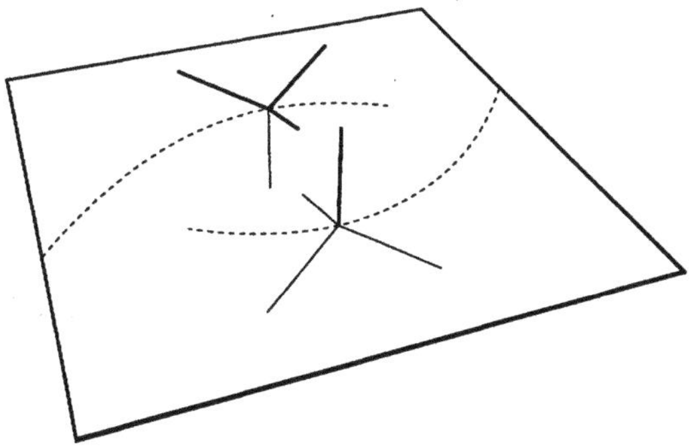

Fig. 12. Possible spin vectors for a modest survey of galactic encounters. Each galaxy can have one of four orientations.

Table 1 lists the new survey of close parabolic encounters. There are eight equal-mass encounters, covering half of the space of disk orientations; this sample is 50% complete. Encounters A–D, including two cases (B and C) with co-planar but counter-spinning disks, bring each possible disk orientation into play once. They are similar to encounters 1–4 and 5–8 of B92: specifically, encounters 1–4 used the same orbits and disk angles as encounters A–D, but employed King (1966) model bulges and halos instead of the present Dehnen (1993) models, while encounters 5–8 were wider analogs, with pericentric separation $r_{\rm p} = 0.4$, again using the older galaxy models. The four new encounters E–H favor skew pairings over co-planar ones. Also new are the eight three-to-one encounters. These are variations on encounters A–D, obtained by allowing either galaxy to be the more massive – a 25%-complete sample of 2×16 possibilities. The smaller galaxy model in these $\mu = \frac{1}{3}$ encounters is scaled by factors of $\mu^{1/2}$ in radius and $\mu^{1/4}$ in circular velocity, thus obeying a Tully & Fisher (1977) relationship with constant M/L ratio.

Initial conditions for all of of the new encounters were realized with $N = 131072$ bodies each. Equal-mass encounters used 65536 bodies per galaxy, apportioned between bulge, disk, and halo as described in § 2.6. In the 3:1 encounters the number of bodies per galaxy was proportional to galaxy mass. The calculations were run with the new "geometric" tree code mentioned in § 2.3; accelerations calculated with this code have relative errors of $\sim 0.25\%$. Following tests with isolated galaxies, I adopted a smoothing length of $\epsilon = 0.01$ length

Table 1. Parameters for self-consistent encounter survey

1:1 encounters										*3:1 encounters*									
ID	i_1	ω_1	i_2	ω_2	ID	i_1	ω_1	i_2	ω_2	ID	i_1	ω_1	i_2	ω_2	ID	i_1	ω_1	i_2	ω_2
A	0		71	30	E	0		71	90	A_1	0		71	30	A_2	71	30	0	
B	-109	90	71	90	F	-109	-30	71	30	B_1	-109	90	71	90	B_2	71	90	-109	90
C	-109	-30	71	-30	G	-109	30	71	-30	C_1	-109	-30	71	-30	C_2	71	-30	-109	-30
D	-109	30	180		H	-109	90	180		D_1	-109	30	180		D_2	180		-109	30

units and a leap-frog time-step of $\Delta t = 1/128$ time units. This represents a compromise guided by the goals of the experiment. A smaller ϵ would have provided better small scale dynamics at the cost of a smaller time-step *and* more bodies to maintain accuracy and smoothness; this cost seemed too high. As it was, each time step took ~ 5.4 minutes on a single IBM SP2 processor. To run the entire sample of sixteen encounters for $t = 9$ time units (1152 time steps) required ~ 70 days of processor time; typically eight encounters were run at once, each on one node.

Several more experiments were run as far as time $t = 4.5$ to assess numerical errors and sampling effects. Two experiments started with exactly the same initial conditions as encounter A: run A_Δ used a time-step $\Delta t = 1/256$, or half the standard value; run A_θ used an opening angle $\theta = 0.7$, doubling the force calculation accuracy. Encounter $\mathrm{A_N}$, in contrast, used twice as many bodies ($N = 262144$) with the standard time-step and force calculation parameters. Runs A and $\mathrm{A_N}$ conserved energy almost equally well, with peak-to-peak variations of $\sim 0.5\%$ and a net drift of $\sim 0.2\%$ between $t = 0$ and $t = 4.5$. Refining the time-step reduced the peak-to-peak variation by half but left the net drift, while improving the force calculation decreased the net drift by half but left the peak-to-peak variation. Thus force calculation and integration errors violate energy conservation by comparable amounts; neither error seems dominant.

Yet conservation tests can't tell the whole story – a simulation may conserve energy and momentum very well yet depart from the collisionless system it purports to represent. So all variants of encounter A contained $N_\mathrm{t} = 2048$ massless test particles, distributed like the luminous matter and started from exactly the same positions and velocities in all four cases. The divergence of these particles provides a way to compare numerical errors with the uncertainties due to finite particle number. Separations grew roughly exponentially, as expected in unstable systems. But run $\mathrm{A_N}$ diverged from the other three significantly faster than runs A, A_Δ, and A_θ diverged among themselves. This implies that finite N effects are the leading source of deviations from a true solution of the CBE with smooth initial conditions.

Encounter Overviews. Fig. 13 shows the earlier stages of the evolution of encounter A. In this simulation, one disk lies in the orbit plane and spins in the same direction as the two galaxies pass one another; the other disk is inclined by $\sim 71°$. Thus both disks, and particularly the first, undergo *direct* passages and rapidly develop dramatic tidal features; indeed, both are quite strongly distorted by the third frame, only an eighth of a time unit after their first passage at $t_\mathrm{p} = 1$. Striking tidal tails are visible in the very next frame, and subsequently grow ever-longer as unbound material escapes to infinity. Meanwhile, a bridge extends from the in-plane disk to its inclined partner, resulting in an appreciable transfer of material which envelops the companion in the fifth frame. After this excitement, however, the last three frames depict a lull in the activity as the galaxies linger near their new apocenter and start to fall back together. This falling-back is of course a consequence of tidal friction between the galaxies, and in particular between their dark halos (§ 6.2).

Similar stories may be told about the color views of encounters D, F, and C_2 presented in Plate 1. Here halo material is shown in red, bulge stars in yellow, and disk stars in blue. Besides demonstrating the ubiquity of orbit decay, these images illustrate the development of a wide variety of tidal features.

All eight 1:1 encounters are shown half a time unit after pericenter in Fig. 14. Each of the eight disk orientations appears once in each column. At this early stage in the interactions, the response of each disk is relatively independent of its partner's orientation, as one can see by matching up corresponding disks. It can be hard to guess the dynamical state of an interacting system from a single image; compare the striking tails of encounter A with the rather ambiguous wisps of encounter B. Both underwent equally close passages 0.5 time units ago, yet the latter system might easily be interpreted as a wider passage if not a passage pending. The present viewing angle – face-on to the orbit plane – doesn't display the tidal features of the inclined disks to best advantage. But even in other views, several of these systems might not qualify for inclusion in Toomre's (1977) sequence of interacting galaxies because they don't display unambiguous *pairs* of tidal tails.

Other mass ratios produce further variations. Fig. 15 compares the eight three-to-one encounters with their four equal-mass counterparts. Across each row, disk angles are held fixed, while mass ratios[2] vary from 3:1 to 3:3 to 1:3 from left to middle to right. Trends of morphology with mass ratio are apparent – massive galaxies extract more extended tidal features from their partners, as TT72 showed with their Figs. 4 and 5. Note in particular the slender bridge and tail extracted from the low-mass disk in encounter C_2, also seen in Plate 1. This long-lived bridge might appear to contradict TT72's rule that low-mass companions build better bridges, but actually it remains visible for so long only because it's viewed almost edge-on in this projection.

To better display tidal features, Fig. 16 views individual disks from the equal-mass encounters and massive partners from the three-to-one encounters face-on to their original spin planes. The left-hand columns report direct passages ($i \leq 71°$) while the right-hand columns report their spin-reversed counterparts. All of these initially-identical disks have undergone parabolic encounters with pericentric separations $r_{\mathrm{p}} = 0.2$ length units. Those in the top row, with inclinations of $i = 0°$ and $180°$, repeat the point of TT72's Figs. 1 and 2 – direct passages produce much more impressive bridges and tails – and add some hint of the *self-consistent* response, which will be discussed shortly. The next three rows report on the $\omega = -30°$, $30°$, and $90°$ passages, respectively; the left-hand columns show the $i = 71°$ cases, while the right-hand columns show those with $i = -109°$.

Among these inclined passages, pericentric argument has important effects on the tidal response. The ring-like galaxies with off-center bulges in the row just below the top show a family resemblance regardless of inclination, as do

[2] Here the equal-mass case is labeled "3:3" as a reminder that the total mass also varies, with the equal-mass case 50% more massive than the others.

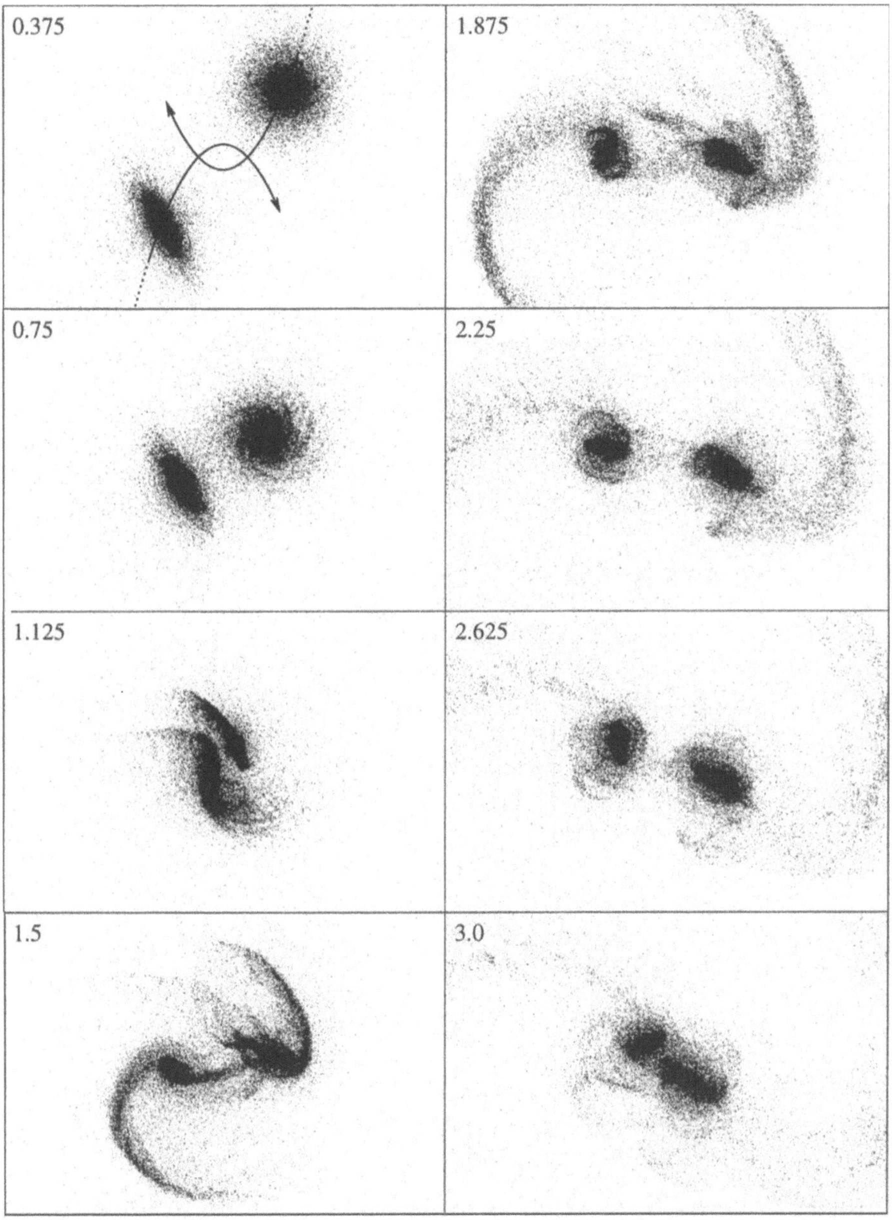

Fig. 13. Evolution of encounter A, viewed face-on to the orbital plane. Time since the start of the simulation is shown in the upper left of each frame; one time unit is about one rotation period at a radius of $3\alpha^{-1}$. The first frame also shows the parabolic paths of the approaching galaxies. Only luminous material is shown.

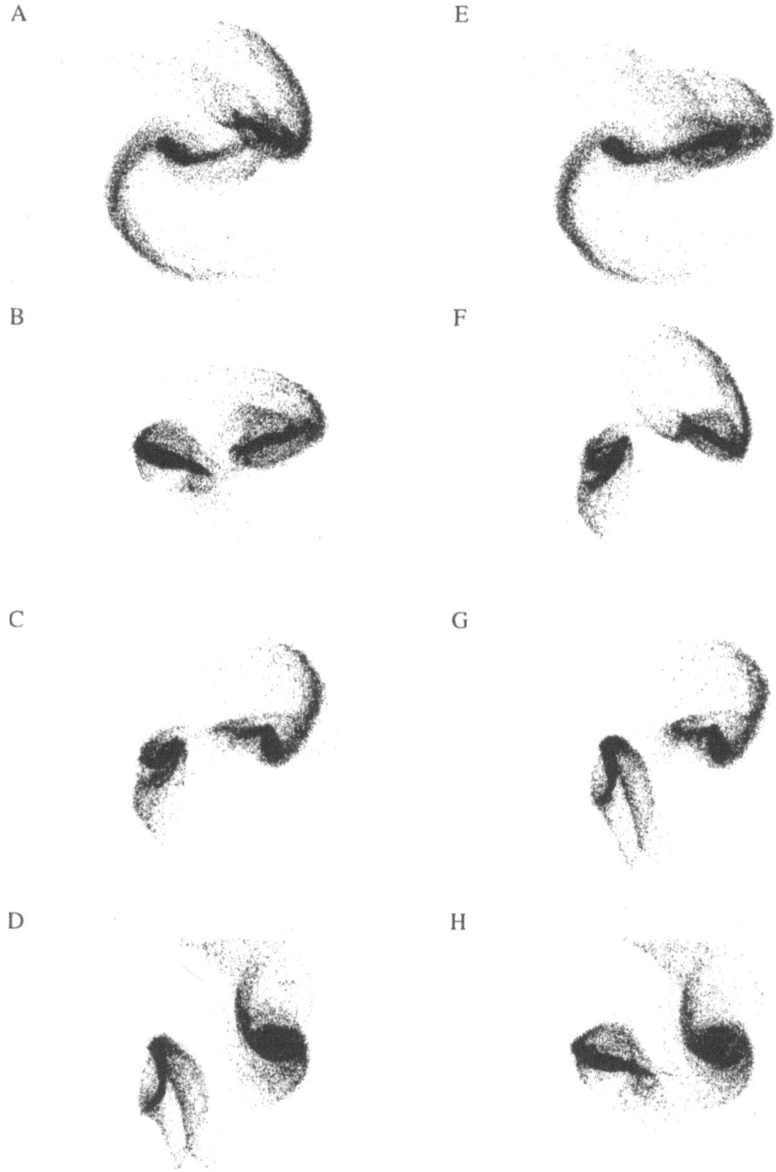

Fig. 14. Encounters A–H at time $t = 1.5$, all viewed face-on to the orbital plane. Only disk material is shown.

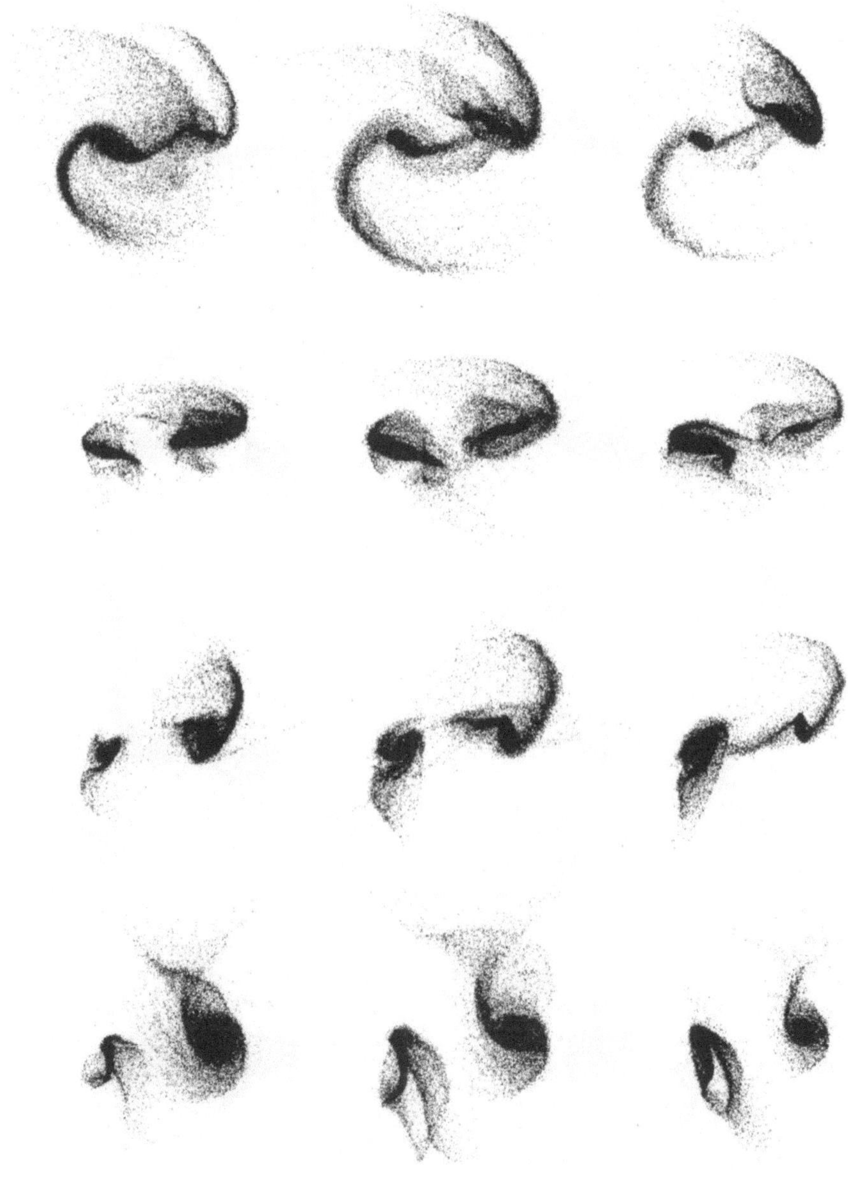

Fig. 15. 3:1 and corresponding 1:1 encounters at time $t = 1.5$, all viewed face-on to the orbital plane. Only disk material is shown.

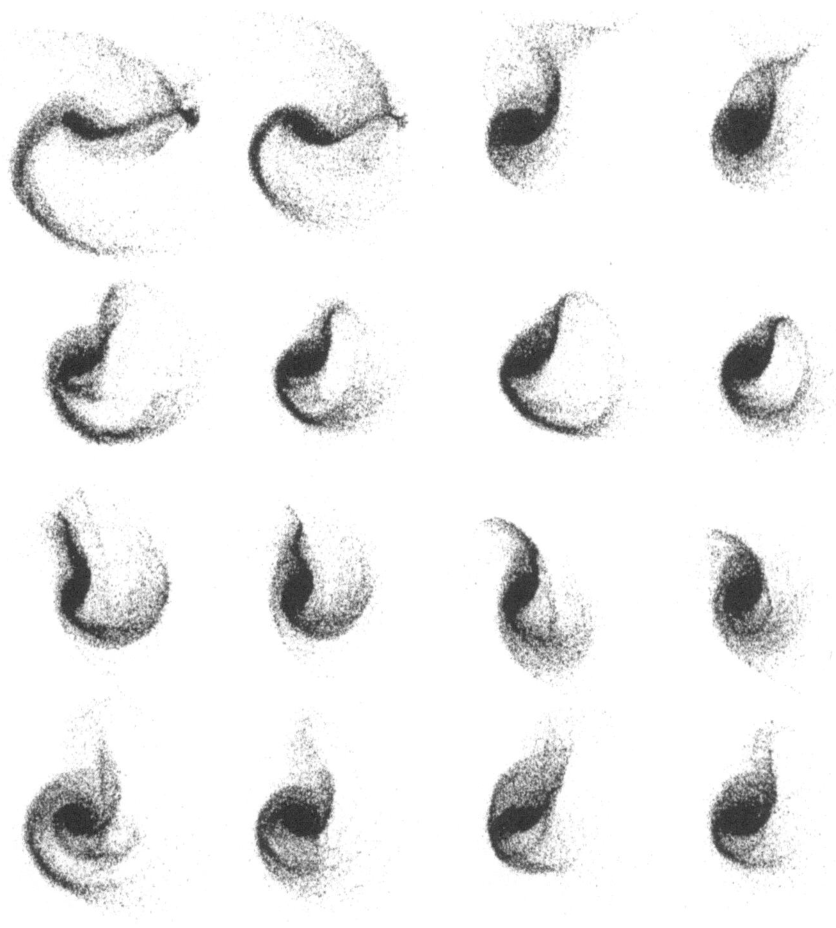

Fig. 16. A survey of disk encounters, viewed along initial spin axes at $t = 1.5$. **Left:** Direct passages ($i \leq 71°$). **Right:** Retrograde passages ($i \geq 109°$). On each side, the first column shows the disk from the equal-mass version, and the second shows the more massive disk from the three-to-one version.

the more **S**-shaped systems in the next row down. How does the change in ω create this difference in morphology? Expanding density-wave rings can result from high-inclination passages through disks (Lynds & Toomre 1976, Theys & Spiegel 1977), and the ring-like results of the $\omega = -30°$ passages resemble the off-center collisions exhibited by Toomre (1978). But the distinctly less ring-like products of the $\omega = 30°$ passages show that not all plunging passages create rings. Evidently a further requirement for ring-making is that the target disk be relatively undisturbed when the companion passes through; if the target is first shaken up by strong tides at pericenter then the ring morphology is suppressed.

Finally, the disks in the bottom row suffered early, rather distant plane crossings followed by closer, almost-polar pericentric passages ($\omega = \pm 90°$). Like the similar (though twice as wide) encounter shown in Fig. 4 of B92, these galaxies exhibited early signs of tidal damage, yet developed quite slowly thereafter.

4.3 Bars and Spirals

The bar instability in self-gravitating disks may be understood as a positive feedback effect (Toomre 1981): a leading spiral wave is swing-amplified as differential rotation shears it into a trailing spiral, which then tunnels across the center of the disk to be reborn as a leading spiral ready for further amplification. After several feedback cycles, the wave seems to "saturate" the disk, producing a stable, long-lived bar. In the B/D/H models used for these studies (§ 2.6), bars are suppressed by lowering the gain of the swing amplifier with a heavy halo and raising the central wave barrier with a cuspy bulge. Nonetheless, these apparently stable disks can develop bars as well as striking "grand-design" spirals when subject to tidal perturbations.

Fig. 17 illustrates the evolution of three tidally perturbed disks. The top row of frames shows the response following a direct ($i = 0°$) passage of an equal-mass companion. Not surprisingly, a strong tidal response starts even before pericenter, and the disk is already distorted into an **S**-shaped pattern in the first frame. In later frames the disk material is organized into a elongated central bar with a rotation period of $\sim$ 1.5 time units. Weaker perturbations produce less extreme responses. The middle row shows the result of an $i = 71°$, $\omega = 30°$ passage by a low-mass companion; this disk first develops an **S**-shaped spiral, and somewhat later a distinct central bar with trailing spiral arms. Finally, the bottom row shows the consequences of an $i = -109°$, $\omega = 90°$ passage by a low-mass companion. This even weaker perturbation produces a grand-design spiral pattern with arms which may be traced outward through almost one complete revolution. The last frame hints that this galaxy too is slowly forming a bar, although less dramatic than those produced in the stronger encounters; the straight segments of the spiral pattern at time $t = 2.625$ may be due to interference between leading and trailing spiral waves (cf. Fig. 12 of Toomre 1981), implying that waves are tunneling through the center of this once-stable disk.

Bars are a fairly common consequence of strong interactions (Noguchi 1987); of the 16 different passages studied here, 5 three-to-one and all 8 equal-mass cases exhibit well-defined bars by time $t = 2.625$. Nonetheless, prompt bar-making seems to require perturbations of finite strength. Violent interactions, like the top one in Fig. 17, form bars immediately; apparently the tidal perturbation is strong enough to saturate the disk directly. Weaker perturbations, on the other hand, may require some swing amplification before a bar can develop. Even gentler interactions, like the $r_{\rm p} = 0.4$ example illustrated in Fig. 1 of Barnes (1990b), give rise to two-armed spiral density waves which slowly wind up but don't produce bars in the time available before the next passage.

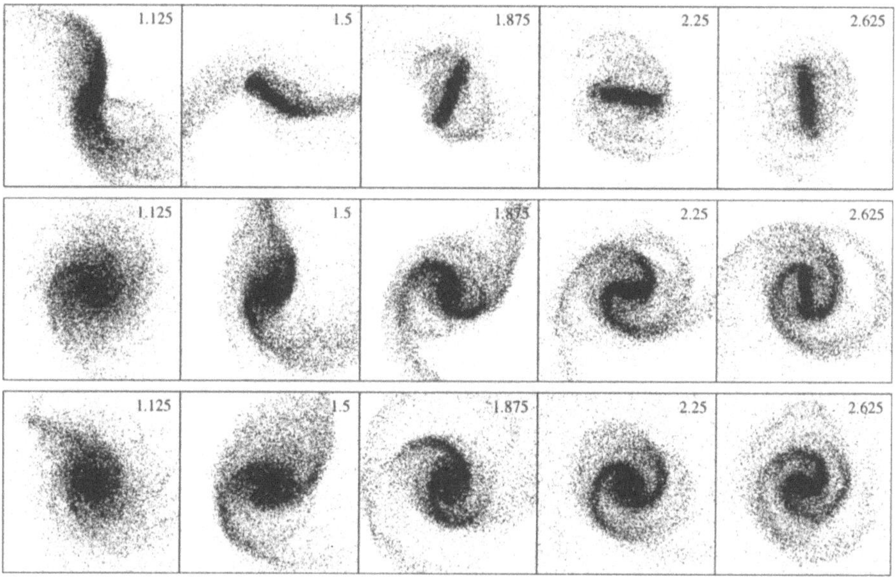

Fig. 17. Closeups of three disks from Fig. 16, showing the development of tidal bars and spiral patterns. Time since the start of the calculation is shown in the upper right of each frame; pericenter occurred at $t_{\rm p} = 1$.

4.4 Tidal Dwarf Galaxies

Self-gravity can also create structure in tidal tails. Zwicky (1956) suggested that the debris of tidal interactions could form dwarf galaxies; observational evidence for such dwarfs is covered in § 2.2 of François Schweizer's lectures. The formation of bound objects in tidal tails may be compared to the collapse of structure in an expanding universe (Gerola, Carnevali, & Salpeter 1983); an even closer analogy is Jeans' long-abandoned tidal theory for the origin of the solar system!

Fig. 18 shows some 23 clumps in the tails of a merged pair of disk galaxies; the encounter parameters are similar to those used in encounter A, though this simulation includes gas. These clumps were identified by a algorithm which links neighboring particles (Barnes & Hernquist 1992b, BH96); all are gravitationally bound and fit within their present tidal radii. Most of these objects are gas poor, and gas plays no significant role in their origin. The largest object, however, is gas-rich; a video shows the stellar component gently collapsing from a bit of spiral structure in the pre-encounter disk, while gas is swept into the dwarf's potential well by converging flows in the nascent tidal tail (BH96).

Only low-inclination encounters yield objects massive enough to hold onto gas with a nominal temperature of 10^4 K. The tail on the right in Fig. 18, produced by an $i = 71°$ passage, has only a handful of loosely-bound clumps. As Fig. 16 shows, inclined passages yield more diffuse tails; out-of-plane forces during such passages may also increase the velocity dispersion of the tidal material. These factors raise the Jeans mass in the tails and thereby suppress the collapse of bound objects.

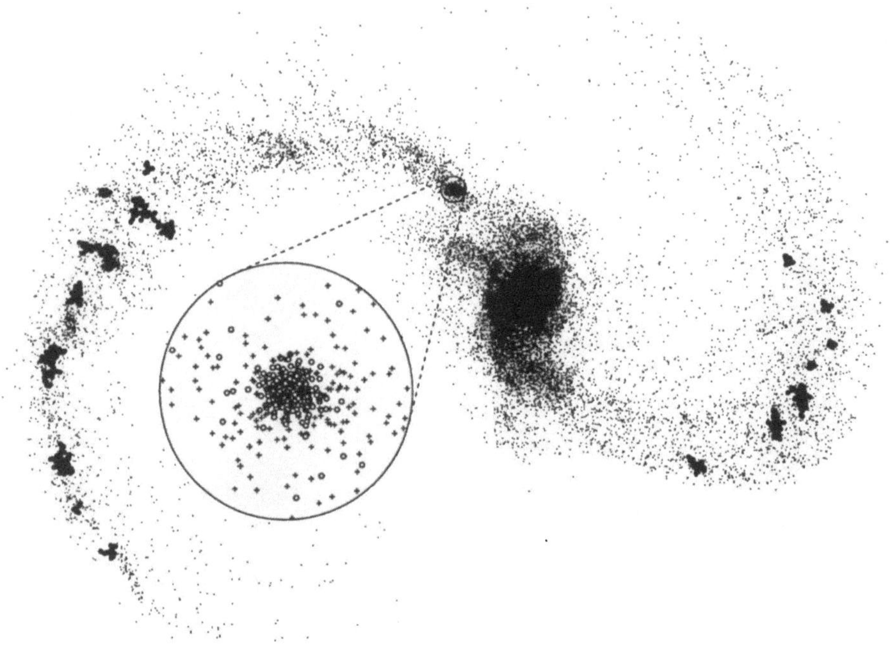

Fig. 18. Bound objects in the tails of a merger remnant (Barnes 1994). This N-body/SPH simulation is further discussed in § 8.3. Only disk and bulge particles are shown, with those belonging to bound objects plotted as filled circles. The insert shows a 12×enlargement of the most massive object; gas particles are plotted as open circles, stellar particles as crosses.

Similar N-body results are given a very different interpretation by Elmegreen, Kaufman, & Thomasson (1993); they focus on the collapse of gas clouds as the primary process in forming bound structures in tidal features. Most of the candidate dwarf galaxies in tidal tails are indeed gas rich, but this may be because gas-rich objects are more easily detected. "The Superantennae", an IR-luminous interacting system, contains *nine* distinct clumps spread along two tidal tails some $260\,h^{-1}$ kpc long (Mirabel, Lutz, & Maza 1991, Colina, Lipari, & Macchetto 1991); the wide range of $B-V$ colors of these clumps hint that some may be gas-poor.

5 Self-Consistent "Lookalikes"

Complementing surveys of tidal encounters are detailed models of specific interacting systems. In simulating a real pair of galaxies, three angles are needed to specify the projection onto the plane of the sky, bringing the formal parameter count to *eleven* – plus the parameters needed to describe the innards of the participants. This plethora of parameters sometimes leads to the grumble that any observation whatsoever can be fit by an N-body simulation. It is undeniably true that this vast parameter space can give rise to a great range of configurations, but this is just the way nature is. What is perhaps more interesting is that interacting systems can often be well-described by models with reasonable and astrophysically sensible parameters.

Starting with TT72's reconstructions of four classic systems, N-body modelers have reproduced the appearances and kinematics of many such objects. As algorithms and computers have become more powerful, a growing fraction of these simulations have included the self-gravity of the participants, extending and generalizing the results of test-particle models. Self-gravity plays a key role in the decay of galaxy orbits and in inflows of interstellar material – two processes which seem relevant to the origins of starbursts in interacting galaxies. This chapter presents self-consistent models of three of TT72's subjects: NGC 4038/9, NGC 4676, and M 51. A self-consistent model of NGC 7252 by Hibbard & Mihos (1995) is described in § 3.1 of François Schweizer's lectures in this volume.

5.1 Getting the Feel of the Antennae

One favorite subject for N-body modelers is "The Antennae" (NGC 4038/9). TT72 showed that the spectacular crossed tails of this system could be reproduced by a symmetric encounter of two inclined disks. The simplicity of these initial conditions made NGC 4038/9 an appealing target for self-consistent simulation with newly-developed tree codes; Fig. 19 compares an optical photograph of this system with a slightly updated version of the simulation presented by Barnes (1988). Patterned on the Toomres' original model, this system resulted from an elliptical, fairly wide encounter ($e = 0.6$, $r_{\rm p} = 0.5$) of two bulge/disk/halo galaxies with equal inclinations and pericentric arguments ($i = 60°$, $\omega = -30°$). Following this rather slow passage, the inclined disks slung tidal tails far above the orbital plane, where they can be seen in projection as crossing one another. The most significant difference between this model and earlier test-particle simulations was the self-consistent treatment of orbit decay, which shortened the post-passage orbit period by about 75%. Thus Fig. 19 is viewed at a somewhat earlier time than TT72's Fig. 23, with the two galaxies about to undergo a second and much-closer passage instead of loitering just past apocenter. The Antennae galaxies may in fact be commencing such a passage, but their line-of-sight velocity difference appears to be only $\sim 40\,{\rm km\,s^{-1}}$. In contrast, the galaxies in this model are approaching each other with a line-of-sight relative velocity comparable to their internal rotation speeds. Better velocity data and more numerical

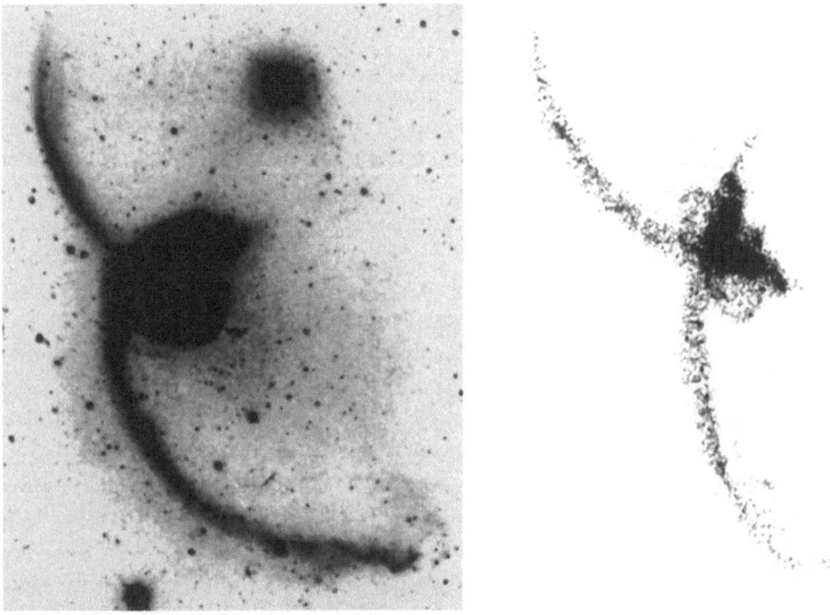

Fig. 19. The Antennae, NGC 4038/9. North is up, west is right. **Left:** Deep optical image (courtesy François Schweizer). **Right:** Self-consistent N-body simulation (after Barnes 1988).

modeling are needed to decide if the Antennae galaxies are actually hurtling back together almost transverse to our line of sight, or drifting past each other slowly in projection.

5.2 Sneaking Up on the Mice

On account of their long tails, the two galaxies making up NGC 4676 were dubbed the "Playing Mice" by Vorontsov-Vel'yaminov (1958). Fig. 20 shows on the left a deep optical image, overlain by contours of HI. The straightness of the bright northern tail indicates that we are viewing the aftermath of a nearly-direct passage roughly edge-on, while the arching curve of the fainter southern tail is consistent with a more face-on view of a higher-inclination disk. The fairly equal lengths of the two tails suggests that the galaxies involved had roughly equal masses. Along with long-slit data showing that both "hulks" rotate internally with north receding, these considerations led TT72 to a simple model for this system. It's worth noting that the Toomres' model, far from fitting all available data, contradicted Burbidge & Burbidge's (1961) claim that the southern hulk is receding with respect to the northern at $\sim 100\,\mathrm{km\,s^{-1}}$; Stockton (1974) reexamined the velocities of both galaxies and found the overall sense of motion predicted by TT72. Gilbert & Sellwood (1993) used Stockton's data in constructing a self-consistent model of the Mice.

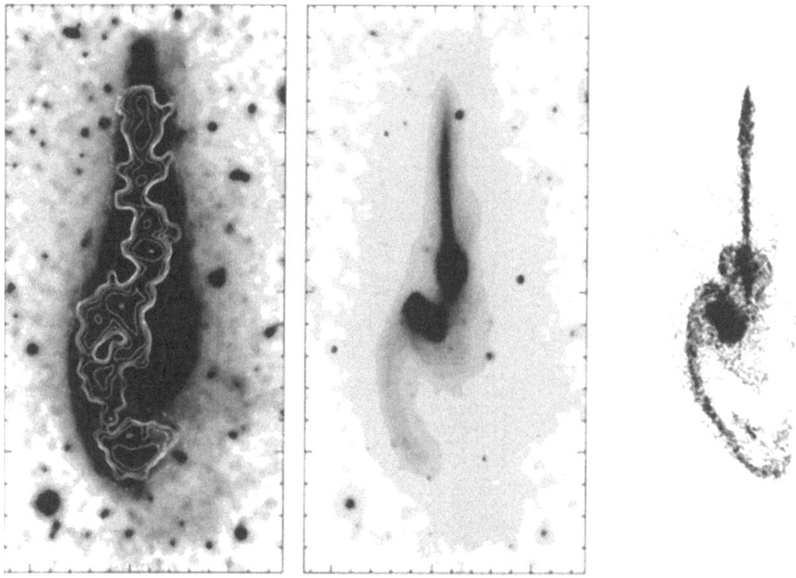

Fig. 20. The Mice, NGC 4676. North is up, west is right. **Left:** Deep optical image and HI contours. **Middle:** Optical image, stretched to show the bodies and tails. **Right:** Self-consistent N-body simulation (Hibbard & Barnes, in preparation).

New velocity data from observations in HI (Hibbard & van Gorkom 1996), CO (Yun & Hibbard, in preparation), and H_α (Mihos, Bothun, & Richstone 1993) are now available. These data are, once again, consistent with the kinematics expected for a fairly direct passage of two disks. John Hibbard and I recently began a project to construct a self-consistent model using these new data. This "work in progress" illustrates a systematic search of parameter space.

We began by considering the initial orbit of the galaxies. TT72 adopted an elliptical orbit ($e = 0.6$) with a rather short period, largely because their restricted three-body simulations did not include orbit decay, while Gilbert & Sellwood used an initially parabolic orbit. As TT72 noted, the latter option is more plausible, and there is no sign of the previous passages expected with an elliptical orbit. We thus ran four experiments: one elliptical orbit with parameters $e = 0.6$, $r_{\rm p} = 0.5 = 6/\alpha$, and three parabolic orbits with pericentric separations of $r_{\rm p} = 5/\alpha$, $4/\alpha$, and $3/\alpha$, where $\alpha^{-1} = \frac{1}{12}$ is the exponential disk scale length. In these four experiments the disks were tilted as in TT72's model, with inclinations $i_{\rm A} = 15°$ and $i_{\rm B} = 60°$ and arguments $\omega_{\rm A} = \omega_{\rm B} = -90°$, where the labels A and B refer to the northern and southern disks, respectively.

All four experiments, evolved about one time unit past pericenter, produced configurations which superficially resemble the Mice. Choice of viewing time proved somewhat subjective; one constraint is the projected length of the tails, which should be 25 to 30 times α^{-1}. The viewing angle is constrained by requiring that tail A appear thin and straight, and that the bodies of the galaxies

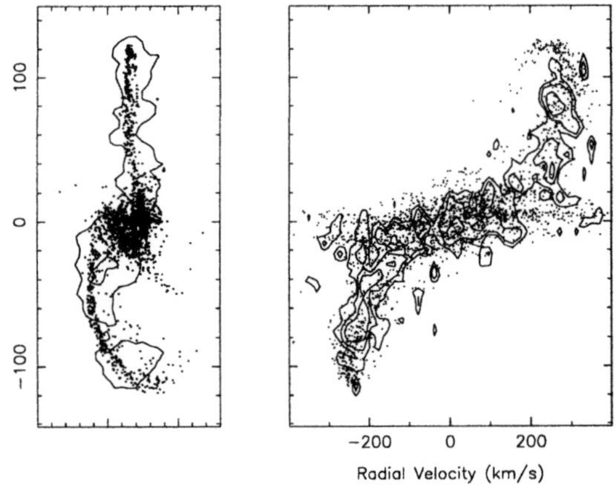

Fig. 21. Neutral-hydrogen contours and N-body simulation of The Mice (Hibbard & Barnes, in preparation). **Left:** View on the sky plane; the vertical scale is in seconds of arc. **Right:** Observations and simulation projected onto the declination versus line-of-sight velocity plane.

appear close in projection. As in the model of the Antennae just described, the post-passage orbit periods were greatly shortened by dynamical friction. Thus the initially elliptical orbit placed the two hulks too close together in space; this match was also marred by a good deal of splattered material and by doubled velocity profiles along tail A. These flaws all seemed less serious in the parabolic encounters. Varying the pericentric distance $r_{\rm p}$, we found that the closer encounters produced systems with a smaller line-of-sight velocity difference between the two hulks, less conspicuous bridge arms, and more massive tails – all changes for the better. Consequently we adopted the parabolic orbit with $r_{\rm p} = 0.25 = 3/\alpha$ for further experiments.

Second, we systematically varied the inclinations $i_{\rm A}$ and $i_{\rm B}$ of the two disks in steps of 10°. The original values $i_{\rm A} = 15°$ and $i_{\rm B} = 60°$ gave tail B too high an arch, and left the east-west separation of the two hulks smaller than observed. Increasing $i_{\rm A}$ implies a corresponding change in the optimal viewing angle, which we used to increase the east-west separation of the two hulks. Reducing the angle between the galaxies, $i_{\rm A} + i_{\rm B}$, had the effect of decreasing the arch of tail B, and slightly increasing its projected velocity towards the observer, both welcome improvements. We thus adopted $i_{\rm A} = 25°$ and $i_{\rm B} = 40°$.

Third, we varied the pericentric arguments $\omega_{\rm A}$ and $\omega_{\rm B}$. Neither disk is highly inclined, so these angles were not strongly constrained. Changing $\omega_{\rm B}$ by $\pm 30°$ shifted the point of maximum projected curvature along the length of tail B; we adopted $\omega_{\rm B} = 60°$ to obtain a "hook" at the end of this tail matching the observed clump of HI emission. Changing $\omega_{\rm A}$ rotates the plane containing the optimal point-of-view; by setting $\omega_{\rm A} = -30°$, we obtained a more end-on view of both tails. This improved the match to the tail velocities, which – up until this point – had fallen considerably short of the mark. Fig. 21 presents views and line-of-sight velocities for this model; the agreement seems satisfactory.

In sum, we follow Gilbert & Sellwood's choice of an initially parabolic orbit, simply because elliptical orbits decay too soon, leaving the tails too little time to grow before the second passage. We also agree, to within 5°, with their favored values for $i_{\rm A}$ and $i_{\rm B}$. We prefer a *closer* encounter, with a pericentric separation of only three disk scale lengths, essentially to give the tails the heft of those observed in the Mice. However, without a systematic study using a range of mass models it is difficult to place a lower bound on the ellipticity e; it seems premature to claim that we have uniquely constrained the parameters of NGC 4676.

5.3 What Happened to the Whirlpool?

The vivid spiral of M 51 (NGC 5194), first described by Lord Rosse in 1845, typifies "grand design" spiral galaxies. But M 51 is hardly a normal galaxy – apart from its splendid spiral, its distorted outer isophotes indicate that it is interacting with its neighbor NGC 5195. There are also hints that the star-formation rate in M 51 has been boosted by this interaction. These considerations make M 51 a very interesting subject for tidal interaction modeling. Despite much effort, however, no really satisfactory model is yet available.

Deep optical images show that both galaxies in this system are distorted (Zwicky 1959, Arp 1965, van den Bergh 1969). M 51 itself has a "bridge" which stretches northward across the face of its neighbor, and a fainter counter-arm emanating from the southwest side of the disk. NGC 5195 is surrounded by extended faint luminosity, including plumes extending to the west-northwest and to the southeast. The relation of these features to the better-known spiral is shown on the left in Fig. 22, where contours of faint emission (Burkhead 1978) overlay an optical image.

These distortions, along with the projected position and line-of-sight velocity of the companion, led TT72 to model this system as the result of a rather recent passage. In a slightly revised model by Toomre (1978), the companion galaxy NGC 5195, assigned one-third the mass of M 51, passed through the plane of the larger galaxy some $2 \cdot 10^8$ years ago at a position of about two o'clock (twelve o'clock being north) and an inclination of $-75°$; it now lies behind M 51. The relative orbit and present appearance of this model are shown on the right in Fig. 22. Rather nicely reproduced are the bridge and counter-arm of M 51, as well as the companion's recession velocity of $\sim 120\,{\rm km\,s^{-1}}$ with respect to M 51. Assuming that companion was also disk-like and about equally inclined to the orbit, the model mimics the faint plumes associated with NGC 5195 and predicts the general sense of this galaxy's observed rotation (Schweizer 1977).

But despite these successes, the model did not reproduce the swirling spiral which earned M 51 the nickname "Whirlpool". TT72 suggested that some exceptional spirals might owe their features to tidal excitation of density waves. This suggestion gained plausibility following Toomre's (1981) discussion of swing amplification in differentially-rotating disks (see § 4.3). It appeared that self-consistent calculations would be needed to test the hypothesis that M 51's spiral is tidal in origin.

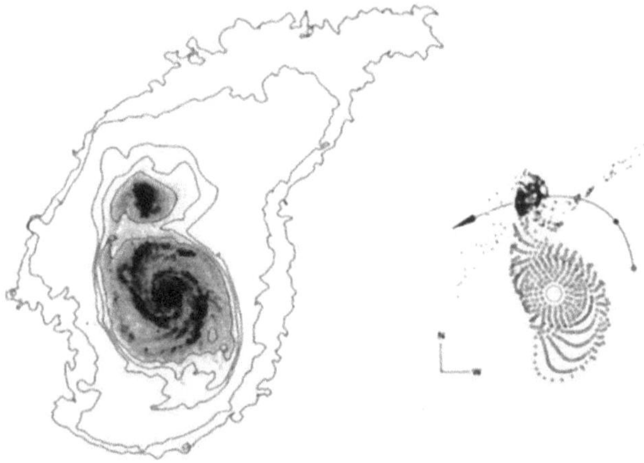

Fig. 22. The Whirlpool, M 51, and its companion, NGC 5195. North is up, west is right. **Left:** Optical contours (Burkhead 1978) superimposed on an optical image (courtesy Tod Boroson). **Right:** Restricted three-body model (Toomre 1978).

Yet another spur to model-building was the discovery of a tail-like HI feature associated with M 51 (Rots et al. 1990). This structure, some 90 kpc in length, is plotted on the left in Fig. 23, with the optical spiral again included for scale. The general appearance of this feature suggests that it is tidal in origin, though it is somewhat curious that most of the tail has no optical counterpart and that double-peaked velocity profiles are observed where it joins the outer body of the galaxy.

Asked to model M 51 for the 1989 Heidelberg meeting, Lars Hernquist and I attempted a self-consistent simulation (Hernquist 1990c). Our goal was to reproduce *both* the spiral pattern within M 51 and the extended HI tail with a tidal interaction. We used a bulge/disk/halo model for M 51 but approximated NGC 5195 with a King (1966) model; thus our modeling did not treat the faint features associated with the companion. We noted that $\sim 10^8$ years is too little time to produce the HI tail, and favored a longer interval between the passage of NGC 5195 and the present. With a longer wait after passage, we were obliged to shift the place where the companion crosses the plane of M 51's disk from two to about four o'clock. Increasing the mass of the companion elongated the tail and sharpened the induced spiral. Thus our best model used a half-mass companion, a south-westerly crossing, and two to three times as long a wait after passage. This longer time allowed the tail to develop and the tidal spiral to wind up into something vaguely resembling the Whirlpool. We were pleased to find that strong streaming motions along the spiral density wave could produce the **S** and **H** shapes Rots et al. (1990) had noted in the channel maps of M 51's disk. But even these efforts were flawed: compared to M 51, the tail seemed underdeveloped, the bridge arm too long, and the spiral pattern rather confused.

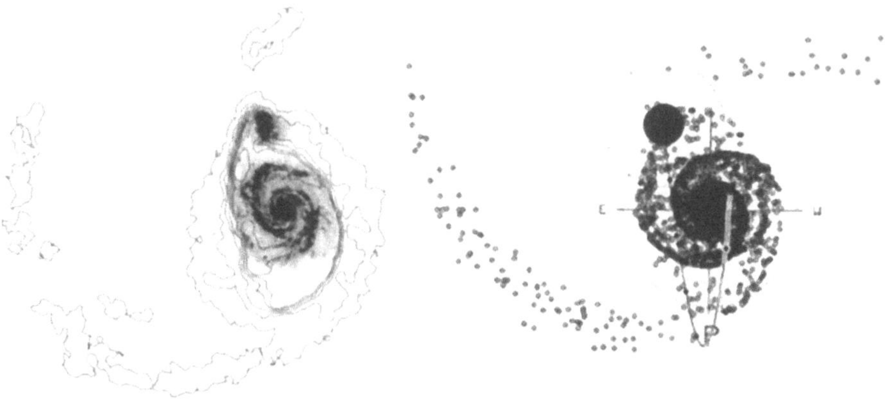

Fig. 23. The Whirlpool. **Left:** Neutral hydrogen contours (Rots et al. 1990) superimposed on an optical image (courtesy Todd Boroson). **Right:** A simple dynamical model imitating the neutral-hydrogen tail and inner spiral (Toomre 1994 Brouwer lecture).

These and other loose ends recently inspired Alar Toomre to revisit the problem of modeling this system. His elegantly minimalistic model of M 51 consists of a rigid Plummer sphere and a self-consistent Kuzmin (1956) disk of equal mass; the companion galaxy NGC 5195 is modeled by a rigid Plummer sphere of the same *total* mass. This rather massive companion diminishes the bridge and lengthens the tail, improving the match with M 51. With a passage through the disk plane at six o'clock, the tidal tail and the induced spiral are both mimicked rather nicely, as shown on the right of Fig. 23. Remarkably, this reconstruction owes rather little to swing amplification; with strong tides from an equal-mass companion, even a kinematic test-particle simulation would have produced a good spiral!

However, this model also has problems. The mass it ascribes to NGC 5195 may not be unreasonable if this galaxy has a hefty dark halo, though such a massive companion will cause rapid orbital decay, which is not properly simulated with rigid potentials. But the companion's passage through the disk of M 51 at about six o'clock seems to rule out any simple *tidal* explanation for its associated faint plumes; following such a passage, the debris from the companion spread out north-south, instead of roughly east-west as observed. The velocity field along the HI tail is likewise a puzzle; the plane of the tail seems tilted with respect to M 51's disk by about 40° (Rots et al. 1990), which is not easy to explain with a tidal interaction.

What do we have to do to solve the puzzle of M 51? Perhaps we are viewing the product of not one but *two* successive passages. A second passage by a tenth-mass companion can "touch up" M 51's spiral, but this in itself does little to resolve the problems noted above, and requires an almost circular orbit with an implausibly short period (Howard & Byrd 1990). With a more massive companion and a first passage at seven or eight o'clock it seems possible to approximate the tidal damage to NGC 5195 while producing at least something of M 51's HI tail (Salo & Laurikainen 1996). But such double passages seem intrinsically less plausible; they will have to show compelling advantages over single-passage models, and so far none have done so.

5.4 Unresolved Issues

The interacting galaxies modeled in this volume are a mixed bag. NGC 4676 (§ 5.2) and NGC 7252 (§ 3.1 of François Schweizer's lectures) seem to be success stories; though neither model is perfect, both manage to fit the data tolerably well with *sensible* initial conditions. NGC 4038/9 (§ 5.1) will probably also turn out well, though the model may require an elliptical orbit (TT72, Barnes 1988). On the other hand, M 51 (§ 5.3) remains perplexing; we may not have all the pieces of this puzzle.

As yet, nobody has produced a *unique* model of any of these systems; there is considerable freedom in choosing halo profiles and initial orbits. Dubinski, Mihos, & Hernquist (1996) argue that halos of much more than ten times the luminous mass of galaxies would prevent the long tails of NGC 4038/9 and similar galaxies from forming. At face value, this seems to rule out the halos expected in an $\Omega = 1$ CDM cosmology (e.g. Navarro, Frenk, & White 1996); however, the simulations with massive halos didn't use cosmologically consistent initial conditions, so the argument is not yet iron-clad.

Finally, it's possible that non-gravitational forces play a larger role than realized. Several galaxies have faint optical and HI features which don't coincide. NGC 520 has a faint, gas-poor tidal tail and an apparently unrelated ring of gas (Hibbard & van Gorkom 1996). Arp 299 has roughly parallel but offset tails of neutral hydrogen and star-light (Hibbard & Yun, in preparation). The forces responsible may be nothing more exotic than ram pressures due to hot gas in quiescent halos (Moore & Davis 1994, Sofue 1994) or starburst-driven outflows (e.g. Heckman et al. 1996). Still, it seems unwise to dismiss out of hand the possibility that some real surprises may yet await modelers of interacting galaxies.

6 Mechanics of Merging

Interactions would rate but a footnote in the theory of galactic structure if galaxies encountered each other elastically and caromed off to drift again through empty space. Even the first theoretical studies of tidal interactions, such as Holmberg's (1941) numerical work (§ 1.2) and Alladin's (1965) analytic investigation of hyperbolic encounters between spherical galaxies, emphasized that this was not the case. Still, the terrible strength of inelastic effects in slow encounters remained largely unrecognized. TT72 estimated that the orbital eccentricity of the Antennae had decreased from $e = 0.8$ to $e = 0.5$ after the most recent passage. But early studies did not include massive dark halos, which dramatically increase the encounter cross-sections of galaxies (Toomre 1977, White 1978).

Likely outcomes of orbit evolution were also mentioned in early work. Zwicky (1959) noted that repeated encounters between bound galaxies might lead to "considerable disruption of both systems" or "total mutual capture". TT72 took this line of reasoning further, mentioning some half-dozen objects each resembling "a single luminous ball, from which protrude several tentacles or filaments", and suggesting that older remnants would resemble elliptical galaxies. While they and Toomre (1977) showed that the expected *number* of old merger remnants is consistent with the observed number of ellipticals, it was not then possible to test the hypothesis that tumbled-together spirals would resemble elliptical galaxies. As it turns out, objects like the one shown in Fig. 24 closely resemble the recent merger remnants that François Schweizer describes in his § 3 of this volume.

6.1 Tidal Drag

Apart from some rather special systems (e.g. Sridhar & Nityananda 1990), analytic descriptions of encounters between galaxies are possible only in limiting cases where the effects can be treated as perturbations (BT87, Ch. 7.1, 7.2). One such limit aries in fast encounters; another in encounters between galaxies of very different masses. In either case, tidal interactions pump energy into galaxies at expense of their relative motions.

Fast Encounters. When galaxies collide with speeds much higher than their internal velocity dispersions, the orbital motions of individual stars during the encounter can be neglected. In this limit, the encounter's effects are to increment the velocity of each star, relative to the center of its galaxy, by an amount $\delta\mathbf{v}(\mathbf{r})$ which depends only on position. Though such fast encounters probably occur only in rich galaxy clusters, the underlying simplicity of the problem creates considerable scope for analytic or quasi-analytic work.

For fast encounters, the orbits of the galaxies may be approximated by straight lines; relative to the center of galaxy 1, let galaxy 2's position be given by the vector $(0, b, Vt)$ where b is the impact parameter and V is the relative velocity. If the galaxies do not overlap during their passage, they act on each

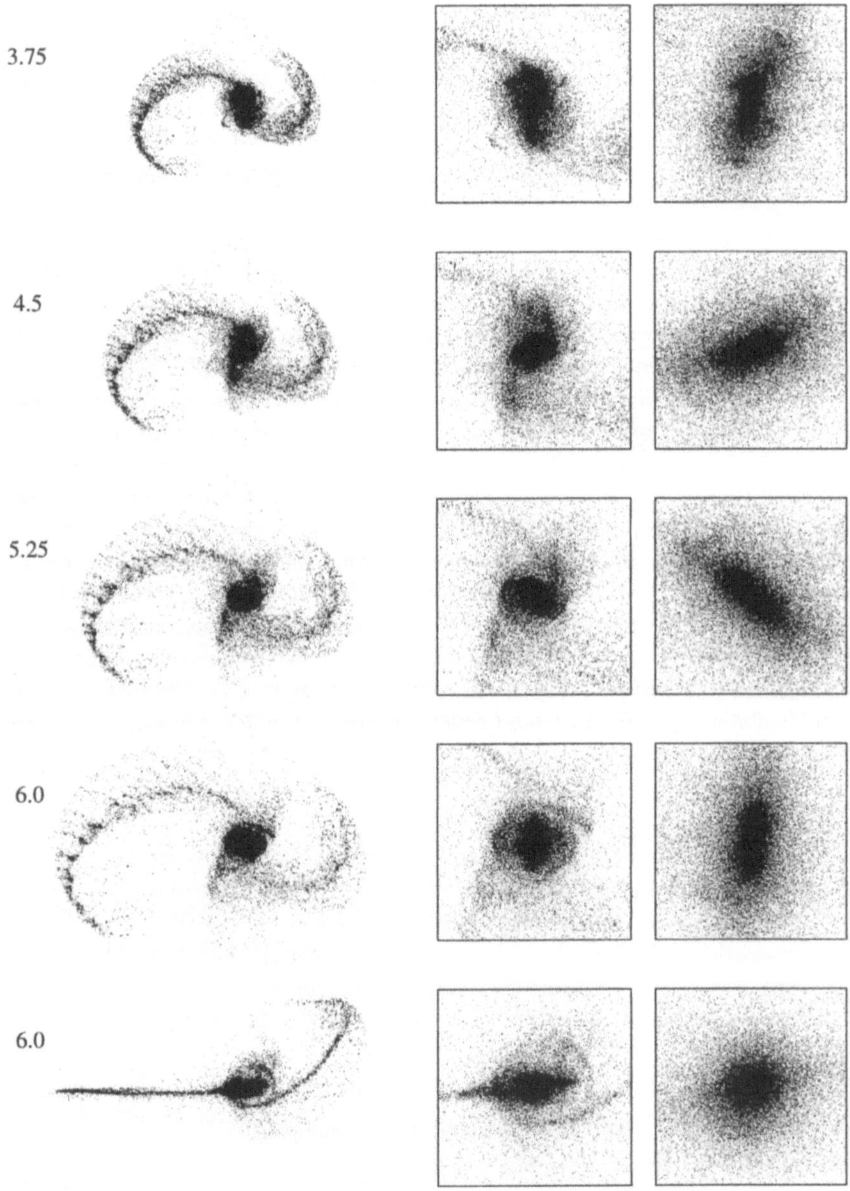

Fig. 24. Structure of the merger remnant from encounter A (see Fig. 13) at the times indicated at left. The first four views are face-on to the orbit plane, while the bottom view is edge on. **Left:** Overviews of the disk material. **Middle:** Twice-enlarged views of the stellar (bulge + disk) material. **Right:** Five-times further enlarged views of the stellar material.

other like point masses, and by analogy with (19), a star at position $\mathbf{r} \equiv (x, y, z)$ relative to the center of galaxy 1 suffers an impulsive velocity change, directed toward the point $(0, b, z)$, of magnitude $2GM_2/b'V$, where M_2 is the mass of galaxy 2 and $b' \equiv \sqrt{x^2 + (y - b)^2}$. Some of this change, however, refers to the net motion of the entire galaxy; subtracting this part and keeping only leading terms in an expansion around the point $(0, 0, 0)$ yields the velocity impulse

$$\delta\mathbf{v}(\mathbf{r}) = \frac{2GM_2}{b^2V}(-x, y, 0) \tag{30}$$

with respect to galaxy 1.

Equation (30), which describes the damage suffered by a galaxy as a result of a fast, distant encounter, has a number of uses (c.f. BT87, Ch. 7.2). For example, when combined with a dynamical rule for the escape of unbound stars, it may be used to estimate mass loss during tidal encounters (Aguilar & White 1985). More to the present purpose, the internal energy gained by a galaxy during an encounter is given by

$$\Delta E = \frac{1}{2}\int \mathrm{d}\mathbf{r}\, \rho(\mathbf{r})|\delta\mathbf{v}(\mathbf{r})|^2 \,, \tag{31}$$

where $\rho(\mathbf{r})$ is the galaxy's mass density (Spitzer 1958, Alladin 1965). The quantity ΔE is positive-definite, implying a one-way transfer of energy from the relative motion of the two galaxies to their internal degrees of freedom. If the energy absorbed by the participants exceeds the energy of their initially-unbound orbit then the galaxies will be bound together as a result of the encounter; however, this is only likely in relatively slow passages, which are not adequately described by the impulse approximation.

Dynamical Friction. The alternate limit where one galaxy is much smaller than the other is often discussed using Chandrasekhar's (1943) theory of dynamical friction. This theory figures widely in models of orbital evolution in interacting galaxies, including studies of "cannibalism" among cluster galaxies (Ostriker & Tremaine 1975), accretion of satellites by the Milky Way (Tremaine 1980), and the fate of the Magellanic clouds (Murai & Fujimoto 1980). However, it's not always recognized that this formalism, when applied to interacting galaxies, is an incomplete approximation; some applications may have overstepped its bounds of validity.

Chandrasekhar's approach to dynamical friction parallels his treatment of two-body relaxation, but the "test star" is replaced by a body with mass M much greater than the mass m of the field stars. Moreover, the impulse approximation is replaced with an exact solution for the hyperbolic relative orbit of this body and passing field stars. Consequently, the massive body acquires a parallel velocity δv_{p} as well as a transverse velocity δv_{t} (§ 2.5) as a result of each encounter; in its pre-passage frame of reference, δv_{p} lies along the field star's initial velocity vector (BT87, § 7.1). If the massive body is at rest with respect to the population of field stars, the impulses $\delta\mathbf{v}$ from successive encounters are uncorrelated and the net effect is a random walk not very different from that discussed in § 2.5.

But suppose the massive body has velocity $\mathbf{v}_M$ with respect to the field stars; then the impulses it receives are partly correlated and accumulate linearly, rather than quadratically as in a random walk. For field stars with a homogeneous and isotropic phase-space mass distribution function $f = f(|\mathbf{v}|)$, integrating the massive body's acceleration over all encounters with impact parameters less than $b_{\max}$ gives

$$\frac{d\mathbf{v}_M}{dt} = -16\pi^2 G^2 \ln(\Lambda) M \frac{\mathbf{v}_M}{v_M^3} \int_0^{v_M} dv\, v^2 f(v) , \tag{32}$$

where $\Lambda \simeq b_{\max} v_M^2 / GM$. This equation describes a force of "dynamical friction" which opposes the massive body's motion and slows it down with respect to the field star distribution. In the limit $M \gg m$ assumed here, this deceleration is proportional to M but depends only on the mass density of field stars and not their individual masses.

Results similar to (32) can be derived in several different ways. One way is to calculate the wake created by a massive body moving through a sea of field particles; the gravitational attraction between the wake and the massive body may then be evaluated using the "Jeans swindle" (BT87, § 5.1), which ignores the potential of the uniform background. This approach shows that the acceleration is linear in M simply because the amplitude of the wake is also proportional to M – at least in the limit that the massive body creates only a small perturbation on the overall density.

Application of (32) to the orbits of satellite galaxy is complicated by several uncertainties (e.g. White 1983a, Bontekoe & van Albada 1987). Satellite galaxies are not point masses, so momentum transfer from field stars is reduced in encounters with impact parameter b less than the satellite's radius; likewise, the maximum impact $b_{\max}$ needs to be more carefully related to the primary galaxy's radius. But such corrections to the factor $\ln(\Lambda)$ do not address the real worries concerning (32), which are that it neglects the periodic motions of the satellite and field stars and ignores the self-consistent response of the primary. In some sense, (32) is only dimensionally correct; for example, it predicts *zero* drag if the satellite orbits outside a primary with a finite radius, whereas numerical experiments show a finite drag.

These shortcomings of the dynamical friction "formula" have led to several efforts to develop and test better models of sinking satellites. One improvement is to sum over all orbits in the primary which resonate with the satellite (Weinberg & Tremaine 1984). This approach makes explicit the transfer of orbital energy and momentum to internal degrees of freedom within the primary galaxy, but does not include self-consistent effects. To address the latter requires solving the CBE, either via semi-analytic methods (Weinberg 1989) or an all-out N-body attack (Hernquist & Weinberg 1989). It is encouraging that semi-analytic and N-body methods yield similar results where they can be compared. But a good description of the orbital evolution of a satellite with one-tenth the mass of its primary is still lacking; the non-linear response provoked by satellite is beyond the scope of the analytic work, but too subtle to be simulated accurately for currently feasible values of N.

6.2 Orbit Decay

For slow encounters and mass ratios of order unity, self-consistent N-body simulations provide a fairly reliable way to study orbit decay. The basic physical mechanisms can be illustrated by encounters between simple spherical systems. In a head-on collision, each galaxy is gravitationally compressed as it passes through its partner, and rebounds thereafter, tossing stars onto loosely-bound orbits; the energy required to do this comes from the relative motion of the two galaxies (Toomre 1974, van Albada & van Gorkom 1977, Miller & Smith 1980). In an off-axis encounter, the decay process is dominated by bodies which orbit within their respective galaxies in the same direction as the two galaxies orbit each other; these are promoted onto wider orbits, extracting both energy and angular momentum from the galaxies' relative motion (White 1978, Roos & Norman 1979, Villumsen 1982, 1983, Navarro 1990). Whether the encounter is head-on or off-axis, the net result is to excite internal motions within the galaxies at the expense of their relative orbit, which therefore decays.

Spherical Systems. Besides sidestepping the complexities of rotating disks, simulations of encounters between spherical systems can use realistically "cuspy" models and high resolution. The models to be used here each consist of two components: a core[3] described by a Jaffe (1982) profile, and a halo described by a $\gamma = 0$ Dehnen (1993) profile; thus

$$\rho_{\rm c} \propto \frac{1}{r^2(r+a_{\rm c})^2}\,, \qquad \rho_{\rm h} \propto \frac{1}{(r+a_{\rm h})^4}\,. \tag{33}$$

In model units, the core has mass $M_{\rm c} = 0.25$ and scale radius $a_{\rm c} = 0.125$, and the halo has mass $M_{\rm h} = 1$ and scale radius $a_{\rm h} = 0.1$; these match the overall luminous-to-dark mass ratio and halo profile of the model disk galaxy described in § 2.6. Following the procedures outlined in that section, the core and halo were realized with truncated Gaussian velocity distributions, using 2^{14} bodies for the former and 2^{15} for the latter. The smoothing length was $\epsilon = 0.006125$, or $\sim 61\%$ of the length used in the disk galaxy simulations; the time-step was reduced to $\Delta t = 1/256$ in order to follow motions more accurately on small scales. Tests show that the initial conditions are near equilibrium and that the core's r^{-2} density profile is represented down to radii as small as $\sim 0.1\epsilon$.

Fig. 25 reports the orbital trajectories followed in two different encounters of core/halo models. Dotted curves represent the initial parabolic orbits, which lead to pericentric passages at $r_{\rm p} = 0.2$ and 0.4 length units; for comparison,

[3] The core represents luminous material residing within a halo of dark matter (White & Rees 1978); despite its name, it is neither wholly contained within a finite radius or constant-density as $r \to 0$.

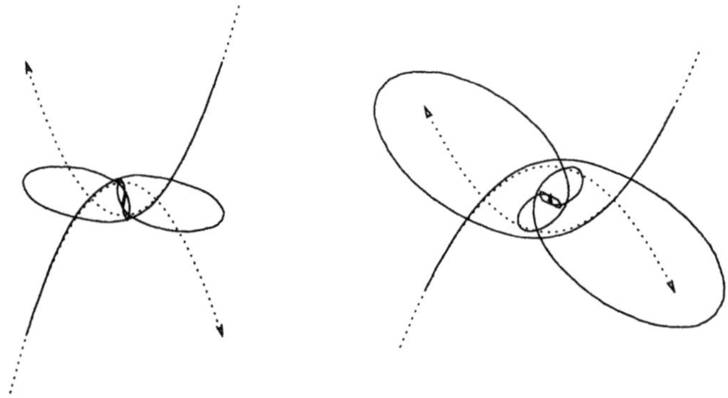

Fig. 25. Orbit trajectories in (left) $r_p = 0.2$ and (right) $r_p = 0.4$ encounters of core/halo galaxies. Dotted lines are the initial parabolic orbits.

the galaxy models have half-mass radii of $r_{1/2} \simeq 0.226$ length units. The actual trajectories, shown by the solid curves, were obtained by following the centroids of the most tightly-bound bodies in each core. In deeply interpenetrating passages like these, the galaxies can't "turn the corner" as sharply as two point masses (Farouki & Shapiro 1982). This happens simply because the galaxies are extended objects; to the extent that they remain spherical, the mass at radii greater than their current separation contributes nothing to the acceleration of their centers. But the fact that both of these pairs fall back together for further passages shows that they interact as *deformable* rather than merely extended objects – rigid potentials might not follow parabolic trajectories, but would escape to infinity.

Orbit decay is due to the transfer of orbital energy and momentum to internal degrees of freedom within each galaxy. It's easier to keep track of the angular momentum since gravitational energy is not localized; Fig. 26 therefore plots specific angular momenta $j(t)$, computed with respect to the orbital center, in the $r_p = 0.2$ experiment. The solid and long-dashed curves show the angular momenta of the halo and core components, respectively. These mirror one another since their linear combination $0.2 j_c + 0.8 j_h$ is the total specific angular momentum. After the first passage at $t = 1$, and again after the second passage just before $t = 4$, there is a transfer of angular momentum from the core to the halo. But the net decline of the core's angular momentum is rather modest, amounting to only $\sim 30\%$ of its initial value; much of this angular momentum remains as spin in the merger product.

More dramatic evidence for tidal transfer to internal degrees of freedom emerges on plotting the specific angular momenta of tightly-bound parts of the galaxy cores. As § 6.4 will show, successive passages and even mergers largely preserve the relative ordering of particles by binding energy; thus core bodies sorted by binding energy in the initial conditions can be used to quickly identify tightly-bound subsets at later times. The short-dashed and dotted curves show $j(t)$ of the most tightly-bound half and quarter of the core, respectively.

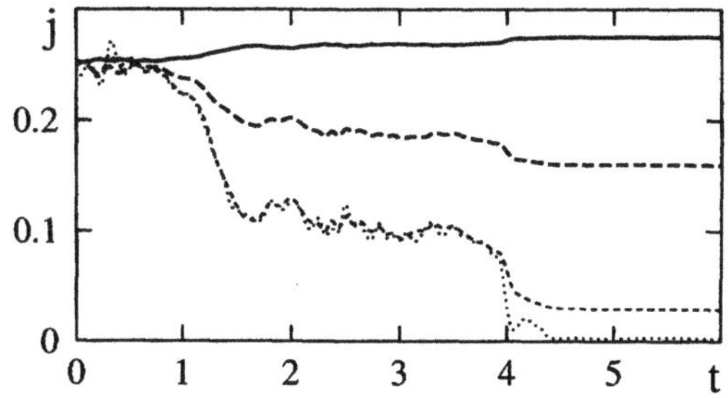

Fig. 26. Transfer of angular momentum in a parabolic encounter of core/halo galaxies with $r_{\rm p} = 0.2$ length units. The solid and long-dashed curves show the specific angular momentum $j(t)$ of the halo and core, respectively. The short-dashed and dotted curves show $j(t)$ for the most tightly-bound half and quartile of the core.

These two subsets, and especially the latter, lose significant amounts of angular momentum after each passage – as they must since they wind up concentrated toward the center of the merger remnant.

The process which enables the most tightly-bound quartile to get rid of $\sim$ 98% of its orbital angular momentum is more complex than plain old dynamical friction. The total torque acting on this material can be decomposed into two terms, one which results from forces between galaxies, the other from forces within galaxies. Even while the galaxies are deeply interpenetrating, the orbital torques on the most tightly-bound quartile due to forces between galaxies are almost nil; instead, the forces which rob this material of its angular momentum are due to interactions *within* each galaxy (B92).

These results suggest the following picture of orbital decay in compound systems. Initially, as the extended halos interact, tidal forces transfer orbital angular momentum to internal degrees of freedom; in other words, the halos start to spin. More tightly-bound components, however, feel no significant tidal forces at this stage and retain their full complement of orbital angular momentum; therefore, they soon get ahead of their own halos. Strong gravitational forces pull these components back toward the centers of their respective halos; these forces oppose the orbital motion of the tightly-bound material and rob it of orbital angular momentum *without imparting spin*. Meanwhile, the halos no longer have enough orbital angular momentum to maintain their initially parabolic orbit. At a later date, then, the entire process of encounter, halo spin-up, and momentum transfer is repeated until the galaxies finally merge.

Disk Systems. Rotating disk components don't fundamentally alter the description of orbital decay. As White (1979) showed with simulated mergers of rotating spherical systems, orbits decay more rapidly if the spin and orbital angular momenta are parallel, and more slowly if they are anti-parallel. This is a simple consequence of the stronger tidal responses produced in direct, as opposed to retrograde, encounters (§ 4.2); direct passages pump more orbital energy into internal motions.

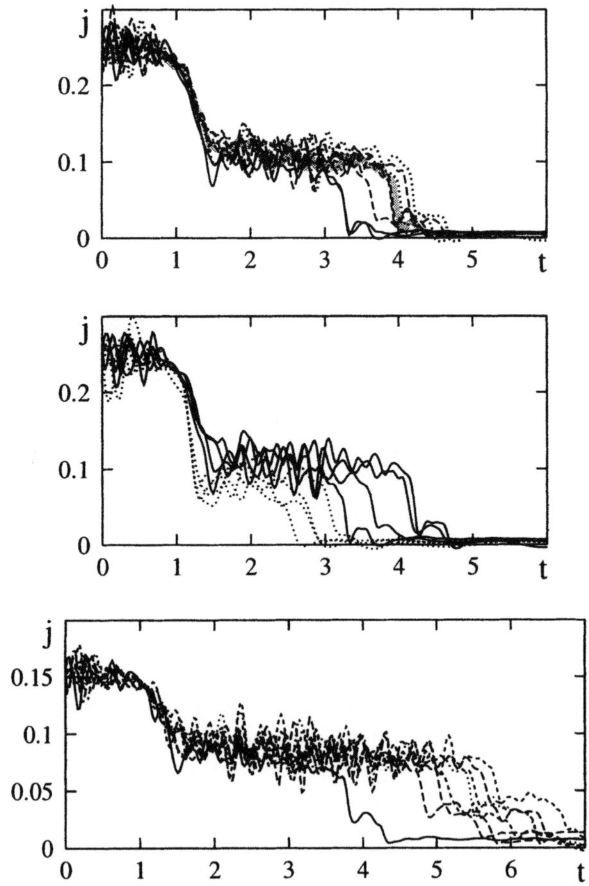

Fig. 27. Orbit decay in disk galaxy encounters. Plotted as functions of time are total specific angular momenta of the most tightly-bound bulge quartiles of each experiment. **Top:** Equal-mass encounters. Line style indicates encounter type: solid for direct encounters (A, E), dashed for inclined encounters (B, C, F, G), dotted for retrograde encounters (D, H); grey shows the core-halo encounter. **Middle:** Encounters A, B, C, and D (solid lines), compared with counterparts from B92 (dotted lines) which used King (1966) model bulges and halos. **Bottom:** Three-to-one encounters. Line type indicates the inclination of the primary disk: solid for $i = 0°$ (A_1), long dash for $i = 71°$ (A_2, B_1, C_1), short dash for $i = 109°$ (B_2, C_2, D_2), and dotted for $i = 180°$ (D_1).

The top panel of Fig. 27 reports orbit decays in the eight equal-mass encounters, and compares them to the $r_{\rm p} = 0.2$ experiment with spherical galaxies. Each curve shows the total specific angular momentum $j(t)$ of the most bound quartile of the bulge material. As in the spherical case just described, these Lagrangian volumes accurately track the centers of the galaxies. The line style indicates the encounter type: solid lines for encounters with direct ($i = 0°$) disks, dotted lines for those with retrograde ($i = 180°$) disks, dashed lines for those between inclined disks; the broad grey curve repeats the results for the most tightly-bound core quartile from the spherical galaxy experiment. Disk orientation clearly influences the rate of orbit decay: the times between first and second passages, indicated by sharp drops in $j(t)$, are about 50% longer for the retrograde cases than for the direct cases. This is consistent with White's (1979) results. The effects of the disks are, nonetheless, relatively small; most of the disk-disk encounters decay at about the same rate as the equivalent encounter between spherical galaxies. And there's no sign that any of the disk galaxies are gaining orbital angular momentum at the expense of their spins, which is not too surprising since the orbits possess most of the angular momentum anyway.

The middle panel of Fig. 27 compares orbit decays in encounters A, B, C, and D with the similar encounters 1, 2, 3, and 4 from B92. The bulges and halos used in the earlier calculations are based on King (1966) models, and the halos in particular, though finite in radius, are more extended than the $\gamma = 0$ Dehnen (1993) model halos used in the latest experiments. The galaxies used in the new calculations have total half-mass radii $r_{1/2} \simeq 0.225$, compared with $r_{1/2} \simeq 0.25$ for the older models. This modest reduction in cross-section causes a significant difference in orbit decay rate, increasing by about a third the interval between first and second passages.

Orbit decays in the eight three-to-one encounters are reported in the bottom panel of Fig. 27. Here, the line style indicates the inclination of the *primary* disk, which might be expected to strongly effect the decay rate. The results largely confirm this expectation, though some interesting exceptions occur. For example, encounter B_1 takes longer to decay than encounter B_2, even though the former has a less-inclined primary. And encounter D_1, with a retrograde primary, is by no means the slowest to decay; evidently the stronger interaction of an in-plane encounter partly compensates for the retrograde passage. On the whole, the 3:1 encounters decay at about two-thirds the rate of their 1:1 counterparts. This seems consistent with the relationship between secondary mass and drag force implied by (32), though the comparison can't be taken too far since the 1:1 encounters have higher pericentric velocities due to their deeper potential wells.

6.3 Violent Relaxation

Repeated and ever-closer passages between bound pairs of galaxies can have only one outcome – the merger of the participants. The details of this process can be investigated numerically. But the theory of violent relaxation, due to Lynden-Bell (1967), provided insight – and created expectations – about the outcome of mergers.

Violent relaxation is different from two-body relaxation (§ 2.5); instead of being scattered by each other, stars are scattered by the net gravitational field $\Phi(\mathbf{r}, t)$. For example, the specific binding energy of a star, $E \equiv \frac{1}{2}|\mathbf{v}|^2 + \Phi(\mathbf{r}, t)$, changes as follows:

$$\frac{\mathrm{d}E}{\mathrm{d}t} = \frac{1}{2}\frac{\mathrm{d}|\mathbf{v}|^2}{\mathrm{d}t} + \frac{\mathrm{d}\Phi}{\mathrm{d}t} = \mathbf{v}\cdot\frac{\mathrm{d}\mathbf{v}}{\mathrm{d}t} + \frac{\partial\Phi}{\partial t} + \mathbf{v}\cdot\nabla\Phi = \left.\frac{\partial\Phi}{\partial t}\right|_{\mathbf{r}(t)} . \tag{34}$$

Thus the total change in the binding energy of the star between times t_1 and t_2 is obtained by integrating $\partial\Phi/\partial t$ along the star's trajectory,

$$E(t_2) - E(t_1) = \int_{t_1}^{t_2} \mathrm{d}t \left.\frac{\partial\Phi}{\partial t}\right|_{\mathbf{r}(t)} . \tag{35}$$

If the potential is static ($\partial\Phi/\partial t = 0$) then E is constant, but in a time-dependent potential each star's binding energy varies in a complex way. For example, in a fast, head-on encounter of two identical galaxies the depth of the potential well briefly doubles as the galaxies coincide. A star on an elongated orbit in one of the galaxies will become less tightly bound if it falls into the well when it is deepest and climbs out after it becomes shallower again; conversely, the star will become more tightly bound if it happens to be climbing out when the well is at its deepest. Typically, some stars will be falling in while others are climbing out, so stars with the same initial binding energy become spread out as a result of the encounter.

Because the scattering is driven by the fluctuating mean field $\Phi(\mathbf{r},t)$, the stellar distribution function $f(\mathbf{r},\mathbf{v},t)$ obeys the Collisionless Boltzmann Equation (5). An important consequence is that the phase-space density along any stellar orbit $(\mathbf{r}(t),\mathbf{v}(t))$ is conserved. The proof is simple:

$$\frac{\mathrm{d}f}{\mathrm{d}t} \equiv \frac{\partial f}{\partial t} + \dot{\mathbf{r}}\cdot\frac{\partial f}{\partial \mathbf{r}} + \dot{\mathbf{v}}\cdot\frac{\partial f}{\partial \mathbf{v}} = 0 , \tag{36}$$

where the second equality is established by substituting the equations of motion $(\dot{\mathbf{r}},\dot{\mathbf{v}}) = (\mathbf{v}, -\nabla\Phi)$ and recognizing the result as (5). Thus no matter how violently the potential fluctuates, the outcome is still constrained by an infinite number of conserved quantities – one for each possible orbit.

Lynden-Bell (1967) attempted to derive the most probable endpoint of the violent relaxation process by maximizing the entropy of the system subject to the constraint implied by (36). This approach turned out to be inconclusive; the resulting maximum-entropy configuration has infinite mass. Still, the expectation remained that remnants should be rather thoroughly homogenized by the merging process. If merging was violent enough to completely mix up the different components of the original galaxies, obliterating all memory of each star's initial energy and angular momentum, merger remnants would be too diffuse and too rapidly rotating to resemble elliptical galaxies (Ostriker 1980). But a detailed examination of the later stages of galactic encounters shows that violent relaxation ends well before such complete mixing is achieved.

6.4 Final Encounters

As a pair of galaxies spiral in towards merger, several dynamical effects are simultaneously active. In mergers of roughly equal-mass galaxies, orbits decay

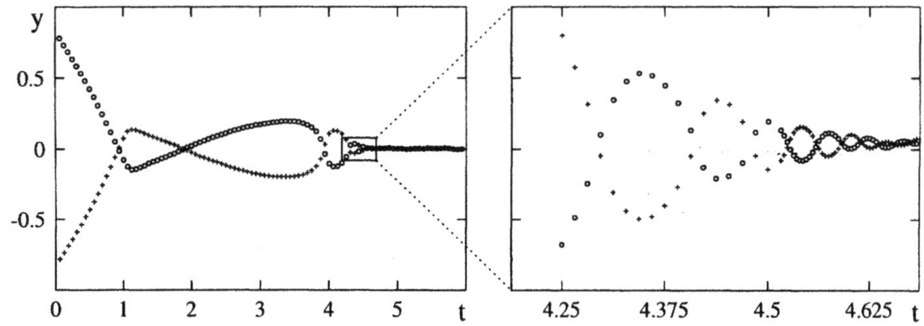

Fig. 28. Multiple passages following a close parabolic encounter of core/halo galaxies. The y coordinate of the center of each core is plotted as a function of time t. First and second passages take place at $t = 1$ and $t \simeq 3.94$, respectively. Later passages, shown in the inset at right, occur with ever-increasing frequency.

on a dynamical time-scale $\sim (G\bar{\rho})^{-1/2}$; thus as material at higher average densities $\bar{\rho}$ comes into play the pace of the passages picks up. Most of the stars and dark matter tidally stripped from either galaxy remain bound to the pair, forming a common envelope which becomes the body of the merger product. The stripped galaxies continue to interact strongly with each other as they draw closer together; they also experience dynamical friction as they move through their common envelope. Plumes or stubby tails of stars shed by rotating disks as they plunge past each other help to populate elongated orbits in the merger remnant. Finally, as the passages get closer, dense regions of interstellar material interact directly, dissipating energy and collapsing to even higher densities; the dynamics of dissipative mergers will be discussed in § 8.2.

Simulations with spherical systems exhibit several basic aspects of this process (e.g. van Albada & van Gorkom 1977, Toomre 1977, White 1978, 1979, Villumsen 1983). Fig. 28 illustrates the decreasing amplitude and increasing tempo of passages as the two "cuspy" core/halo galaxies in the $r_{\rm p} = 0.2$ experiment (§ 6.2) fall together. This plot takes advantage of the fact that in the second and subsequent passages the galaxies basically shuttle back and forth along the y-axis (Fig. 25). Remarkably, these two cores undergo at least *eight* passages, each closer and more nearly head-on than the last, before the amplitude of their relative orbit falls to a value of order the smoothing length. At later times the amplitude continues to decay, but the interval between passages remains roughly constant since both cores are confined within the region in which Plummer smoothing (12) creates a nearly-constant density $\bar{\rho}$. With less smoothing, the tempo of the passages would continue to increase as the amplitude decayed, and the final stages of the merger would be quicker and more violent. Yet as these orbital tracks show, the merger is almost over before smoothing effects begin to seriously perturb the dynamics; thus the body of the remnant should be only slightly affected by smoothing.

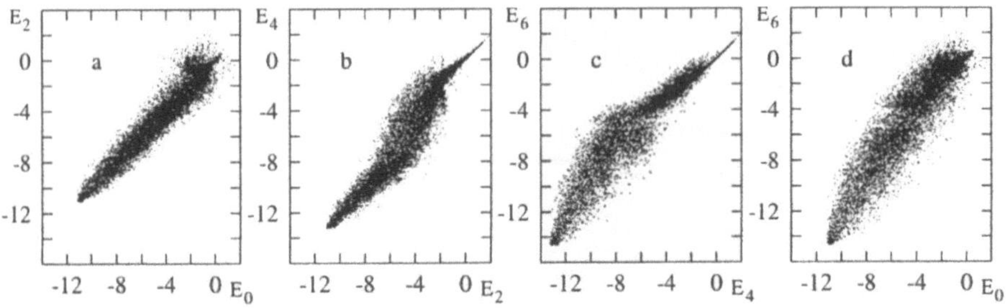

Fig. 29. Stages in violent relaxation in a close parabolic encounter of core/halo galaxies. Each scatter plot show specific binding energies at different times. (a) Change in binding energy between $t = 0$ and $t = 2$. (b) Change in binding energy between $t = 2$ and $t = 4$. (c) Change in binding energy between $t = 4$ and $t = 6$. (d) Total change in binding energy between $t = 0$ and $t = 6$.

The repeated passages of this system redistribute stars in binding energy as expected with violent relaxation. This process is illustrated in Fig. 29. Here panel (a) shows E for each body at times $t = 0$ and $t = 2$. The point-distribution is distinctly wider for the less tightly-bound bodies, which have been scattered during the first passage. Tightly-bound bodies, having orbital radii smaller than r_{p} and orbital periods short compared to the time-scale $\Phi(\partial\Phi/\partial t)^{-1}$ on which the gravitational field changes, are relatively undisturbed by the first passage, though two-body relaxation causes some diffusion in energy space. Changes in binding energy between times $t = 2$ and $t = 4$ are shown in panel (b); this time interval includes the second passage, which leaves the extremes of the binding-energy distribution relatively undisturbed but spreads out the midrange. Effects of the third and subsequent passages are shown in panel (c); here the most tightly-bound bodies are scattered, while those further out are basically unaffected by the inner turmoil. Finally, panel (d) shows the cumulative effects of all passages between $t = 0$ and $t = 6$; perhaps most striking is the strong correlation between initial and final binding energy which persists in even this rather drawn-out merger. The survival of such a correlation, noticed by White (1979), is evidence of *incomplete* violent relaxation (see White 1987).

Rotating disks increase the range of morphologies seen in final encounters. If the first passage was relatively gentle, the disks may arrive at second pericenter fairly undisturbed and still capable of throwing off tidal filaments; such a merger is shown in Fig. 30. Here two disks, each dressed with a certain amount of tidal debris from their first passage, have a second pericenter at $t \simeq 4.5$, briefly separate, and fall back together at $t \simeq 5.25$. Each passage adds new strata of stars to the merger remnant, much in the form of tail-like structures flung outward on bound, elongated orbits. If the first passage was closer, one or both galaxies may develop strong bars (§ 4.3). In such cases the outcome may depend on the exact phases of the bars in later passages; while the "hotter" bars don't throw off such well-defined tidal structures, they may still be quite effective at scattering stars onto elongated orbits as they churn together.

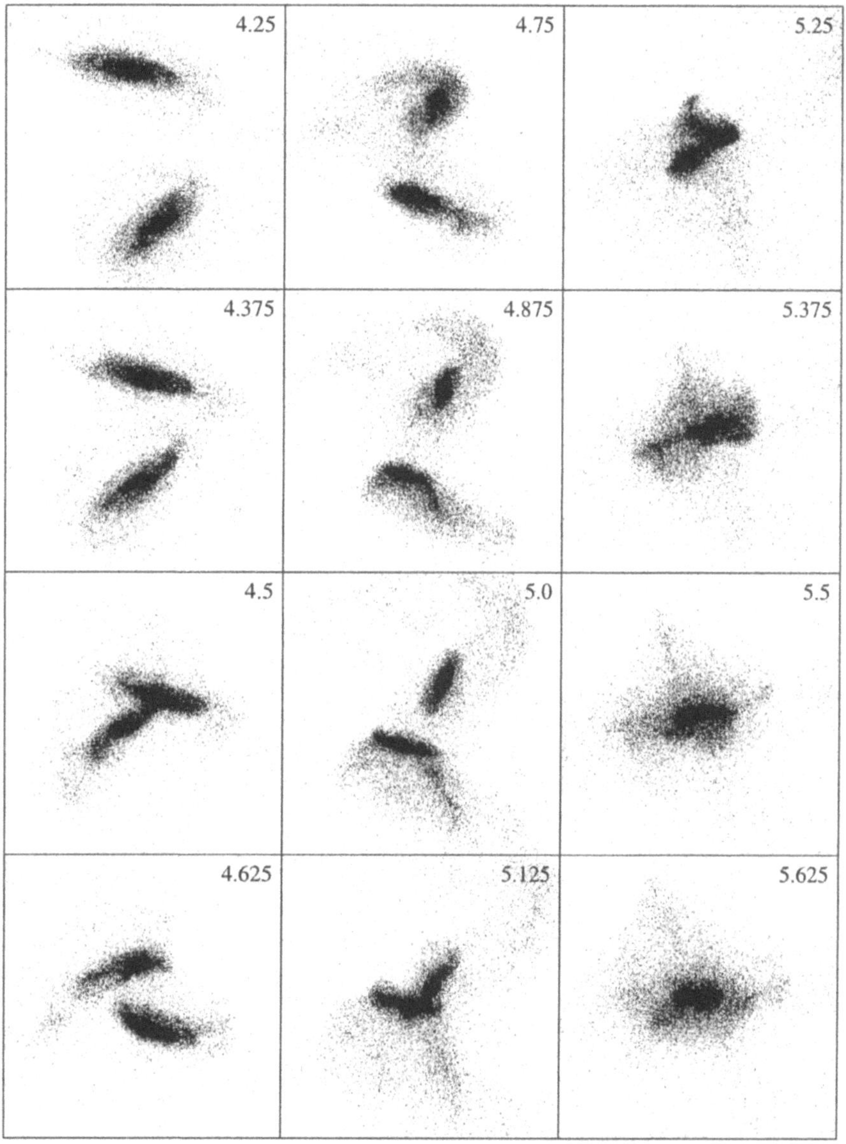

Fig. 30. Final passages and merger of two inclined disk galaxies (B92). This is a wide ($r_{\rm p} = 0.4$) version of encounter B, with King (1966) model bulges and halos.

7 Remnant Structure

If galaxies could speak, what stories they would tell! The history of a galaxy is written in its structure; to revisit the past of a system described by the collisionless Boltzmann equation, one need only integrate the dynamical equations backwards[4] instead of forwards. van Albada & van Gorkom (1977) applied this technique to a simulated remnant and saw it divide into two galaxies, which between them captured every particle before separating along a time-reversed version of their initial orbit; such reversible behavior is guaranteed by a leap-frog integrator (§ 2.4). But we can't play this game with real galaxies; to read the past lives of a galaxy, one must understand how its memories are coded into its present structure.

7.1 Phase Mixing

Phase-space density is conserved along orbits (§ 6.3), so violent relaxation – even if continued to completion – is subject to important constraints. Collisionless dynamics pushes elements of phase fluid around, but leaves their densities unchanged. This implies that the Gibbs-Boltzmann entropy,

$$S = -\int \mathrm{d}\mathbf{r}\, \mathrm{d}\mathbf{v}\, f \ln(f) \,, \tag{37}$$

is actually *conserved*; in this sense, violent relaxation is adiabatic (White 1987).

But as a violently relaxed system settles down to a new equilibrium state, an initially localized blob of phase fluid will be mapped into an ever-more convoluted volume. As Fig. 31 shows, the distribution eventually become so tightly-wound that the observable properties of the system are adequately described by the *coarse-grained* distribution function $F(\mathbf{r}, \mathbf{v}, t)$, which is a smoothed version of the fine-grained distribution. Unlike the latter, however, F can't be reversed; it lacks the fine details which are "unscrewed" by integrating backwards. In terms of information content, F clearly has more entropy than f. Following Tolman (1938), Tremaine, Hénon, & Lynden-Bell (1986) show that for *any* convex function $C(F)$ the quantity

$$S' = \int \mathrm{d}\mathbf{r}\, \mathrm{d}\mathbf{v}\, C(F) \,, \tag{38}$$

increases during phase mixing, provided that the initial distribution varies slowly as a function $\mathbf{r}$ and $\mathbf{v}$; this provision holds at only one instant, so S' does not define an arrow of time (White 1987). Since $C(F)$ can be any convex function, the specific choice $C(F) = -F \ln(F)$ doesn't have a special status, and there is no reason to expect a unique endpoint of violent relaxation which maximizes the Gibbs-Boltzmann entropy.[5]

---

[4] Strictly speaking, this is only possible if no dissipative component is present; otherwise one must supply the time-reversed potential of the dissipative material.

[5] James Binney, in "A short sermon on entropy" (de Zeeuw 1987, p. 269), disagrees, arguing that the Gibbs-Boltzmann entropy *does* have a special status and the apparent freedom to pick an arbitrary convex function $C(F)$ exists only because of neglect of prior constraints. As far as I know this issue has not been resolved.

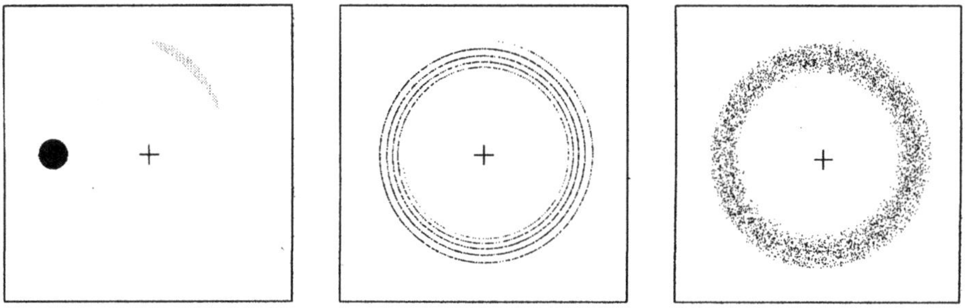

Fig. 31. Phase mixing and the coarse-grained distribution function, illustrated for a 2-D phase space. On the left are the initial conditions; the phase points are packed into a small circle. Also shown in grey is the image of this circle shortly thereafter as the phase flow carries it clockwise around the origin. In the middle is the situation after a number of orbits; the circle has now been mapped into a tightly-wound spiral. On the right is a representation of the coarse-grained distribution function at the same time; local averaging smooths out the spiral pattern.

While it may be impossible to define a unique endpoint, we can still make some generalizations about the outcome of phase mixing. Let $m(F)\mathrm{d}F$ be the mass at phase-space densities between F and $F + \mathrm{d}F$, and define the functions

$$M(F) = \int_F^\infty \mathrm{d}F' \, m(F') \quad \text{and} \quad V(F) = \int_F^\infty \mathrm{d}F' \, \frac{m(F')}{F'} \,, \tag{39}$$

Since the functions $M(F)$ and $V(F)$ are monotonic, one can eliminate the argument F and define $M(V)$, which is the maximum mass enclosed in any region of phase-space, connected or not, with volume V. Suppose that at time t_0 the distribution function varies slowly with $\mathbf{r}$ and $\mathbf{v}$, so that $F \simeq f$. Then the specific phase-space region V_0 containing the maximum mass has fairly simple outline. But at any other time t_1 before *or* after t_0, the region V_1 which contains that same mass will have a complex outline, which will be lost when f is smoothed to construct F. No amount of gerrymandering will recapture the material formerly enclosed by V_0; consequently $M_1(V) < M_0(V)$. Tremaine, Hénon, & Lynden-Bell (1986) express this by saying that the system is *more mixed* at t_1 than at t_0.

An important consequence of mixing is that the maximum value of the coarse-grained distribution function, $F_{\max}$, is bounded from above by its initial value; it can never increase as a result of merging. In a merger, the relationship between the initial and final $F_{\max}$ depends on the structures of the participants. Hot stellar systems merge with relatively little decrease in $F_{\max}$ (e.g. Duncan, Farouki, & Shapiro 1983, Carlberg 1986). On the other hand, a cold, rotating disk can't be converted into a hot system without a good deal of mixing; simulations show $F_{\max}$ decreasing by an order of magnitude when disks merge (B92).

Fine Structure. Another important consequence of phase mixing is the gradual fading and eventual disappearance of "fine structures" in merger remnants.

Recently-formed remnants like NGC 7252 and NGC 3921 have a wide variety of not-so subtle features, including shells, plumes, and loops of star-light (see § 3 of François Schweizer's lectures). In a potential which has by and large settled down to equilibrium, such structures exist because stars have correlated orbital phases, and they become more convoluted with time as stars with different periods drift out of phase. Eventually the convolutions become so intricate that they are no longer visible. This is *precisely* the mixing process illustrated in Fig. 31.

Examples of fine structures are evident in Fig. 24 (see also Barnes 1988, B92). The top row shows the system just after the galaxies have merged; strong shell-like features are visible in the highly-enlarged plot on the right. By the next time plotted, $t = 4.5$, most of this obvious small-scale structure has phase-mixed away, and the remnant looks much smoother. This smoothness is partly an illusion; though they are hard to see against the grainy background of the N-body model, phase-wrapped shells can be rendered visible by high-lighting particles as they reach their apocenters (Hernquist & Spergel 1993). The intermediate-scale views in the middle also show that the central stellar ellipsoid is surrounded by an irregular stellar distribution with plume-like features; other examples are visible in the last frames of Fig. 30. These features seem to be relics of the violent tidal interactions of merging systems; composed of stars with a wide range of orbit periods, they fade quickly once the merger is complete.

Infall of material from the tidal tails creates the loops seen at later times in Fig. 24. At time $t = 6$, a loop associated with the tail from the direct disk extends below and to the left of the remnant in the face-on view; another, associated with the tail from the inclined disk, can be seen circling the body of the remnant in the edge-on view. Late return of tail material, continuing for $\sim 10^9$ yr or more, makes phase-mixing a drawn-out process. The rate of infall can be estimated as follows: let $N(E)\mathrm{d}E$ be the number of stars with specific binding energies between E and $E + \mathrm{d}E$. Tails contain stars on both bound and unbound orbits, so $N(E)$ is continuous across the escape energy $E = 0$. If the potential at large radii is approximately keplerian, the relationship between binding energy and orbit period t_p is $E \propto t_\mathrm{p}^{-2/3}$. Then the number of stars with periods between t_p and $t_\mathrm{p} + \mathrm{d}t_\mathrm{p}$ is proportional to $N(E)(\mathrm{d}E/\mathrm{d}t_\mathrm{p})\mathrm{d}t_\mathrm{p}$, or $t_\mathrm{p}^{-5/3}$ in the limit of large t_p. Stars from the tails fall back into the remnant roughly one orbit period after the merger, so the infall rate declines like $t^{-5/3}$ at late times (B92).

7.2 Characteristic Scales

Before looking into details of phase-space structure, it's worth a detour to examine the factors which fix the dimensions of merger remnants. An argument invoking only conservation laws and the virial theorem predicts that the characteristic scales of merger remnants are closely tied to those of their progenitors (Hausman & Ostriker 1978, White 1983b). Suppose that two identical systems, each of mass M_i and binding energy E_i, fall together along a parabolic orbit and merge with no mass escaping; the orbit's binding energy is zero, so the merger product has mass $M_\mathrm{f} = 2M_\mathrm{i}$ and binding energy $E_\mathrm{f} = 2E_\mathrm{i}$. Assuming that the virial theorem holds for the individual progenitors and for their remnant, it follows that the virial velocity dispersion $V_v \equiv \sqrt{-2E/M}$ is unchanged by the merger, while the virial radius $R_v \equiv -GM^2/2E$ is doubled. This argument, only slightly elaborated, has been used to constrain the putative progenitors of elliptical galaxies (Ostriker 1980, Veeraraghavan & White 1985).

The assumption of zero mass loss is violated in any encounter which produces tidal tails escaping to infinity. But simulations of such encounters show only modest amounts of mass escaping (e.g. B92). As the last panel of Fig. 29 shows, the escaping material, which comes from the loosely-bound envelopes of the progenitors, is only slightly unbound; left behind in this example is something like 91% of the total mass and 104% of the total binding energy. Allowing for these escapers revises the expected virial velocity and radius of the merger product upward by 6% and downward by 20%, respectively.

However, the above argument only holds for the *entire* system; gravitational energy is not localized, so the binding energy of a specific component, such as the luminous material, can't be defined. If merger remnants were homologous with their progenitors – that is, if they were simply scaled-up versions – then this argument would work for the entire system and for any lagrangian component. But merging of spherical galaxies violates homology; the characteristic phase-space density $F_0 \propto G^{-3}(-E)^{3/2}M^{-7/2}$, defined using the total mass M and energy E, falls to one-quarter of its initial value in a parabolic merger of identical galaxies with no mass loss, while the maximum phase-space density $F_{\rm max}$ probably drops by less than a factor of two (Duncan, Farouki, & Shapiro 1983, Carlberg 1986). Thus merger remnants tend to be more concentrated than their progenitors.

Evidence for non-homology in merging appears in Table 2, which lists median radii for the luminosity (bulge and disk), dark halo, and total mass of remnant B and its progenitor galaxies. Values for the remnant are given twice, once for all particles and once for those with binding energy $E < 0$; some 99%, 91%, and 92% of the light, halo, and mass, respectively, remain bound to the remnant. As in earlier studies (e.g. B92), the median radius of the total mass is fairly close to predicted value, while that of the light is smaller, and that of the halo is larger. Merging increases the radial segregation of the luminous component with respect to the dark matter; the ratio their median radii goes from ~ 0.37 in the progenitor to ~ 0.25 in the bound remnant. This comes about because the luminous

Table 2. Median radii $r_{\rm m}$ of merger remnant and progenitor

	light	halo	mass
Remnant B	0.200	0.996	0.677
... bound	0.195	0.776	0.552
Progenitor	0.133	0.360	0.280

material – being concentrated toward the centers of the progenitors initially – isn't much disturbed until the galaxies have undergone substantial orbital decay. Consequently the luminous stuff winds up tightly bound at the center of the remnant, while the dark matter, having provided the orbital braking, acquires a more extended distribution (Barnes 1988).

What implications does this dynamical segregation of luminous material have for the optical properties of merger remnants? To begin with, merger remnants have higher surface brightnesses than the systems they were built from, whereas homologous merging predicts that the characteristic surface brightness should fall by a factor of 2. For remnant B, the median projected radius of the bound luminous stuff is $R_{\rm m} \simeq 0.148$, as compared to 0.127 for one of its progenitors viewed face-on, so the average surface brightness within this radius has increased by a factor of 1.46.

Dynamical segregation of luminous material also affects the velocity dispersions of merger remnants; more mass is packed into a smaller radius, so the velocity dispersion must go up (Farouki & Shapiro 1982). In parabolic mergers of composite bulge/disk/halo systems, the velocity dispersions of those particles belonging to the bulges are $\sim 40\%$ higher in the final remnant than in the initial galaxies (Barnes 1988, B92). Moreover, recent merger remnants have velocity dispersions only slightly lower than normal cluster ellipticals of the same luminosity (Lake & Dressler 1986; see § 3.3 of François Schweizer's lectures). Mergers of present-epoch disk galaxies seem capable producing objects with characteristic scales consistent with those of elliptical galaxies; these remnants may be slightly more extended than cluster ellipticals, but seem to be in good agreement with slightly more heterogeneous samples including field ellipticals.

7.3 Radial Profiles

For the remnant of a major merger – as for any hot stellar system – the trend of the distribution function with binding energy has a strong influence on the radial density profile. Indeed, for a spherical and isotropic configuration this influence takes the form of a strict one-to-one relationship (BT87, § 4.4). Most remnants are not exactly spherical and isotropic; nonetheless, the interpretation of radial profiles is informed by knowledge of the coarse-grained distribution function.

To begin with, if the remnant has a fairly isotropic core with constant density $\rho_{\rm c}$ and velocity dispersion $\sigma_{\rm c}$, the maximum phase-space density is finite and approximately $F_{\rm max} \sim \rho_{\rm c}\sigma_{\rm c}^{-3}$. The converse is also true; that is, a finite $F_{\rm max}$ implies that a constant-density core exists. In fact, the core parameters of violently relaxed systems are sensitive to the maximum phase-space densities present in the initial conditions; generally, the higher the initial $F_{\rm max}$, the higher the final $F_{\rm max}$, and the smaller and denser the final core (May & van Albada 1984). To get an object with a bona-fide cusp, even the rather gentle cusp of a de Vaucouleurs profile, the maximum phase-space density must actually be *infinite*.

Spherical systems merge with relatively little radial mixing (White 1979), so the density profiles of the initial galaxies are largely preserved in the merger

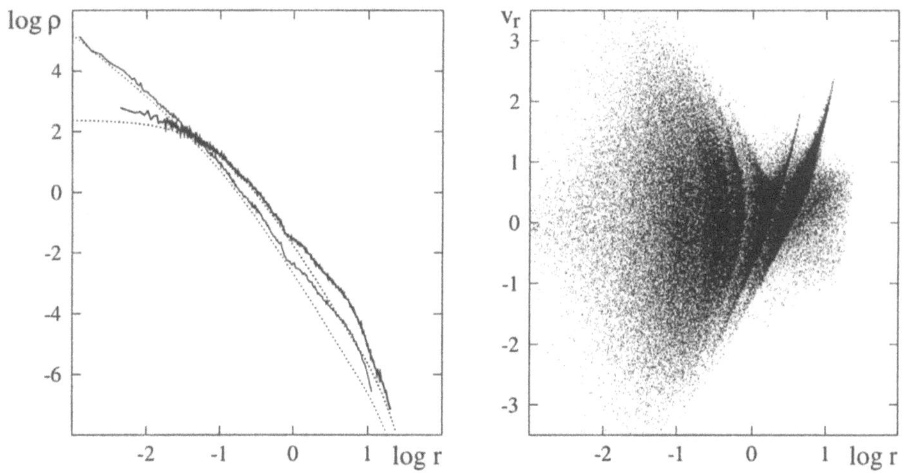

Fig. 32. Left: Density profiles for the core and halo components of a remnant produced by the merger of two spherical galaxy models. **Right:** Radial velocity v_r plotted against log radius for this remnant.

remnants. This is shown on the left in Fig. 32, where density profiles of the remnant produced by a merger of two core/halo galaxies (§ 6.2) are compared with initial profiles. Initially, the core component followed a Jaffe (1982) profile, with a logarithmic slope of -2 at small radii and a steeper slope of -4 at large radii, while the halo followed a $\gamma = 0$ Dehnen (1993) profile, with a constant-density core and a logarithmic slope of -4 at large radii; these profiles are indicated by the dotted lines. The solid lines show the actual density profiles measured about 1.5 time units after the galaxies merge. Out to radius $r \simeq 1$, the merged core profile parallels its initial profile; in particular, the -2 power-law slope of the inner core is essentially unchanged. Evidently, the high phase-space densities required to support steep inner cusps are robust with respect to merging.

The merged halo profile closely matches the initial halo profile between $r \simeq 0.03$ to 1, but at smaller radii it shows a modest positive gradient to the innermost point measured at $r \simeq 0.005$. This shallow slope appears to be due to an inconsistency in the initial conditions; the constant-density core of a $\gamma = 0$ Dehnen (1993) model can't be realized in the logarithmic potential well of a Jaffe (1982) model with a halo distribution function $f(E)$ which is a monotonic function of binding energy. At smaller radii the halo profile should level off because smoothing in the force calculation prevents $\Phi(r)$ from actually diverging as $r \to 0$.

Beyond radius $r \simeq 1$ the core and halo profiles deviate almost identically from their initial power-law slopes. This is not a permanent feature; the local dynamical time-scale increases with radius, and $r \simeq 1$ is the current division between the phase-mixed interior and the still unmixed exterior of the remnant. This is evident from the right-hand part of Fig. 32, where radial velocity v_r is plotted against radius r; the interior is relatively smooth and symmetric with respect to the sign of v_r, while the exterior contains two coherent and approximately parallel streams of material, consisting of particles scattered onto less-bound orbits by the first and second passages (§ 6.4). As Aguilar & White (1985, 1986) show, such an unmixed surge of stars creates a "bump" on the luminosity profile which travels outward, leaving a new profile behind as stars reach apocenter, turn around, and phase-mix into the galaxy. Such bumps, also created by fast tidal interactions, may explain the distended profiles of some early-type galaxies with nearby companions (Kormendy 1977).

The power-law slope of the relaxed envelope can be understood with a simple analytic argument (Jaffe 1987, White 1987). As in § 7.1, let $N(E)$ be distribution of stars in binding energy E, and note that this function is continuous across the escape energy $E = 0$. Once phase-mixing is complete the characteristic binding energy of the stars at radius r is $E \propto -GM/r$, where the enclosed mass M is roughly constant for large r. Therefore $N(E)(\mathrm{d}E/\mathrm{d}r)\mathrm{d}r$ is proportional to the number of stars at radii between r and $r + \mathrm{d}r$, and the stellar density is

$$\rho(r) \propto \frac{1}{r^2} N(E) \frac{\mathrm{d}E}{\mathrm{d}r} \propto \frac{1}{r^4} , \tag{40}$$

where the second relation follows because $N(E)$ is slowly changing near $E = 0$. This asymptotic r^{-4} slope is a common feature of the Jaffe (1982), Hernquist (1990b), and Dehnen (1993) models.

Roughly speaking, the spherically-averaged mass profiles of disks are also preserved by merging. A thin, rotating disk has a much higher phase-space density than an isotropic spherical system with the same mass profile $M(r)$; moreover, in a disk with constant scale-height and velocity ellipsoid shape the maximum phase-space density actually *increases* with radius. But the coarse-grained phase-space density is knocked down by a large factor in the process of scrambling rotating disks to make a pressure-supported spheroid. Mergers of disk systems without bulges produce remnants with rather large cores (Hernquist 1992); anticipating this point, Carlberg (1986) had already argued that only the very brightest ellipticals, which have the largest cores, could be produced by non-dissipative mergers of purely stellar disks. This conclusion is strengthened by the recent finding that luminous elliptical galaxies actually have shallow power-law slopes at small radii (Crane et al. 1993, Jaffe et al. 1994, Byun et al. 1996).

The left-hand panel of Fig. 33 presents density profiles of the remnant produced by a wide ($r_p = 0.4$) version of encounter B with King (1966) model bulges and halos (B92). The merged bulges yield a density profile which levels off at small radii. This contrasts with the profile shown in Fig. 32, which has nearly the same power-law slope as the initial galaxies. However, mergers between disk galaxies with realistically cuspy bulges should yield remnants with steep inner profiles. The density profile of disk particles levels off at small r as expected since the initial disks had finite F_{max}. At larger radii, the exponential fall-off of the initial disks has been erased by violent relaxation, and everywhere apart from the center the disk material dominates the luminosity density. Finally, the

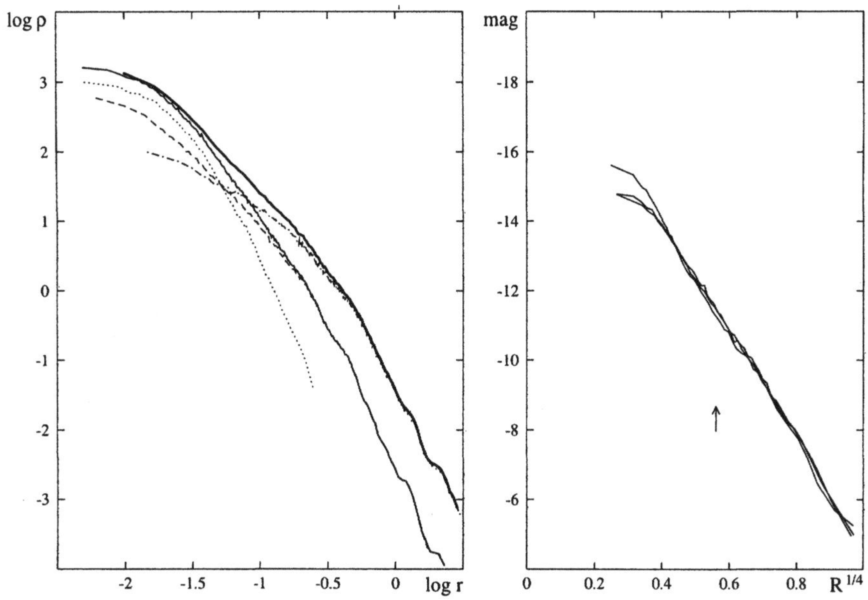

Fig. 33. Density profiles for the remnant in Fig. 30 (B92). **Left:** Spherically-averaged space density profiles; dotted, dashed, and dot-dashed lines refer to material from the bulges, disks, and halos of the original galaxies. **Right:** Surface-densities of luminous particles, projected along three orthogonal directions; the arrow indicates the projected half-light radius.

halo component behaves much as in the core/halo merger, closely tracking its initial profile at most radii. The halo's contribution to the total mass density increases from roughly 10% at $r = 0$ to 50% at $r = 0.08$ and to 90% at $r = 1$. Thus luminosity and mass are still quite segregated in this remnant.

Projected surface brightness profiles, constructed by assigning all bulge and disk particles equal luminosity and averaging in circular apertures, are also shown in Fig. 33. The right-hand panel plots $\mu \equiv -2.5 \log(\text{surface brightness})$ against $R^{1/4}$, where R is the projected radius; in such a plot, a de Vaucouleurs (1948) or $R^{1/4}$ law is a straight line. As in other studies of disk-galaxy merging, the remnant approximates a $R^{1/4}$ law over a fair range of radii. This outcome is often blamed on violent relaxation, which does produce profiles resembling a $R^{1/4}$ law from sufficiently cold and irregular initial conditions (e.g. van Albada 1982, McGlynn 1984). But as the above discussion makes clear, the violence of disk-galaxy mergers falls short of completely rearranging the luminosity distribution; the final match to a de Vaucouleurs law owes a good deal to the fact that even the *initial* luminosity profiles of the bulge + disk used in the simulations are not all that different from $R^{1/4}$ law-profiles (B92). Incomplete violent relaxation has smoothed the transition between the bulge and disk profiles – and replaced the exponential cutoff of the outer disk with a more gradual $\rho \propto r^{-4}$ profile. Compared to the initial conditions used in violent relaxation studies, disk-galaxy mergers are too "hot" to automatically produce $R^{1/4}$ profiles; that such mergers typically do produce such profiles is partly due to the presence of material with high phase-space densities in the initial galaxies (e.g. White 1987, Vedel & Sommer-Larsen 1990).

Fitting a straight line to the right-hand plot yields an extrapolated central surface brightness $\mu_0 \simeq -20.0$ (in highly arbitrary units!) and an effective radius of $R_e \simeq 0.1$ length units. If the initial disks are scaled to the Milky Way (§ 2.6), the projected effective radius is $R_e \simeq 4$ kpc. This is consistent with observations of recent merger remnants such as NGC 7252 (Schweizer 1982), NGC 3921 (Schweizer 1996), and Arp 220 (Wright et al. 1990).

7.4 Shapes and Kinematics

Axial ratios of the remnants produced by the 16 encounters in Table 1 are reported in Fig. 34. Here b and c are derived from the moment of inertia tensor of the more tightly-bound half of the luminous particles; the lines are contours of constant triaxiality $T \equiv (1 - b^2)/(1 - c^2)$. Remnant shape is sensitive to the initial orientations, internal structures, and mass ratio of the victim galaxies. The flattest remnants result when the disks lie in the orbital plane, producing objects with extremely anisotropic velocity dispersions; Hernquist, Heyl, & Spergel (1993) find that the flattening of such remnants is constrained by bending or "firehose" instabilities (e.g. Toomre 1967, Fridman & Polyachenko 1984). When the disks are inclined to the orbital plane, stars are scattered more effectively, producing rounder remnants. Mergers of bulgeless galaxies produce flatter remnants than equivalent encounters between galaxies with bulges as well as dark halos; on the other hand, bulge spin has no significant effect on remnant shape (Hernquist 1992, 1993c). Finally, mass ratio has a strong influence on remnant shape; the products of 3:1 mergers are distinctly more oblate than those produced in 1:1 encounters.

The connection between disk inclinations and remnant shapes is further explored in Fig. 35. The remnant on the left, formed from two disks each inclined by 71°, exhibits boxy isophotes rather as envisioned by Binney & Petrou (1985), while the remnant on the right, produced by a zero-inclination encounter, has distinctly disky isophotes. Not all mergers can be recognized by boxy isophotes; indeed, some even appear *both* boxy *and* disky, depending on radius and viewing angle (Governato, Reduzzi, & Rampazzo 1993, Heyl, Hernquist, & Spergel 1994). Investigations of isophote shape in merger simulations are hampered by poor statistics; compare the 10^5 particles in a typical N-body experiment with the 10^{10} photons captured in a CCD frame. It's also unclear if the non-elliptical features seen in some simulations last for many Gyr; slow evolutionary effects which might weaken such features are largely beyond the scope of present simulations.

To examine the degree of rotational support, the parameter $\lambda' \equiv J/J_{\rm max}$, where $J_{\rm max} \equiv \sum_i m_i |\mathbf{r}_i| |\mathbf{v}_i|$, was computed for the more bound half of the luminous particles of each remnant. This parameter takes a maximum value of $\lambda' = 1$

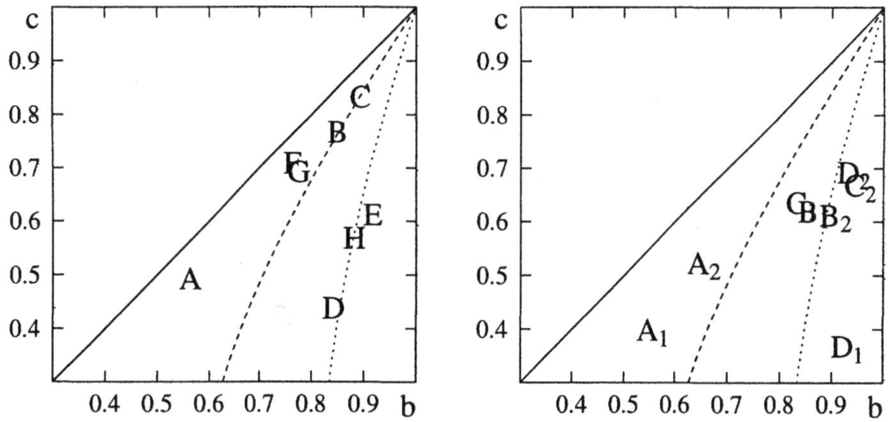

Fig. 34. Axial ratios of remnants from mergers of bulge/disk/halo galaxies. Solid, dashed, and dotted lines are contours of triaxiality $T = 1$, $\frac{2}{3}$, and $\frac{1}{3}$, respectively. **Left:** equal-mass mergers; **right:** three-to-one mergers.

Fig. 35. Edge-on views of two simulated merger remnants, with isophotal contours spaced 0.5 mag apart. On the left is the result of an encounter between two inclined disks, while on the right, both disks lay in the orbital plane (after Hernquist 1993c).

for a cold rotating disk. Fig. 36 plots λ' against the minor axis ratio c. Disk orientation, mass ratio, initial galaxy structure, and initial orbit all influence the angular momentum a remnant retains. Three-to-one remnants rotate rapidly; all have λ' values greater than that of the most rapidly rotating equal-mass remnant. The various equal-mass remnant samples generally overlap one another, but those produced using the present B/D/H galaxy models (§ 2.6) rotate about twice as fast as remnants produced using King (1966) model bulges and halos; compare remnants A–D with the otherwise equivalent remnants 1–4. Evidently the more compact halos used in the present simulations, besides delaying orbit decay (§ 6.2), are also a bit less effective at absorbing angular momentum. Finally, wide ($r_{\mathrm{p}} = 0.4$) encounters produce remnants rotating more rapidly than close ($r_{\mathrm{p}} = 0.2$) ones (B92); compare remnants 1–4 with 5–8.

These points aside, the basic message of Fig. 36 is simply that flattening and rotation are not well-correlated; objects with similar c span a wide range of λ', and vise versa. Rotation may impose a *lower* limit on flattening as hinted by

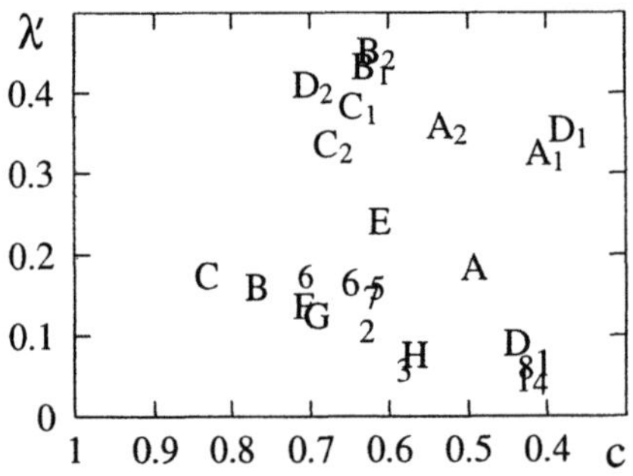

Fig. 36. Minor axis ratio c plotted against angular momentum parameter λ'. Remnants 1–4 and 5–8 were produced by close ($r_p = 0.2$) and wide ($r_p = 0.4$) encounters from B92; larger symbols are reruns with larger N. Remnants A–H and A_1–D_2 are the products of the equal-mass and three-to-one encounters from Table 1.

the lack of points in the upper-left corner of this plot. Thus remnants like D_2 may be flattened by rotation, but most of these objects – like bright elliptical galaxies – owe their shapes to velocity anisotropy (e.g. Binney 1976, Aarseth & Binney 1978). On the whole, the range of flattenings and the lack of rotational support both strongly identify remnants of equal-mass mergers with bright elliptical galaxies.

Given how little angular momentum many of these remnants posses, it's not surprising that some have significant misalignments between their spin and minor axes. Fig. 37 compares misalignments in the two sets of equal-mass remnants from B92 and from Table 1. Here ψ_X, ψ_Y, and ψ_Z are the angles between the spin axis and the major, intermediate, and minor axes, respectively. A remnant with $\psi_Z = 0$ rotates about its minor axis, while one with $\psi_X = 0$ rotates about its major axis. These two plots are distinctly different. The first sample (left) includes a number of objects with large misalignments, and some fall quite far from the $\psi_Y = 90°$ line. The second sample (right) is much better aligned, and in no case is ψ_Y very different from 90°. The larger misalignments seen in the left-hand panel are partly a consequence of the slower rotation of these objects; only a little angular momentum about another axis is needed to create a large misalignment in slowly-rotating remnants 2, 3, 4, and 8. On the other hand, the misalignment angles for the 3:1 sample, not shown here, are small; all have $\psi_Z \lesssim 15°$, and more than half have $\psi_Z \lesssim 5°$. Their good alignments are consistent with a significant role for rotational flattening in unequal-mass merger remnants.

Projected misalignments between the spin and minor axes of elliptical galaxies span the full range from 0° to 90°. However, this range is not uniformly populated; there is a large peak in the distribution of misalignment angles at $\sim 0°$, and a much smaller peak at $\sim 90°$ (Franx, Illingworth, & de Zeeuw 1991). The wide range of misalignments in remnants 1–8 seemed inconsistent with the observed distribution; consequently B92 suggested that dissipationless equal-mass

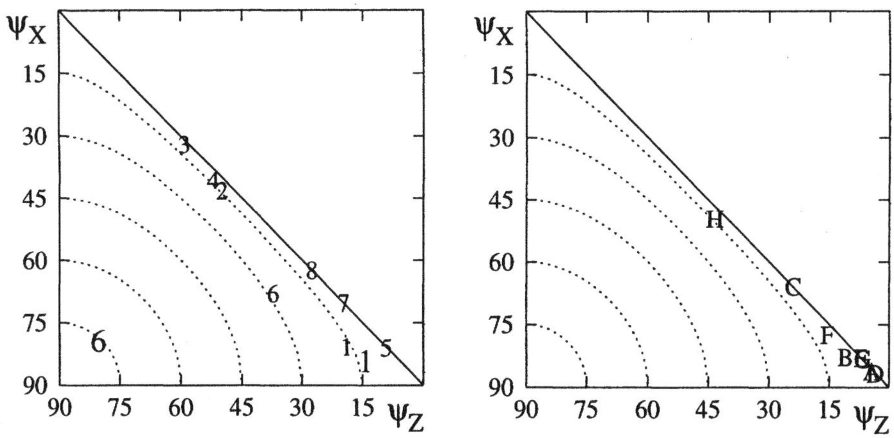

Fig. 37. Intrinsic misalignments of merger remnants. Dotted lines are contours of $\psi_Y = 15°, \ldots, 75°$, while the solid line represents $\psi_Y = 90°$. **Left:** Remnant sample from B92; see caption of Fig. 36. **Right:** Remnants of equal-mass encounters from Table 1.

mergers might have difficulty in reproducing the small misalignments found in most elliptical galaxies. On the other hand, the smaller misalignments of remnants A–H are more consistent with the observations. Evidently initial galaxy structure has a significant impact on the outcome of merging encounters, and misalignment and rotational support should be considered together in comparing simulated merger remnants and real elliptical galaxies.

7.5 Orbit Structure

To understand how misalignments arise, it helps to adopt another point of view, and regard a galaxy as composed of *orbits* rather than stars. In a triaxial potential with a constant-density core and no figure rotation, most orbits can be classified into one of several major orbit families, each parented by a set of stable closed orbits (Schwarschild 1979, de Zeeuw 1985). One set of stable closed orbits circulate about the minor axis; these give rise to the "Z-tube" family. Another set of closed orbits circulating about the major axis give rise to the "X-tube" family.[6] There are no stable orbits circulating about the intermediate axis. However, there are stable closed orbits which shuttle along the major axis and these give rise to the "box" family.

While angular momentum is not an integral of motion in a static triaxial potential, the time-averaged angular momentum of a particle on a tube orbit is nonzero since the orbit maintains a definite sense of circulation about its axis. Averaging over time is equivalent to averaging over orbital phase. Thus the

[6] Actually, there are *two* X-tube families, each parented by a distinct set of closed orbits (de Zeeuw 1985), but this distinction isn't critical for the present application.

stars on Z-tube orbits are responsible for the Z component of a triaxial system's angular momentum, and likewise those on X-tube orbits are responsible for the X component. The net spin vector is therefore constrained to lie somewhere in the plane containing the major and minor axes, implying $\psi_Y = 90°$ (e.g. Levison 1987, de Zeeuw & Franx 1991); Fig. 37 shows that most remnants obey this constraint.

Classification Algorithm. Inspection of individual trajectories shows that many particles follow orbits belonging to one of the major families described above. If the system has a sufficiently steep central potential well, minor orbit families may also be present, including the "boxlet" families described by Miralda-Escudé & Schwarschild (1989). To analyze orbit populations, a fast and reliable algorithm for classifying orbits into families is required. Tube orbits are easily recognized since exactly one component of their angular momentum does not change sign with time (e.g. B92), while boxlets have time-averaged positions offset from the origin (Spergel, private communication). A simple method which classifies major orbit families *and* boxlets is described here.

As input data, the orbit classifier requires the trajectories of particles over some tens of orbital periods. In principle, trajectories could be extracted directly from an N-body simulation. But since different particles may have very different orbit periods, it's quite expensive to run the system long enough to collect the data needed for a comprehensive analysis; moreover, there may be no point in classifying every particle's orbit when a representative sample will often do just as well. Instead, the gravitational potential is "frozen" some time after the merger remnant has settled down, and individual orbits are followed in the frozen potential. It's convenient to expand the potential in Cartesian moments (e.g. White 1983a, Hernquist & Barnes 1990). This expansion retains the overall structure of the potential but suppresses higher spatial frequencies; here the expansion is truncated at quadrupole order, approximating the N-body force field to $\sim 1\%$.

The next step is to set up a coordinate system (X, Y, Z) aligned with the major, intermediate, and minor axes of the remnant. These axes are conveniently defined by the eigenvalues of the moment of inertia tensor for the mass contained within some equipotential surface; here a surface enclosing the inner $\sim 12\%$ of the mass is used.

To classify a particle's orbit, follow its trajectory and record each crossing of the principal coordinate planes. This effectively reduces the trajectory to a string of symbols drawn from the set $\{X, Y, Z\}$. Typically it's enough to accumulate about a hundred such crossings, though in ambiguous cases up to a thousand may be necessary. This symbol string is then tested against various patterns characteristic of different orbit families.

Fig. 38 illustrates the relationship between orbit family and crossing pattern. Tubes, which maintain a definite sense of circulation about the origin, yield simple strings of alternating crossings. Boxlets, such as the "Banana", "Fish", and "Pretzel" families, produce more complex crossing strings, each composed

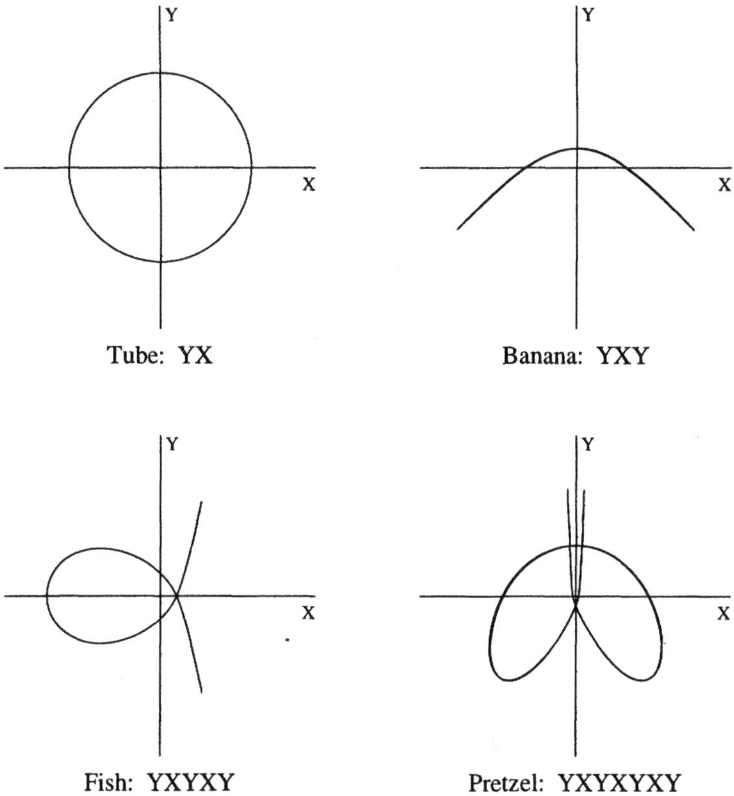

Fig. 38. Parent orbits for several families and the plane crossing patterns they produce (Fulton & Barnes, in preparation). Note that crossing the X plane is equivalent to crossing the Y axis, and vise versa. Only the tube orbit has a definite sense of circulation.

of a basic phrase repeated ad infinitum. On the other hand, true box orbits produce crossing strings without any stable pattern.

Because the initial phase of the orbit may be undetermined, the symbol string starts at an arbitrary point; the routine which compares this string with a stored pattern must allow for this before rejecting the match. And in three-dimensional systems the crossings which identify an orbit's family will be interspersed with crossings of the third plane. For example, a Z-tube orbit might yield the crossing string "$ZYXZYZXYZXYZ\ldots$", and this string must first be projected down to two dimensions by deleting the Z symbols before it can be recognized as a tube. In practice, projections along all three axes are made to take an inventory of possible orbits; this recognizes not only X-tubes, but also such exotic creatures as "Y-Fish", which look like fish orbits when viewed along the Y axis.

This scheme is very fast since it "boils down" a trajectory to a compact string of symbols. A simple extension enables it to recognize orbit families which plunge almost exactly through the center instead of avoiding it; such orbits produce strings in which a specific pair of crossing symbols occur in arbitrary order since the trajectory passes slightly to one side or the other of the origin. Finally, an open-ended version which searches for arbitrary periodic patterns in two *or* three dimensions may be useful in discovering new orbit families (Fulton & Barnes, in preparation).

Equal-Mass Merger Remnants. Applied to the products of merger simulations, orbit classification helps illuminate the relationship between initial conditions and remnant structure. Such an analysis enlarges and supports the results on shapes and kinematics. It also points to "merger signatures" which might distinguish well-mixed remnants of equal-mass mergers from other violently-relaxed systems.

Fig. 39 shows populations of major orbit families for the remnants of equal-mass encounters in Table 1. Each histogram, constructed from a uniform one-in-eight sample of the 65536 luminous particles in each remnant, shows orbit population as a function of specific binding energy E; solid, short-dashed, and long-dashed lines represent boxes, X-tubes, and Z-tubes, respectively. In this plot, boxlets are lumped together with boxes.

Some common features are seen in all these remnants: box orbits dominate the most tightly-bound particles, X-tube orbits occur mostly at intermediate energies, and Z-tube orbits are strongly favored among loosely-bound particles. Since box orbits support triaxial density distributions, while Z-tubes support oblate ones, this change in orbit population with binding energy is also reflected in the shapes of merger remnants, which tend to become more oblate with increasing radius (B92). This general trend originates in the process of orbit decay; particles stripped during the first passage tend to retain a good deal of orbital angular momentum and so wind up on tube orbits, while tightly-bound particles participate in the later and nearly head-on passages and tend to be left on box orbits.

Not unexpectedly, the inclinations of the initial disks significantly influence the resulting orbit populations, with the largest Z-tube populations resulting from in-plane ($i = 0°$ or $180°$) mergers, and significant X-tube populations occurring in mergers between inclined ($i = 71°$ or $109°$) disks. Orbit population tends to correlate with triaxiality; Z-tubes are more prominent in the more oblate remnants, while boxes and X-tubes are favored in strongly triaxial or prolate remnants.

The relationship between orbit population and angular momentum content is illustrated in Plate 2, which presents scatter plots of the binding energy E and the Z component of the specific angular momentum j_Z for particles on Z-tube orbits. Color codes galaxy of origin, with blue for particles from the galaxy with the smaller inclination angle i. Fig. 40 presents similar plots of E and j_X for X-tube orbits.

From these figures it's clear that different encounters produce a variety of remnant structures. Simplest is remnant A, in which both red and blue points

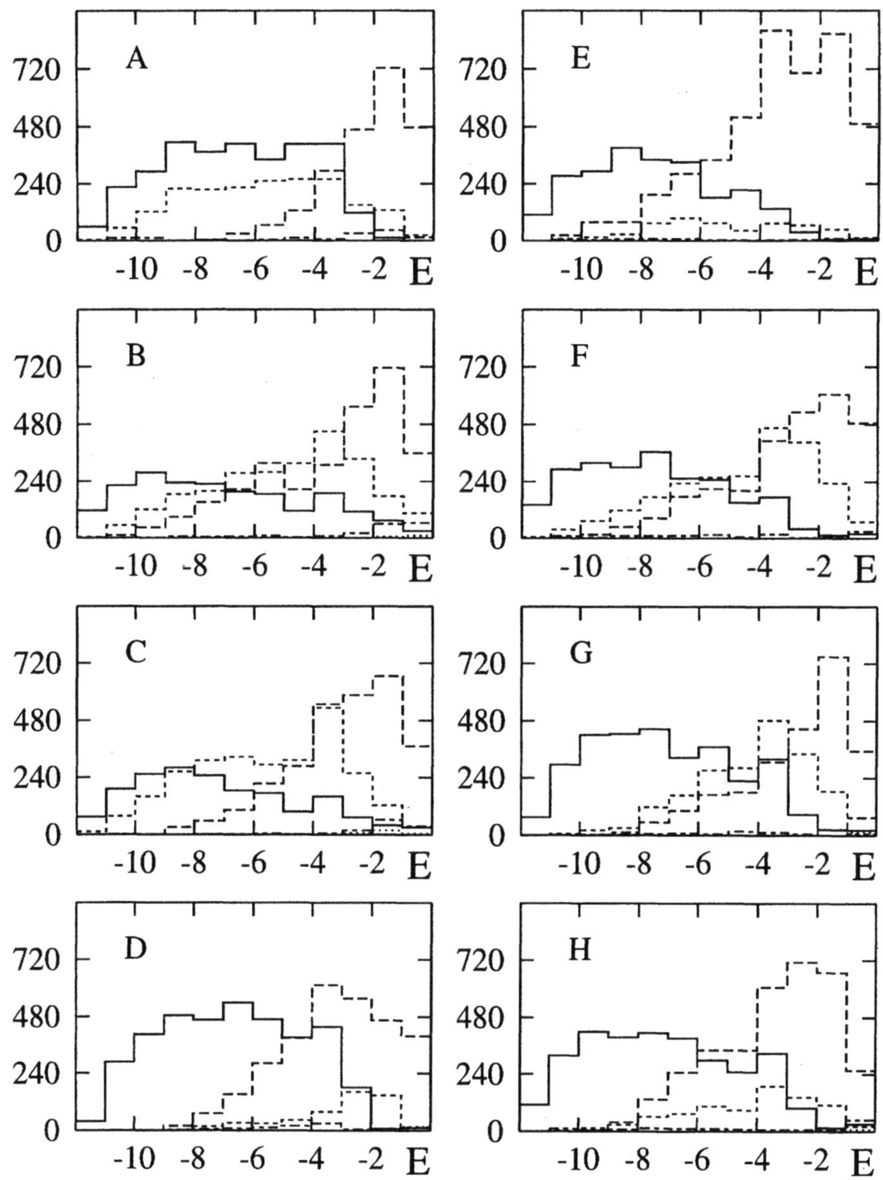

Fig. 39. Orbit populations for remnants of equal-mass mergers. Each plot shows orbit classifications for a uniform one-in-eight sample of luminous particles. The horizontal axis is the specific binding energy $E \equiv \Phi + 0.5v^2$. Solid, short-dashed, and long-dashed histograms show box, X-tube, and Z-tube populations, respectively; other histograms are uncertain classifications.

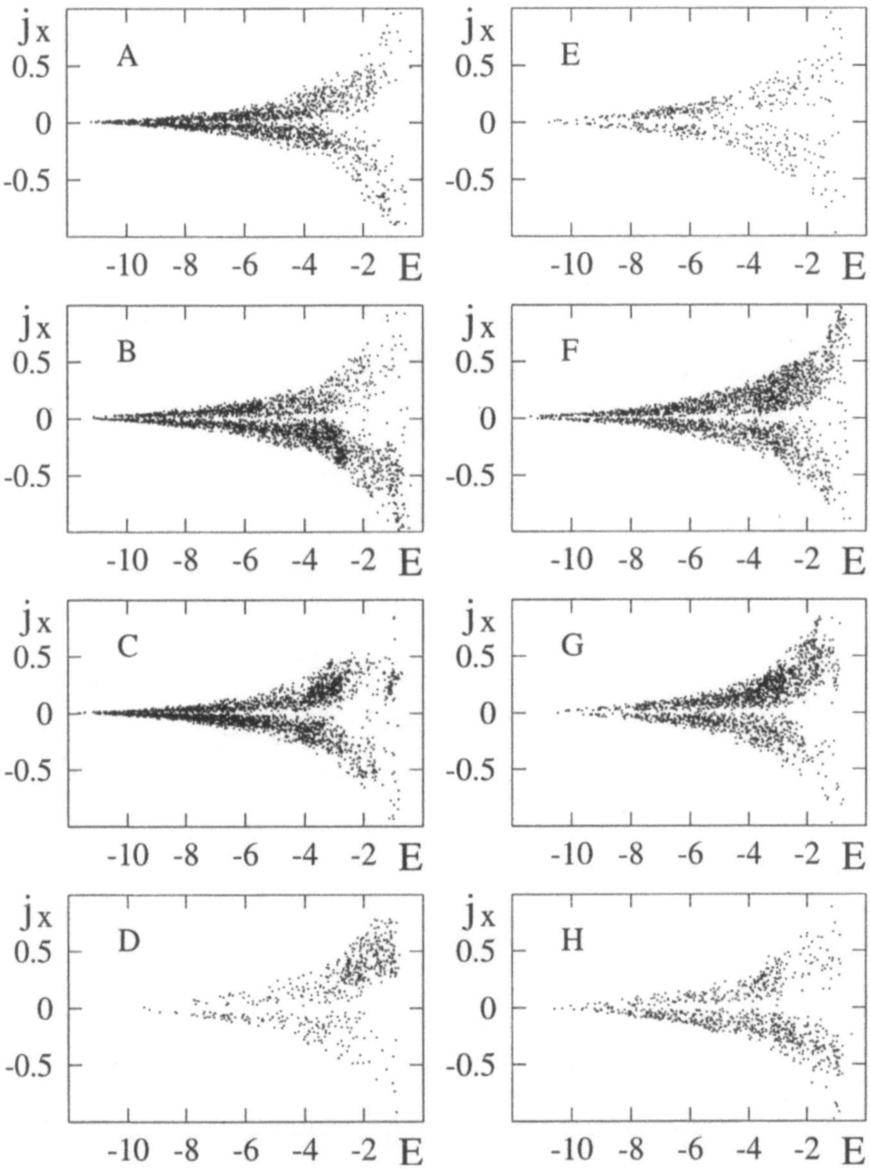

Fig. 40. Specific binding energy E and specific angular momentum j_X for particles on X-tube orbits in equal-mass merger remnants.

have fairly smooth and similar distributions; the particles with the greatest j_Z at a given E generally come from the in-plane ($i = 0°$) disk, but on the whole both disks are fairly well-mixed. The other remnants are more complex, showing either large-scale segregation between particles from different disks or small-scale features. While the latter tend to phase-mix away over time, asymmetries with respect to the j_Z or j_X axes and gross differences in the distributions of particles from the two disks persist indefinitely.

The circulation of the Z-tube orbits is a robust memory of the initial disk orientations. For example, remnants A and E have many direct ($j_Z > 0$) Z-tube orbits from the direct ($i = 0°$) disks; these two objects rotate more rapidly than the other equal-mass remnants plotted in Fig. 36. Remnants B, C, F, and G have large populations of direct Z-tubes from the less inclined ($i = 71°$) disks in these encounters; these remnants also have retrograde ($j_Z < 0$) populations preferentially coming from the $i = 109°$ disks. Finally, remnants D and H have nearly equal direct and retrograde Z-tube populations neatly divided by disk of origin, with the retrograde orbits originating in the $i = 180°$ disks.

The X-tube orbits shown in Fig. 40 also encode memories of disk kinematics. In the remnants produced by encounters D, E, and H, these orbits are populated almost exclusively by particles from the inclined disk; particles from the in-plane disks are not efficiently scattered into X-tube orbits. Remnant A seems to owe its large and relatively well-mixed population of X-tubes to its highly flattened and rather prolate shape, which makes a large volume of phase-space available to such orbits. In contrast, encounters B, C, F, and G, in which both disks inclined, produce remnants with marked segregations about the j_X axis as well as various "clumps" of particles from one disk or another. Such clumps probably come about as the churning gravitational field of the merging galaxies scatters coherent tidal features into X-tube orbits; as they are well-localized in energy, they will persist even after phase-mixing is complete.

Several remnants, in particular C and H, have tightly-bound X-tube populations asymmetrically distributed with respect to the j_X axis. It's no coincidence that these two remnants are the most misaligned of those plotted on the right in Fig. 37. Remnant H, as already noted, is a rather slow rotator. Remnant C, on the other hand, is a fairly rapid rotator and comes by its $\psi_Z \simeq 22°$ misalignment through a strong excess of X-tube orbits with $j_X < 0$. Strong rotators with even more extreme misalignments can probably be produced by adjusting the encounter parameters; for example, a prolate configuration rotating about its *major* axis might result from a face-on encounter between two co-rotating disks.

Misalignments and counter-rotating populations are potential signatures of mergers between disk galaxies. Some of these features may also be produced by other kinds of mergers. However, the angular momenta of the retrograde particles in encounters D and H are fairly unique signs of massive disks. In such remnants, particles from these disks counter-rotate with respect to those from the bulges and halos (B92). Such "kinematic segregation" effects may be relevant to galaxies like NGC 5128, where planetary nebulae show significant rotation, while the system of globular clusters does not (Hui et al. 1995).

Three-To-One Merger Remnants. Fig. 41 presents orbit populations for remnants of the three-to-one encounters listed in Table 1. Like the remnants of equal-mass mergers, these objects show a general trend from tightly-bound box orbits to loosely-bound Z-tube orbits. Consistent with their relatively oblate shapes, they have more Z-tubes orbits and fewer boxes. Most are also quite poor in X-tube orbits, and this can't be blamed on shape alone; only remnants A_1, A_2, and C_1, the first two of which are actually quite prolate, have anything like the X-tube populations found in the equal-mass mergers. Evidently a smaller fraction of material is thrown onto X-tube orbits in the inclined, unequal-mass mergers.

Another striking difference between these remnants and the products of equal-mass mergers becomes evident on plotting E and j_Z for Z-tube orbits as shown in Plate 3. In equal-mass remnants, particles of a given energy are spread over a wide range of specific angular momenta. But in remnants B_1, B_2, C_1, and D_1 the particles from the primary disks, shown in blue, have instantaneous $|j_Z|$ values within a narrow range just below the maximum possible at that E. The orbits of these particles are not as circular as those in the initial disks, but they are much closer to circles than the Z-tubes found in equal-mass merger remnants. Viewed edge-on, these objects show rapid rotation, with mean streaming velocities up to 75% of the circular speed.

In several cases, particles from the primary disks remain not only on relatively circular orbits but also close to the original disk plane. This is most evident – and least unexpected – in remnants A_1 and D_1, where the primary disk initially lay in the orbit plane; as the right-hand part of Fig. 35 already shows, in-plane encounters may yield rather disky remains. Several other remnants including B_1 and possibly B_2 and C_2 also show traces of disks, though incomplete phase-mixing at $t = 9$ time units may account for some of this morphology.

Thus in about half of the three-to-one encounters, the rapid rotation and disk-like morphology of the remnants seem inconsistent with the known characteristics of bright elliptical galaxies. However, these objects could well be classified as S0 galaxies.

Boxlet Orbits. The possible consequences of significant populations of boxlet orbits are incompletely understood. Ordinary box orbits can be quite thin, extending far more along the major axis than along the intermediate and minor axes; thus highly flattened triaxial density distributions can be constructed by superimposing box orbits. Boxlet orbits, however, aren't as thin, and superpositions of boxlets may be unable to reproduce very triaxial figures. Thus the replacement of ordinary boxes with boxlets in galaxies with steep central cusps may force such galaxies to assume fairly oblate shapes (e.g. Merritt 1997).

Orbit analysis in merger remnants identifies a modest fraction of what seem to be boxlets; some are shown in Fig. 42. These classifications are a bit uncertain since they partly depend on the details of the expansion used for the frozen potential. Moreover, particles in self-consistent N-body simulations are constantly being scattered in and out of boxlet families. When the remnant produced by

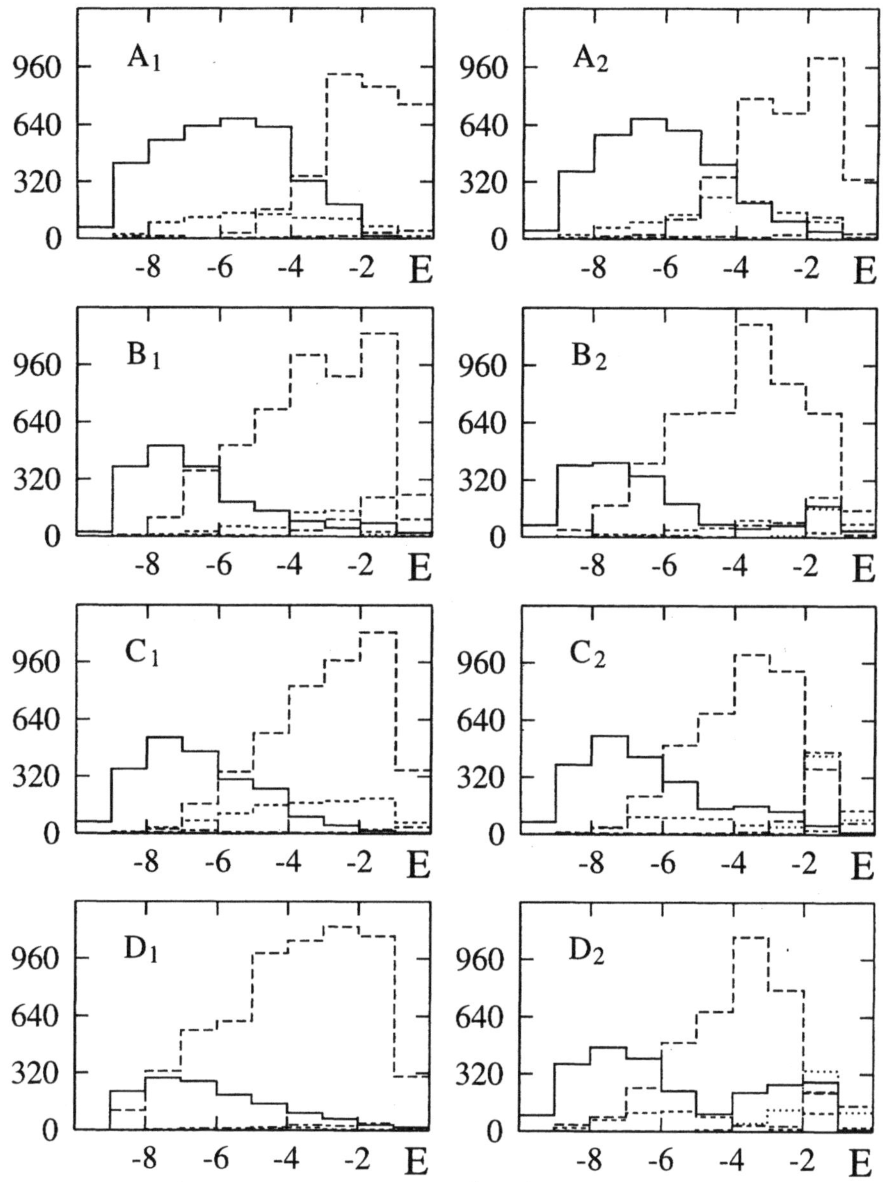

Fig. 41. Orbit populations of three-to-one mergers. As in Fig. 39, a one-in-eight sample is shown. Solid, short-dashed, and long-dashed histograms again represent boxes, X-tubes, and Z-tubes; other histograms represent minor misclassifications arising because of persistent axial twists in the outer parts of remnants B_2, C_2, and D_2.

encounter A is analyzed separately at times $t = 6$ and $t = 9$, the energy distributions and memberships of the box, X-tube, and Z-tube families are fairly stable; the modest changes observed may be blamed on two-body relaxation (e.g. van Albada 1987), and on "after-shocks" of the merger at $t \simeq 3.5$. But over the same interval the memberships of the various boxlet families change drastically; for example, while the total population of fish orbits is fairly constant between times $t = 6$ and $t = 9$, relatively few particles remain on fish orbits throughout this interval. At best, the results of the classification procedure may indicate how many boxlets a hypothetical, partly-smoothed version of the system might support.

With this caveat in mind, Table 3 list populations of Z-axis boxlets as percentages of all luminous particles. Remnants are sorted by intermediate axis ratio b. Only remnants with more than 0.5% boxlets are listed, and columns are left blank if the entry is less than about 0.05%. Y-axis boxlets, not listed, are less numerous but show similar patterns. There are no analogous "X-axis boxlet" orbits since orbits in the YZ-plane are unstable.

Different simulations may produce different boxlet populations. As particles diffuse in and out of boxlet orbits due to two-body scattering, each family will reach an equilibrium with the phase-space density in its vicinity. Thus the steady-state population of a boxlet family should scale with the phase-space volume that family occupies; in these remnants, which all have shallow central cusps further smoothed during force calculation, axial ratio governs boxlet phase-space volume. With parenting orbits remaining in the $Z = 0$ plane, the existence of Z-boxlets will depend on the intermediate axis ratio b. For example, remnants A and A_1 are the only ones with fish orbit populations, probably because these remnants are the only ones flat enough ($b \lesssim 0.6$) to support fish orbits, which oscillate 1.5 times faster in Y than in X. And yes, we have no bananas in any of these remnants because none of these remnants are elongated enough to raise the Y-to-X frequency ratio to 2. But available phase-space volume may not be the only factor influencing boxlet populations. Suppose, for example, that most box orbits in a remnant are relatively long and narrow; these box orbits will be

Table 3. Boxlet populations in merger remnants.

ID	b	5:3	3:2	7:5	4:3	5:4	6:5	7:6
A	0.56	0.12	1.36	0.52	0.43			
A_1	0.56		2.69	0.69	0.71	0.09		
A_2	0.66			0.50	0.88	0.17		
F	0.76			0.09	0.56	0.33		0.07
G	0.78			0.44	1.41	0.33	0.54	
D	0.84			0.39	1.25	0.77	0.20	0.11
H	0.88				0.21	0.30	0.07	0.11
C	0.89				0.28	0.37	0.07	

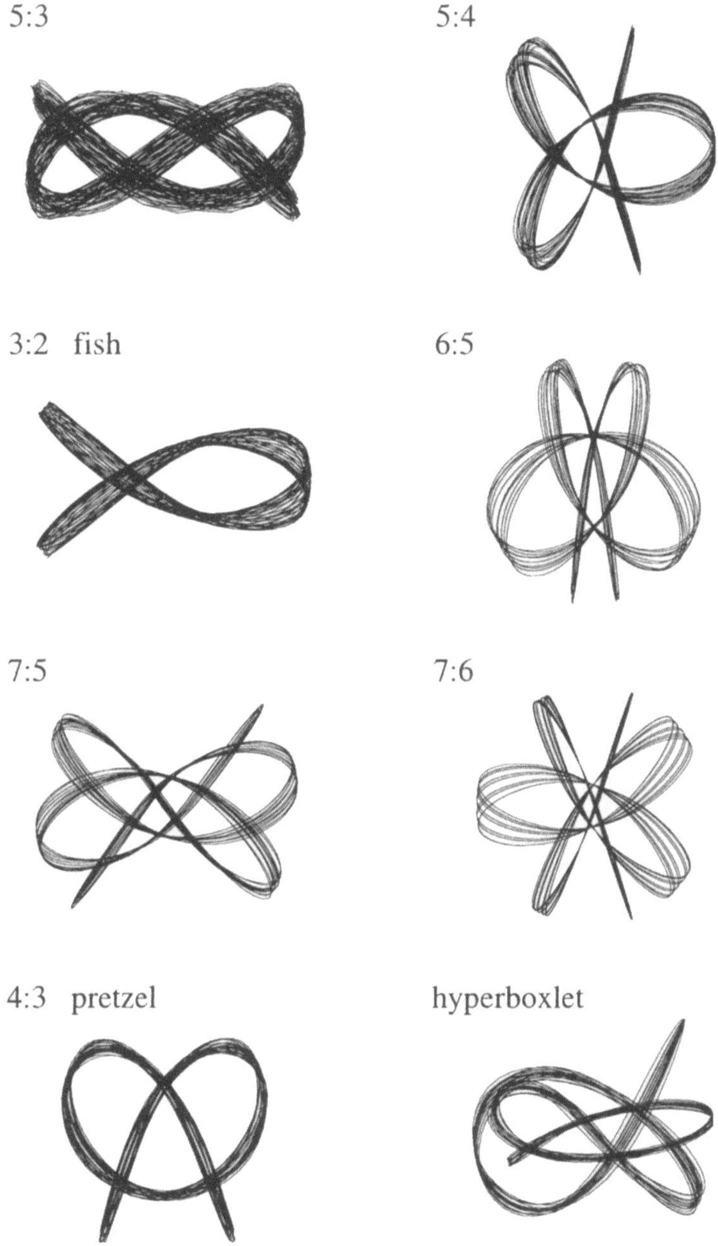

Fig. 42. Boxlet orbits found in simulated merger remnants. The left-hand column shows the low-order fish and pretzel boxlets and two centrophilic orbits; the 5:3 orbit lies between the banana and fish orbits, while the 7:5 lies between the fish and pretzel orbits. The right-hand column shows three higher centrophobic orbits, and a 3-D resonance or "hyperboxlet".

slow to diffuse into pretzels and higher resonances which extend nearly as far in Y as in X. Thus the density of particles on orbits near each boxlet family must also play a role in populating boxlets.

The rather small boxlet populations in these remnants, always less than 10% of all box orbits, are evidently consistent with their moderate flattenings. It seems likely that flatter remnants could be constructed, though as some point violent bending instabilities may develop. But in more realistically concentrated remnants the boxlet fraction may increase, and at some point boxlet orbits could constrain remnant shapes.

In real mergers, boxlet orbits may be populated by stars scattered off potential fluctuations during violent relaxation. Such fluctuations have no way to "pick out" trajectories which eventually lead to boxlet orbits; they sow stars widely across phase-space, and some stars land on boxlets. But like tube orbit families, boxlet families occupy phase-space volumes consisting of several disjoint regions. For example, fish orbits may point in either direction, but a left-pointing fish never turns around and points to the right. Some encounter geometries could produce remnants with more fish orbits pointing in one direction than the other, and such remnants might retain a "fishy" asymmetry even after phase-mixing. This admittedly hypothetical scenario does imply that some forms of fine structure may persist indefinitely. On the other hand, boxlets could also be populated as slow changes in the potential fill boxlet families from a well-mixed sea of box orbits. This would produce symmetric boxlet populations, with fish pointing equally often in either direction, and no significant asymmetries would result.

8 Gas Dynamics in Mergers

Including even modest amounts of gas can markedly alter the outcome of merger simulations. Unlike the collisionless components, gas evolves *irreversibly* once shocks develop. Shocked gas can't return to its initial state; in contrast to the stars, the gas rapidly forgets its past history. Dissipative effects may help explain why elliptical galaxies have such a limited range of kinematic structures; dissipation generally creates attractors in dynamical systems, and galaxies are no exception.

In fact, dissipative effects are needed to explain how galaxies form at all. Collisionless stellar dynamics only decreases the coarse-grained phase-space density of a galaxy; it can never pack more stars into a given phase-space volume than are present initially. Dissipation allows gas to become concentrated in regions of high space density. Stars formed from this gas will – at least initially – have relatively small velocity dispersions and thus high phase-space densities. The subsequent collisionless evolution of these stars may well produce the range of coarse-grained phase-space densities populated in galaxies. Many aspects of this general picture are active in dissipative encounters and mergers of disk galaxies.

As pressure forces on cool interstellar material are normally weak compared to gravity, tidal features such as bridges and tails generally contain similar distributions of gas and stars. Unlike the stellar components, however, streams of gas can't pass through each other; the velocity field must be single-valued everywhere. As a result, gas and stars become separated. For example, galaxies can be swept clean of gas in deeply interpenetrating collisions (Spitzer & Baade 1951). Perhaps more commonly, nearly-direct encounters can produce tidal features with the gas concentrated along a narrow ridge-line bisecting the stellar distribution (e.g. Wallin 1990). But the most general mechanism for separating gas and stars involves the response of gas-rich disk galaxies to tidal perturbations (e.g. Noguchi 1988, Hernquist 1989, Barnes & Hernquist 1991).

8.1 Inflows in Perturbed Disks

Fig. 43 compares gaseous and stellar components of a tidally perturbed disk (Barnes & Hernquist 1991, BH96). These frames come from a version of encounter 1 (§ 4.1) simulated with combined N-body/SPH code (Hernquist & Katz 1989). The disk inclinations and initial orbit are identical to those of encounter A, the galaxies have King (1966) model bulges and halos, and 10% of the disk mass is gaseous. The first pair of frames show the pre-encounter disk with noise-driven spiral structure, while the second pair show the other galaxy's passage at time $t = 1$. After this passage the gas and stars briefly define "S-shaped" spiral patterns, but by $t = 1.375$ both components have developed strong bars. The gas bar then sinks toward the center and settles into an elongated oval ring, roughly aligned with the stellar bar but only one-third as long.

From this figure and from the video of this calculation (BH96) it's evident that the gas which sinks to the center does so without much gas-dynamical interaction with other material. The proof is presented in Fig. 44. Here, the top

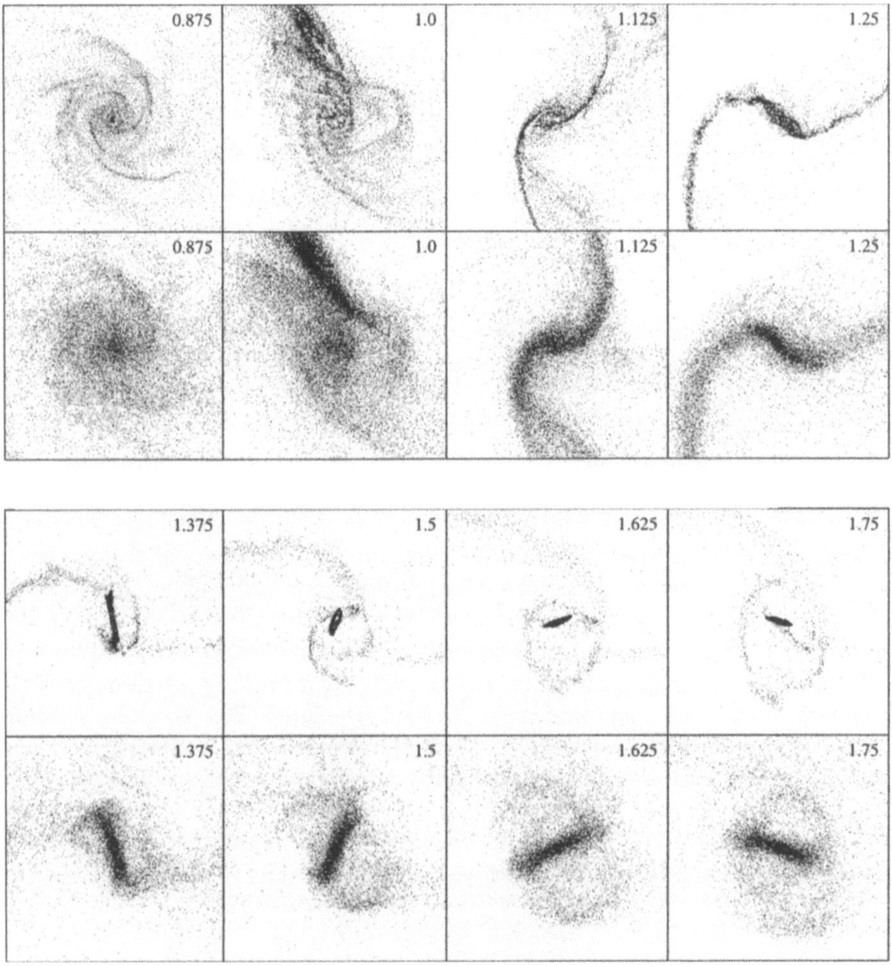

Fig. 43. Response of a gas-rich disk galaxy to a strong tidal perturbation. This is the $i = 0$ disk from the SPH version of encounter 1. Time in simulation units appears in the upper right of each frame. At each time, the upper plot shows the gas, while the lower plot shows the stellar disk.

panel shows the specific angular momentum, j, of the Lagrangian volume of gas which has collected within a radius of 0.03 length units of the center of the direct disk at time $t = 1.75$. Initially most of this gas lies within 0.1 length units of the center, though some comes from a leading two-armed spiral which extend several times further out, and this outlying gas carries a good fraction of the initial angular momentum. During the time interval shown in this figure, the total angular momentum of the gas drops by a factor of ~ 30. The rate of change of j with respect to time, $\mathrm{d}j/\mathrm{d}t$, found by differentiating the curve in the upper panel, is plotted as a smooth curve in the middle panel of Fig. 44.

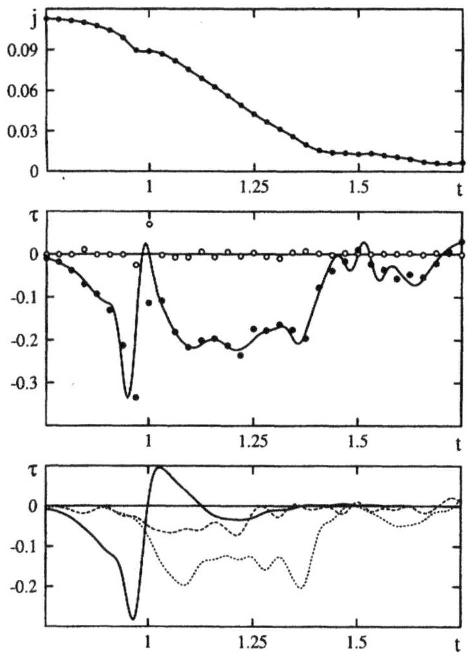

Fig. 44. Angular momentum transfer from central gas (Barnes & Hernquist 1991, BH96). **Top:** Specific angular momentum of the gas which collects at the center of the direct ($i = 0$) disk; the smooth curve is a spline fit through the data points. **Middle:** Specific torques acting on this gas. Here the smooth curve is the derivative of the curve plotted above. Filled circles are gravitational torques on the gas due to the rest of the system, while open circles are torques due to gas-dynamic forces. **Bottom:** Components of the gravitational torque; the solid line is the torque due to the other galaxy, the dashed line is the torque due to the bulge and halo, and the dotted line is the torque due to the disk and bar.

Also shown are the gravitational and gas-dynamic torques acting on this gas as filled and open circles, respectively. Close agreement between $\mathrm{d}j/\mathrm{d}t$ and the gravitational torque makes it quite clear that the angular momentum of the gas is being extracted by *gravity* (Noguchi 1988, Combes, Dupraz, & Gerin 1990, Barnes & Hernquist 1991). Only near first pericenter is the non-gravitational torque at all significant, and here angular momentum is added to the gas as a result of direct interaction between the interpenetrating disks. At other times, the net gas-dynamic torque on this material is basically zero.

Sources of gravitational torques are disclosed in the bottom panel of Fig. 44. Here the solid line represents the torque due to the *other* galaxy, while the dotted and dashed lines represent the torques on the gas due to its *own* disk and spheroid, respectively. The torque directly due to the other galaxy is initially negative but changes sign at the passage ($t = 1$); on the whole it brakes the rotation of the gas, but only by a modest amount. But as the disk develops a tidal bar after $t = 1$ the net torque between the in-flowing gas and its own galaxy takes over. This torque is dominated by the gravitational force between the gas and stellar bars. It arises because the gas bar slightly leads the stellar bar as seen in Fig. 43, especially at time $t = 1.375$ when the gas bar is very narrow and the torque is large.

Though gravity actually extracts the angular momentum from the gas, gas-dynamic forces create the situation in which gravity can act. After all, no inflow occurs if the same material is evolved collisionlessly! Gas-dynamic forces, especially ram pressures in highly-supersonic shocks within the gaseous bar, redistribute angular momentum within the material sinking to the center. Such shocks, being irreversible, lend the system an arrow of time; gas evolves rapidly as long as strong shocks are present.

This simulated inflow stops when the gas settles onto closed, non-intersecting orbits aligned with the bar; once the gas has found such orbits its subsequent evolution is much slower. At this stage, about half of the disk's inventory of gas is concentrated within a radius of 0.03 length units ($\sim$ 1 kpc). The factors governing this length scale are not entirely clear. In isolated barred galaxies the behavior of the gas depends on the existence of inner Lindblad resonances; if the pattern speed of the bar is low enough to permit such resonances then gas is trapped on closed x_2 orbits perpendicular to the bar (e.g. Simkin, Su, & Schwarz 1980, Combes 1994), while if no resonances are present the gas flows inward more effectively (e.g. Friedli & Benz 1993, Athanassoula 1994). The bar in this simulation, like those in simulations of isolated disk galaxies (Sellwood 1981, Sparke & Sellwood 1987), initially rotates too fast to permit inner Lindblad resonances, and no sign of the associated x_2 orbit family is seen in videos of the gas flow (BH96). It's plausible that rapid gas inflow is a transient process associated with bar formation, and the rate and extent of inflow may depend on details of the encounter.

Bar formation is not the only way to transport gas to the centers of interacting systems; encounters between disk galaxies and small satellites may also produce substantial central concentrations of gas (Hernquist 1989, Mihos & Hernquist 1994). In such cases the inflow of gas is correlated with the development of high-amplitude spiral structure as the satellite gets dragged in to the primary. The efficiency of this process depends on the shape of the primary's rotation curve; a rapidly rising rotation curve tends to reduce the amount of accumulated gas.

8.2 Merging Encounters

Plate 4 illustrates the dynamics of cool and hot gas in two similar merger simulations, both started with the same initial conditions as encounter 1. To provide context, the frames on the left show the stellar component, and the boxes within these frames indicate the regions shown in the other columns. The experiment shown in the middle included radiative cooling, while the one on the right allowed only adiabatic cooling. These will be referred to as "radiative" and "adiabatic" models, respectively. The former serves as a proxy for the cold and warm phases of the ISM, which are too cool to generate large-scale pressure forces, while the latter might describe a hot phase, too tenuous to cool rapidly.

Both kinds of gas, responding to gravity, follow the bridges and tails evident in the stellar distribution. But gas streams can't pass through each other, so interstellar material flowing along the bridge from the direct ($i = 0°$) galaxy to its inclined ($i = 71°$) companion stops at the center of that galaxy instead of populating the stellar plume seen to the right at $t = 1.75$. After this time, the cool and hot phases both diverge, albeit in different directions, from the stellar distribution.

The hot gas is perhaps simplest to discuss. In this simulation, it starts out at $T = 10^4$ K, but shocks associated with spiral arms and early tidal features heat it to several times this temperature by first passage. When the disks interpenetrate, a small amount of gas is heated to virial temperatures by high-velocity shocks. Large-scale shocks in the perturbed disks then heat most of gas to $T > 2 \cdot 10^5$ K. Thus by $t = 1.75$ each galaxy is dressed in a thick disk of hot gas enveloping the stellar distribution; cooler gas is confined to the tidal tails, while a small amount of very hot gas is produced as the bridge impinges on the inclined disk. At second passage ($t = 2.25$), these gas disks run head-on into each other, creating an extensive shock front which heats the gas to $T > 2 \cdot 10^6$ K. In the last row of frames, this shock-heated gas has formed a pressure-supported gaseous halo, roughly as extended as the stellar distribution but already smoother and more regular.

Early stages in the evolution of the cool gas were described in the last section; dynamical interactions with tidally-induced bars rob gas of its spin and drive it inwards. By $t = 1.75$, over half of all the gas in these galaxies is concentrated in two elongated oval rings with densities $\gtrsim 10^3$ model units ($10^2\,\mathrm{cm}^{-3}$). Moreover, each of these narrow ovals is surrounded by a roughly circular disk or ring of low-density gas. This material, originally part of the tidal tails, has been re-accreted by the galaxies; an accretion stream and the ring it builds up are seen in the last three frames of Fig. 43. During the second passage, which commences around $t = 2.25$, the re-accreted material interacts both gravitationally and gas-dynamically, while the central ovals remain relatively undisturbed.

Fig. 45 details what happens to the cool gas in the second and later passages. The most immediate consequences of the second passage are to temporarily increase the spin of the cloud in the direct disk, and to slew the rotation axis of the cloud in the inclined disk counterclockwise by $\sim 45°$. But orbit decay following the second passage reduces the orbital angular momentum of the nuclei by about an order of magnitude; thus the third passage ($t \simeq 2.563$) is very nearly head-on, with the gas clouds plowing into each other at $\sim 500\,\mathrm{km\,s^{-1}}$. Radiative cooling almost instantly dissipates the kinetic energy of this collision. Nonetheless, the gas clouds retain some momentum and manage to separate slightly before falling back together. By the last frame, they have coalesced, forming a single massive gas cloud at the center of the merger remnant.

While details of this encounter depend on the disk inclinations and pericentric separation, the general outcome is similar in other cases. Fig. 46 summarizes the

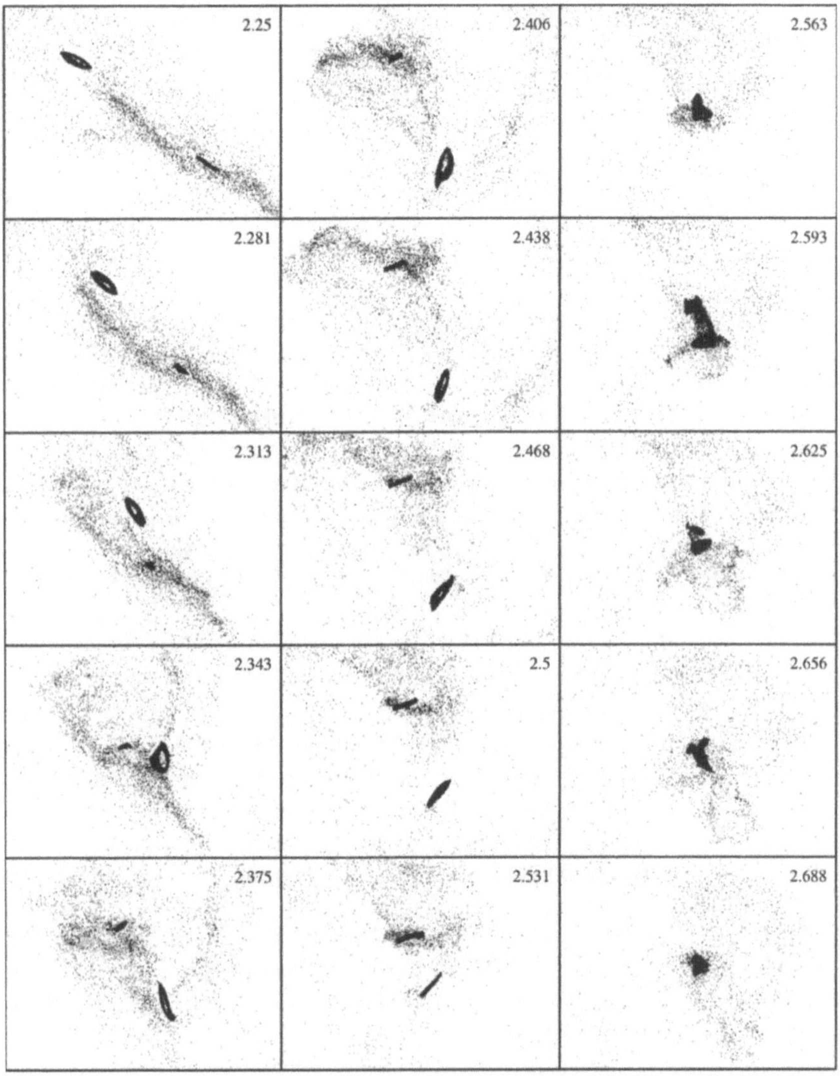

Fig. 45. Final passages and merger of gas-rich disk galaxies in an SPH version of encounter 1 (BH96). These frames, each 0.4×0.3 length units, show only the gas.

dissipative evolution of four radiative simulations (BH96). Two are N-body/SPH versions of encounters 1 and 4, resembling encounters A and D respectively; the other two are N-body/SPH versions of encounters 6 and 7, which are wider ($r_p = 0.4$) variants of encounters B and C. All were run with King (1966) model bulges and halos. The heavy descending curve in each frame shows the binding energy lost to radiation. The energy lost can be quite significant; for example, the SPH version of encounter 4 gives up some 14% of its *total* initial binding energy, or several times the initial kinetic energy of the gas. Much of the energy

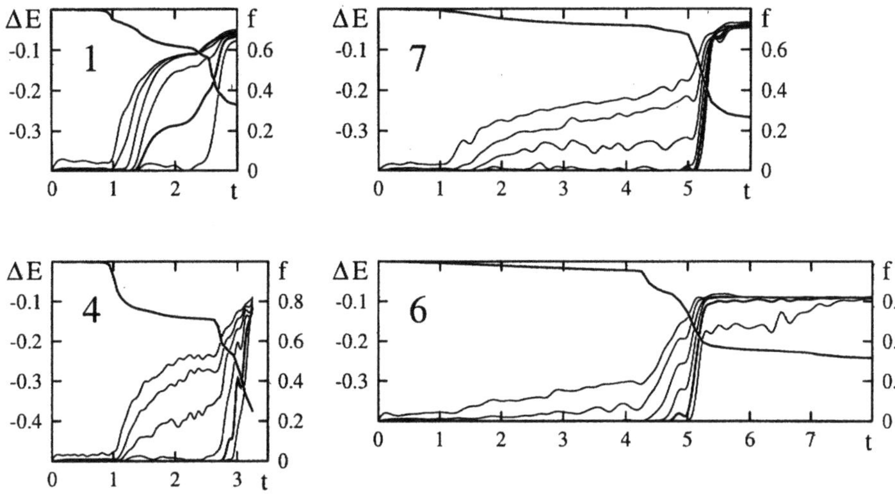

Fig. 46. Energy dissipation and density evolution for various encounters (BH96). Descending curves show changes in binding energy due to radiative losses; for comparison, the total binding energy of each encounter is $E \simeq -2.81$. Ascending curves show gas fractions above densities $\rho = 10^2$, $10^{2.5}$, 10^3, $10^{3.5}$, 10^4, and $10^{4.5}$; a heavy line is used for $\rho = 10^4$, which corresponds to $\sim 10^3\,\mathrm{cm}^{-3}$.

lost comes from potential energy liberated as the gas falls into and deepens the potential wells. Falling inward, the gas also becomes denser; the ascending curves show fractions of cool gas at densities above thresholds logarithmically spaced between 10^2 and $10^{4.5}$ model units. The tempo of dissipation and density increase varies from one run to another. Only in the SPH version of encounter 1 does gas at densities of 10^4 model units appear before the second or third passage; the weaker bars and spiral patterns produced in the other encounters are less effective at cramming gas into the centers of disk galaxies. In the end, however, all four simulations yield comparable amounts of high-density gas; evidence, albeit crude, for an attractor which "forgets" details of the initial conditions.

8.3 Remnant Structure

Fig. 47 shows the continued evolution of the remnant of encounter 1. The initial conditions are much like those used in encounter A, and Fig. 47 may be compared with Fig. 24, and even more closely compared with Fig. 11 of B92, which like this calculation employed King (1966) model bulges and halos. The large-scale tidal features shown here are very similar to those in the other figures. On the other hand, the structure and kinematics of the central stellar remnant are quite different; in this case the stars relaxed to form a fairly oblate spheroid, while the N-body versions produced tumbling prolate remnants. Possible explanations for this difference will be discussed in § 8.4.

Within the stellar body of the remnant, gas falling in from the tidal tails builds up a warped disk, ~ 0.4 length units in radius, containing about 20% of all

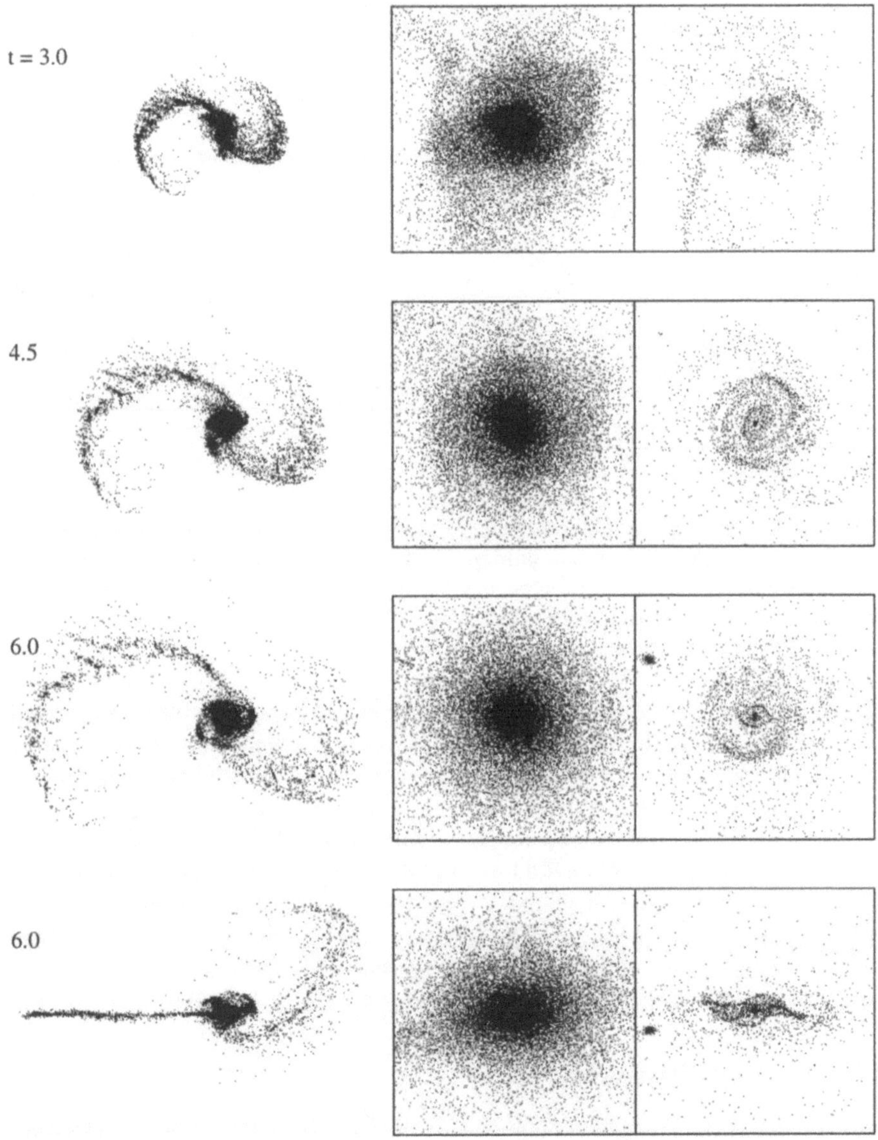

Fig. 47. Evolution of the merger remnant produced by the radiative SPH version of encounter 1 (BH96). The first three views are face-on to the orbital plane, while the last one is edge-on. **Left:** The large-scale stellar distribution at the times indicated. **Middle:** Enlargements of the central stellar distribution; each frame is 0.8×0.8 length units. **Right:** Similar enlargements showing the gas.

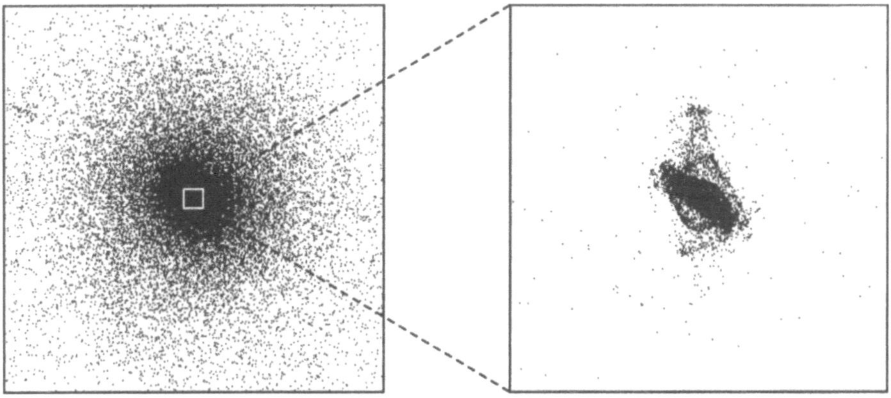

Fig. 48. Stars (left) and gas (right) in the merger remnant produced by the radiative SPH version of encounter 1 at time $t = 6$ (Barnes & Hernquist 1991, BH96). The stellar view is 0.8×0.8 length units, while the view of the gas is enlarged by a factor of 20. Over 60% of all the gas in this remnant lies in this dense central blob.

the gas in the simulation. Another $\sim 20\%$ of the gas remains in the tails awaiting reaccretion or escape. But most of the gas, some 60%, collapses to form a dense central cloud, much as in earlier sticky-particle calculations by Negroponte & White (1983). This compact blob of gas, shown in Fig. 48, formed by the collision of the massive oval rings in Fig. 45; during this collision, much of the angular momentum supporting the rings canceled out, increasing the gas density by a hundred-fold. Scaling the initial disks to the Milky Way, this central cloud has a radius of $\sim 100\,\mathrm{pc}$, a mass of $\sim 5 \cdot 10^9\,\mathrm{M}_\odot$, and a density of $\gtrsim 10^3\,\mathrm{M}_\odot\,\mathrm{pc}^{-3}$. Such densities, if realized in stars, are quite sufficient to account for the central regions of early-type galaxies (Carlberg 1986, Vedel & Sommer-Larsen 1990, Kormendy & Sanders 1992).

The gas distribution in a merger remnant depends on its thermodynamic history. At densities characteristic of cold and warm gas in disk galaxies, radiative cooling is very efficient and shock-heated gas cools down quickly. As described in § 3.2, we therefore imposed a lower cutoff of $10^4\,\mathrm{K}$ on the cooling curve to prevent runaway cooling from "curdling" the simulated gas; most of the gas hovers near this cutoff. Consequently, pressure forces are small in simulations including cooling processes, and the results are very similar to those obtained with the temperature fixed at $10^4\,\mathrm{K}$ (BH96). This point is illustrated in Fig. 49, which shows distributions of gas density in the remnants from three different N-body/SPH simulations of encounter 1 at time $t = 3$. Solid and dashed histograms report simulations with radiative processes and with T fixed at $10^4\,\mathrm{K}$, respectively; the isothermal gas attains slightly higher densities, but by and large these distributions are quite similar. On the other hand, the adiabatic simulation produced a very different result; as the last panel on the lower right of Plate 4 already shows, the hot gas forms an extended, pressure-supported atmosphere instead of collapsing to high densities.

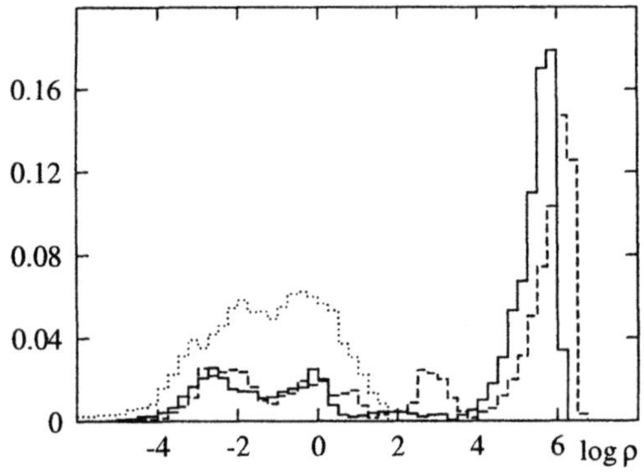

Fig. 49. Histograms of gas density for merger remnants from various SPH versions of encounter 1 (BH96). The solid line shows results for a run with radiative cooling; the dashed line shows results for isothermal gas at $T = 10^4$ K, and the dotted line shows results with only adiabatic cooling.

Hot Gas: Profiles and Kinematics. Radial density and temperature profiles of the hot gas from the adiabatic version of encounter 1 are shown in Fig. 50. The upper panel presents densities from SPH interpolation (points) and spherical averaging (heavy solid line). These measurements are in good agreement out to $r \simeq 1$; beyond that radius much of the gas is concentrated in the tidal tails, and its density is thus underestimated by spherical averaging. Also shown in this panel is the spherically averaged density of the disk *stars* (dotted line), which initially had the same distribution as the gas but evolved collisionlessly. The gaseous and stellar distributions are still pretty similar, though the former seems slightly more extended. As noted by Navarro & White (1993), adiabatic gas may be slightly heated relative to the stellar distribution because the latter can interpenetrate; thus an orbital phase difference develops which transfers kinetic energy to the gas. The lower panel shows the temperature profile of the gas. Within the body of the remnant the temperature is fairly flat at a few million K, matching the velocity dispersion of the stellar distribution. Beyond radius $r \simeq 1$ is material which has adiabatically cooled to $T < 10^4$ K by expanding with the tidal tails.

If the gas can't dissipate, its kinematics soon become very regular; even as early as time $t = 3$ shown in Fig. 51 the velocity field is consistent with rotation about a single axis aligned with the net angular momentum of the remnant. The motion of the hot gas is simpler than that of the stellar component (§ 7.3, 7.4), which is perhaps only natural since the velocity field of the gas must be single-valued, while stellar velocities may span a wide range at each point.

Scaling the initial galaxies roughly to the Milky Way, the remnant has $\sim 8 \cdot 10^9\,\mathrm{M}_\odot$ of hot gas at densities of $\lesssim 1\,\mathrm{cm}^{-3}$. The cooling time at $r \simeq 4\,\mathrm{kpc}$ (~ 0.1 length units) is about 10^8 yr; if only a fraction f of the gas is heated to virial temperatures then the cooling time increases by f^{-1}. This provides some post-hoc justification for including adiabatic gas in merger simulations. However,

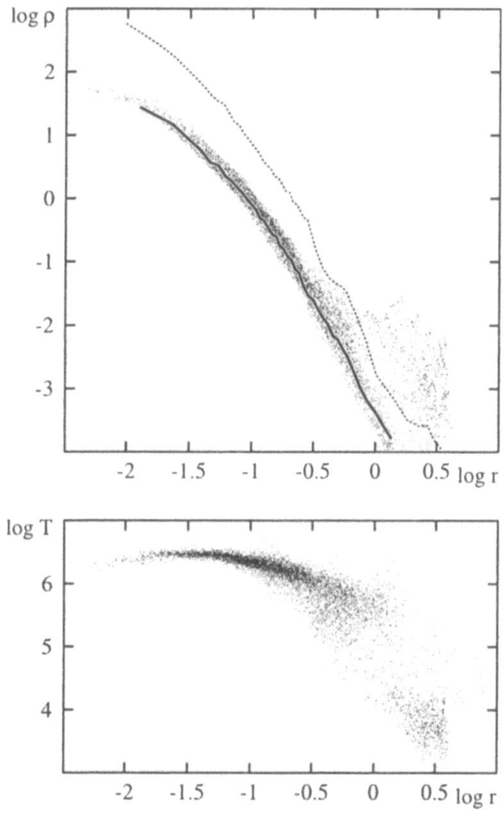

Fig. 50. Density and temperature profiles for the remnant with adiabatic cooling only, measured at time $t = 3$ (BH96). **Top:** Density of gas (solid curve and points) and disk stars (dotted curve). **Bottom:** Temperature of gas particles.

cooling times are significantly shorter for gas at 10^5 K, and it's quite possible, at intermediate stages before the final encounter, that much of the gas could cool off between shocks; if such cooling is effective, much of the available mechanical energy is lost to radiation.

The extended atmosphere of hot gas in this simulation invites comparison with the X-ray gas observed in interacting systems such as The Antennae (Read, Ponman, & Wolstencroft 1995, Sansom et al. 1996) and merger/starburst remnants like Arp 220 (Heckman et al. 1996). In The Antennae, extended soft X-ray emission indicates the presence of $\sim 10^9\,M_\odot$ of gas at $T \simeq 4 \cdot 10^6$ K; much of this gas is associated with the galactic hulks, while a small fraction seems to be part of a bipolar outflow. The bodies of these galaxies contain $\sim 2 \cdot 10^9\,M_\odot$ of molecular gas (Stanford et al. 1990) and $\sim 10^9\,M_\odot$ of atomic hydrogen (van der Hulst 1979), so the hot gas is about 25% of the total inventory. Compared to the simulation discussed above, the hot gas in The Antennae has been heated *earlier*, perhaps even before the two galaxies collide head-on, and possibly more

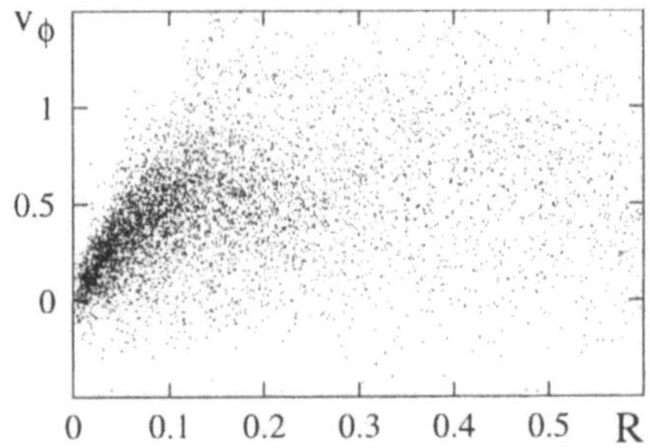

Fig. 51. Azimuthal velocity plotted against cylindrical radius for hot gas from the simulation with only adiabatic cooling.

energetically. In Arp 220, X-ray emission extending to several times the optical radius is produced by $\sim 5 \cdot 10^8\,\mathrm{M}_\odot$ of gas at $T \simeq 10^7\,\mathrm{K}$. This gas may be too hot to remain in Arp 220's potential well, and indeed there is optical evidence for a bipolar outflow (Heckman, Armus, & Miley 1987). Once again, it seems that some additional energy source is needed to heat the gas; it's energetically impossible for purely mechanical heating to drive a bulk outflow.

Hot gas in interacting galaxies and mergers may affect other forms of interstellar material. § 5.4 has mentioned possible dynamical effects of hot gas. In addition, such material might ionize atomic gas returning from the tidal tails; this could explain why the body of NGC 7252 is poor in neutral hydrogen even though a fresh supply is constantly flowing in from the tidal tails (Hibbard et al. 1994, Hibbard & Mihos 1995). The hot gas may also implode molecular clouds within the body of the remnant as suggested by Jog & Solomon (1992); see § 8.2 of Rob Kennicutt's lectures in this volume. In the experiment shown in Fig. 50, the hot gas pressure at $r \simeq 4\,\mathrm{kpc}$ is $\sim 2 \cdot 10^5\, k_\mathrm{B}\,\mathrm{K\,cm^{-3}}$, where k_B is Boltzmann's constant. This is comparable to the internal pressures in GMCs; more detailed studies are needed to see if hot gas can actually implode GMCs throughout the merging hulks before adiabatic expansion and radiative cooling lower the ambient pressure. Such implosions offer a possible way to trigger galaxy-wide starbursts and form the young star clusters seen throughout the bodies of recent merger remnants (see § 3 of François Schweizer's lectures).

Cool Gas: Kinematics. Unlike the material which collapses to the center, gas which first dissipates late in the merging process is likely to share the general sense of rotation of the stellar remnant. An example is shown in Fig. 52. This warped disk of gas, glimpsed roughly edge-on in the bottom-right frame of Fig. 47, is here viewed along the minor axis of the merger remnant. Gas particles close to the equatorial plane are plotted as points, while those more than 15° above or below the plane are shown as crosses and circles respectively. This reveals a *leading* spiral pattern characteristic of a warped disk precessing in an oblate potential. In angular momentum space the warped disk appears as a helix which joins smoothly onto the gas falling in from the tidal tails; thus the warp is driven by the continued accretion of gas following the merger.

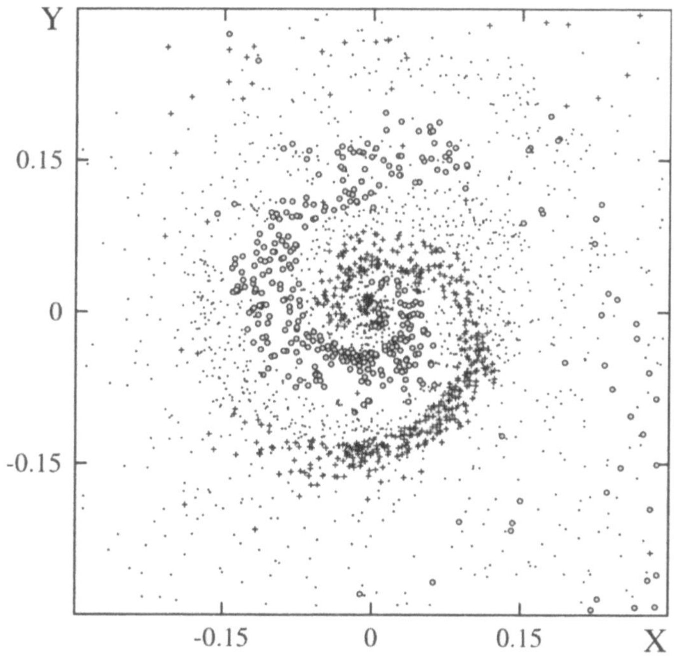

Fig. 52. Face-on view of the warped gas disk in the radiative SPH version of merger remnant 1 (BH96). The disk rotates clockwise. Particles significantly above or below the disk plane are plotted as crosses or circles, respectively.

The disk in this remnant is reminiscent of the warped dust disk in the early-type galaxy NGC 4753 (Steiman-Cameron, Kormendy, & Durisen 1992), though its tilt is more extreme: as much as 40°, as compared to $\sim$ 15° for NGC 4753. Steiman-Cameron et al. blame the disk in NGC 4753 on the accretion of a gas-rich satellite galaxy. In the remnant described here the disk forms, warp and all, by continued accretion of misaligned gas from the tidal tails. This it seems possible that some warped disks in early-type galaxies may have formed not by accretion of low-mass companions but rather as a by-product of mergers between comparable progenitors. A galaxy worth mentioning in this context is NGC 5128, also known as Centarus A (van Gorkom et al. 1990, Nicholson, Bland-Hawthorn, & Taylor 1992, Schiminovich et al. 1994); though its warped disk is often attributed to the accretion of a small companion, the disturbed structure revealed at low light levels hints at a more violent origin (e.g. Schweizer 1986).

In radiative merger simulations some fraction of the gas experiences rapid dissipation, often coinciding with increasing gas density. This material loses a large fraction of its initial angular momentum, as indeed it must if it's to fall to the centers of the interacting galaxies and eventually to the center of the merger remnant. What little angular momentum it does retain may not be well-aligned with the angular momentum of the rest of the remnant.

A dramatic example is provided by the nested pair of gas disks in Fig. 53 (Hernquist & Barnes 1991). At the time $t = 6$ shown here, the outer disk, while rather diffuse, can be traced out to a diameter of ~ 0.2 length units; edge-on it's only 0.01 units thick, and exhibits a modest warp. In the bottom panel, which shows velocities seen by an observer viewing the disks edge-on, the outer disk and gas at larger radii together trace out a nearly flat rotation curve with a peak-to-peak amplitude of ~ 4 velocity units. The inner disk, which appears "burnt-out" in the upper panel, is an oval ring with dimensions of 0.023×0.015 length units; edge-on it is just 0.002 units thick. This component produces the narrow swath of points slanting across the bottom panel, with a amplitude just half that of its outer partner. Thus the two disks are *counter*-rotating; at the instant shown here their spin axes are separated by 165°.

This remarkable state of affairs arose in the radiative N-body/SPH version of encounter 6, in which the two disk galaxies with initially antiparallel spin vectors have a wide, nearly polar encounter. The counter-rotation in the remnant is not, however, a direct result of this initial antiparallelism; both inner and outer disks contain material from both galaxies. The inner disk, comprising $\sim 60\%$ of the total gas, is largely composed of material which collected in the central regions of the two galaxies between their second and third passages. The outer disk, which at time $t = 6$ comprises $\sim 20\%$ of the gas, comes from material initially at somewhat greater radii. Because the outer disk suffered little dissipation until late in the encounter, it more or less shares the angular momentum distribution of the rest of the remnant. The inner material, on the other hand, lost a great deal of angular momentum, and it's perhaps natural that what little it retains is not very well-correlated with its motions; still, it's quite remarkable to find it rotating in the *opposite* direction!

8.4 Dissipation and Stellar Backlash

As noted above, the *stellar* remnant in Fig. 47 differs in one striking respect from similar N-body versions: it's more oblate than its collisionless counterparts. Cosmological simulations show a similar effect; purely collisionless models produce strongly triaxial halos (e.g. Barnes & Efstathiou 1987, Frenk et al. 1988, Dubinski & Carlberg 1991), while models including gas produce rounder and more oblate halos (e.g. Katz 1991, 1992, Katz & Gunn 1991, Udry 1993, Dubinski 1994). It would be very interesting if merger remnants were systematically "rounded out" by dissipative effects. Elliptical galaxies seem to be limited to a small range of stereotypical structures; for example, most have well-aligned rotation and spin axes (Franx, Illingworth, & de Zeeuw 1991) though misalignments of up to 90° are possible outcomes of violent relaxation (§ 7.4). Lower-luminosity ellipticals in particular seem to be nearly oblate systems (e.g. Trembaly & Merritt 1996) which would naturally explain their small misalignments; this seems consistent with the suggestion that fainter ellipticals have undergone more dissipation (e.g. Kormendy 1990).

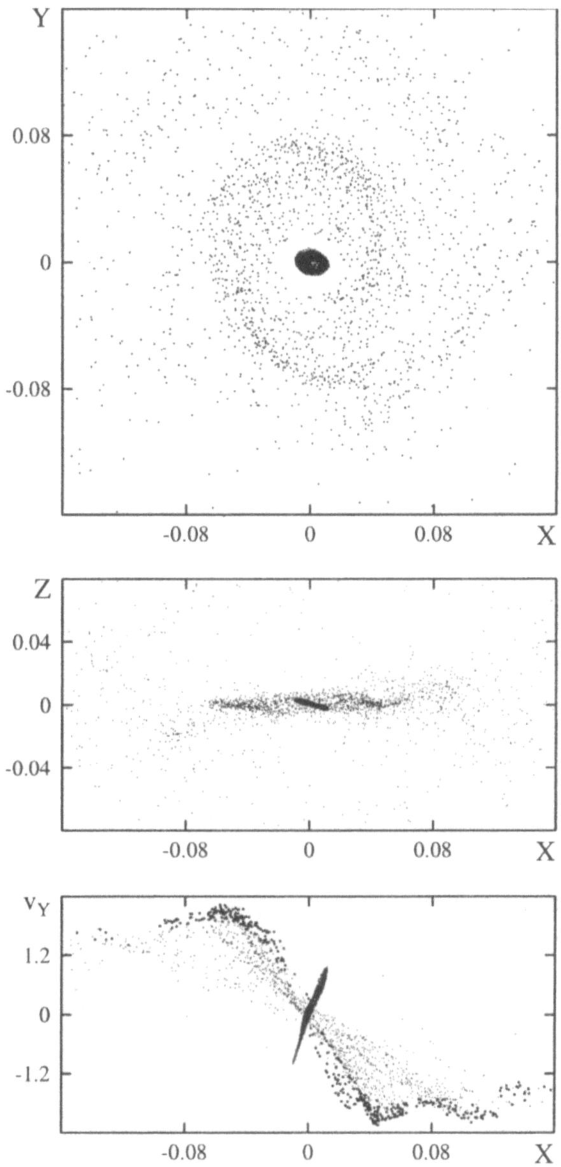

Fig. 53. Counter-rotating gas disks in the remnant produced by a radiative SPH version of encounter 6 (BH96). **Top:** Face-on view; the solid oval at the center is the inner disk, while the loose spiral pattern belongs to the outer disk. **Middle:** Edge-on view of both disks. **Bottom:** Line-of-sight velocities seen from the edge-on projection; points in the outer disk within 0.025 length units of the Y-axis are plotted as small filled circles to highlight the shape of the rotation curve. The counter-rotating inner disk produces the dark streak rising left-to-right across the middle of this plot.

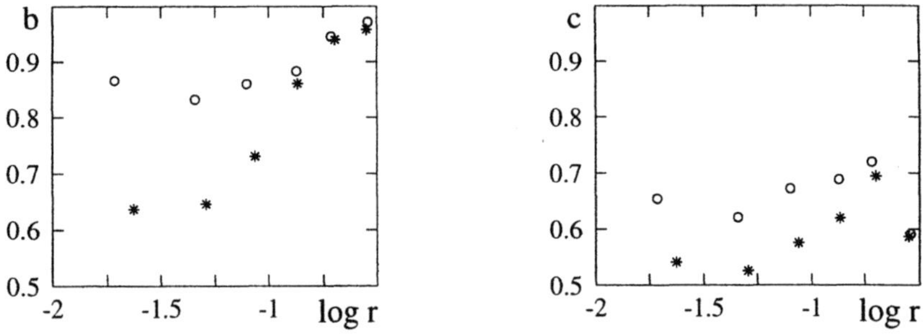

Fig. 54. Ellipticity profiles for remnants from the N-body and SPH versions of encounter 1, plotted using stars and circles respectively. **Left:** Intermediate axis ratio. **Right:** Minor axis ratio.

There are at least two general ways in which dissipative material might influence the dynamics of the stellar component; both rely on the tendency of gas to accumulate in the central regions of merging galaxies. First, as described in § 8.1, inflows during the earlier stages of encounters build up massive central rings of gas, deepening the potential wells of individual galaxies and binding their centers more tightly. Consequently the central regions are more robust and take longer to completely merge, effectively prolonging the violent relaxation phase. Second, as shown in § 8.4, a large amount of gas may wind up in the center of a merger remnant, where it again deepens the potential well. In deeper potential wells a larger fraction of box orbits become boxlets and a smaller range of orbital shapes is available to self-consistently generate the remnant's potential.

In seeking to isolate the effects of gas dissipation it's important to exclude other influences. One possible influence is that N-body and SPH simulations often use different smoothing kernels in gravitational force calculations. BH96 discuss a pair of simulations designed to eliminate this uncertainty. One calculation is the radiative SPH version of encounter 1 already discussed, while the other starts from *exactly* the same initial conditions but treats all particles as collisionless. These will be called the "SPH" and "N-body" calculations, respectively. Both compute gravity using SPH smoothing.

While the stellar remnant produced in this specially-matched N-body calculation is not as flattened and prolate as the remnant of encounter A, it's still flatter and more triaxial than the remnant from the SPH calculation, as shown in Fig. 54. Here collisionless bulge and disk particles have been sorted by binding energy into eight equal bins. Net axial ratios for each bin are plotted as functions of the r.m.s. radius, excluding the two outermost bins which were not yet in equilibrium at time $t = 6$. As the resulting figure shows, out to $r \simeq 0.1$ the N-body remnant has a much shorter intermediate axis than its SPH counterpart; it's also somewhat flatter along the minor axis. What is the origin of this difference?

The central gas rings deepen the potential wells of the individual disks by $\sim 10\%$ at time $t = 2$. Among other things, the deeper potential well in the direct disk seems to increase the pattern speed of its bar by about 50% (BH96). But

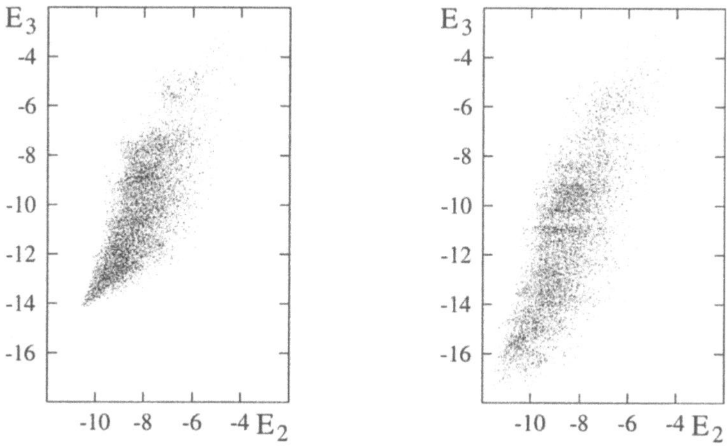

Fig. 55. Scatter plots of specific binding energies of bulge particles at times $t = 2$ and $t = 3$. **Left:** N-body calculation. **Right:** SPH calculation.

these deeper wells don't significantly increase the level of violence in the final encounter. As noted in § 6.3 and shown in Fig. 29, violent relaxation erodes correlations between initial and final binding energies of individual particles. Fig. 55 shows changes in specific binding energy for bulge particles over a time period which straddles the last passages and merger. At first glance the SPH scatterplot appears less correlated than its N-body counterpart, but this is mostly an illusion. Both scatter plots yield identical values of ~ 0.73 for Spearman's rank-order correlation coefficient r_s. On the whole, bulge particles were not shuffled much more effectively in binding energy as a result of the fluctuating gravitational fields due to the merging gas clouds. The most tightly-bound octile seems bit more shuffled in the SPH version, but the effect is small compared to the dramatic difference in shapes illustrated in Fig. 54.

If increased violence during relaxation can't explain the nearly oblate form of the SPH remnant, perhaps the effects of a deeper remnant potential well are to blame (e.g. Gerhard & Binney 1985). The massive blob of gas at the center of the SPH version (Fig. 48), by deepening the potential well by about 20%, does have measurable effects on the collisionless components. Fig. 56 shows orbit family distributions for the N-body and SPH remnants as functions of binding energy E. Consistent with its more oblate shape, the SPH remnant has a larger population of Z-tubes and a smaller population of boxes than its N-body counterpart. As BH96 note from a comparison of these remnants with a more heavily-smoothed version, remnants with deeper potential wells tend to have fewer box orbits, more Z-tubes, and more oblate figures. But how does potential well depth control shape? The central gas blob in the SPH remnant directly adds a nearly-radial component to the force field, which may increase the phase-space volume available to tube orbits. Alternately, the deeper potential well may favor boxlet orbits over the the more elongated box orbits needed to support a highly triaxial system.

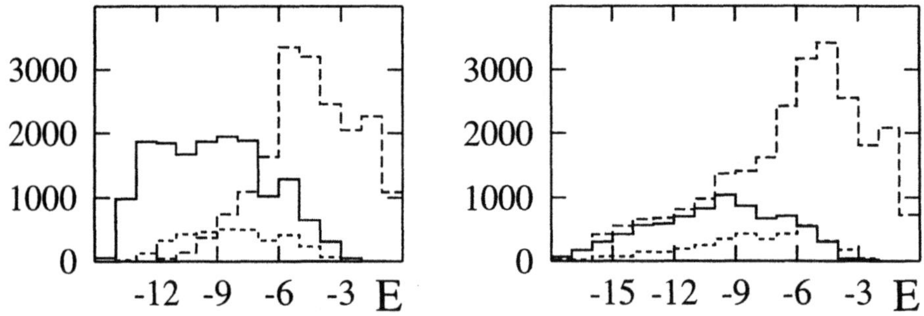

Fig. 56. Orbit populations in two versions of remnant 1. The horizontal axis is the specific binding energy. Solid, short-dashed, and long-dashed curves show populations of box, X-tube, and Z-tube orbits, respectively. **Left:** N-body calculation. **Right:** SPH calculation.

These alternatives may be tested by assaying the phase-space volumes available to box and tube orbits in the potentials of different remnants. In particular, the SPH and N-body potentials can be compared to a composite generated by a mass distribution with the density profile of the former and the ellipticity of the latter; this chimera was constructed by rescaling the radius of each particle in the N-body calculation to its radius in the SPH version. Phase-space volumes of orbit families were estimated by classifying a sample of initial coordinates $(\mathbf{r}_i, \mathbf{v}_i)$ drawn from an isotropic distribution with the same distribution in binding energy as the luminous particles. By this estimate both remnants have low X-tube populations compared to the available phase-space volumes, as do others studied by B92. In the N-body and SPH potentials, volumes associated with orbit families mirror the trends seen in Fig. 56 – box orbits give way to Z-tube orbits as the potential well deepens. On the whole, boxlet orbits seem relatively unimportant; they occupy no more than $\sim 2.5\%$ of the volume in any of these potentials, consistent with the small populations described in § 7.5.

The orbit phase-space volumes in the composite are intermediate between those of its parents, but closer in overall pattern to the SPH potential. While the N-body potential supports only box orbits within its rather "puffy" core, the SPH and composite potentials both permit all three major orbit families at almost all binding energies. With a more limited selection of box orbits, it may be impossible for systems with the composite's flattening and density profile to exist self-consistently; even if self-consistent versions could be constructed, some finesse would probably be required. Merging of equal-mass galaxies is unlikely to sow stars across phase-space with enough finesse to create objects *both* centrally concentrated and highly triaxial. Deepening the potential well shifts the boundaries between *major* orbit families (BH96), and this shift in turn explains trends of shape with potential depth.

Further experiments could test the general prediction that deeper potential wells, however created, lead to more tubes and fewer boxes. For example, pure

N-body mergers of galaxy models with a range of central density profiles might show that remnants with steeper central cusps are typically more oblate – a result with implications for observed trends of shape and kinematics with luminosity profile.

8.5 Galaxy Transformation and the Arrow of Time

Rapid gas inflows during galaxy collisions are crucial for the transformation of disk-disk mergers to ellipticals: inflows fuel the starbursts which apparently build the central regions of early-type galaxies (Schweizer 1987, 1990, Kormendy 1990, Kormendy & Sanders 1992). Such inflows are seen in simulations of barred disk galaxies (e.g. Athanassoula 1994), satellite accretions (Hernquist 1989), and tidal interactions (Noguchi 1988) as well as the equal-mass mergers discussed above. While gravitational torques are the proximate cause of these inflows, the deeper reasons for this generic one-way behavior pose an interesting puzzle (e.g. BH96).

An isolated self-gravitating gas sphere will radiate while contracting along a sequence of near-equilibrium configurations. This is an irreversible, entropy-producing process; less energy can be extracted from the system after than before. In rotating systems the same one-way behavior is slowed but not staved off by conservation of angular momentum. For example, gas in isolated accretion disks drifts inward on a time-scale governed by the outward transport of angular momentum via kinematic viscosity (see Papaloizou & Lin 1995). Once again less work can be extracted from the system after the gas has spiraled in; inflows in accretion disks are irreversible. And likewise, tidal interactions in gas-rich galaxies give rise to rapid inflows by creating situations favorable for angular momentum transport. The net effects of such a transformation – mass inward, angular momentum outward – are similar to the effects of viscous dissipation in isolated disks, though the action takes place on a dynamical rather than a viscous time-scale.

Gas inflows in tidal interactions resemble irreversible inflows in other situations. One might conjecture that all these inflows are manifestations of the second law of thermodynamics. According to this view, gravitational torques on gas disks are simply enabling agents by which angular momentum transport and entropy production take place. The dramatic collapse of gas in mergers (e.g. Fig. 45) seems to be governed by similar considerations. Although it's hard to abstract underlying physics from such messy systems, it's evident that shocks play a large role in the dynamical evolution of the dissipative material. Until the potential of the remnant settles down, gas flows are likely to intersect, and the resulting shocks demand irreversible behavior. The inflow of the gas and its accumulation within the centers of gas-rich merger remnants is thus a corollary of its dissipative nature.

As gas densities increase, further irreversible behavior is likely; in particular, the gas may fragment, collapse, and form stars. Star formation adds a source term, $\dot{f}$, to the right-hand side of the collisionless Boltzmann equation:

$$\frac{\partial f}{\partial t} + \mathbf{v} \cdot \frac{\partial f}{\partial \mathbf{r}} - \nabla \Phi \cdot \frac{\partial f}{\partial \mathbf{v}} = \dot{f} \,. \tag{41}$$

Thus fine-grained phase-space densities are no longer conserved; instead, the present value of $f(\mathbf{r},\mathbf{v})$ is the integral of $\dot{f}$ along the entire trajectory leading to $(\mathbf{r},\mathbf{v})$. Phase-space density can be built up slowly by extended star formation or created suddenly in a starburst; the former may describe the formation of disk galaxies, while the latter seems to be the mode observed in luminous infrared systems (e.g. Sanders & Mirabel 1996).

As yet we don't know in detail how star formation lays down newly formed stars in phase-space. Kennicutt (1989, also § 8.1 of his lectures in this volume) argues that star formation is a threshold phenomenon, "kicking in" if the gas surface density is high enough to approach Toomre's (1964) local instability criterion. Systems at the threshold have modest rates of star formation, consuming about 1% of their gas per 10^8 yr; formation rates increase non-linearly above this threshold, culminating in catastrophic starbursts which may consume $\gtrsim 50\%$ of the available gas in $\lesssim 10^7$ yr. Such star formation events can create very high phase-space densities; for example, newborn open clusters have central phase-space densities $F_{\rm max} \simeq \rho_0/(2\pi\sigma^2)^{3/2} \simeq 10^{1\pm1}\,{\rm M}_\odot\,{\rm Myr}^3\,{\rm pc}^{-6}$. Typical star formation events may yield phase-space densities perhaps two orders magnitude lower than those required to create open clusters.

Coarse-grained phase-space densities in galaxies are often orders of magnitude lower still. Most of the stars in galactic disks occupy a relatively narrow range of phase-space densities; while the detailed distribution is model dependent, typical disks have median densities $F_{\rm med} \simeq 10^{-6\pm1}\,{\rm M}_\odot\,{\rm Myr}^3\,{\rm pc}^{-6}$, with central phase-space densities perhaps an order of magnitude lower (Carlberg 1986, Lake 1989). Elliptical galaxies, in contrast, contain a wide range of phase-space densities. Carlberg (1986) found peak phase-space densities $F_{\rm max} \simeq 10^{-5\pm0.5}\,{\rm M}_\odot\,{\rm Myr}^3\,{\rm pc}^{-6}$ for ellipticals with $M_{\rm B} = -20$, with a steep power-law dependence on luminosity: $F_{\rm max} \propto L^{-2.35}$. These values are *underestimates* since elliptical galaxies with power-law cusps formally attain infinite phase-space densities. However, characteristic phase-space densities in ellipticals, estimated from $F_0 \simeq 3M/(8\pi r_{\rm g}^3 \langle v^2\rangle^{3/2})$, are at least two orders of magnitude lower than Carlberg's peak values (White 1987). Thus while ellipticals have small amounts of high phase-density material, the bulk of their stars actually lie at phase-densities *lower* than those found in spirals (Vedel & Sommer-Larsen 1990; see also Carlberg 1986, Lake 1989).

Transforming spiral galaxies into ellipticals entails forming a central cusp of stars with high phase-space densities while mixing the bulk of the stars to lower phase-space densities. Dissipative inflows triggered by interactions and mergers collect the raw materials needed to build high-density cusps; meanwhile, violent relaxation and phase-mixing reduce the coarse-grain phase-space densities of existing stars. One direct consequence of the mixing process is the gradual disappearance of the "fine structures" in young elliptical galaxies, described by François Schweizer in his § 6 of this volume. The arrow of time defined by the fading of fine structure is partly a consequence of the limited information we can obtain from observations; stellar dynamics itself is reversible. But star formation provides time-asymmetric boundary conditions for the collisionless Boltzmann equation, lending a sense of direction to galactic transformation. The "delayed formation of some elliptical galaxies" proposed by TT72 thus involves reversible *and* irreversible processes; *both* are crucial for galaxy formation.

Acknowledgements.

I thank my co-authors, Rob Kennicutt and François Schweizer, for many stimulating discussions during the preparation, presentation, and aftermath of these lecturers. I also thank the organizers of the SAAS-FEE winter school, Daniel Friedli, Louis Martinet, and Daniel Pfenniger, for inviting me to the school and for their help and hospitality.

John Hibbard kindly allowed me to present our work on The Mice in advance of publication and generated several of the figures reproduced here. Lars Hernquist also gave me free access to unpublished simulations and commented on several drafts of the written lectures. Alar Toomre generously supplied software, slides, and images of his Whirlpool simulations and provided much helpful feedback on the text.

Tod Boroson, Martin Burkhead, and Arnold Rots generously allowed me to reproduce their images and plots of The Whirlpool galaxy. Eliza Fulton computed several boxlet orbits. Cathy Ishida provided written notes from the winter school which were helpful in writing this report. Anjana Kar and Joel Welling of the Pittsburgh Supercomputing Center swiftly generated the videos I showed in the lectures. Naryan Raja provided expert technical help with the Web pages. Karen Teramura redrew several of the figures and created the composite images of The Whirlpool.

Conversations with participants at the Aspen 1996 Summer Workshop on *Galaxy Interactions at Low and High Redshift* expanded my perspectives on many of the topics presented here, and I thank my colleagues for openly sharing ideas and the Aspen Center for sponsoring the workshop.

This work was supported in part by NASA through grant NAG 5-2836.

I'm grateful to the Pittsburgh Supercomputing Center for past support and for continued access to data and video facilities.

Through the use of the Maui High Performance Computing Center, this research was sponsored in part by the Phillips Laboratory, Air Force Materiel Command, USAF, under cooperative agreement number F29601-93-2-0001. The views and conclusions contained in this document are those of the author and should not be interpreted as necessarily representing the official policies or endorsements, either expressed or implied, of Phillips Laboratory, the U.S. Government, The University of New Mexico, or the Maui High Performance Computing Center.

References

Aarseth, S.J. 1985, Multiple Time Scales, ed. J.U. Brackbill & B.I. Cohen (Orlando: Academic Press), p. 377

Aarseth, S.J. & Binney, J.J. 1978, MNRAS 185, 227

Ahmad, A. & Cohen, L. 1973, J. Comp. Phys. 12, 389

Aguilar, L.A. & White, S.D.M. 1985, ApJ 295, 374

Aguilar, L.A. & White, S.D.M. 1986, ApJ 307, 97

Alladin, S.M. 1965, ApJ 141, 768

Appel, A.W. 1985, SIAM J. Sci. Stat. Comput. 6, 85

Arp, H. 1965, ApJS 14, 1

Arp, H. 1971, Science 174, 1189

Athanassoula, E. 1994, Mass-Transfer Induced Activity in Galaxies, ed. I. Shlosman (Cambridge: Cambridge University Press), p. 143

Athanassoula, E. & Sellwood, J.A. 1986, MNRAS 221, 213

Barnes, J.E. 1986, The Use of Supercomputers in Stellar Dynamics, ed. S. McMillan & P. Hut (Berlin: Springer Verlag), p. 175

Barnes, J.E. 1988, ApJ 331, 699

Barnes, J.E. 1990a, J. Comp. Phys. 87, 161.

Barnes, J.E. 1990b, Dynamics and Interactions of Galaxies, ed. R. Wielen (Berlin: Springer-Verlag), p. 186

Barnes, J.E. 1992, ApJ 393, 484 (B92)

Barnes, J.E. 1994, The Formation of Galaxies: Proceedings of the V Canary Islands Winter School of Astrophysics, ed. C. Munoz-Tunon & F. Sanchez (Cambridge: Cambridge University Press), p. 399

Barnes, J.E. 1997, The Nature of Elliptical Galaxies, ed. M. Arnaboldi, G.S. Da Costa, & P. Saha (San Francisco: Astronomical Society of the Pacific), p. 469.

Barnes, J. & Efstathiou, G. 1987, ApJ 319, 575

Barnes, J.E. & Hernquist, L. 1991, ApJ 370, L65

Barnes, J.E. & Hernquist, L. 1992, Nat 360, 715

Barnes, J.E. & Hernquist, L. 1996, ApJ 471, 115 (BH96)

Barnes, J. & Hut, P. 1986, Nat 324, 446

Barnes, J.E. & Hut, P. 1989, ApJS 70, 389

Barnes, J. & White, S.D.M. 1984, MNRAS 211, 753

Benz, W. 1990, Numerical Modeling of Nonlinear Stellar Pulsations, ed. J.R. Buchler (Dordrecht: Kluwer), p. 269

Binney, J.J. (1976), MNRAS 177, 19

Binney, J. & Tremaine, S. 1987, Galactic Dynamics (Princeton: Princeton University Press) (BT87)

Binney, J. & Petrou, M. 1985, MNRAS 214, 449

Bontekoe, Tj.R. & van Albada, T.S. 1987, MNRAS 224, 349

Burbidge, E.M. & Burbidge, G.R. 1961, ApJ 133, 726

Burbidge, E.M., Burbidge, G.R., & Hoyle, F. 1963, ApJ 138, 873

Burkhead, M.S. 1978, ApJS 38, 147

Byun, Y.-I., Grillmair, C.J., Faber, S.M., Ajhar, E.A., Dressler, A., Kormendy, J., Lauer, T.R., Richstone, D., & Tremaine, S. 1996, AJ 111, 1889

Carlberg, R. 1986, ApJ 310, 593

Chandrasekhar, S. 1942, Principles of Stellar Dynamics (Dover: New York, 1960)

Chandrasekhar, S. 1943, ApJ 97, 255

Clutton-Brock, M. 1972a, Ap&SS 16, 101

Clutton-Brock, M. 1972b, Ap&SS 17, 292

Colina, L., Lipari, S., & Macchetto, F. 1991, ApJ 379, 113

Combes, F. 1994, The Formation of Galaxies: Proceedings of the V Canary Islands Winter School of Astrophysics, ed. C. Munoz-Tunon & F. Sanchez (Cambridge: Cambridge University Press), p. 317

Combes, F., Dupraz, C., & Gerin, M. 1990, Dynamics and Interactions of Galaxies, ed. R. Wielen (Berlin: Springer), p. 205

Combes, F. & Gerin, M. 1985, A&A 150, 327

Condon, J.J., Helou, G., Sanders, D.B., & Soifer, B.T. 1993, AJ 105, 1730

Colella, P. & Woodward, P.R. 1984, J. Comp. Phys. 54, 174

Crane, P., Stiavelli, M., King, I.R., Deharveng, J.M., Albrecht, R., Barbieri, C., Blades, J.C., Boksenberg, A., Disney, M.J., Jakobsen, P., Kamperman, T.M., Machetto, F., Mackay, C.D., Paresce, F., Weigelt, G., Baxter, D., Greenfield, P., Jedrzejewski, R., Nota, A., & Sparks, W.B. 1993, AJ 106, 1371

Dejonghe, H. 1989, ApJ 343, 113

Dehnen, W. 1993, MNRAS 265, 250

de Vaucouleurs, G. 1948, Ann. d'Astrophys 11, 247

de Zeeuw, T. 1985, MNRAS 216, 273

de Zeeuw, T. 1987, Structure and Dynamics of Elliptical Galaxies (I.A.U. Symp. 127), ed. T. de Zeeuw (Dordrecht: Reidel)

de Zeeuw, T. & Franx, M. 1991, ARA&A 29, 239

Draine, B. & McKee, C. 1993, ARA&A 31, 373

Dubinski, J. 1994, ApJ 431, 617

Dubinski, J. 1996, New Astronomy 1, 133

Dubinski, J. & Carlberg, R.G. 1991, ApJ 378, 496

Dubinski, J., Mihos, J.C., & Hernquist, L. 1996, ApJ 462, 576

Duncan, M.J., Farouki, R.T., & Shapiro, S.L. 1983, ApJ 271, 22

Dyer, C.C. & Ip, P.S.S. 1993, ApJ 409, 60

Efstathiou, G., Lake, G., & Negroponte, J. 1982, MNRAS 199, 1069

Elmegreen, B.G., Kaufman, M., & Thomasson, M. 1993, ApJ 412, 90

Eneev, T.M., Kozlov, N.N., & Sunyaev, R.A. 1973, A&A 28, 41

Evrard, A.E. 1988, MNRAS 235, 911

Farouki, R.T. & Shapiro, S. 1981, ApJ 259, 103

Farouki, R.T. & Shapiro, S. 1982, ApJ 259, 103

Franx, M., Illingworth, G., & de Zeeuw, T. 1991, ApJ 383, 112

Frenk, C.S., White, S.D.M., Davis, M., & Efstathiou, G. 1988, ApJ 327, 507

Fridman, A.M. & Polyachenko, V.L. 1984, Physics of Gravitating Systems (Springer-Verlag, Berlin)

Friedli, D. & Benz, W. 1993, A&A 268, 65

Gerhard, O. 1981, MNRAS 197, 179

Gerhard, O. & Binney, J.J. 1985, MNRAS 216, 467

Gerola, H., Canrevali, P., & Salpeter, E.E. 1983, ApJ 268, L75

Gilbert, S.J. & Sellwood, J.A. 1993, Numerical Simulations in Astrophysics, ed. J. Franco, S. Lizano, L. Aguilar, & E. Daltabuit (Cambridge: Cambridge University Press), p. 131

Gingold, R.A. & Monaghan, J.J. 1977, MNRAS 181, 375

Gingold, R.A. & Monaghan, J.J. 1982, J. Comp. Phys. 46, 429

Governato, F., Reduzzi, L., & Rampazzo, R. 1993, MNRAS 261, 379

Greengard, L. & Rokhlin, V. 1987, J. Comp. Phys. 73, 325

Hausman, M. & Ostriker, J. 1978, ApJ 224, 320

Heckman, T.M., Armus, L., & Miley, G.K. 1987, AJ 92, 276
Heckman, T.M., Dahlem, M., Eales, S.A., & Weaver, K. 1996, ApJ 457, 616
Hernquist, L. 1987, ApJS 64, 715
Hernquist, L. 1989, Nat 340, 687
Hernquist, L. 1990a, J. Comp. Phys. 87, 137
Hernquist, L. 1990b, ApJ 356, 359
Hernquist, L. 1990c, Dynamics and Interactions of Galaxies, ed. R. Wielen (Berlin: Springer), p. 108
Hernquist, L. 1992, ApJ 400, 460
Hernquist, L. 1993a, ApJS 86, 389
Hernquist, L. 1993b, ApJ 404, 717
Hernquist, L. 1993c, ApJ 409, 548
Hernquist, L. & Barnes, J.E. 1990, ApJ 349, 562
Hernquist, L. & Barnes, J.E. 1991, Nat 354, 210
Hernquist, L. & Katz, N. 1989, ApJS 70, 419
Hernquist, L. & Ostriker, J.P. 1992, ApJ 386, 375
Hernquist, L. & Spergel, D.N. 1993, ApJ 399, L117
Hernquist, L. & Weinberg, M.D. 1989, MNRAS 238, 407
Hernquist, L., Heyl, J.S., & Spergel, D.N. 1993, ApJ 416, L9
Heyl, J.S., Hernquist, L. & Spergel, D.N. 1994, ApJ 427, 165
Hibbard, J.E., Guhathakurta, P., van Gorkom, J.H., & Schweizer, F. 1994, AJ 107, 67
Hibbard, J.E. & Mihos, J.C. 1995, AJ 110, 140
Hibbard, J.E. & van Gorkom, J.H. 1996, AJ 111, 655
Hockney, R.W. & Eastwood, J.W. 1988, Computer Simulation Using Particles (New York: Adam Hilger)
Hohl, F. 1971, ApJ 168, 343
Holmberg, E. 1941, ApJ 94, 385
Howard, S. & Byrd, G.G. 1990, AJ 99, 1798
Hui, X., Ford, H.C., Freeman, K.C., & Dopita, M.A. 1995, ApJ 449, 592
Hut, P., Makino, J., & McMillan, S. 1995, ApJ 443, L93
Jaffe, W. 1982, MNRAS 202, 995
Jaffe, W. 1987, Structure and Dynamics of Elliptical Galaxies (I.A.U. Symp. 127), ed. T. de Zeeuw (Dordrecht: Reidel), p. 511
Jaffe, W., Ford, H.C., O'Connell, R.W., van den Bosch, F.C., & Ferrarese, L. 1994, AJ 108, 1567
Jenrigan, J.G. 1985, Dynamics of Star Clusters (I.A.U. Symp. 113), ed. J. Goodman & P. Hut (Dordrecht: Reidel), p. 275
Jog, C.J. & Solomon, P.M. 1992, ApJ 387, 152
Kalnajs, A.J. 1978, Structure and Properties of Nearby Galaxies (I.A.U. Symp. 77), ed. E.M. Berkhuijsen & R. Wielebinski (Dordrecht: Reidel), p. 113
Katz, N. 1991, ApJ 368, 325
Katz, N. 1992, ApJ 391, 502
Katz, N. & Gunn, J.E. 1991, ApJ 377, 365
Kennicutt, R. 1989, ApJ 344, 685
King, I.R. 1966, AJ 71, 64
Knuth, D.E. 1969, The Art of Computer Programming, Vol. 2. Seminumerical Algorithms (Reading: Addison-Wesley)
Kormendy, J. 1977, ApJ 218, 333
Kormendy, J. 1990, Dynamics and Interactions of Galaxies, ed. R. Wielen (Berlin: Springer), p. 499.

Kormendy, J. & Richstone, D. 1995, ARA&A 33, 581
Kormendy, J. & Sanders, D.B. 1992, ApJ 390, L53
Kuijken, K. & Dubinski, J. 1995, MNRAS 277, 1341
Kuzmin, G. 1956, Astron. Zh. 33, 27
Lake, G. 1989, AJ 97, 1312
Lake, G. & Dressler, A. 1986, ApJ 310, 605
Laplace, P. 1820, A Philosophical Essay on Probabilities, trans. F.W. Truscott & F.L. Emory (New York: Dover, 1951)
Larson, R.B. & Tinsley, B.M. 1978, ApJ 219, 46
Levison, H. 1987, ApJ 320, L93
Lindblad, B. 1961, Problems of Extragalactic Research (I.A.U. Symp. 15), ed. G.C. McVitte (New York: Macmillan), p. 146
Lindblad, P.O. 1960, Stockholm Obs. Ann. 21, no. 4
Lucy, L. 1977, AJ 82, 1013
Lynden-Bell, D. 1967, MNRAS 136, 101
Lynds, R. & Toomre, A. 1976, ApJ 209, 382
Makino, J. 1990, J Comp Phys 87, 148
Mair, G. Müller, E., Hillebrandt, W., & Arnold, C.N. 1988, A&A 199, 114
May, A. & van Albada, T.S. 1984, MNRAS 209, 15
McGlynn, T.A. 1984, ApJ 281, 13
McKee, C.F. & Ostriker, J.P. 1977, ApJ 218, 148
Merritt, D. 1997, The Nature of Elliptical Galaxies, ed. M. Arnaboldi, G.S. Da Costa, & P. Saha (San Francisco: Astronomical Society of the Pacific), p. 32
Mihos, C. & Hernquist, L. 1994, ApJ 425, L13
Mihos, J.C., Bothun, G.D., & Richstone, D.O. 1993, ApJ 418, 82
Miller, R.H. & Smith, B. 1980, ApJ 235, 421
Mirabel, I.F., Lutz, D., & Maza, J. 1991, A&A 243, 367
Miralda-Escudé, J. & Schwarschild, M. 1989, ApJ 339, 752
Monaghan, J.J. 1992, ARA&A 30, 543
Monaghan, J.J. & Gingold, R.A. 1983, J. Comp. Phys. 52, 374
Moore, B. & Davis, M. 1994, MNRAS 270, 209
Murai, T. & Fujimoto, M. 1980, PASJ 32, 581
Navarro, J.F. 1990, MNRAS 242, 311
Navarro, J.F., Frenk, C.S., & White, S.D.M. 1996, ApJ 462, 563
Navarro, J.F. & White, S.D.M. 1993, MNRAS 265, 271
Negroponte, J. & White, S.D.M. 1983, MNRAS 205, 1009
Nelson, R.P. & Papaloizou, J.C.B. 1993, MNRAS 264, 905
Nelson, R.P. & Papaloizou, J.C.B. 1994, MNRAS 270, 1
Nicholson, R.A., Bland-Hawthorn, J., & Taylor, K. 1992, ApJ 387, 503
Noguchi, M. 1987, MNRAS 228, 635
Noguchi, M. 1988, A&A 203, 259
Olson, K.M. & Dorband, J.E. 1994, ApJS 94, 117
Ostriker, J.P. 1980, Comments Astrophys. 8, 177
Ostriker, J.P. & Tremaine, S. 1975, ApJ 202, L113
Ostriker, J.P. & Peebles, P.J.E. 1973, ApJ 186, 467
Papaloizou, J.C.B. & Lin, D.N.C. 1995, ARA&A 33, 505
Pfleiderer, J. 1963, Zs. f. Ap. 58, 12
Pfleiderer, J., Siedentopf, H. 1961, Zs. f. Ap. 51, 201
Pikel'ner, S.B. 1968, ARA&A 6, 165
Plummer, H.C. 1911, MNRAS 71, 460

Porter, D. 1985, PhD Thesis, University of California, Berkeley
Press, W. 1986, The Use of Supercomputers in Stellar Dynamics, ed. S. McMillan & P. Hut (Berlin: Springer Verlag), p. 184
Quian, E.E., de Zeeuw, P.T., van der Marel, R.P., & Hunter, C. 1995, MNRAS 274, 602
Quinn, P.J. 1982, PhD Thesis, Australian National University
Read, A.M., Ponman, T.J., & Wolstencroft, R.D. 1995, MNRAS 277, 397
Roos, N. & Norman, C. 1979, A&A 76, 75
Rots, A.H. & Shane, W.W. 1975, A&A 25, 25
Rots, A.H., Bosma, A., van der Hulst, J.M., Athanassoula, E., & Crane, P.C. 1990, AJ 100, 387
Saha, P. & Tremaine, S. 1994, AJ 108, 1962
Salmon, J.K. 1990, PhD Thesis, California Institute of Technology
Salmon, J.K. & Warren, M.S. 1993, J. Comp. Phys. 111, 136
Salo, H. & Laurikainen, E. 1996, unpublished manuscript
Sanders, D.B. & Mirabel, I.F. 1996, ARA&A 34, 749
Sansom, A.E., Dotani, T., Okada, K., Yamashita, A., Fabbiano, G. 1996, MNRAS 281, 48
Schiminovich, D., van Gorkom, J.H., van der Hulst, J.M., Kasow, S. 1994, ApJ 423, L101
Schwarzschild, M. 1979, ApJ 232, 236
Schwarzschild, M. 1993, ApJ 409, 563
Schweizer, F. 1977, ApJ 211, 324
Schweizer, F. 1982, ApJ 252, 445
Schweizer, F. 1986, Structure and Dynamics of Elliptical Galaxies, ed. T. de Zeeuw, (Dordrecht: Reidel), p. 109
Schweizer, F. 1987, Nearly Normal Galaxies, ed. S.M. Faber (Berlin: Springer Verlag), p. 18
Schweizer, F. 1990, Dynamics and Interactions of Galaxies, ed. R. Wielen ((Berlin: Springer Verlag), p. 60
Schweizer, F. 1996, AJ 111, 109
Sellwood, J.A. 1981, A&A 99, 362
Sellwood, J.A. 1987, ARA&A 25, 151
Simkin, S.M., Su, H.J., & Schwarz, M.P. 1980, ApJ 237, 404
Singh, J.P. 1993, PhD Thesis, Stanford University
Sofue, Y. 1994, ApJ 423, 207
Sparke, L.S. & Sellwood, J.A. 1987, MNRAS 225, 653
Spitzer, L. 1942, ApJ 95, 329
Spitzer, L. 1958, ApJ 127, 17
Spitzer, L. 1978, Physical Processies in the Interstellar Medium (New York: Wiley)
Spitzer, L., Baade, W. 1951, ApJ 113, 413
Sridhar, S. & Nityananda, R. 1990, Dynamics and Interactions of Galaxies, ed. R. Wielen (Berlin: Springer), p. 375
Stanford, S.A., Sargent, A.I., Sanders, D.B., & Scoville, N.Z. 1990, ApJ 349, 492
Steiman-Cameron, T.Y., Kormendy, J., & Durisen, R.H. 1992, AJ 104, 1339
Steinmetz, M. 1996, Dark Matter in the Universe, in press
Steinmetz, M. & Müller, E. 1993, A&A 268, 391
Stockton, A. 1974, ApJ 187, 219
Sugimoto, D., Chikada, Y., Makino, J., Ito, T., Ebisuzaki, T., & Umemura, M. 1990, Nat 345, 33

Syer, D. & Tremaine, S. 1996, MNRAS 282, 223
Tashpulatov, N. 1969, Soviet Astr.–AJ 13, 968
Tashpulatov, N. 1970, Soviet Astr.–AJ 14, 227
Theys, J.C. & Spiegel, E.A. 1977, ApJ 212, 616
Tolman, R.C. 1938, The Principles of Statistical Mechanics (Clarendon, Oxford)
Toomre, A. 1964, ApJ 139, 1217
Toomre, A. 1967, Notes from the Geophysical Fluid Dynamics Summer Program (Woods Hole Oceanographic Institute: Woods Hole), p. 111
Toomre, A. 1974, The Formation and Dynamics of Galaxies, ed. J.R. Shakeshaft (Reidel: Dordrecht), p. 347
Toomre, A. 1977, The Evolution of Galaxies and Stellar Populations, ed. B. Tinsley & R. Larson (New Haven: Yale University Obs.), p. 401
Toomre, A. 1978, The Large Scale Structure of the Universe, ed. M.S. Longair & J. Einasto (Dordrecht: Reidel), p. 109
Toomre, A. 1981, The Structure and Evolution of Normal Galaxies, ed. M. Fall & D. Lynden-Bell (Cambridge: Cambridge University Press), p. 111
Toomre, A., & Toomre, J. 1972, ApJ 178, 623 (TT72)
Tremaine, S.D., Ostriker, J.P., & Spitzer, L. 1975, ApJ 196, 407
Tremaine, S.D., Hénon, M., & Lynden-Bell, D. 1986, MNRAS 219, 285
Tremaine, S. & Weinberg, M.D. 1984, MNRAS 209, 729
Tremblay, B. & Merritt, D. 1996, AJ 111, 2243
Trimble, V. 1987, ARA&A 25, 425
Tully, R.B. & Fisher, J.R. 1977, A&A 54, 661
Udry, S. 1993, A&A 268, 35
van Albada, T.S. 1982, MNRAS 201, 939
van Albada, T.S. 1987, Structure and Dynamics of Elliptical Galaxies (I.A.U. Symp. 127), ed. T. de Zeeuw (Dordrecht: Reidel), p. 291
van Albada, T.S. & van Gorkom, J. 1977, A&A 54, 121
van den Bergh, S. 1969, Ap. Letters 7, 117
van der Hulst, J.M. 1979, A&A 71, 131
van der Hulst, J.M. & Hummel, E. 1985, A&A 150, L7
van Gorkom, J.H., van der Hulst, J.M., Haschick, A.D., & Tubbs, A.D. 1990, AJ 99, 1781
Vedel, H. & Sommer-Larsen, J. 1990, MNRAS 245, 637
Veeraraghavan, S. & White, S.D.M. 1985, ApJ 296, 336
Villumsen, J.V. 1982, MNRAS 199, 493
Villumsen, J.V. 1983, MNRAS 204, 291
Vorontosov-Vel'yaminov, B.A. 1958, Soviet Astr.–AJ 2, 805
Vorontosov-Vel'yaminov, B.A. 1961, Problems of Extragalactic Research (I.A.U. Symp. 15), ed. G.C. McVitte (New York: Macmillan), p. 194
Wallin, J.F. 1990, AJ 100, 1477
Weinberg, M.D. 1989, MNRAS 239, 549
Weinberg, M.D. 1993, ApJ 410, 543
White, S.D.M. 1976, MNRAS 174, 467
White, S.D.M. 1978, MNRAS 184, 185
White, S.D.M. 1979, MNRAS 189, 831
White, S.D.M. 1982, Morphology and Dynamics of Galaxies, ed. L. Martinet & M. Mayor (Sauverny: Geneva Observatory), p. 291
White, S.D.M. 1983a, ApJ 274, 53

White, S.D.M. 1983b, Internal Kinematics and Dynamics of Galaxies, ed. E. Athanassoula (Dordrecht: Reidel), p. 337.
White, S.D.M. 1987, Structure and Dynamics of Elliptical Galaxies (I.A.U. Symp. 127), ed. T. de Zeeuw (Dordrecht: Reidel), p. 339
White, S.D.M. & Rees, M.J. (1978), MNRAS 183, 341
Wright, A.E. 1972, MNRAS 157, 309
Wright, G.S., James, P.A., Joseph, R.D., & McLean, I.S. 1990, Nat 344, 417
Yabushita, S. 1971, MNRAS 153, 97
Zang, T.A. 1976, PhD Thesis, Massachusetts Institute of Technology
Zasov, A.V. 1968, Soviet Astr. 11, 785
Zwicky, F. 1956, Ergeb. Exakten Naturwiss. 29, 344
Zwicky, F. 1959, Hdb. d. Phys. 53, 373.

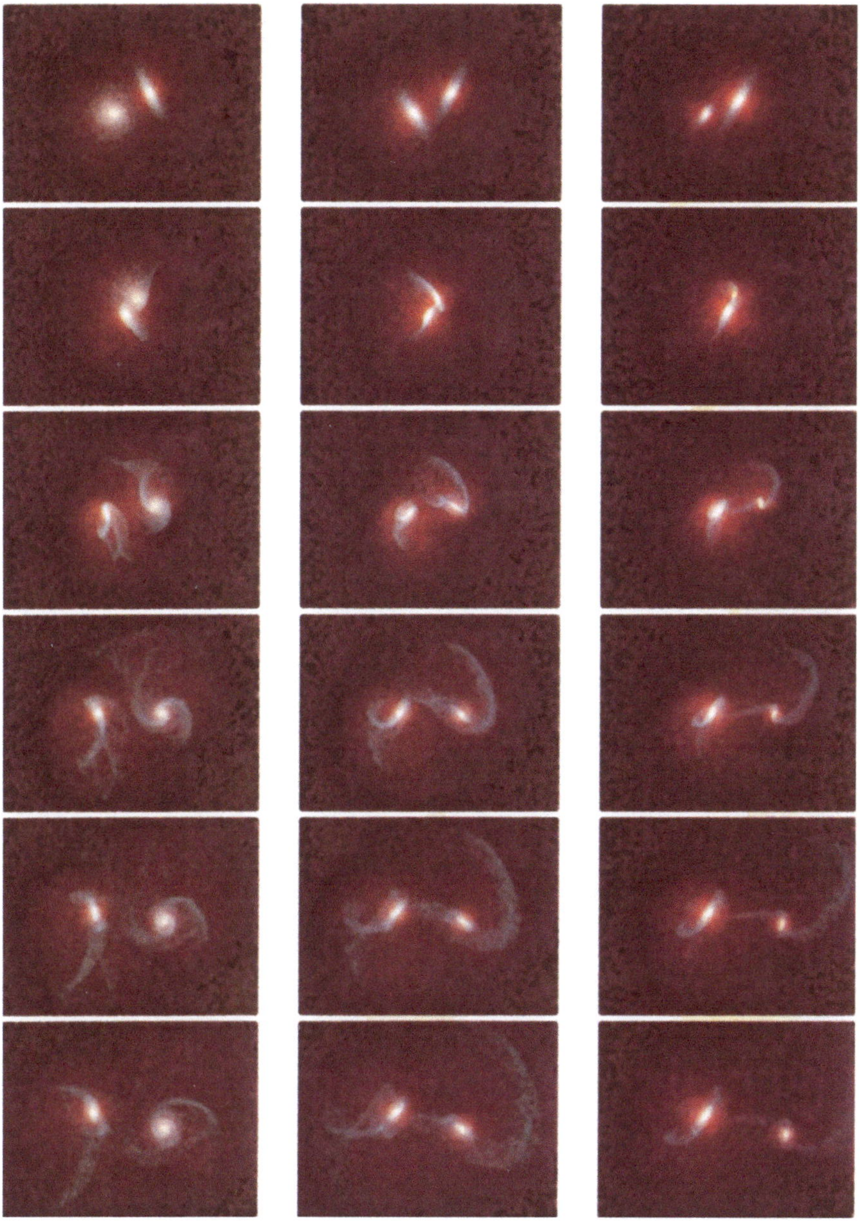

Plate 1. Selected frames from three encounter videos. Each system is shown at six equally-spaced times between $t = 0.75$ and 2.625. **Left:** Encounter D. **Middle:** Encounter F. **Right:** Encounter C_2.

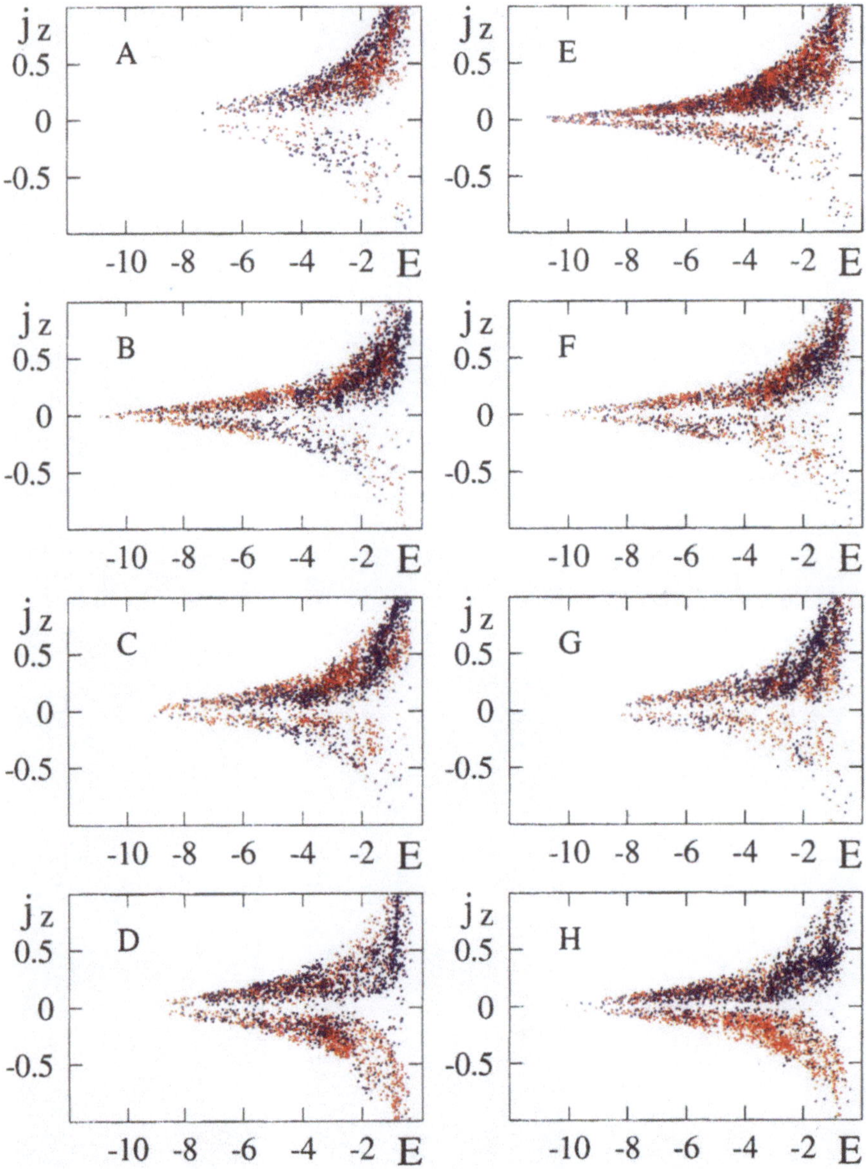

Plate 2. Specific binding energy E and specific angular momentum j_Z for particles on Z-tube orbits in equal-mass merger remnants. Color indicates galaxy of origin, with blue used for the galaxy with the *smaller* inclination i.

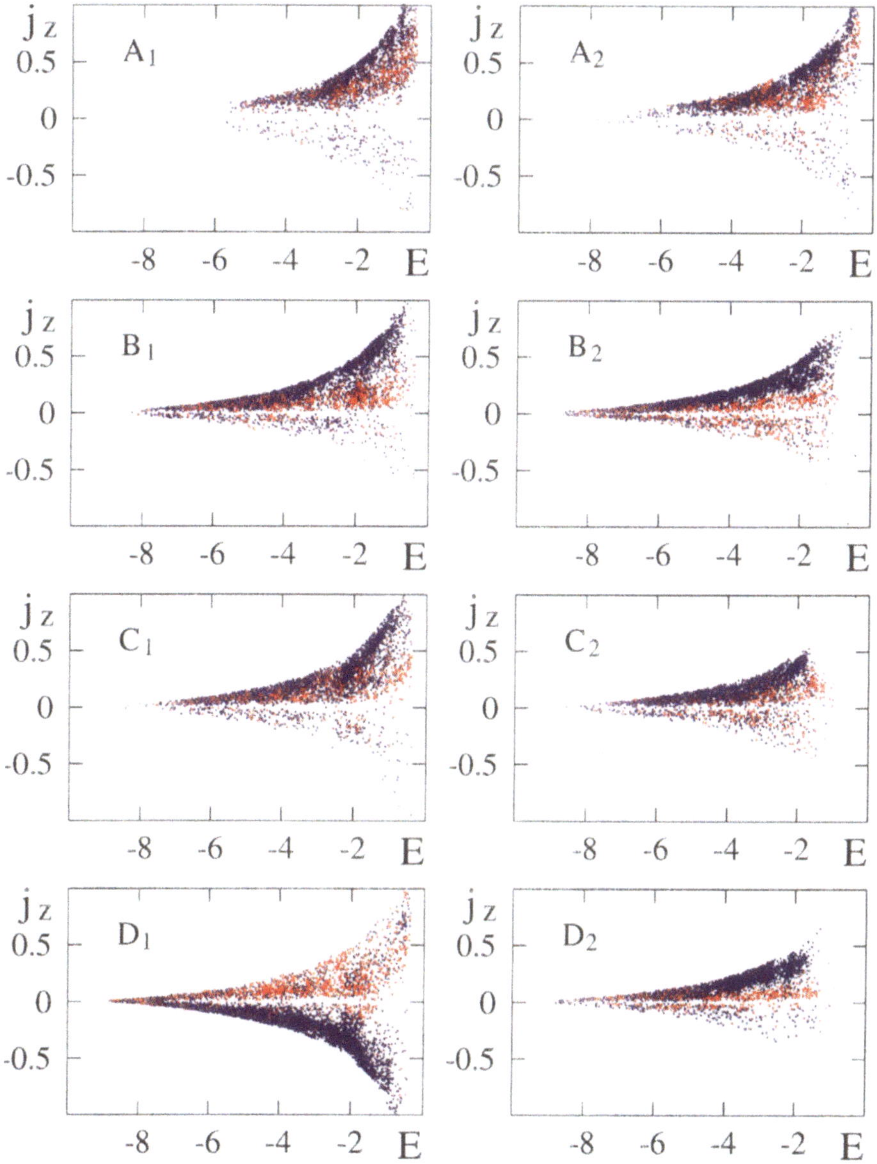

Plate 3. Specific binding energy E and specific angular momentum j_Z for particles on Z-tube orbits in third-mass merger remnants. Color indicates galaxy of origin, with blue used for the more massive galaxy.

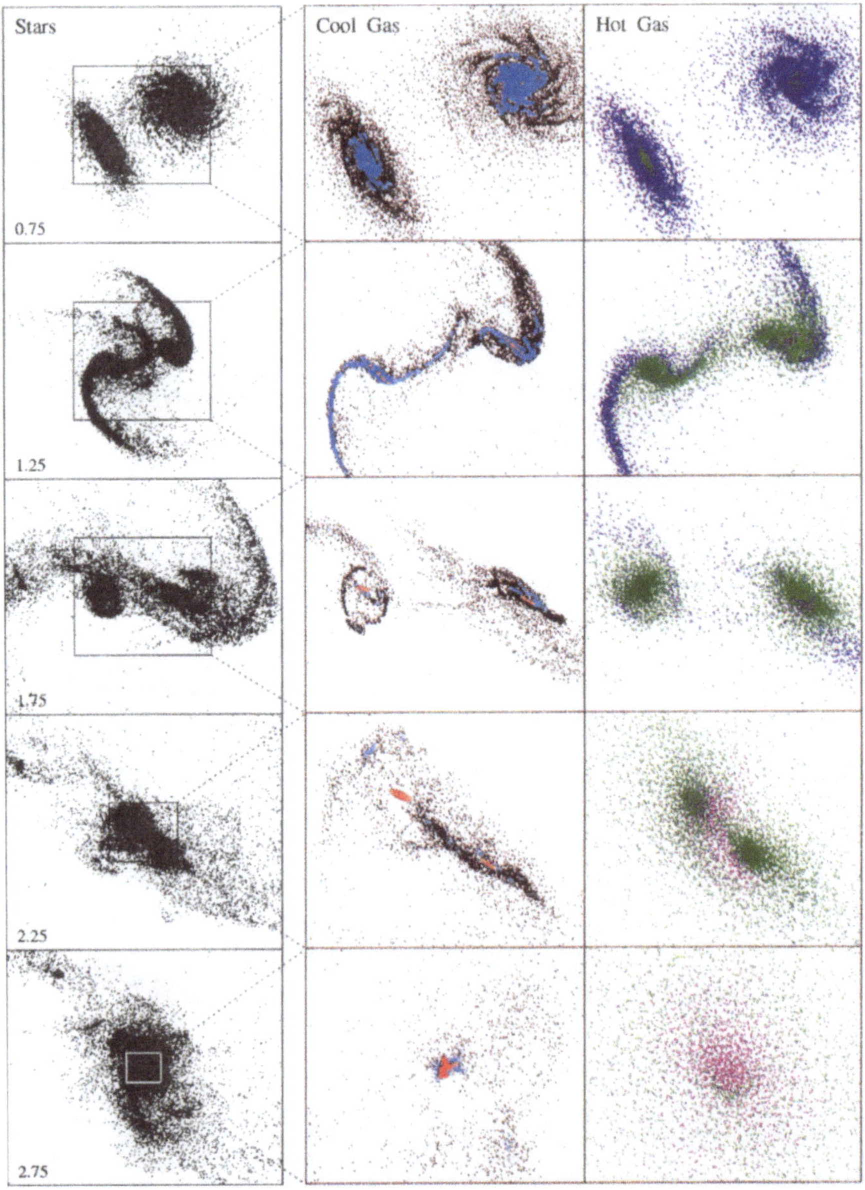

Plate 4. Stars, cool interstellar material, and hot gas in a merger of spiral galaxies. **Left:** Stellar distribution; each frame is 2×2.4 length units, with times since the start of the simulation given at the lower left. **Middle:** Cool gas, enlarged with respect to the stellar frame as shown; here color codes local gas density, with black, light blue, and red indicating successive factor-of-hundred increases up to $\sim 10^2\,\mathrm{cm}^{-3}$. **Right:** Hot gas, shown on the same scale as the cool gas; here color codes temperature, with dark blue, green, and purple indicating factor-of-ten increases up to $\sim 2 \cdot 10^6\,\mathrm{K}$.

Subject Index

MIX
Papier aus verantwortungsvollen Quellen
Paper from responsible sources
FSC® C105338

If you have any concerns about our products,
you can contact us on
ProductSafety@springernature.com

In case Publisher is established outside the EU,
the EU authorized representative is:
Springer Nature Customer Service Center GmbH
Europaplatz 3, 69115 Heidelberg, Germany

Printed by Libri Plureos GmbH
in Hamburg, Germany